CW01183586

Hans J. Ebert, Johann B. Kaiser, Klaus Peters

Willy Messerschmitt - Pioneer of Aviation Design

The History of German Aviation

Willy Messerschmitt -
Pioneer of Aviation Design

by Hans J. Ebert,
Johann B. Kaiser,
Klaus Peters

translated by Ray J. Theriault and Don Cox

Schiffer Military/Aviation History
Atglen, PA

Photo Credits

The majority of photos appearing in this book were taken over the decades by the company photographers of BFW/Messerschmitt AG, EWR, and MBB. These included Margarete Thiel in the 'thirties and 'forties, Barbara Heissig in the 'seventies and 'eighties, Dorothea Trautmann, Irmgard Großhauser (Haßlacher) and Josef Pfeffer in Munich and Ottobrunn, and Joachim Gruner in Manching.

In addition, the following organizations have also kindly supplied photographs published within these pages: Aero Club Bamberg/Harth Archives (9), Aérospatiale, Paris (1), Akaflieg/TU Munich (1), Richard P. Bateson (2), Werner Blasel (6), Etienne Boegner (1), H. Bueschel (1), Bundesarchiv, Bildstelle, Koblenz (3), Deutsches Museum, Sondersammlung, Munich (11), DFVLR-AVA, Göttingen (1), Hans J. Ebert (4), Fédération Aéronautique Internationale, Paris (2), "Flugsport" (2), FMR-Fahrzeug und Maschinenbau Regensburg (4), Hispano Aviacion S.A., Seville (17), Israel Aircraft Industries, Tel Aviv (1), Gero von Langsdorff (8), Lockheed Aircraft Corporation, Burbank (1), "Luftwissen" (2), R. P. Lutz (1), Gero Madelung (2), Willy Messerschmitt (10), Musée de l'Air, Paris (6), National Air and Space Museum (Smithsonian Institute), Washington, D. C. (8), Klaus Peters (4), Kurt Schnittke (5), Börje Sjögren (1), "Der Spiegel" (1), Mano Ziegler (7). All remaining photos were provided from the Messerschmitt and MBB archives.

The publisher wishes to thank the MBB company and the Messerschmitt Foundation for their generous support.

Translated from the German by Don Cox

Copyright © 1999 by Schiffer Publishing, Ltd.
Library of Congress Catalog Number: 99-65169

All rights reserved. No part of this work may be reproduced or used in any forms or by any means—graphic, electronic or mechanical, including photocopying or information storage and retrieval systems—without written permission from the copyright holder.

Printed in China.
ISBN: 0-7643-0727-4

This book was originally published under the title,
Willy Messerschmitt-Pionier der Luftfahrt und des Leichtbaues
by Bernard & Graefe.

We are interested in hearing from authors with book ideas on related topics.

Published by Schiffer Publishing Ltd.
4880 Lower Valley Road
Atglen, PA 19310 USA
Phone: (610) 593-1777
FAX: (610) 593-2002
E-mail: Schifferbk@aol.com
Visit our web site at: www.schifferbooks.com
Please write for a free catalog.
This book may be purchased from the publisher.
Please include $3.95 postage.
Try your bookstore first.

In Europe, Schiffer books are distributed by:
Bushwood Books
6 Marksbury Ave.
Kew Gardens
Surrey TW9 4JF
England
Phone: 44 (0)208 392-8585
FAX: 44 (0)208 392-9876
E-mail: Bushwd@aol.com
Free postage in the U.K. Europe: air mail at cost.
Try your bookstore first.

CONTENTS

Introduction by Ludwig Bölkow	4
Introduction by Gero Madelung	5
Preface	7

Willy Messerschmitt - Aircraft Designer, Aviation Pioneer, and Researcher — 9

Reducing Drag — 9
Reducing Design Weight — 10
Simplifying Construction — 12

Young Messerschmitt (1913-1923) — 14

From the Family Journals	14
Starting Point	15
Beginnings - Harth-Messerschmitt Sailplanes	18
Flugzeugbau Messerschmitt Bamberg	32
The Powered Gliders	36

From Engineer to Businessman (1924-1934) — 40

The Step to Powered Aircraft	40
M 17 Two-Seater Light Plane (1925/26)	40
Design history and characteristics	40
M 17's contests and races	42
M 18 Feeder Service Commercial Plane (1925-1930)	44
Situation in the mid-'twenties	44
M 18a four-seat commercial plane (1925/26)	44
M 18b five-seat commercial plane (1927-1930)	46
M 19 Competition and Sportplane (1927)	48

From Bamberg to Augsburg (1927) — 52

BFW M 20 All-Metal Commercial Plane (1927-1931)	56
M 21 Trainer	62
M 22 Twin-Engined Biplane (1930)	63
M 23 Trainer and Sportplane (1928-1931)	64
M 23b	67
M 23c Sportplane and the 1930 Challenge de Tourisme Internationale	69
M 23b Floatplane (1930/31)	70
M 24 Eight-Seat Commercial Plane (1929)	71
M 18 Follow-On Developments (1929)	72
M 18c (1929/30)	73
M 18d commercial plane (1929/30)	73
M 25 Aerobatic Plane (1929)	75
M 26 High-Speed Commuter Plane (1929/30)	75
M 27 Trainer and Sportplane (1930/31)	77
M 28 Mailplane (1931/32)	78
Organization of the Developmental Team	80
RVM Subsidies	81
Bankruptcy (1931)	81
M 29 Competition and Sportplane (1931/32)	82
Description	83
M 30, the Elektron M 26 (1929)	85
M 31 Sportplane and Trainer (1932)	85
M 32 Trainer Project	86
M 33 Kitplane (1932)	87
M 34 Long-Range or Antipode Plane (1932)	88
Forced Settlement (1933)	89
M 35 Trainer and Aerobatic Plane (1932/33)	89
M 36 Commercial Plane for Romania (1934)	

Collapse and Rebirth (1935) — 93

Restructuring the Aviation Industry	93
Bf 108 *Taifun* Four-Seat Commuter Plane	96
Development of the Bf 108	96
The Solution: Monocoque Construction and Technical Characteristics of the Bf 108	98
The Bf 108 in the 1934 Challenge de Tourisme Internationale	100
How the Bf 108 became the "*Taifun*"	100
Move to the B-series	101
Enthusiasm for the *Taifun*	102
Bf 108C record-setting variant	103
D-series for the Luftwaffe	104
Me 208 - Bf 108 nosewheel variant	104
Continued production of *Taifun* in France	105
Willy Messerschmitt and his *Taifun*	106
Taifun accomplishments	106
Bf 108 *Taifun* variants	107

Me 108 - Postwar Projects in Germany — 107

Bf 109 - Messerschmitt's Entry into Military Aviation	110
Bf 109 Goes into Production	114
Bf 109 Becomes a Sensation	116
Bf 109V-13 Sets World Record	117
Bf 109E with the 30 liter DB 601 Engine	118
Bf 109T for the Aircraft Carrier *Graf Zeppelin*	120
Bf 109 - Special Developments	121
Bf 109 - Two-Seat Trainer	121
Bf 109 with Radial Engine	122
Bf 109F - the Perfect One-Oh-Nine	123
Bf 109G - the Most-Produced Variant	125
A British Test Pilot Assesses the Bf 109G	125
Performance Boost Using the MW 50 and GM-1	128

5

Various Types of Special Evaluation	130
Bf 109 High-Altitude Fighter Projects: Me 409, Me 155, P 1095, Me 109H	131
Bf 109K - Final Large-Scale Production Variant	133
The Bf 109 Goes International	134
Postwar Production - the Czech 109	134
Spanish License Production of the Bf 109G	135
Bf 109 Variants	136
Me 209 Record-Breaking Airplane	137
The "New" Me 209	142
Me 209/309 Performance Table	146
Me 309 - Successor to the 109	147
Bf 110 - A Heavy Fighter Defined	150
Development of the Bf 110	151
Bf 110 in Combat	154
Bf 110 Achieves Success as a Nightfighter	157
Bf 110 Production	158
Five Episodes with the "Political" Bf 110	160
Bf 110 Variants	163
Me 210 - A Major Setback	164
First Me 210s Delivered	170
Me 210 Production Halted	171
Consequences for Willy Messerschmitt	174
Longer Fuselage and More Powerful Engine Change the Me 210 into the Me 410	174
Me 310 and Follow-On Developments	176
Me 210/410 Variants	178
Search for Replacement Heavy Fighters and High-Speed Bombers	178
Projects by A. Lippisch and H. Wurster	178
Tailless or Conventional?	180
Swept Wings and Butterfly Tails	182
Aircraft Using the Modular Assembly Principle - The P 1090 Project	184
Streamlining and Commonality	185
Commonality Even for Jet Aircraft	186
From BFW to the Messerschmitt Company	187
Establishment of the BFW-GmbH Regensburg	191
Bf 109 Large-Scale Production at WNF	195
F 2 Production Ring for Messerschmitt Series Production	196
The Fighter Staff Swings Into Action	197
Comparison Data from Messerschmitt Aircraft Production	197
Messerschmitt Aircraft Production from 1934 to 1945	199

Special Role Developments - From High Altitude Strategic Reconnaissance Platforms to the "Giants"	199
Bf 161 - Tactical and Strategic Reconnaissance Plane	199
Bf 162 - 500 km/h High Speed Bomber	201
Bf 163 - Competitor for the *Storch*	203
Bf 164 (P 1053) - Aircraft for a Round-the-World Flight	205
Bf 165 - Strategic Bomber Project	206
Me 308 - Me C 164, Commercial Airliner	207
Me 261 - The 11,000 km Airplane	208
Me 264 - The 15,000 km Airplane	212
The RLM Vacillates	213
Four Engines or Six?	214
The Me 264 Flies	216
The Last of the Futuristic Projects	218
The Me 321 and Me 323 Giants	220
Heavy Lift Gliders for Operation *"Seelöwe"*	220
Me 321 Assembly in Leipheim and Obertraubling	223
Problems with Glider Towing	224
Transport Gliders in Operation	225
Projects using Steam Turbines and Pulse-Jet Engines	226
The Me 323 Heavy Lift Transport	226
Me 323 Production	228
Me 323 - Technology and Operations	228
Me 321 and Me 323 Variants and Projects	231
Messerschmitt Jets	232
Me 262, First Operational Jet Fighter	232
The P 1065 Pursuit Fighter Project	232
The P 1065 Officially Becomes the Me 262	233
Me 262 Straight Wing Aerodynamic Trials	234
Assembly and First Flights of the Me 262V-1	234
Evaluation with Junkers Jet Engines	235
Luftwaffe Officers Enjoy Flying the Me 262	237
The Long Road to Series Production	238
Momentous Involvement	239
Flight Testing Continues	243
Impetus From the Fighter Staff	244
Me 262 In Combat	246
Me 262 Follow-On Developments	248
High Speed Development Projects	251
Transferring Me 163/Me 262 Technology To Japan	254
Overview of Me 262 Variants	255
The Lippisch-Messerschmitt Me 163 Rocket Fighter	255
Prehistory Up To The DFS 194	255
Project X Becomes The Me 163	256
The 1000 km/h Speed Barrier Is Shattered	258
The Long Road to the Operational Me 163B	258
Erprobungskommando 16 For The Me 163B	265
Readying for Combat	265
Lippisch Leaves the Messerschmitt Company	266
Further Development Leads to the Me 263	266
Me 163/263 Variants	269
Me 328 - Pulse-Jet Powered Multi-Role Aircraft	269
Flight Testing the Me 328 in Hörsching	269

Manufacturing Program for 300 Aircraft	272
Me 328 SO For Kamikaze-Type Operations	273
The Messerschmitt P 1101 Swing-Wing Fighter	275
History of the Swept Wing	275
Planned Increase of Mach Number Leads to P 1101	276
P 1101 Configurations for the Requirement	276
P 1101V-1 Under Construction and as Pattern for Postwar Aircraft	280
The Americans Arrive	281
The P 1101 Becomes the Bell X-5	281
Swing-Wing Types Throughout the World Designed on the Basis of German Research	282
Enzian Remote-Controlled Anti-Aircraft Missile	283
Messerschmitt Projects from P 1000 To P 1112	286
Projects with Official or Unofficial RLM Type Numbers	292
Messerschmitt Propeller Developments (Me P)	292
Type Names	292

The War's End 293

The Americans and French "Capture" the Project Bureau	293
Internment	293
Messerschmitt's Nephew Obtains First Postwar Work	294
New Beginnings	295
Windmills for Bavaria's Energy Needs	295
Neue Technik GmbH Founded	296
Innovative Prefab Homes	296
Postwar Struggles to Keep Aviation Industry Alive - Messerschmitt's "Japanese Funds"	299
The End of Augsburg	301
Messerschmitt Falls Under Trust Company	301
Dismantling Destroys Capital	302
Sewing Machines Bring Initial Success	303
First Postwar Aeronautical Inroads	304
New Concepts for Engine and Automobile Construction	306
P 511 Mid-Size Automobile	306
Messerschmitt's Bubble Cars in Full-Scale Production	310
Vespa Built Under License - Contacts With Alfa-Romeo	313
Minicars Follow Bubble Cars	313
Technical Office Keeps Busy With the *Kurvenleger*	315
Messerschmitt in Spain	316
Aircraft Development Again	316
Close Cooperation with Hispano Aviacion	317
HA 100 Piston-Engine Trainer	317
Comparing the HA 100 with the T-28A	320
HA 200: First Post-1945 Jet Development	321
HA 300 Supersonic Fighter Developed	326
Messerschmitt Reaches Pinnacle of Fighter Development In Egypt	331

Saeta Becomes Al Kahira	331
Engine Troubles Delay the HA 300	332
HA 300 - Lightest Supersonic Fighter in the World	337
Signs of Renewed Aircraft Development	340
Aviation Development Begins Anew in 1955	340
Licenses, Programs, Cooperative Efforts	341
Fouga Magister License-Built	341
License Production of Fiat G.91R3 Ground Attack Plane	342
F-104G Starfighter Most Significant License Program	343
Parts for the Bell UH-1D	344
Fuselages for the Sikorsky CH-53G	344
Transall C-160 Center Wing Sections	344
Gleaning "Know-how" from the F-4 Phantom II	345
Casa C-101	346
Tornado European Combat Aircraft	346
VJ 101: World's First Supersonic VTOL Plane	347
Project Competition for Interceptor	347
Two Prototypes Built	349
VJ 101C-X2 Tests STOL Capabilities	352
VJ 101D Follow-On Development	353
German-American Weapons System	355
The Road to Merger	355
Messerschmitt's Legacy: The "Rotor-Jet" Heliplane	356
Competing Developers Reveal Advantages of Messerschmitt's Concept	357
Projects	365
In Praise of the Messerschmitt Foundation	378
Messerschmitt-Bölkow Merger Paves the Way for Restructuring the Aviation and Space Industry	380
70th Birthday Celebrated as Sign of Messerschmitt-Bölkow Merger	384
75th Birthday in Auditorium of Deutsches Museum	385
80th Birthday: "Messerschmitt" - More Than Just a Name for Technological Products	387
Epilogue	390

Appendix

- Photo Credits	4
- Coworkers and Contemporaries	391
- Company Organization/Overview	396
- Aircraft Data	398
- Sources and Suggested Reading	403
- Abbreviations	408
- Index by Subject	409
- Index by Persons	411
- The Authors	415

Introduction

A memorable event from when I worked with Professor Willy Messerschmitt perhaps best portrays his successful organizational creativity, a unique ability which is repeatedly manifest in the great designs addressed within the pages of this book.

A moment in time from 1941 illustrates one of the key aspects of this creative talent.

The Messerschmitt AG was contracted to evaluate the Bf 109 using a so-called butterfly tail—a combination elevator and rudder produced by arranging these control surfaces into a V form. The improvement in drag seemed enticing. However, we aerodynamics engineers, particularly those in the flight characteristics group, had a few reservations due to the stable neutral position necessary when approaching a target.

With wind tunnel testing completed, the contract was awarded. After awhile, the company's contracting office began criticizing us in the project section—we had overall responsibility for the project—about the delays. I complained to Messerschmitt that work wasn't progressing because we were overloaded. "Give me a few of your documents on the support spar, the empennage and the aerodynamic data. I'll think about how we can work through this on Sunday," he replied.

Monday morning I was called in. Messerschmitt handed me 28 pieces of drafting paper, some of them stuck together, covered with scribblings from a blue pencil, and for the most part with measurements noted. "I just drew up a few things, see what you can do with it." I took the papers to Lorenz Bosch's empennage team in the design office. Together with the stress analysts we'd called over, we came to the conclusion that the documents had been so well drawn up, calculated and measured that a modern testing workshop could have used them to construct the tail without the need for any further drawings. All we needed was a draftsman. Two days later the blueprints were turned over to the shop.

That was the enthusiastic and unusually creative Willy Messerschmitt. The VJ 101's landing gear from the 'sixties is another example of his special gift, one which he retained until well into an advanced age. During the many years of cooperative work we learned to value another side of his engineering skills: after clear, sensible observation of a problem, he had the resolve to subsequently work the problem through and then apply the solution, taking into account extreme demands for minimal weight and surface area. When necessary, Willy Messerschmitt even fought for his convictions to the point where he would draw up a detailed parts list. For me and many others, these were years of learning, traces of which we are continually discovering after all this time.

An example of his engineering team's abilities was the Me 321 transport glider, a project which also revealed how the team, with him at its helm, was able to develop a design in short order. At the time, we were able to go from contract award to first flight of this—for us seemingly gigantic—airplane in less than four months.

Messerschmitt's restless personality simply was unable to sit quietly in the years following the collapse. After several more or less successful ventures with pre-fabricated houses, bubble cars, and automobiles, it wasn't long before he was again working on aircraft development as a technical advisor to the Spanish company of HASA. He eagerly jumped at this opportunity, one which ultimately resulted in the creation of the HA 100 piston-engined trainer, the HA 200 jet trainer, and the HA 300 supersonic fighter.

As part of the EWR during the 'sixties and 'seventies, he played a role in the development of the MRCA (later Tornado) and in the Airbus design, making many recommendations. However, all the while his dream was in reality focused on a VTOL commercial airliner. Applying his typical radical thinking approach, Messerschmitt went beyond his ongoing ideas he'd been working on since the mid-'fifties with organizations such as the Bell Company in the U.S. His "Rotor-jet" concept envisioned helicopter-like rotors shutting down during horizontal flight, when turbojet engines would then take over and provide the means of propulsion. Because of drag, the rotor blades would then fold back and retract into nacelles. However, he underestimated the technical difficulties and the requisite financial means to support the project. The complex Rotor-jet concept revealed his willingness to tackle even the most extremely challenging technical problems.

Those of us who had the opportunity to work with Professor Messerschmitt are grateful for what we've learned from him—and continue to carry with us—about creative fantasy, attainable reality, and the power of persuasion when applying sound research findings.

Dipl.-Ing. Dr.-Ing. E. h. mult. Ludwig Bölkow

Introduction

Four years after the end of the First World War the 25-year-old graduate engineer Willy Messerschmitt set up his own company and, in doing so, became a direct competitor to the well established aircraft designers and companies of the day. His ten years of instruction in sailplane development, provided under the tutelage of *Regierungsbaumeister* Friedrich Harth, was unconventional given the state of technology at the time. Extremely lightweight design in conjunction with good aerodynamic form and controllability became the bases for the successful metamorphosis of glider into sailplane. The subsequent introduction, application and continued development of these principles with regard to powered flight carried over into the jet age and became hallmarks of Messerschmitt's life work as a creative engineer.

As a businessman in the aviation field, he also masterfully displayed a knack for developing marketable aircraft; the sport of aviation in the 'twenties was punctuated by numerous publicly sponsored races, and Messerschmitt's customers not infrequently placed quite well in these events when flying his early lightplane designs.

In addition to the ability to carry an optimal load and have good cruising speeds, the economically feasible commercial airliner also had to be able to withstand inclement weather conditions and provide passengers with better protection in the event of a crash, features only a metal plane could offer. Thus it was that by 1925 the young engineer/businessman was hard at work developing the smooth-skinned metal alloy construction with the intent of garnering the contract for the M 18 commuter airliner. Marketable planes could not be too expensive, however, and Messerschmitt—not unlikely due to his beginnings marked by an ability to craft things by hand—was a master in the fine art of producing easy-to-build designs. This talent enabled him, in the interest of bettering his products' performance, to be one of the first to successfully employ what was then considered to be complicated subsystems: automatic slats, Fowler flaps, and the cantilever single-strut undercarriage. In 1934 the Bf 108 Taifun, an all-metal design using these flap and landing gear systems, became the prototype of today's private planes with its pioneering max/min speed performance characteristic.

The quintessential light fighter concept proved to be both marketable as well as pioneering when it took shape in 1935 as the Bf 109, a design in which Messerschmitt combined a high performance engine with the technological advances already incorporated into the Taifun. In doing so, the resulting product (which had what was then considered high wing loading) did not initially meet the *Luftwaffe* leadership's fixed ideas laid down in the prototype competition, ideas which primarily focused on dogfighting maneuverability. However, Messerschmitt and other visionary personalities were able to convincingly argue that, given the state of modern technology, it was entirely possible to design a high speed transport that could simply walk away from a fighter design which stressed maneuverability over speed.

The marriage of high performance (in 1937 a variant of the Bf 109 broke the world's speed record) with a well-conceived design ensured that the Bf 109 (later Me 109) would become the *Luftwaffe*'s standard fighter.

With few exceptions, the majority of Messerschmitt's other developments also proved to be marketable and saw large production orders. These also included larger aircraft, particularly the Me 321 glider transport which—due to the war's pressures—went from initial design work to first flight in the unbelievably short time of just four months, and to this day serves as a shining example of the creative power of Messerschmitt and his team. The first jet fighter produced in significant numbers (1944 to 1945) was the Me 262, a design which achieved worldwide recognition for its ability to leave all other fighters of the day in the dust. This design heralded the departure from piston-driven aircraft throughout the world.

Messerschmitt's next developmental step occurred in 1945 with the practical development of the prototype P 1101 fighter, a design whose sweep-wing concept—fundamental technology for today's jet age—was based on initial wind tunnel testing by the team back in 1939/40. Although the war prevented the plane from being flight tested, recognition of the technology embodied in the prototype was a further impetus for the rapid onset of the jet age.

Because of the postwar conditions prevalent in Germany, it wasn't until nearly twenty years later that Messerschmitt's association with the jet age became apparent once again with the flight testing of his HA 300 supersonic lightweight fighter and VJ 101 VTOL plane. These projects were followed by concept studies for the later MRCA Tornado and the trend-setting Airbus design.

For his part, Messerschmitt himself had received the highest recognition for his pioneering work in aviation back in 1953 when he was invited to the U.S. as the guest of honor at a celebration commemorating 50 years of powered flight.

Prof. Dipl.-Ing. Gero Madelung

Foreword

The eight decades of the *Luftwaffe*'s colorful history can be broken down into five main periods:

A number of aircraft manufacturers emerged during the buildup and industrialization phase from 1908 to 1914. These manufacturers initially built French aircraft designs under license, or were chiefly influenced by French products. Large scale production of warplanes during the First World War saw these companies undergo major expansion.

Following the end of World War One, the victorious Allies levied a ban on construction, bringing ruin to many aircraft companies or forcing them to diversify into other product lines. Once the ban was lifted (although military aircraft still fell under the restriction), brave young engineers set up new aviation businesses. These firms (with much encouragement in the form of State-sponsored competitions) over the next ten years developed numerous prototypes, with a handful entering small scale production. A few of these developments even became international trendsetters.

The political upheaval in Germany in 1933 brought with it the creation of a new military and *Luftwaffe* and an unprecedented boom in aeronautical research and development. This resulted in fundamentally new aerodynamic principles and radical new approaches to construction methodology. Jet engines began appearing in the mid 'thirties, adding an entirely new dimension to the aviation industry. Eventually 350,000 people would be employed in this field and, despite working under extremely harsh wartime conditions in the final years of the conflict, would produce many different types in significant numbers (35,000 Me 109s, 20,000 Fw 190s, 15,000 Ju 88s). By the end of the war Germany's aviation program had risen to become the world's indisputable leader from a technological standpoint.

The complete destruction of Germany's aviation industry in 1945 resulted in a ten-year forced hiatus. Aircraft manufacturers—those which had survived—were once again compelled to diversify into non-aviation products. Many engineers quickly and easily found employment abroad. In 1955 the Federal Republic of Germany began to exercise its sovereignty, taking on responsibility for its own defense, and surviving aircraft companies began planning accordingly. Maintenance work on first-generation combat aircraft inherited *in situ* from NATO partners served to provide the initial foundation for new programs. Government contracts influenced the formation of joint ventures. Trainers and transports, followed by combat aircraft, were built under license. Development of a family of VTOL aircraft saw Germany's aviation industry again reach prominence on the international scene; it was now capable of sustaining itself as a partner in the great cooperative programs so common in today's European aviation industry.

By the mid 'sixties Germany's aviation industry was beginning to restructure itself—again under the influence of the Federal government. Focke-Wulf and Weserflug joined to become the "Vereinigte Flugtechnische Werke" (VFW) in 1963. Heinkel cast in its lot in 1964, and when Holland's Fokker joined in 1969 the first European partnership was formed: VFW-Fokker. In 1968, Messerschmitt and Bölkow merged in southern Germany. In 1969 the Hamburger Flugzeugbau was acquired and Messerschmitt-Bölkow-Blohm (MBB) established. In 1981 MBB absorbed the Vereinigte Flugtechnische Werke, from which Fokker had broken off previously. With Daimler Benz AG throwing its hat into the ring in 1985, there began a comprehensive restructuring of the aviation and space industry in Germany. The Deutsche Aerospace was founded in 1989 by combining Daimler-Benz's interests with those of MTU, Dornier, and shares from AEG. Takeover of MBB at the end of 1989 led to an enterprise which, like the approach pioneered in neighboring England, France, and Italy, combined the key national-level system management functions in the various divisions of the aviation branch. This was a prerequisite, not only for the company's long-term successful position in the world's competitive marketplace, but also for securing Germany's role in both European and global cooperative efforts, as well as joint-venture companies.

Willy Messerschmitt not only experienced this tumultuous history of German aviation for over six decades, he played a major role in influencing it. This book is an attempt to portray that. It is founded on meticulously researched files (which not only convey facts, but to a lesser extent also reveal something of the circumstances), as well as numerous conversations and interviews (some of which were recorded) with contemporary witnesses—whose numbers have dwindled considerably during the years it has taken to prepare this book. Despite the utmost care, the material contained within these pages must remain heterogeneous, even fragmentary in places, because of the various levels of information and lack of documentation. This is particularly true with regards to the scope of this book. In order to provide an easier-to-understand overview of Messerschmitt's work, the material has been compiled with an emphasis on content over chronology.

The authors wish to thank those individuals without whose discussion, advice, and varying kinds of support this book would not have been possible:

Heinrich Beauvais, Dr.-Ing. E. h. Ludwig Bölkow, Ferdinand Brandner (†), Heinz Braun, Dr. Johannes Broschwitz, Wolfgang Degel, Oskar Friedrich, Josef Fuchshuber, Kyrill von Gersdorff, Werner Göttel, Werner Heinzerling, Helmut Kaden (†), Josef Kemper (†), Günter Klug, Arnulf Krauß, Gero von Langsdorff, Wolfgang Mach, Ello Madelung (†), Prof. Gero Madelung, Jürgen Mataré, Hans Justus Meier, Prof. Dr.-Ing. Elmar Messerschmitt, Alfred Nerud (†), Willy Radinger, Julius Reicherter, Kurt Schnittke, Karl Seifert, Dr. Hans-Heinrich Ritter von Srbik (†), Dr. Bruno Storp, Mano Ziegler (†), and Horst Zocher, as well as the Bayerische Staatsbibliothek Bamberg, the Bundesarchiv (Militärarchiv) Freiburg, the Stadtarchiv Augsburg, the Institut für Zeitgeschichte, Munich, the Imperial War Museum, London, the Musée de l'Air, Paris, and the Deutsches Museum, Munich. Where possible, originals were used for the drawings. Where this was not possible and new drawings had to be created, these were provided by Horst Fechner, Theo Lässig, Theodor Mohr and Helmut Stützer.

Munich, winter 1991/92

Hans J. Ebert, Johann B. Kaiser (†), Klaus Peters

Willy Messerschmitt - Aircraft Designer, Aviation Pioneer, and Researcher[1]

The name of Willy Messerschmitt first generally became associated with aircraft design as a result of Harth's endurance flight in the S 8 glider.[2] The S 8, as was typical of all Harth's and Messerschmitt's gliders, made use of "wing controls"; i.e. tipping the wings from a pivot point on the fuselage side changed the angle of attack, which in turn affected the craft's altitude. Aileron control was accomplished by twisting the wingtip rib sections around the rounded outer spar. The basic concept behind the idea was to take advantage of short-term wind changes, i.e. gusts, with the aid of the flexible wing. In actual fact, on several occasions Harth succeeded in taking off from a standing start without the need for outside assistance.

In understanding Professor Willy Messerschmitt's background, it is important to know that while in his early, formative, years he had the opportunity to work with an experienced professional who set before him design tasks which deviated markedly from the norm. Just a few years later we find a young Messerschmitt as a loner on the brink of his own meteoric rise in the world of powered flight.

Taking a look at the entire line of his creations today, from the S 15 powered glider through the Me/HA 300 jet fighter (totaling 37 basic types, not taking into account the numerous variants of many of his designs), there are three separate and distinct principles employed by Messerschmitt in his designs which stand out:

reducing drag
reducing design weight
simplifying construction

Reducing Drag

Nearly all of Messerschmitt's designs were monoplanes, some high-wing, some low-wing. Messerschmitt (like Junkers) had recognized early on that: a) two wings (i.e. a biplane) exposed to the flow of air produce more drag than one single wing, even if the single wing has equal

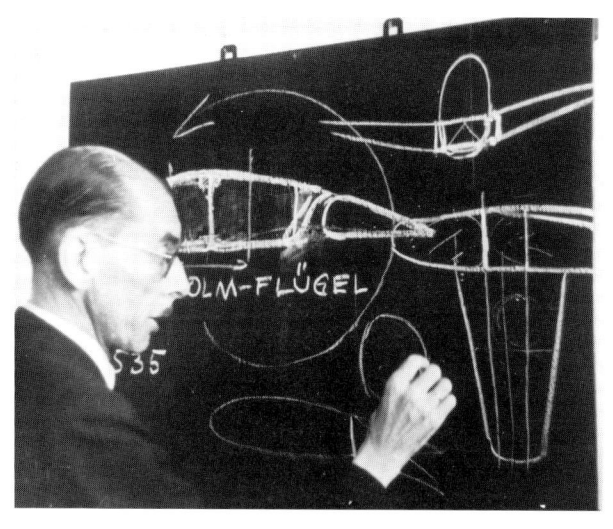

Regierungsbaumeister Prof. Dipl.-Ing. Julius Krauß (1894-1972) here explains the advantages associated with Willy Messerschmitt's single-spar wing design. From 1924 to 1945, Krauß had worked at the Udet Flugzeugbau, BFW and ultimately Messerschmitt AG. In 1942 he was awarded a professorship as a faculty member of Munich's Technische Hochschule machine engineering department, at the same time becoming chairman of the institute's aircraft branch.

load capacity of the two, and b) thick wing profiles are structurally more sound, and in order to reduce this drag effect he became a diehard proponent of the cantilever monoplane idea. The few biplane construction designs he produced were created expressly for customers partial to that design.

The M 20 was publicly displayed for the first time at the 1928 International Aviation Exhibition in Berlin. It was a portend of great things to come in the world of aviation: a 12-seat, all-metal airliner; cantilever single-spar tapered wing with an aspect ratio of 10; additional load equating to 47% of its take-off weight; less than 90 km/h stalling speed, with a maximum speed conservatively estimated at 175 km/h with a 500 horsepower engine. It came as quite a surprise when during test flights the plane actually reached a maximum speed of 220 km/h, and thereby a speed range of almost 2.5. At a time when the concept of increasing lift

---

[1] Taken from the anniversary publication "40 Jahre Messerschmitt-Flugzeugbau 1923-1963" authored by Prof. Julius Krauß and published in Augsburg in 1963 in honor of Willy Messerschmitt's technical products. Later republished as "Messerschmitt - 50 Jahre Flugzeugbau von 1923 bis 1973 - Messerschmitt-Bölkow-Blohm" (Ottobrunn 1973), and "Messerschmitt - Sechs Jahrzehnte Flugzeugbau - Messerschmitt-Bölkow-Blohm" (Ottobrunn 1978).

[2] Compare with W. Messerschmitt: "Der 21-Minutenflug auf dem Heidelstein am 13. September 1921" in the magazine "Flugsport" 1921, no. 21.

through the use of flaps and other means had not yet been explored this was an unusually good value.

Several foreign developers (including Dewoitine in France) subsequently picked up on the ideas embodied in the M 20. In any case, by this time the name of Messerschmitt had already become synonymous with the concept of clean, aerodynamic design workmanship.

Considering that, with regard to frictional drag, it was important to keep an airplane's surfaces exposed to the flow of air to a minimum, Messerschmitt constantly worked at keeping the size of a given airplane to a minimum. Since the external covering is what forms the aircraft's surface and accounts for 50% or more of the aircraft's structural weight, in addition to the aerodynamic advantages any reduction in dimensions would also involve a considerable savings in weight, benefitting the plane's flight performance in turn.

For reasons of stability, reducing the size of the wing naturally meant shortening the fuselage considerably, as well as reducing the size of the empennage.

Messerschmitt's standard practice involved a methodical, systematic hunt for an aircraft design which limited the amount of additional drag. In this respect the undercarriage became a particular focus of his attention, since in an age before the "disappearing" undercarriage even the most favorable design—a spatted, streamlined, triple-strut system for each side—had a major negative impact on aircraft performance. In this case Messerschmitt created the "single-strut" landing gear, consisting of nothing more (aside from the wheel) than a single rigid, non-twisting, non-bending strut housing the shock absorption device (first used in 1931 on the M 29 series). This innovation later proved to be particularly suited for the retractable landing gear, since it incorporated a minimum of moving parts required to draw in the undercarriage, as for example on a low-wing monoplane (Me 108, Me 109, etc.).

It is well known that the varying characteristics of ambient airflow over the wings and fuselage produce interference drag where the flows come in contact with each other; this interference drag can be a significant factor in low wing designs, especially when the wing/body junction angle is less than 90 degrees.

So-called "fillets" are used to reduce this type of drag. These are curved metal sections which round out or blend in the area between the inner wing and the fuselage, and on airliners can often be quite large so that their drag reducing effect is seriously offset by an increase in frictional drag. For example, on the Me 262 Messerschmitt dispensed with the idea of a fillet, instead designing the fuselage cross section in this area as a triangle (base on the bottom) with curved corners, so that a distinctly shallow angle resulted between the upper wing and the fuselage sidewall and thereby—for the most part—avoided interference drag without necessitating additional frictional drag.

Improvement in aircraft performance is a natural byproduct of any reduction in drag. In two well-attended lectures presented before the *Deutsche Akademie der Luftfahrtforschung* (German Academy for Aeronautical Research) in 1937 and 1938, Messerschmitt shared his thoughts on the possibilities of increasing flight performance.[3] His observations were not exclusively reserved for advancing airframe design, but also addressed engine development, and in so doing took the view of distributing the requisite engine power into many small units side-by-side under the wings to avoid increasing drag.

In a memorandum to the *Deutsche Akademie der Luftfahrtforschung* in 1942 he discussed the "Development of the Concept of the Flying Wing and Swept Wing Design for Increasing Flight Performance," this coming at a time when other countries were still completely unaware of employing the swept wing concept as a means of raising the critical Mach number.

One particular characteristic of Messerschmitt's own personal makeup is reflected in one of the previously mentioned writings, which contains his commentary on the professional ethics of the aircraft designer:

"The designer sees not only the aircraft flying today, and he sees not only the aircraft currently being built or in the first stages of design, but instead looks further into the future; what is reality to the general public today is already obsolete in the designer's mind. Long before the completion of a new aircraft type he knows what could have been done to improve it. This tragedy of our profession is also the driving force that spurs us to ever newer creations, and we are fortunate to be Germans, to have the full understanding that our work will never be finished."

Reducing Design Weight

Messerschmitt's designs are a rich source of demonstration material to illustrate the consistent use of the principles of light construction.

The first and foremost principle: reduction in the size of the object to an acceptable minimum while meeting all other requirements, something which has already been discussed. For aircraft this chiefly applies to the weight of the rigid materials (skin + stringer + ribs), which accounts for 50% (spar wing) to 80% (monocoque wing) of the structural weight. Thanks to their low weights, Messerschmitt designs, assiduously adhering to these principles, were con-

---

[3] Writings of the Deutsche Akademie der Luftfahrtforschung, vol. 1: W. Messerschmitt "Die Entwicklung der Flugleistungen," and vol. 31: W. Messerschmitt "Probleme des Schnellflugs."

stantly the best proposals in requests for tender in which the customers placed equal emphasis on payload, engine and minimum flight performance.

The second principle of light construction calls for the multiple exploitation of parts and sub-assemblies on hand. For instance, this refers to the open spaces generated by the external shell being drafted to absorb part of the load, as with the long-range Me 261 where its entire fuel capacity was carried in the close-riveted sealed wings to give it an operational range of approximately 11,000 kilometers. It also means making use of an available structural section for several jobs.

One example is the monospar wing with torsion-resistant leading edge. On this wing the leading edge, an aerodynamic necessity, is made so torsion resistant that, together with the spar, it forms a rigid tube. The leading edge as well as the spar frame then both absorb the shearing stress from the torque, but this is offset by the stress from the lateral forces, whereas the spar flanges inherit the longitudinal force from the strain of bending.

This typical light construction methodology, standardized with wood materials in glider aircraft construction, was first introduced in light metal by Messerschmitt in his M 18 and utilized in subsequent models. This clearly wasn't just a simple "translation" from wood to metal; rather, the requisite stringers and support spars used in the metal construction concept resulted in changes in stress points, which necessitated detailed research, particularly with regard to the danger of instability of the thin-walled structural elements.

In this context, it is interesting to note that a variant of Messerschmitt's metal monospar wing can be found on the Vickers Viscount airliner, among other types (the Viscount wing utilizes a main spar for flex support, two support spars acting as front and rear torsion tube boundaries, and closely spaced ribs in place of stringers).

Another example of multiple exploitation is found in the fuselage former construction employed on several of Messerschmitt's designs. In general, formers have the role of displacing lateral forces, supporting the stringers (reducing their flex length), and providing overlap attachment points for the individual metal skin panels of the fuselage.

For these different functions, Messerschmitt found a brilliant, weight-reducing and inexpensive solution in which he dispensed with the former as an independent element, instead flanging off the edges of the metal skin panels to act as a replacement for the formers (first used with the Me 108).

The concept of multiple exploitation also brings to mind a stroke of genius by the German cast steel industry which, like engine manufacturers and many in the armament industry, demonstrated their wholehearted acceptance of Messerschmitt's ideas. This was the forward fuselage structure of the Me 109, which had to fulfill three roles simultaneously, namely: the attachment point for the single strut undercarriage, lower attachment point for the engine support frame, and the attachment point for the forward support spar in the monospar wing.

The third light construction principle states: exploiting the aircraft's dimensions, i.e. using available space. One example from Messerschmitt's designs was locating the main spar inside the monospar wing at the thickest point of the wing's profile. It is easy to see that there is an optimal total height for any flexible support structure comprised of cap strips and rib webs. However, this height is far greater than that permitted by the aircraft's physical limitations. All the more important for utilizing the available dimensions.

Another example also focuses on the metal monospar wing: landing flaps and ailerons consume up to 1/3 to 1/4 of the wing chord. Comparing the torsional rigidity of a fully skinned wing to that of a wing covered only to a fixed trailing edge, which varies in chord along the length of the wing, the designer finds that the best ratio at a given point of the fixed trailing edge is somewhere between 60% to 75% of the wing chord, depending on the type of wing profile. The skinning called for in later Messerschmitt designs, covering the main spar as far as the edge where the landing flaps and ailerons connected, can be chalked up as an optimal solution, a fact which, upon careful investigation of structural details of Messerschmitt designs, is borne out time and time again.

Though there are many other basic principles besides those already mentioned, they cannot be covered here due to space limitations. The fourth and last basic principle of light construction to be addressed is the requirement to get the most out of the materiel already referenced. Without a doubt, it is this principle which causes the greatest amount of consternation in light material construction. In many cases, materiel research is completely unable to supply the concept with the strength values for a piece or component subjected to operational conditions (endurance strength, design strength, creep limitation, and so forth). Therefore, this knowledge must be acquired through relevant, albeit limited testing and, in the end, those involved must simply hope that the lengthy (and costly) multi-stage stress test program with the prototype airframe for determining the life of the aircraft won't cause any significant reworking of the design. For the monospar wing the situation here is a critical one, since it is almost impossible to achieve "fail safe" conditions. Instead, the focus is placed on what is known as "safe life."

Independent of this, the strength of an aircraft type still hinges on the the field of statics, or stress analysis. When designing metal aircraft, there is always some element of doubt with regard to the support strength of the thin-walled structural elements, and Messerschmitt initially employed the method of determining the strength based on a "calculated weight" 20% higher than actual anticipated takeoff weight. For example, a flying weight of 5,500 kp was applied to the M 20, although it had an actual takeoff weight of 4,600 kp. This approach, while it may in general go a long way toward compensating for uncertainties, offered no guarantees that potentianl localized problems, e.g. instability of stringers, could be totally overcome.

On later aircraft prototypes Messerschmitt paved the way for "weak point research" which, although hotly contested initially—mainly at the proving ground—is today even making inroads in foreign aircraft construction. At the time, German design loads (i.e. safe load) were expected to have a safety factor of 1.8 for production aircraft, while V-types (prototypes) only required a 1.65 safety factor. Messerschmitt demanded from his employees accurate stress data for his prototypes and expected them to find any weak areas cropping up when the safety factor was raised from 1.65 to 1.8 through load testing on the finished product. During the test phase, these mostly localized weaknesses were overcome through temporary structural reinforcements, which would then later be incorporated into the production blueprints using a more permanent fix.

In this manner, all of Messerschmitt's aircraft types, in attaining the level of strength prescribed for production aircraft, succeeded in approaching close to the ideal concept of "an object of equal strength." And in so doing, his designs were kept to their minimum possible structural weight, a particularly important aspect in light of the weight multiplier (1 kp of additional weight equates to an 8 to 17 kp increase in the overall aircraft weight, depending on aircraft type).

Simplifying Construction

Several examples of this concept have already been illustrated in the preceding sections. The following lines are taken from a report based on the enemy's analysis of a captured Me 109:

"Investigation into the production of this aircraft clearly shows the great successes achieved by prioritizing simplicity, functionality of design and ease of operation. The systematic use of strong metal panels enables the number of stiffeners to be reduced. Particular merits of the design include the liberal use of cast and molded parts—used to advantage to replace smaller components in large-scale production, the mastery of casting techniques using magnesium-based alloys—evidenced by its abundant usage, and the interchangeability of parts, which seems to have been well-resolved if examination of airframes after a dogfight is anything to go by. (from "Luftwissen," 1940, page 423)

Messerschmitt the Pioneer naturally had to take into account numerous setbacks as he embarked into uncharted territory. However, they were always of the sort that spawned scientific research, the consequences of which invariably benefitted general aviation. Breaks in the wing ribs and rudder on the M 20 prompted a rework of the design load in relationship to pressure distribution along the length of the profile chord; swelling on the wing stringers of the M 24 in gusty weather resulted in Küßner and Kaul carrying out an extensive series of flight test series to evaluate the effects of gust forces on an aircraft; the M 27's abnormal spin characteristics prototype provoked a whole series of wind tunnel tests with generally noteworthy results; porpoising of the M 29 prototype, equipped with an all-moving tailplane, initiated a series of tests for understanding the problems of pitch stability on such aircraft. The simultaneous cross-pollination between production and research played a major role in maintaining Germany's leading position in aviation for more than a decade.

Messerschmitt himself often only mentioned his role in passing, such as in his article "German Contributions to Aviation Development" (Interavia 1953, Number 12), in a speech presented before the "Instituto Superior Tecnico Lissabon" entitled "Development of Aircraft Construction," excerpts of which appeared in "Messerschmitt Nachrichten" Nos. 11 and 12, 1962, as well as No. 1, 1963. Glancing over all of Messerschmitt's creations, the reader should not ignore his other developments, most of which stemmed from times of crises. Messerschmitt's creative spirit is readily apparent in all his works, including windmills, prefabricated housing, automobiles, engines, transmissions, bubble cars, and sewing machines, and it should come as no surprise to discover that by mid-1963 Professor Messerschmitt held 283 personal patents, and the Messerschmitt AG yet another 470 patents.

A more detailed look at those Messerschmitt types not already mentioned will be provided in the following chapters. However, some of his other, newest designs are worthy of note here; these were shown as models at the 1962 Paris Aero Salon to the delight of visitors at the exhibition. It was especially the Me 308 which caught their eyes, a follow-on design to the trusty old Me 108 recognized by aviation buffs as a modern commuter and business plane featuring an increase in the number of seats from 4 to 6, a pressurized cockpit, nose gear in place of the earlier tailwheel (already planned for the Me 208, a new development of the *Taifun* dating from 1941 and 1942), twin-jet engines in place of the centrally mounted piston engine, 780 km/h cruising speed at an altitude of 11,000 m instead of 265 km/h near sea level, and modernized equipment.

Prof. Willy Messerschmitt here congratulates *Prof.* Julius Krauß, who for many years was one of his closest colleagues, in Munich on 3 August 1969 on the occasion of the latter's 75th birthday.

Young Messerschmitt (1913-1923)

From the Family Journals

Wilhelm Emil Messerschmitt (nickname and signature: Willy) was born on June 26, 1898, in Frankfurt am Main. He was the second son from the second marriage of Ferdinand Baptist Messerschmitt and had four siblings: a brother, Ferdinand, 14 years his senior (born in 1884); brother Rudolf, three years younger than Willy (nickname Bubi, born in 1902); and sisters Maria (nickname Maja, born in 1901) and Elisabeth (nickname Betti, and later Ello, born in 1907).

Initially his father had aspirations of becoming an engineer, and to this end studied mechanics at the polytechnical center in Zürich, Switzerland. His older brother Andreas was expected to eventually take over the family's wine shop on Lange Straße in Bamberg. Although he indeed worked in the wine trade, he had little interest in assuming the mantle of responsibility (he had emigrated to the United States), and so his brother Ferdinand was forced to give up his occupation and take over the family business; in 1906 he therefore moved with his entire family to Bamberg and took over his father's operations.

From 1908 onward Willy Messerschmitt attended the first classes held at the new high school in Bamberg. He was an intelligent young man, with musical as well as scientific interests. Good role models for Messerschmitt during this period of his life were his two uncles: Pius Ferdinand Messerschmitt, a particularly successful nature painter living in Munich-Solln, and Johann Baptist Messerschmitt,

The Baptist Ferdinand Messerschmitt family in Bamberg, circa 1911 (l. to r.): Rudolf, youngest son (1902-1983), mother Anna Maria, nee Schaller († 1943), father Baptist-Ferdinand (1858- 1916), daughters Ello (1907-1990) and Maja (1901-1985), eldest son Ferdinand (1884-1958) and Wilhelm (Willy, 1898-1978).

18

The Messerschmitt wine merchant's shop, Willy Messerschmitt's home during his early years, in Bamberg at Lange Straße 41. This is also where the first aircraft designs were built. Today the building houses a restaurant by the same name.

who had a PhD and worked as a surveyor in Switzerland. These two occupations, seemingly so different, were indistinguishable to Willy Messerschmitt; he both designed and painted with considerable talent and devotion, and ever since learning to read had devoured everything he could get his hands on relating to science and technology. That same year he witnessed the Zeppelin "*Bodensee*" at Friedrichshafen, and the following year (1909) was a visitor to the colossal International Aviaton Exhibition in Frankfurt am Main, where he experienced firsthand the rapid growth of technology, and in particular aviation. Such stimuli sparked in Messerschmitt a passion that would determine the course of his life's work.

The 12-year old began designing rubber-band powered model aircraft, and even at this early stage went through the process of going from complicated biplanes to monoplane layouts, the latter at first being elaborate constructions which quickly evolved into radically simplified designs.

Nobody was happier than his father Ferdinand, who saw in his son what he himself had once given up in order to fulfill a sense of family duty. Therefore, in the autumn of 1910, he pulled his son Willy out of high school and sent him to the Bamberg secondary school.

Starting Point

Bamberg, autumn 1913: the 33-year old government architect Friedrich Harth and a handful of assistants had been experimenting with hand-made, wing controlled sailplanes (technically known as gliders in today's terminol-

Messerschmitt's practical aviation began with the construction of various aircraft models, initially based on available plans; this is his first original design, "Eindecker I," from January 1913.

19

The student Willy Messerschmitt poses with one of his compressed air powered models in a field outside the city of Bamberg (circa 1913).

ogy) for three years. These experiments took place on the Ludwager Kulm, a flat knoll approximately 20 kilometers north of Bamberg, not far from the small hamlet of Scheßlitz. Around this time a young Willy Messerschmitt joined this group, whose goal was to develop an aircraft "that could rise into the air without the aid of anything other than the wind and remain aloft, floating freely and safely" (quote from Harth's journal). They wanted to build the sailplane.

Friedrich Harth had already performed flap-wing flight tests while a *Gymnasium* student in Landshut in 1897/1898. Running into the wind produced noticeably strong resistance when flapping, and this reinforced in him the belief that, by flapping powerfully enough, the weight of a man could become airborne. However, a full ten years would pass before Harth would once again entertain any serious thoughts of flight. The most likely reason was because, at this time, no serious person would publicly engage in such "tom-foolery" as flying if he didn't want to be made a fool of. Peter Riedel called this the spirit of the times, and considered it the greatest hindrance to the development of the aircraft's "heavier-than-air" concept. In his book "Start in den Wind," he states:

"The spirit of the times—and this will come as a great surprise to us today—represented the greatest hindrance of all when it came to finding our way into the air. It restrained many people from becoming involved in flying who otherwise had a burning desire to experiment. Famous scientists on both sides of the ocean repeatedly declared that flight by a "heavier-than-air" craft would be totally impractical, now that travel using the "lighter-than-air" concept had been possible for some time with hot-air balloons and, later, gas balloons. Furthermore, the press took great delight in mocking those few aeronautical researchers who remained undaunted in their pursuits. Engineers feared for their reputation when they seriously and publicly became involved in matters of aviation. In order for us today to understand the difficult path traveled by the first aviation pioneers, it is important that we are able to visualize the prejudices which the brilliant research and engineering achievments of the last century's final decades were, remarkably, able to overcome.

It was a total surprise for me when I discovered, while sorting through contemporary documents for source material, how strange it was that the general prejudice of the European community suddenly changed and reversed itself to become an enthusiasm for flying which bordered on hysteria. It was Wilbur Wright's first flight in Europe on the 8th of August, 1908, near the city of LeMans in central France. A flight at a height of thirty or forty meters, lasting no longer than a single minute and forty-five seconds!"

By this time, Otto Lilienthal, "the first flying man", had already been dead for 12 years. It is difficult to understand that aeronautical research, a shining path so brilliantly marked out by Lilienthal, had practically come to a standstill in Germany since his death. Was it his death from flying that had caused this? The saying that a prophet is not welcome in his own land has rarely been more applicable than in this case. In the event, Lilienthal slipped into oblivion, and thoughts of flight were pushed further out of the limelight by the airship builders and their initial successes. On the other hand, it was mainly the Wright brothers in the U.S. who carried on the work of Lilienthal and who publicly stated they owed a debt of gratitude for their fundamental knowledge to this circumstance. In Washington, while Lilienthal was yet alive, the two became familiar with a Lilienthal-designed hang-glider and fastidiously studied Lilienthal's designs, writings, and calculations. The two brothers were also the first to control a flying craft along its lateral axis by applying the concept of wing warping/wing controls to the wing, which up until then had been a rigid structure. Between 1900 and 1902 they carried out successful glider flights while lying prone in homebuilt biplanes using the principle of wing warping before turning to "powered flight" for nearly a decade, and today the Wright brothers are recognized as true pioneers of aviation, the first to successfully fly a heavier-than-air powered aircraft.

Among the many others who rose to the occasion were the American Chanute (an immigrant from France) and Herring. As outspoken pioneers on the heels of Lilienthal, they were the first to fit movable control surfaces on their self-made hang gliders for controlling the yaw and pitch axes and establish the standard directional and altitude control principles still used in aerodynamic flight today.

In Germany, Friedrich Harth was one of those idealists who, in the first years following the turn of the century, embraced the work of the successful Americans. Harth carried out tests in a similar vein, and like many others sacrificed much time, money, and, ultimately, his health, for the cause he'd devoted himself to. As he began to entertain the idea of flying, most probably beginning in 1908, his role models were found exclusively in the Wright brothers. In Würzburg, Harth picked up in 1908/09 at the point where the brothers had left off glider aircraft in 1902. This not only is borne out by surviving photos, but also is proven in a compilation published by Harth entitled: "Aus der Praxis des freien Segelflugs" (Exercises in Free Flight). In it he refers to the first aircraft he built in Bamberg in 1910, the S 1:

"In design, except for the fact that it was a monoplane our first aircraft resembled the well-known glider type built by the Wright brothers. It was fitted with forward control (canard empennage) and made use of the Wright's wing warping principle. The only substantial difference was in the design of the wing, which had a sharp upward sweep to the front."

Harth had relocated to Bamberg in the beginning of 1910, where he was employed as the *Stadtbaumeister*, or municipal government architect. That he also began development of an aircraft immediately after Wilbur Wright's flight in France on August 8, 1908, without access to Lilienthal's writings is evidence that he, too, was a victim of the "spirit of the times." In a letter to Messserschmitt dated September 27, 1915, it is quite clear that this was the first time he'd expressed a desire to study the tests conducted by Lilienthal more closely and therefore wanted to take a look at the documents in the National Library in Munich. This came 19 years after the death of Lilienthal; "the prophet in his own land" had gone unrecognized far too long.

When Harth started, there were hardly more than a dozen people in Germany working in the field of flight,

Friedrich Harth's first sailplane, the S 1 from 1910, showed the influence of the Wright Brothers in its layout. The elevator was located in front of the pilot (canard) and the method of wing warping (wing control) provided lateral control. The contraption, with a wingspan of 12 meters and weighing a meager 56 kilograms, was only able to achieve short hops using the cable tow takeoff method, as seen here in the spring of 1911 on the Ludwager Kulm.

and like him, many of these also went down as pioneers in the developmental history of aviation.

Hundreds of youths of Willy Messerschmitt's generation, enthused with the idea of flight, may have helped build gliders and sailplanes and all kinds of flying machines by occupying themselves just as he did in attics, cellars, sheds, and barns, sawing and cutting and filing and gluing models, then to test them with their own hands. Over the following decades this army of unknowns later generated renowned glider pilots and the famous combat pilots of both World Wars, as well as courageous pioneers of commercial aviation, daring stunt pilots, and a large number of inventors, technologists, engineers, researchers, scientists, aircraft designers and entrepreneurs, who in barely fifty years collectively raised their profession to a standard unparalleled in any branch of technical development (with, perhaps, the exception of automobile construction). Without a doubt, Messerschmitt was one of the chief movers and shakers of this development.

Beginnings - Harth-Messerschmitt Sailplanes

Harth had built and tested his S 2 in the interim; by the time Messerschmitt joined the Harth group to work as an assistant (around September of 1913) the decision to build a third glider had just been made: the S 3. This aircraft, developed for free flight, had a decidedly larger wing than the two previous versions and appeared to incorporatae several advanced concepts.

It was a high-wing plane with an exposed (non-skinned) frame fuselage built around a central skid. The pilot sat just above the skid beneath the forward third of the twistable wing surface; used here for the first time, the mono-skid construction concept quickly became popular on the Rhön and was to become a standard for later trainer gliders.

In addition to the proven concept of wing warping around the central spar, another new feature was to mount the wing so that the pilot could adjust its pitch, changing

The S 2, with its stabilizing surfaces moved aft, if given sufficient wind, was still only capable of unmanned and tethered flights.

The S 3 from 1913 was the first to make use of a truss fuselage design, which tapered to a single centerline skid. As such, the S3 became the prototype for subsequent training gliders in which many a sailplane pilot made his first hop during the 'twenties through the 'fifties. By January 1914 Harth was regularly making flights up to 120 meters. Here, Harth's assistant Baumgartner holds the S 3 on the line. Willy Messerschmitt, who joined the Harth group about this time, photographed this scene on 8 February 1914 on the Kulm.

Demonstration of the wing warping control method which Friedrich Harth employed in all of his designs, seen here on the S 3 in the winter of 1913/14. The control levers, attached to the skid, pulled the control cables in such a manner that when the levers were moved in opposite directions each side of the wing twisted simultaneously, changing the wing's angle of attack and also effecting elevator control.

the wing's angle of attack in flight. This enabled the craft's altitude to be controlled more effectively—a Harth first—and the principle of "wing control" was born. For well over a decade a segment of sailplane flyers considered this to be a prerequisite for dynamic soaring, although their convictions were not borne out in the end.[1] It should be added that the wing's central section was braced on top by a cabane strut, and from below with the support of a tension block, forming a rigid unity with the fuselage in the region of the wing spar. The twisting effect was further enhanced through the extensive use of rattan cane.

The tail consisted of a horizontal rectangular and a vertical, nearly square stabilizing surface; a rudder was left out since Harth felt that even birds get by without one.

Control "on the wing" was accomplished as follows: two control sticks were mounted on the skid. The control lines ran in such a way that moving the sticks in the opposite direction produced the twisting, or warping effect, while moving the sticks in the same direction changed the wing's angle of attack. On December 13, 1913, the craft was taken to the Kulm, assembled, and the first static balance tests were carried out. During week-long testing on flat ground, Harth worked out the correct application of the control mechanisms by trial and error. In the end, Harth became so proficient that he would chat amicably with bystanders while balancing the aircraft so perfectly that, even with strong gusts of wind, the wings would hardly rise or fall more than a hand's breath.

"It seemed the time for free flight had finally arrived. After several short flights, on the 6th of January, 1914, in ice cold weather and with winds blowing up to 15 meters per second coupled with a heavy snow fall, Harth risked an even longer flight. Take-off was accomplished with the use of a rope which had been attached to the skid beforehand.

[1] The concept is being reexamined today, see also X. Hafer/G. Sachs, Flugmechanik. Moderne Flugzeugentwurfs- und Steuerungskonzepte.
[2] 2nd ed., Springer Verlag, 1987.

After travelling only three short meters the aircraft broke the bonds of earth and moved into the wind, easily passing over its point of separation. However, the craft had to be brought down at once because of the limited space of the testing ground and the poor visibility caused by the snow flurries. Nevertheless, a free-flight distance of 120 meters was covered, and this in heavy winds.

The important matter of maintaining balance had therefore been resolved. In the future, the goal would be to increase the distance of the flight."

Thus did Harth describe his 1916 flight in the magazine "Flugsport." Another source, excerpts from Harth's daily journal entries, listed the length of this flight as only 60 meters. In any event, these entries also mention several other successful flights of a similar nature and recalled this particular day as:

"the moment of the birth of free flight with an aircraft completely controlled by the wings."

Harth and his assistants set their sights still higher. They envisioned a new aircraft with a somewhat larger wing surface area, approximately 15-18 m^2, and they were convinced of the need of finding a more suitable testing area, since the top of the Kulm dropped away abruptly and forced the pilots to land prematurely. When one flight ended with the craft "drilling" its nose into the crusty ice, breaking off its tail in the subsequent somersault, it was finally decided to end flight testing on the Kulm once and for all. The last successful flight on a nearly flat piece of land on the Ludwager Kulm took place on 8 February 1914, resulting in a pleasant flight at an altitude of 6-8 meters and a distance of 87.5 meters.

On 5 April 1914 the Rhön was scouted out and a site at Heidelstein near Oberelsbach selected as a suitable area for further testing. On Whitsun, they visited the site once again and thus began construction of a shelter hut.

The reason that Harth and Messerschmitt had not chosen the well-known Wasserkuppe site used since 1912 by students from Darmstadt was most likely not based on competitive thinking by the glider pilots, but rather was more probably due to the fact that the Heidelstein and the town of Bischofsheim was in Bavarian territory, while the Wasserkuppe and the town of Gersfeld was in Hessian country.

For the first time, Messerschmitt was involved in the construction of a full-size aircraft from start to finish when the group began work on the S 4. Reports indicate that this was the first time steel tubes were used in the construction of the fuselage, as well as a larger wing span; further data is unknown. Top priority would have been given to technically precise arrangment of all details, ensuring that Harth could put his faith in the flying contraption's reliability even over longer flights and in strong winds. Late July saw the last pieces of equipment gathered up from the Kulm and, together with the aircraft, loaded onto railcars. The aircraft alone weighed 50.4 kilograms. On July 28, 1914, the transportation to the Rhön began. Harth, Messerschmitt and another colleague by the name of Stürmer carried the heavy boxes from Bischofsheim up to the Heidelstein where they put the aircraft together. Just as they were finishing up—it was the first of August—the war broke out, and a few days later Harth was called up. The sheds at Heidelstein were locked up, and the promising new sailplane and the plans of its designers were left to an uncertain fate. At almost the same time as Harth, his oldest compatriots were also called up to serve their country, leaving Messerschmitt behind as the sole remaining member of the group. He was the only one left able to keep the dream of flight alive. For Harth, the 16-year old Messerschmitt had gone from being an apprentice to his most valued colleague, since for more than a year he'd been right alongside Harth for virtually every flight test and for all aircraft design and repair work. On top of that, he had taken weight off the shoulders of the "master" and contributed his own inspiration and ideas.

At first, the letter exchanges between Harth and Messerschmitt were scanty at best, but Harth eventually requested in a postcard dated the 5th of April, 1915, that Messerschmitt visit the Rhön site and check up on their property. When Messerschmitt then made his way to Bischofsheim and visited the Heidelstein, he found that the shed had been broken into, the door torn open and lying on the ground and the glider destroyed. Harth had hoped to be able to use his leave time for more flight testing, and his disappointment at the news was deeply felt.

But not for long. From Harth's letters to Messerschmitt it was obvious that Messerschmitt had offered to build a new flying machine. This decision is testimony to the growing self-confidence of the 16-year old young man. He took from Harth's workshop anything useful he could find and set to work in a shed on his parents' property on Lange Straße in Bamberg.

Thus was born the first plane Messerschmitt built by himself, the "Harth-Messerschmitt S 5 wing controlled sailplane." From April until August, 1915, there came from Harth a storm of postcards and letters, some up to ten pages, filled with information such as sketches, designs, bits of advice, instructions, and appraisals all addressing the preparation and construction of this craft.

While most of Harth's letters painted an almost morbid despair due to the possibility of yet another summer, 1915, passing without flying, he seems to have suddenly rallied considerably to pep up his young follower in a letter from July the 4th:

"You have a herculean task ahead of you. In building an airplane by yourself, you will find that it will also become quite boring at times. But hang in there. Everything depends on you. Have no doubts that I fully and completely understand your values in devotion, your zeal and your brilliant talent; bear in mind also that this is being done for a great flight. As soon as I have the feeling that the craft is reliable enough I will, after short preparations, trust myself to any altitude and commit my life to helping us achieve success. That being said, there isn't a day that goes by when I don't prepare myself spiritually for soaring. I feel completely at home dealing with the rocking motions the aircraft is subjected to as a matter of course. But I think the vertical control surfaces must be made to move. This would mean a new era of experimentation."

For all the design details known from Harth's letters, we still know nothing of the S 5's dimensions. The aircraft was essentially similar to both of its predecessors: a wing-controlled high-wing monoplane with a welded steel tube fuselage, single skid, with non-moving horizontal and vertical stabilizing surfaces at the aft end of the fuselage. The weight was given as 32 kilograms. The major features of the glider comprised: the use of rattan cane in the outer wing sections; a differential for tipping the wing; cotton fabric covering, saturated in enamelled cellone; and a steel cable running through the craft with forked end turnbuckles. The aircraft had cost 235 Deutschmarks, and the letters subsequently confirmed the assumption that some parts from the destroyed S 4 were recycled for the S 5. Nevertheless, since he had only been on school break since the end of July and prior to that would have had to devote at least a portion of his time for schooling and preparations for his college entrance exams it seems astonishing that the young Messerschmitt was able to complete the aircraft by the time Harth's August 24, 1915, postcard arrived from Munich with the laconic note: "I'll be coming to Bamberg on Friday. I'll have a look at you, and we'll drive immediately to the

Rhön." On 26 August Messerschmitt and his assistant, one of the French prisoners of war assigned to help at the *Weinhaus Messerschmitt*, loaded the new aircraft and arrived in Bischofsheim/Rhön on the 29th of August, with the first test flights being carried out just two days later. As seen in the photo below, signed "1.9.15. Rhön, W.M.," the aircraft initially had a quite normal rudderless non-moving vertical and horizontal tail attached to the aft end of the fuselage. Although journal excerpts daily mention several flights, beginning with a distance of 80 meters on the first day to one of 300 meters at an altitude of 20 meters on September 11, Harth seems not to have been satisified with the aircraft itself, nor with the performance achieved.

As can be seen in the photos and intimations from Harth's later letters, the aircraft was quite tail heavy, and the flights undulated sharply and zig-zagged somewhat,

The S 5, built by Willy Messerschmitt based on written instructions. The glider was photographed in the workshed by Messerschmitt in September 1915 and subsequently flown by Harth.

Friedrich Harth and Willy Messerschmitt (left in each of the pictures) had themselves photographed with the S 5 on 12 September 1915. The man on the right in the pictures is a French prisoner-of-war who was assigned to the Messerschmitt wine shop as a worker and became an assistant to the two flyers.

depending on the wind direction. For this reason they worked feverishly on a new, larger empennage, which they attached on the 12th of September.

One photo above, marked with "Rhön, 12.9.15. W.M." shows this improvisation. In the photo are Harth (with moustache and cap) and the French prisoner of war. In the other similar photo, Messerschmitt traded places with Harth. Also noteworthy are the cables hanging loosely from the cabane, indicating the aircraft is not yet fully assembled. Whether or not the aircraft was ever finished and tested in this form is not known.

Harth had to be back in Munich by the middle of September, and Messerschmitt's summer vacation was also at an end. He started the new school year by attending the 7th class at the *Oberrealschule* in Nuremberg. Following the time spent together on the Rhön Harth drew up further plans for the construction of a somewhat larger, improved aircraft in a letter from September 18, 1915. Surprisingly, he was sent to the Western Front in October. Until the end of the year he was in Flanders, but within just a few short weeks of his arrival he again began sending descriptive letters with plans for the new aircraft. In sketches and descriptions, some of them made to the finest detail, the design for the later S 6 began to come into focus.

Beginning on 1 January 1916, Harth once again found himself at the "flight school" in Munich, where he now had "a beautiful room, even a drafting table, and, moreover, few duties." In short order he'd finished the general plans for the craft and soon even had the detail blueprints for the "iron parts" completed. But then fate caught up with him again: on the 7th of February he once again was sent to the front and experienced "hard times" there. According to a letter from March, Messerschmitt went to the Heidelstein again, found the gate open, but other than the theft of a few woolen blankets found that no damage had been done. Harth gradually resumed penning the design details for the S 6, and at the end of March he was able to announce his return from the front to Munich.

First hesitantly, and then with ever increasing zeal he began to construct the S 6. He ordered materiel, contracted for the parts, wire pulleys, and the steel tube frame, and even found a business, the Beißbarth company, where he could build his aircraft.

In almost one hundred letters and postcards (until July 25, 1916), Harth reported on the progress of his work with the S 6 sailplane, enthusiastically expressing his gratification for working with his hands again and outlining future plans for a small factory where sailplanes with small motors could be built, but then declined Messerschmitt's proposal for entering into such a contract with him with the reason that Messerschmitt was still a minor. This, plus a few other design differences regarding the S 6 led to the first real, albeit temporary, feeling of dissatisfaction between the two dissimilar partners. Harth's nerves seem to have been a bit jarred from his experiences at the front, and Messerschmitt, now 18, was most likely anything but a "yes man."

In late July 1916, Harth informed Messerschmitt that the aircraft was being sent to Bischofsheim, and a few days later the two met to begin new tests on the Heidelstein.

As with former designs, Harth had again dispensed with a rudder; his non-moving tail assembly design, which he called "bird's tail," seemed to him to still offer the best solution despite the setback with the S 5. Furthermore, he wanted to control the craft purely by the wings, as he had expressed already. He seemingly overlooked the fact that

During the summer of 1916 Friedrich Harth achieved somewhat better success with a three-and-a-half minute flight in the S 6. Like its S 5 predecessor, this glider also had a fixed empennage which Harth called a "bird's tail."

his archetypes, the soaring birds, had been blessed with a flexible tail as an important means of control. The flight tests must obviously have been carried out in low winds. Nevertheless, the craft lifted easily off the ground from the almost flat plot of land, climbed slowly upwards and forwards with perfect balance and no difficulty whatsoever in maintaining pitch stability.

Harth succeeded in making his best flight on the 15th of August, with a flight time of 3 1/2 minutes, rising approximately 15 meters from the point of takeoff and crossing a distance of 200 meters. In writings this flight is repeatedly exaggerated, nor is there much mention of the fact that this particular flight ended in a crash landing (when a wire slipped off its pulley), and forced Harth to break off his testing.

Immediately following the flight, however, Harth found the time to summarize his views and the results he and Messerschmitt had achieved in the magazine "Flugsport," 1916, volumes 22/23. Up to this point, he had only sporadically made use of the local press to publish his views.

After 1916 there was a long period of inactivity following the S 6. It was wartime, and all assets were required to defend the homeland. And so great was the military crisis that the development of powered flight took priority,

Friedrich Harth during airflow testing on one of his own wing designs in the summer of 1916.

leaving little room for glider flight, which "only" served the interests of research and sport flying.

Messerschmitt graduated from the *Oberrealschule Nürnberg* in 1917. A few months later he was drafted into the military and began his basic training. He became ill, and from his hospital bed he pulled every lever he could to get himself into the Air Corps following his convalescence. He then began to lobby the military hierarchy in Munich and Berlin to use the glider for military operations. Ultimately, however, he was unable to prevent himself from serving as a recruit with the *Fliegerpionierabteilung Milbertshofen* (explosives) from June 5 to November 8, 1918. Reliable sources there reported that, in front of the whole squad, he once made the remark: "What in the world am I doing here? I'm an aircraft designer." The end of the war allowed him to take up his studies at Munich's *Technische Hochschule* (TH, or technical university). There he was provided ample playing room to resume his work in the field of aeronautics. It was also beneficial that Harth was still actively employed in Munich. By the end of 1916 and after many attempts, Harth had finally succeeded in transferring to a *Fliegerersatzabteilung*, or pilot replacement regiment. Beginning in January 1917 he taught aeronautics in Milbertshofen. Along with Messerschmitt, he worked out the dimensions for a new aircraft, the S 7. This was the first aircraft for which there exists precisely detailed blueprints for individual components, completed by Harth in the timeframe from December 12, 1917, until January 25, 1918. Even a scale three-view drawing has survived. One of the things revealed in these blueprints is a pilot seated somewhat behind the leading edge of the wing, the reason for the tail heaviness which cropped up during subsequent flight testing. For the first time, the aircraft had a rudder placed behind the non-moving horizontal stabilizer which—as with motorized aircraft—was controlled by means of foot pedals, but there was no tailfin. It can be safely said that Harth had this aircraft built at the Bayerische Flugzeugwerke

While Friedrich Harth was out in the field, Messerschmitt continued learning and became more and more confident of his own abilities. In 1915 and 1917 he acquired two of the most important works on aviation at the time. Notice that he still used his proper name of Wilhelm as his signature.

(BFW) in Munich,[2] where he was able to work in his spare time and could influence the quality of the workmanship (Harth had been working as a design engineer at the company since 1917). It was finished in late May of 1918. Since Harth was not given leave, Messerschmitt took the glider to the Rhön alone, where he limited himself to making balancing tests on the ground.

Harth wasn't able to carry out any flight testing until his leave in October of 1918, and it was quickly discovered that the aircraft was quite tail heavy. A test with the wind blowing at 10 meters per second ended with the craft flipping over backwards, damaging the glider to the point where repairs could not be made at the site.

The end of the war arrived, and along with it came the hardships of the postwar period. The "two Bambergers," as they were called by insiders in the sailplane movement, remained unperturbed and set about resuming their work. Even though the S 7 had proved unable to match the performance characteristics of earlier aircraft, it did provide the two with much experience and knowledge which would then be applied to a significantly better machine.

Thus emerged the wing controlled S 8 sailplane.

As it would later play a special role, it is worth examining more closely: the fuselage was designed as an open, welded triangular steel tube structure which ended in a horizontal stabilizer and rudder in tandem. The wing, mounted onto the fuselage in such a manner that it could rock freely, was supported by wires running to a cabane strut above and a skid frame below. It was designed using the monospar concept, with wooden ribs, a plywood leading edge and fabric-covered rear sections. At the center of gravity beneath the spar was the exposed pilot's seat with the dual lever wing control and foot pedals for the rudder.

An extensive array of flight tests was completed with this aircraft throughout the summer and fall of 1920 on the Heidelstein; included among these were a few flights with the the rudder being replaced by spoilers on the wing tips, the wing controlled plane's further step in the direction of a soaring bird's "design concept."

In another step to this end, Messerschmitt built a tailless wing controlled craft, the S 9, on his own. It had an

---

[2] This company had been renamed from the Otto-Flugzeugwerke, München, in 1909 and was again renamed the Bayerische Motorenwerke AG in 1922. It was not associated in any way with the BFW Augsburg company formed in 1926.

Willy Messerschmitt only reluctantly served as a soldier in 1917/18. It was during this period that he most likely had the opportunity to fly in a biplane (right).

The S 7 was the first to be fitted with a rudder. As Harth was not allowed to take leave in order to test fly the craft, Messerschmitt spent his time during May of 1918 carrying out balance testing on the ground.

Below: Harth drew up the blueprints for the S 7 during his time with BFW in Munich; these show some of the wing ribs.

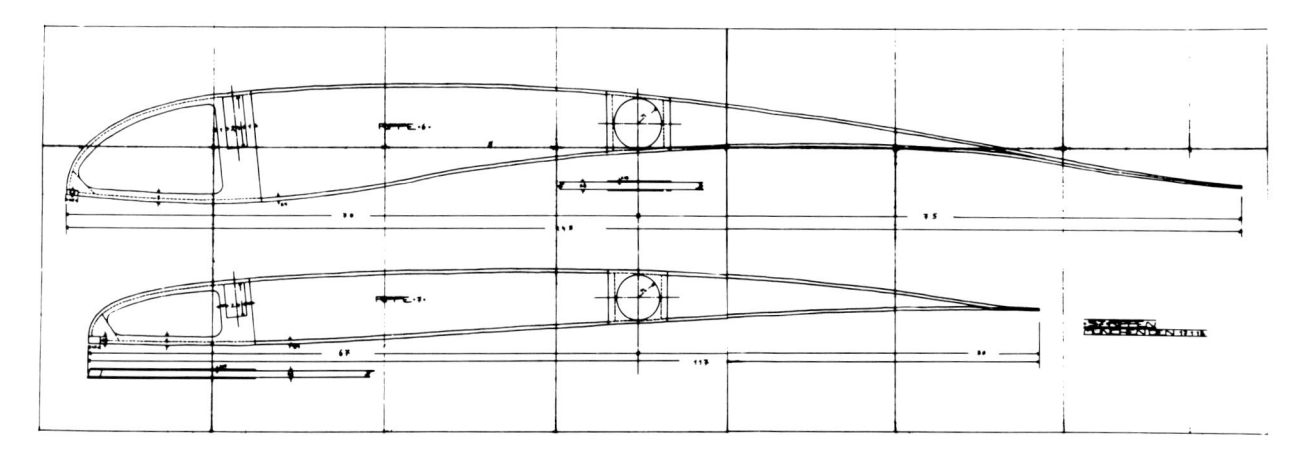

Willy Messerschmitt's letter from 9 July 1919 to the Aerodynamische Versuchsanstalt in Göttingen reveals that Harth and Messerschmitt had reached the limits of their collective knowledge when it came to the development of wing profiles.

aerodynamically designed hull fuselage. The wings were swept forward and the wing tips fitted with drag slats, which produced directional control. In place of a control cable, a torsion tube which ran inside the wing was used for the first time to produce the wing's twisting effect. With the S 9, Harth and Messerschmitt had hoped to take part in the Rhön sailplane competition in 1921, but unforseen problems cropped up in testing this aircraft. The root cause was found to be the problematic stability and the number of new concepts employed, preventing the aircraft from completing its tests in time for the competition.

This setback was initially cause for a disagreement between Harth and Messerschmitt; eventually they reconciled, however, and with the previous year's S 8 attempted to challenge the 15-minute endurance record set by Martens. This was successful: on September 13, 1921, Friedrich

The S 9 from 1921 was Willy Messerschmitt's first original aircraft design. It can be assumed that Messerschmitt had been stimulated to build this forward swept flying wing by other writings of the day. The S 9 was not flown successfully.

The S 8 from 1921 was the biggest success for Harth and Messerschmitt after nearly ten years of research and experimentation.

Harth set a new world record for endurance flights at 21 1/2 minutes on the Heidelstein/Rhön. However, the flight ended in a crash from a high altitude, the aircraft was smashed, and Harth had to be taken to the Fulda hospital. He never fully recovered.

Messerschmitt, now working alone as a result of Harth's accident, found time despite his studies to establish a provisional workshop with a paid work force in Bischofsheim on the Rhön under the company title of Harth und Messerschmitt. He also had time to negotiate with the community of Bischofsheim for a site and wood for their own production shop for sailplanes. He also scouted out flying students for a flight school to be established on the Wasserkuppe, where they would be trained in operating wing controlled sailplanes. The first result of this search produced the student Wolf Hirth, who had responded to the firm after seeing their advertisement in "Flugsport" and who would later go on to become a famous sailplane pilot.

The S 10 student trainer glider, the plans of which had been drawn up by Harth before his crash, may have initially been started by the BFW Munich. Messerschmitt brought the unfinished craft with him to Bischofsheim. There he passed on the managerial reins of the temporary workshop to Wolf Hirth, who oversaw assembly of the first glider and tackled the construction of four more S 10 sailplanes in the hopes that they would be ready for the next Rhön competition. Wolf Hirth outlined in humorous fashion the course of events during this period of time. From

Friedrich Harth in the S 8 during its record flight of 21 1/2 minutes on 13 September 1921.

Map showing the route of the record-setting flight on 13 September 1921 on the Heidelstein.

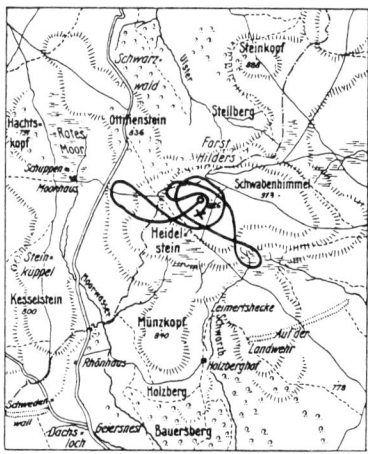

him we also discover that Messerschmitt had simultaneously begun work on an "improved" craft, the first "hull glider"; this would become the S 11.

In the early spring of 1922 newcomer Wolf Hirth taught himself to fly on the first S 10 (a wing controlled plane), then went on to teach several other students so that they would be able to pass their B-class certification before the Rhön competition.

These former students participated for the first time at the glider competition portion of the Rhön competitions in August 1922, where they won many awards and thereby proved the toughness and reliability of the S 10 trainer glider, since virtually all of the flights had been carried out in bad weather.

SEGELFLUGZEUGBAU

HARTH-MESSERSCHMITT

BAMBERG
BISCHOFSHEIM-RHÖN

BAU u. VERTRIEB v. SEGELFLUGZEUGEN
Ausbildung von Flugschülern.

□□

Aeltestes Unternehmen zur Erforschung des Segelfluges
Aus systematischen Versuchen seit 1910 entwickelte,
anpassungsfähige Tragflächen.
Bisherige Erfolge:
1914 Flüge ohne Höhenverlust.
1916 Flug von 3½ Minuten ohne Höhenverlust,
1920 Minutenlange Flüge bis zu 50 m Höhe über der Abflugstelle.
13. September 1921 Welthöchstleistung:
Flug von 21 Minuten bis 150 m Höhe über Abflugstelle
Landung nur 12 m tiefer.

Advertisement carried in the 1/1922 edition of "Flugsport" magazine which resulted in Wolf Hirth and Willy Messerschmitt coming together.

Messerschmitt, however, found little satisfaction with this showing. The Hannover group, with Georg Madelung's *Vampyr*, had achieved the goal Harth and Messerschmitt had long been unsuccessfully striving toward: endurance flights of one hour and longer, thereby proving that an aerodynamic refined aircraft with conventional controls far surpassed any non-powered sailplane. The wing controlled sailplane, though, had a perfidious nature that had to be eradicated. Based on the experiences with the S 9, he constructed his S 11 along the lines of the concepts of the earlier gliders, but with an enclosed fuselage and braced wings with bendable tips, a non-moving tailplane and a standard rudder. Altitude control by tilting the wing and the dual control stick design was retained. And the S 11 was created. Tested by Wolf Hirth, the type repeatedly exhibited undesirable features for which no explanation was immediately forthcoming.

Messerschmitt later reported:

"The S 11 was conceived as an improvement to the S 10, aerodynamically speaking. It did not deviate from the S 10 with regard to the main points of the wing control systems, but differed chiefly in the use of a plywood covered fuselage and by moving the entire control system to the inside of the fuselage and the wings. Because of unfavorable circumstances, poor quality control fittings due to the isolation of the construction location, and the lack of a suitable test pilot, flight testing did not extend beyond short hops. Since...the type revealed poor handling characteristics, which were later traced back to the minor causes mentioned above, the parts planned for the S 11 (wooden fuselage, new wing profile, and the proven cable control from the S 10) became a new aircraft variant..."

This variant would be the S 12. Wolf Hirth, as mentioned, went to the Rhön in early 1922 with Messerschmitt, where he became a student pilot and workshop supervisor for construction of the S 10. He soon became flight instructor and test pilot (crashing several times along the way), and can be considered Messerschmitt's closest coworker from this period of time. It was he who, in test flying the S 11, crashed and damaged his larynx so badly that his voice was never to be normal again. He also relates that at the time he became acquainted with the two *Regierungsbaumeister* Harth and Messerschmitt, the two were having constant "technical quarrels," whereas they otherwise got along splendidly. Harth was the "brake" with his constructive, resolute thoughts, and often had to quell Messerschmitt's rush-into-it temperament, and get back into proper step with the system.

From Harth's standpoint, this system—from the S 8 through the S 10 and continuing on to the S 11—seems to have been the incremental application of minor improve-

Three-view of the S 10 by Willy Messerschmitt, with reference to the dimensions of the S 11.

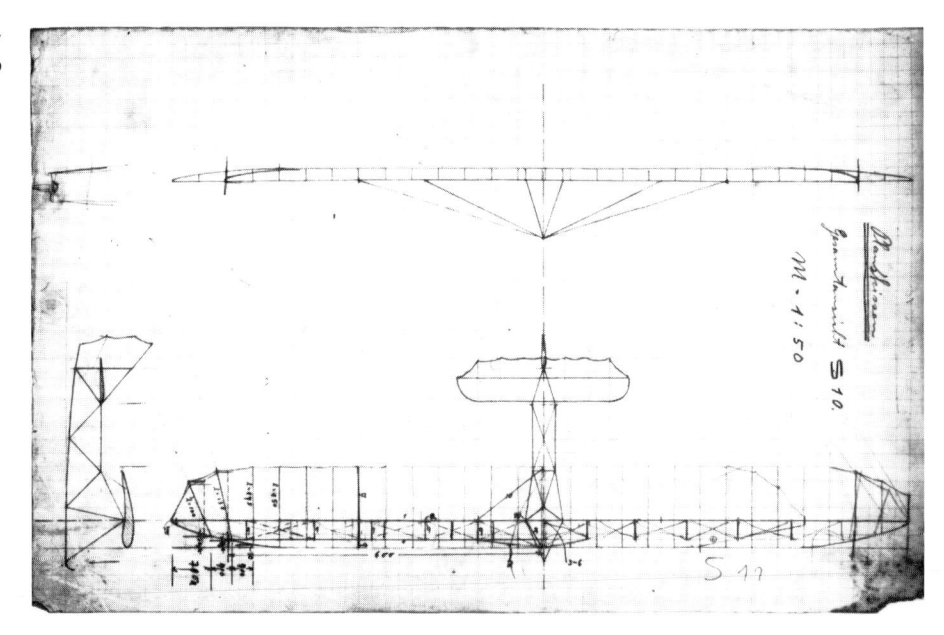

ments, such as moving the control wires into the wings while for the time being retaining the uncovered girder fuselage, not least of which was due to the fact that he was a staunch supporter of self-launching and no friend of the enclosed cockpit.

Messerschmitt had not only gradually developed into the initiater of construction and training in Bischofsheim, but in the meantime appears to have taken over the intellectual lead and thereby become the deciding partner in the team. He was now able to make advances in follow-on developments of several ideas planned for the S 9, such as reducing harmful drag by moving the control cable to the inside, covering the pilot's seat and girder fuselage, shortening or even removing support struts, aerodynamically improving the profile, and of course, simplifying the design as much as possible.

In the magazine "Flugsport," volume 4, 1924, there was a photo with Messerschmitt's description above it that showed; "W. Hirth in the S 17 in May, 1922"; it was in 1922, however, that Wolf Hirth had tested the S 11 and crashed, leading to the immediate construction of the S 12. So there was never a S 17, and the caption obviously was a typographical error.

In any case, the S 12 was an interim step on the way to realizing Messerschmitt's constructive ideas, which by tenaciously following he would strive toward perfection of the wing controlled sailplane from this point on. It seemed that in these efforts Harth, who was probably just as strong

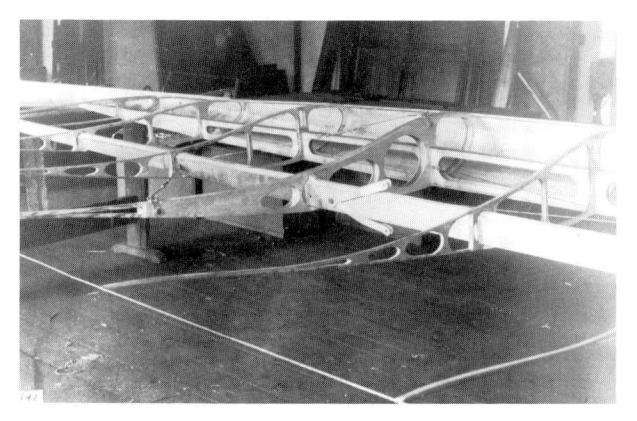

Exposed wing and truss fuselage of the S 10 in the workshop.

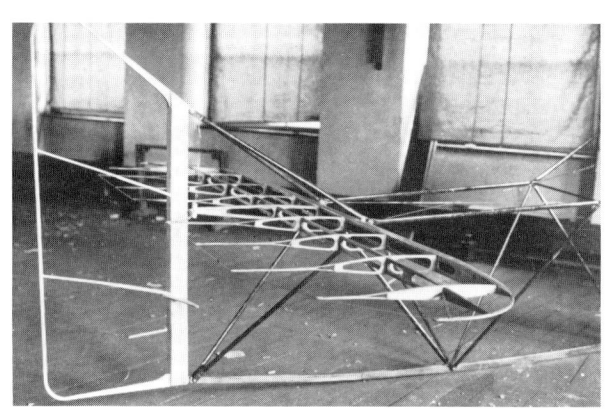

Typical aerial photo of the S 10 around 1922. Aside from the wing control it had the same basic design as later training gliders.

willed, was apparently either unable or unwilling to follow along: the student had outpaced his teacher!

The S 12 appears to have embodied all the improvements Messerschmitt had been working to implement. In addition to the pilot, the fuselage also incorporated all controls, on the wings "the leading edge's sharp ridge disappeared and the rear bottom edge was somewhat more curved," and all tension wires were dispensed with, as well. Supports and braces for pivoting the wing were teardrop shaped, so that the aircraft (which carried on with the proven concept of wing control) not only manifested a remarkably advanced shape for its day, but even today offers quite an amazingly well-balanced, aesthetic design.

Two or three of this aircraft type were available for the Rhön competition of 1922 (entrant numbers 12, 40, and 36, the last of which may have been the S 11 rebuilt as the S 12).

But, starting with Wolf Hirth, there were...:

"a series of crashes before the very eyes of premier German scientists, none of whom could figure out the reason why. As was often the case in the early days, one took the easy route of placing the blame for the accidents at the feet of the pilot. Willy Messerschmitt got upset about things, to be sure, particularly since there were two more accidents over the course of the competition. But he was much more interested in research and probably too little interested in medicine to spend too much unnecessary time dealing with the pilots of these crashes. These logically only interested him in so far that they were able to reliably tell him a rea-

The S 11 from 1922 was the first Messerschmitt design having an enclosed fuselage.

The S 12, incorporating more and more of Willy Messerschmitt's design ideas, stemmed from the S 11. This photo shows the aircraft over the Heidelstein with Wolf Hirth at the controls.

son behind the crash. Then he immediately rolled up his sleeves and plunged back into the task of correcting the design problem.

In the end, six months later it was Messerschmitt himself who, while still a student at Munich's *Technische Hochschule* (Munich Technical Institute), discovered what had been the problem with the design."

Messerschmitt later discussed the situation:

"The S 12 became the victim of an odd set of circumstances, namely the design of a new wing profile controlled by the old wire system.

The wing profile of the S 11 and S 12 had been developed from that of the S 10, but with marked improvement

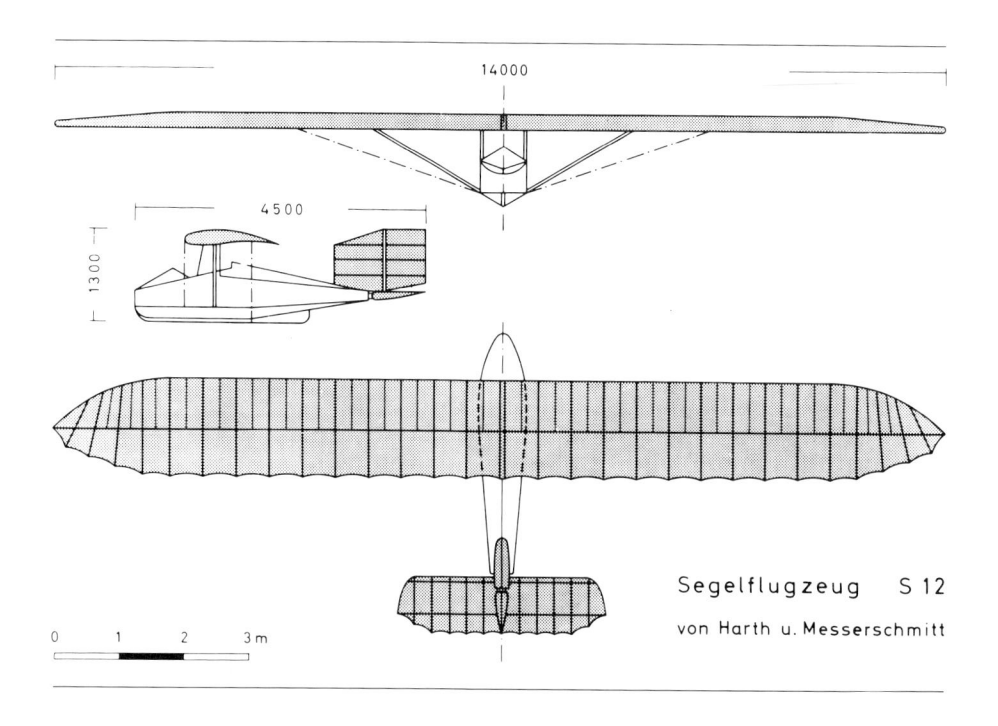

S 12 three-view.

so that other aircraft designers used it later (Konsul-Darmstadt, Hirth also mentions the record-setting *Fafnir, Windspiel, Rhönbussard, Rhönsperber, Moazagotl,* and *Kranich* sailplanes). Since time was of the essence in building the S 12 (because of the impending Rhön competition), it had not been possible to evaluate the profile's characteristics in wind tunnels. It was found that, whereas the S 10's center of pressure (cp) was located directly behind the wing's pivot point 'D,' the cp of the new profile was located approx. 100 mm further aft. As soon as the wing had a negative angle of attack of more than -3 degrees, the circumstance shown in the drawing would occur. The cp, situated in normal flight at M', would drift past B toward M'. At the moment that it passed through B, the wing surface would then flex at this fixed point and the spar, until now under upward stress, suddenly found itself bending downward. This rapid shifting of the wing resulted in a sudden negative climb effect which could not be reversed, since pulling on the elevator control wire would simply increase the bending of the wood. All of the crashes by this aircraft can be traced back to this cause."

While all hopes of the S 12 making a good showing at the 1922 Rhön competition were literally dashed to pieces, some consolation was found that year with the S 10's excellent performance. Over an eight-day period in decidedly poor weather conditions, Harth and Baron von Freyberg successfully flew "entrant number 10" in 16 registered flights without mishap, winning a total of six awards in their class.

Before the Rhön competion, Harth and Messerschmitt may have given the appearance that they were partners, but at the competition it was soon apparent to everyone that they were well on their separate ways.

To a great extent, Messerschmitt had already been working independently with the construction of the S 11 and S 12. These developments and all subsequent designs were shouldered by Messerschmitt alone, i.e. he was responsible for their design, financing, and construction. Harth was only interested in finishing "his" S 10 aircraft on time, although he was unable to personally work on them himself since he had not yet regained full control of his capacities. Following the Rhön competition, Harth gave notice of the partnership's dissolution (in a letter dated September 18, 1922), and the firm "Segelflugzeugbau Harth-Messerschmitt Bischofsheim" was no more. The workshop in Bischofsheim was disbanded, and the flight school on the Wasserkuppe was initially taken over by Wolf Hirth. But then the end of inflation in 1923 resulted in the exclusion of foreign flying pupils (chiefly the enthusiastic Swedes) who lost their exchange rate advantage. The Rhinelander Ludowici intervened in an attempt to support the school out of his own pocket, but was soon forced to give up himself. The *"Weltensegler"* ultimately inherited the school.

In late 1922 the Bischofsheim community called upon Harth and Messerschmitt in turn to take delivery of and pay for the building wood they had ordered, which had now been shipped in. As neither one had the means to pay for this wood, Ferdinand Messerschmitt stepped in to pay for the cost and shipment.

Flugzeugbau Messerschmitt Bamberg

Once Messerschmitt had succeeded in finding the reasons for the setbacks with the S 12 in 1922, he tackled the heavily modified S 13 design with unabated enthusiasm. The S 13 was based on the collective knowledge gained and was the result of a series of studies and detailed calculations. With the exception of the Harth-Messerschmitt wing control principle, this aircraft actually retained little of the characteristics of their earlier designs. It was obvious that with the S 13 Messerschmitt had finally broken free of Harth's influence and stepped on the road leading from glider to sailplane alone.

The S 13 was a wing controlled, high-wing design whose wings were mounted to the fuselage using a plywood-covered plug, enabling them to pivot, and supported by two diagonal braces.

The fuselage was a single skid, plywood hull with enclosed seat located below the forward third of the wings and with the aft section terminating abruptly in an upward sweep. The wing, with the improved, thickened profile of the S 12 (Gö 535), was attached to the upper fuselage at two points in such a manner that it could move. It had a rigid box spar, plus a pliable wooden frame with wood-covered leading edge and cellon fabric covering, with tapered ends.

For the first time, the tailplane could be trimmed in flight via a hand lever, but there was no elevator. The vertical stabilizer was designed as an upward canted parallelogram, 1/3 fin, 2/3 rudder. It was controlled by the feet with cable lines running through the fuselage. Lateral and vertical control were based on the standard concept of wing con-

The most successful Harth-Messerschmitt glider during the 1922 Rhön competition was the S 10, seen here just after a cable-tow takeoff with Baron von Freyberg at the controls.

trol; however, altitude control was accomplished by adjusting the wing's angle of attack in the center third of the wing, while lateral control was the result of twisting the outer wing sections.

This was Messerschmitt's first use of a single joystick to manipulate the control rods and torsion tubes.

During flight testing, the sailplane became uncontrollable and was destroyed in a crash following several good flights due to buckling of a steel control tube which had been manufactured by a subcontractor, whose wall thickness had eluded verification once it was installed. The pilot, Wolf Hirth, suffered a broken pelvis. The roster for the 1923 Rhön competition shows entrant number 55, Count Hamilton, Swede, listed as owner of the S 13, although it never actually participated in the technical acceptance trials or in the contest itself.

On the other hand, the S 14 sailplane was an imitation of its predecessor, with the exception of the new wings having a somewhat smaller span and with the profile of "*Vampyr*," Gö Number 482 (according to a statement by Alexander Lippisch).

Two of these sailplanes were completed just in time for the Rhön competition.

The S 14 proved its effectiveness during the storm flight by Hackmack on 30 August 1923, at the conclusion of the Rhön competition when, with the wind blowing at 20 meters per second, Hackmack attained the highest altitude during the entire competition.

The final and most successful Messerschmitt sailplane was the S 14, designed for the Rhön competition in 1923.

Victor of the 1923 Rhön competition was Hans Hackmack in the S 14, who flew the craft in extremely gusty conditions.

That both plane and pilot had accomplished an outstanding feat during this storm was clearly borne out by no less than the aghast body of the competition board in its flight reports:

"30 August - wind 20 m/s westerly - very gusty. Hackmack was first to take off in the Messerschmitt S 14. The wing controlled monoplane reached an altitude over the take-off point of 303 meters in approximately 2 minutes. After a few loops, Hackmack made a large left turn in the direction of Eube, Oberhausen, and Heidelstein, where he landed behind the forest with minimal breakage to the aircraft. The flight can be considered a step towards solving the overland problem and the best performance of this year's competition.

The flyers were so excited about flights of this type that a real rush to take off arose.

Thomsen took off in the "*Dessau*." The gusts were extraordinarily strong, and the wingtips could be seen to start vibrating and moving up and down. Recognizing the danger, the pilot nosed the craft over toward the ground, but a heavy landing caused the wings to snap downward. Then the third craft, the "*Galgenvogel*," was towed to the starting point. Tracinski moved the horse-drawn carrier into place and...because of the unusually powerful ground gusts, the contraption was thrown onto a knoll and slid down another 20 meters. Next came Standfuß with the "*Erfurter Eindecker*." The gusts shook the plane violently, and it was driven into the Abstroda valley. Up to this point, Standfuß had been able to master any dangerous situation. However, just above the trees a gust struck him so sharply that the wing broke and the fuselage plummeted from a height of approximately 30 meters into a swamp. Farmers rushing to the scene pulled the pilot out of the swamp. He was brought to the Tann hospital where he succumbed to his serious internal injuries later that evening."

The S 14 had not only demonstrated its effectiveness under particularly strong wind conditions which had spelled disaster for three other proven designs, it had also fulfilled a dream Harth and Messerschmitt had been nurturing since 1913. This vision, "to soar like the stormbirds riding a storm," which the two had been unable to reach in ten years of working together, had eluded Harth while being achieved with the flight by Hackmack. Alone, the young Messerschmitt had reached this goal with his S 14 after barely a year's work.

Hackmack won the following with the S 14 in the 1923 Rhön competition:
- Top prize for the altitude of 303 meters (altitude competition);
- 2nd place for the distance competition (distance of the flight);
- the "Albert Böhm prize" for the greatest altitude;
- the medal of honor from the *deutscher Luftfahrtverband* for the highest flight performance during the competition

Only presented twice in history, Messerschmitt received the George Foundation award for his design.

The location of the S 14's construction was first mentioned in the *Messerschmitt Nachrichten*, no. 3/1963 under the title: "Co-workers Discuss the Messerschmitt Flugzeugbau's Beginnings," which read in part:

"Later in 1923 the 25 year old student Willy Messerschmitt founded 'Flugzeugbau Messerschmitt Bamberg' with the help of his brother Ferdinand, who had taken over the family's wine business. This business then set about manufacturing its own airplanes.

The first Messerschmitt sailplanes were built in a shed on his father's property at '41 Lange Straße' in Bamberg. His brother Ferdinand initially took care of the financing. The first sailplanes built here were the wing controlled S 14, which also served as the young student's thesis, and the S 15."

The design of the S 14 also served as Willy Messerschmitt's thesis at the Technische Hochschule Munich in November of 1923.

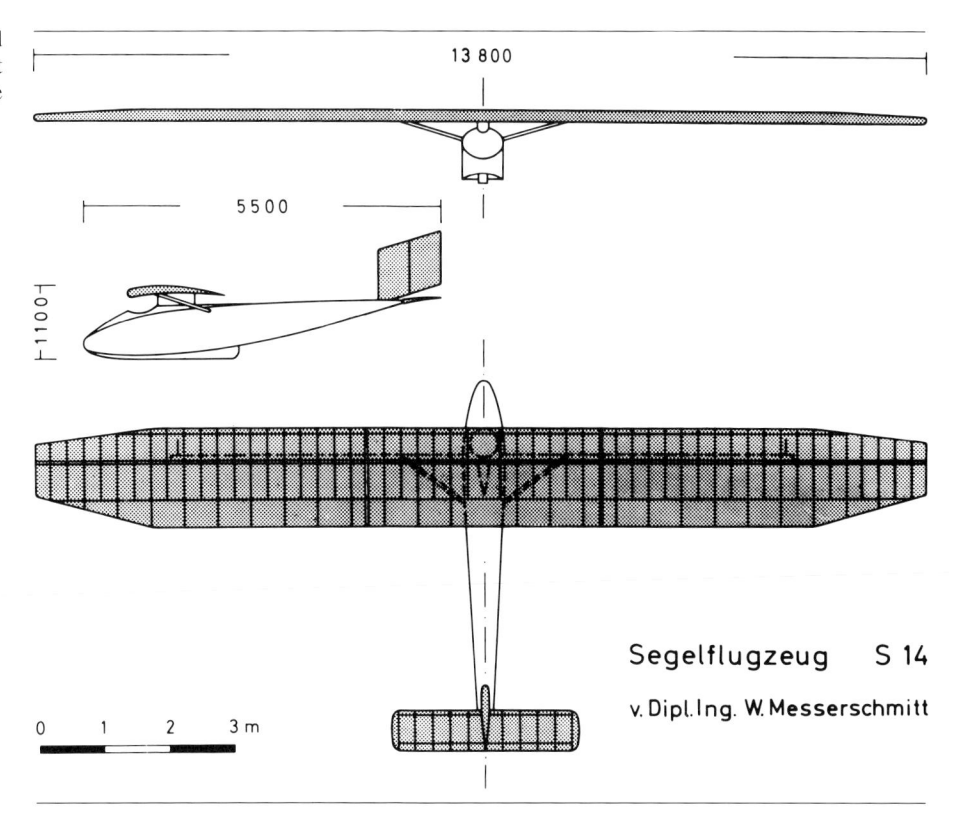

Segelflugzeug S 14
v. Dipl. Ing. W. Messerschmitt

The story of the thesis is perhaps best told in the words of Professor Julius Krauß:

"As mentioned, Messerschmitt received his scientific training in the mechanical engineering and electronics department of Munich's *Technische Hochschule*, where he completed his final exams in late 1923. At that time, the idea of a final thesis had not yet been introduced; instead there were obligatory studies projects for the various subjects, which were met with varying degrees of enthusiasm by the students themselves. Particularly feared for different reasons was the subject of lifting devices. Now young Messerschmitt, who alongside his studies had been busily working on his his S 14, said to himself that an airplane was also a type of lifting device and attempted to submit blueprints for a sailplane design in place of a crane design. At first, he was coldly rejected. Fortunately, the evaluation board included such eminent figures as *Geheimrat* Professor Finsterwalder who, as head of the photogrammetry department and father of a son interested in sailplanes himself, had some familiarity with aviation, and Professor Ludwig Föppl, a senior professor of technical mechanics. Both of these individuals wielded their influence to ensure the eventual acceptance of Messerschmitt's design.

Flugzeugbau Messerschmitt, Bamberg
baut flügelgesteuerte Segelflugzeuge bewährter Konstruktion.
Schulflugzeuge mit Gitterrumpf.
Höchstleistungsflugzeuge m. Sperrholzboot.
Segelflugschule Wasserkuppe.
Nächster Schulkursus 1. Juni 1923.
Anmeldungen an W. Hirth, Wasserkuppe, Gersfeld-Rhön.

One of the first advertisements appearing in the May 1923 edition of "Flugsport" for the new company of Flugzeugbau Messerschmitt, Bamberg and an associated glider pilots' school run by Wolf Hirth.

Willy Messerschmitt was an engineering student at the Technische Hochschule in Munich from 1918 to 1923.

Thus, engineeer-candidate Messerschmitt's difficult construction work on the S 14 paid off in this sense, as well.

Today, where aeronautics departments have long been commonplace and, aside from regular studies, a graduate is only required to submit a thesis selected from a single subject area, as an airplane builder he would have found the road much easier. Nevertheless, his breakthrough into the field of conventional mechanical engineering caused a considerable sensation.

Unencumbered by exam studies, the green 25-year old engineering graduate and businessman was finally able to dedicate himself to aircraft design and use his successes to draw the attention of the international aviation community."

But circumstances beyond his control gave the young entrepreneur little hope in obtaining contracts for building aircraft. Inflation had reached its pinnacle, and billions became worth only pfennigs, and nobody could forsee the end to it. In addition, the number of jobless rose to over 6 million, with many people in Germany finding themselves in the direst of circumstances. It took a great deal of self-confidence to establish and build up a business, all the more so because it had set itself the task of producing aircraft of all things.

Even the family business, which had lost its business leadership when the father passed away in 1916 at an early age (brother Ferdinand was a soldier in the field at that time), was hit hard by these poor economic conditions. In addition to the mother, three of Messerschmitt's siblings had not yet left the nest—Maja, Rudolf, and Ello.

The Powered Gliders

Messerschmitt took his first cautious step with his second S 14 sailplane, for which he had found a buyer in the "*Arbeitsgemeinschaft Unterfranken Würzburg*" after making a few changes to the design. Prior to building light construction aircraft, he developed sailplanes with auxiliary engines which, even then, were already being christened "powered gliders." Next came the S 15, designed in the fall of 1923 and built and tested in the early part of 1924.

The S 15 powered glider was an unbraced high-wing monoplane with the wing set directly on the fuselage.

The fuselage was built as an enclosed plywood hull, with a horizontally tapered aft section, a vertical cut to the nose, and an upholstered seat under the wing leading edge.

The undercarriage consisted of two rubber tires with an axle running through the fuselage, narrow wheel track, and a tail skid at the aft end of the fuselage.

The tail assembly consisted of a vertical fin and supported a rudder, while the horizontal stabilizer, which had originally been a simple horizontal damping surface, now incorporated a normal elevator.

Lateral control was produced by twisting the outer wings over inner push-pull rods and torsion shafts inside the wings, directional control by foot pedals and cables, altitude control also by cable lines, with both altitude and lateral controls receiving their inputs from a single column control stick.

The wing was a three-part wooden frame affair, joined to the fuselage by means of two short posts; the center section wing was rectangular shaped with a non-twisting box spar, while the outer wing sections had freedom to twist and a wood covered leading edge with fabric covering.

The engine was situated immediately in front of the pilot's seat under the instrument panel and had a drive chain with a gear reduction of 1:2.6 for the propeller shaft located directly ahead of the fuel tank. The propeller had two blades, and the fuel tank had a total capacity of twelve liters and was arranged as a gravity feed tank in front of and above the motor.

"Flugzeugbau Messerschmitt Bamberg" built the S 15 power glider in its workshop at 41 Lange Straße in the winter of 1923/24, where both of the S 14s had previously been

During the winter of 1923/24 Willy Messerschmitt developed the S 15 for the 1924 Rhön Gliding Competition, which by then had been expanded to include powered sailplanes.

The aircraft was given a thorough inspection by light aircraft expert Werner von Langsdorff (in cockpit). Also seen in the photograph, taken at the airfield in Bamberg, is company manager Graichen (next to propeller).

built. The wing of the powered glider was somewhat slimmer, the fuselage shape altered to account for the engine and undercarriage installation, but the wing control concept initially remained unchanged from previous models.

Initial taxiing trials took place in the spring of 1924. Carl-August von Schönebeck acted as the pilot. When attempting the common procedure of "pushing" the nose forward to lift the tail off the ground, the aircraft flipped over, damaging the propeller, tail assembly and aft fuselage, though the pilot was uninjured. It was discovered that wing control, at least with regard to controlling the airplane along the pitch axis, was entirely unsuitable and unfit for a motor-driven aircraft taking off. In addition, engine power was found to be inadequate. The powered glider underwent modifications. The wing was fixed in place and could no longer be tilted. For the first time, a Messerschmitt design incorporated an elevator. Lateral control through twisting of the wings was retained, however. The 700 cm³ Douglas engine took the place of the Viktoria engine and was fitted with a quick-change propeller.

Testing resumed in late May with *Hauptmann* Heinz Seywald at the controls. He had flown during the war and was a member of the *Arbeitsgemeinschaft Unterfranken Würzburg* (Lower Franconia Worker's Union-Würzburg), which in fact was a front for the local Reichswehr's *7. Kraftfahrkompanie*, which as the customer had a vested interest in the effectiveness of this powered glider. The "Flugsport" in 1924 carried reports of flying demonstrations being carried out in front of large crowds on the 19th of June, 1924. According to the article the plane had exhibited quite good flight characteristics, its takeoff and climb performance most acceptable and the engine's reliability proven in earlier flights of up to three-quarters of an hour.

Shortly following these demonstrations, the customers in Würzburg took delivery of the aircraft.

Messerschmitt had also been contracted to deliver two more powered gliders to the same customer, one a one-seater and one a two-seater, in time for the 1924 Rhön Competition. These two planes were built between March and July of 1924, though not inside the shop in Messerschmitt's parent's house, which had become too small, but in a rented space in the Murmann Brewery at Jakobsberg in Bamberg, where the firm had set up its new residence. These were:

the single-seater S 16a (Bubi)
the two-seater S 16b (Betty)

Messerschmitt's younger brother and the younger sisters, working at the time as unpaid assistants of the firm, were compensated when their Christian names were "immortalized" onto the two aircraft.

Friedrich Schwarz, who had come to Flugzeugbau Messerschmitt Bamberg on May 1, 1924 (as an apprentice), for the purpose of reworking the S 15's engine, also assisted in fitting engines in the two S 16s, remembers that the latter's dimensions were very similar to those of the S 15 and that the two-seater S 16b had a proportionately larger

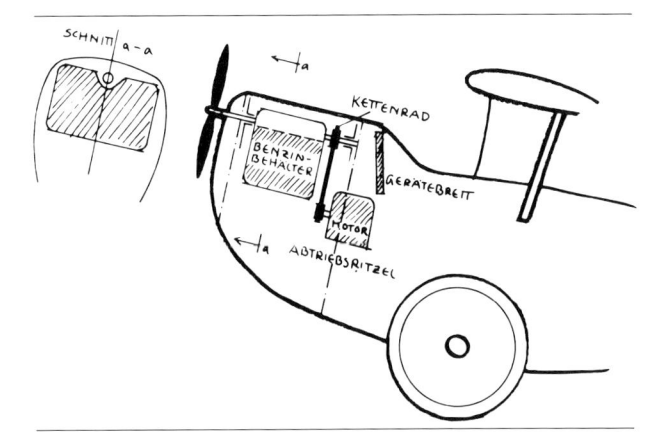

Conceptual sketch for engine installation in the S 15 and S 16. Initial powerplant was the 500 cm³ Victoria engine, having an output of 3.5 to 10 hp (2.5 to 7.36 kW), although the two cylinder Douglas engine was later used, this being rated at 22.6 hp (16.6 kW).

Takeoff of the S 16 during the 1926 Rhön competition. Both participating S 16s dropped out because of engine problems. Thus ended Messerschmitt's future role in the Rhön competitions, as well.

wing span than the S 16a, though the wing load was the same with the pilot's seat toward the nose.

For the first time, a regular paid work force was used to build the two aircraft. The company hired carpenters Rühler, Kistner, Masching, and Ditmar, with the errand boy Konrad acting as everyone's helper, plus engineer Max Graichen as the shop supervisor. Then there was the *Schlosserei Müller* fulfilling the role as a "sub-contractor" for metal fittings, whose agent Kaspar Meinhard was eventually hired full-time by Messerschmitt a year later. Sometime in July 1924 the metalworker Wolfrum joined the group, and it was with this team that the greenhorn businessman Willy Messerschmitt designed and built both S 16s during the early summer of 1924.

The S 16 was no longer "wing controlled." It had normal ailerons, rudder, and elevator. For the first time, Messerschmitt had totally dispensed with the Harth concept, and in future projects never returned to using the idea. This may have been because he felt "wing control" to be unsuitable for engine-powered aircraft, simply out of respect for the domain of his former mentor Harth, or it may have been because he no longer wanted to take the risks with this construction method as he branched into commercial aircraft production. Whatever the reason, the two S 16s were finished in time for the Rhön Competition, but had not been flight tested by the time of the contest. They were shipped virtually "factory fresh" by railcar to Bischofsheim and from there via truck to the Wasserkuppe without having ever made a single flight beforehand.

With *Hauptmann* Heinz Seywald as pilot, apprentice F. Schwarz and the carpenter Kistner made up the competition team. During the competition, the first to have tentatively allowed powered gliders to participate on a trial basis, they were tasked to participate and clear the two aircraft over the technical acceptance hurdle.

During the competition, Seywald flew the single-seat S 16a "Bubi" first. The propeller block broke, causing the propeller itself to fly apart, and he was forced to make an emergency landing which resulted in serious damage to the plane. Upon landing, the plane caught in the trees alongside a wooded path, where it was soon discovered and brought down by the team. But pilot Seywald was missing for a time. He was later found, uninjured, but quite intoxicated; it seems he had decided to celebrate his "birthday" at the Wasserkuppe's cantina..

The same situation is referred to in Peter Riedel's memoirs thusly:

"Willy Messerschmitt's S 16a took off shortly after the Rhön Competition began and made its way into the rainy, gray sky. Pilot Heinz Seywald, an old hand at flying motorized aircraft, steered his machine from the runway into the prevailing westerly wind toward the summit of the Wasserkuppe. 'Yes, that sure is a real powered glider,' I thought, and glanced up at the elegant 14-meter craft, whose propeller spun more slowly than its British Douglas engine thanks to the reduction gearing. It was now flying majestically, slowly along at an altitude of about 40 meters toward the west. Suddenly a hysterical wailing sound came from the engine—and then nothing but silence. Ssssst-sssst-sssst came the sound of the propeller, as it spun freely downward to fall into the fir trees behind the Messerschmitt barracks, followed by other pieces, such as the reduction gear's chain and a few pieces of the cowling.

We stood by with opened mouths and watched with amazement as Seywald calmly made a left turn and returned to the landing strip with the wind at his back...The failure of the S 16a was yet another setback for the sailplane movement. This craft had come so close to being the ideal powered glider that we could have done research into thermal gliding and cloud soaring...that is, if this misfortune hadn't compelled a disappointed *Dipl.-Ing.* Willy Messerschmitt to set his sights on new projects. Ultimately, these projects would make him the leading designer of many successful engine-powered aircraft. Messerschmitt's last engine-less type, the elegant S 14, was piloted at the 1924 competition by *Oberleutnant a. D. C. A.* von Schönebeck, who had flown with the Richthofen squadron during the war. It was to be the last time that a Messerschmitt sailplane would take off from the Wasserkuppe."

Seywald didn't fare much better with the S 16b "Betty." After passing the technical acceptance portion, he took off in the two-seater on the "round-trip long-distance flight to Bad Kissingen." The honor of being the first ever passenger in a Messerschmitt plane went to apprentice F. Schwarz, who after the feverish take-off preparations had enough stamina to observe whether or not the cotter pin on the wing connection fitting and strut, lying directly in his field of vision, would slowly and surely pop out of their holes, or if they would hold: in the preflight haste, two of them had not been properly bent over.

But while still in sight of the Wasserkuppe, the chain between the motor and the propeller shaft broke: emergency landing in an open field, repair the chain, takeoff again from the field; everything went ok and relatively fast. However, after only a short time, the engine "coughed"—another emergency landing. It went smoothly, but this time they took a little longer finding the source of the problem—water in the carburetor; no chance of taking off now. Even the field in which the plane rested was unsuited for takeoff, and the attempt was abandoned.

"Betty"—and what was still salvageable from "Bubi"— were transported via truck to Würzburg, since the "worker's union" had become the owners of both aircraft.

At this Rhön competition, Ernst Udet and his "*Kolibri*" brought home all the laurels for powered aircraft. Peter Riedel noted that this was an injustice, for the *Kolibri* possessed none of a sailplane's characteristics. Seywald received a consolation prize for his passenger flight, and the "Honorary Cup of the Aeroclub" for his participation in the flying events.

Despite this bit of bad luck, Messerschmitt nevertheless achieved what was for him at the time the most important thing: he became a topic of conversation and a force to be reckoned with in the future; his designs were beginning to draw attention.

Prince Heinrich of Prussia, a great promoter of German aviation in its infancy and a pilot himself, inspects the S 16 together with pilot Heinz Seywald.

From Engineer to Businessman (1924-1934)

The Step to Powered Aircraft

After powered gliders, Messerschmitt turned away from sailplanes altogether and initially devoted his efforts to building light planes. At that time, the definition of light plane embraced the construction of trainer aircraft and sportplanes possessing low weight, low engine performance, and low purchase and operating costs. Airplanes and flying would both have to become less expensive, thereby making them more accessible to a larger number of people. And just as abruptly as he'd rejected Harth's stubborn insistence on the wing warping principle, Messerschmitt also turned his back on the Rhön.

Yet there was hardly a powered aircraft event in Germany where his designs were not competing, and indeed leading these competitions in increasing numbers. That Messerschmitt wanted to develop motorized aircraft from this point on was obvious in his use of "M" for the designator (in place of "S," which had stood for "sailplane"). The fact that he did not begin his numbering system with 1 following his breakup with Harth, as had most of the designers evolving from glider development, is further evidence that Messerschmitt not only wanted to demonstrate that he considered his connection with Harth an important part of his personal development in becoming an aircraft designer, but that he perceived it as a calling and an obligation to advance those ideas which he and Harth had worked for and initiated: developing an "economical" powered aircraft using the knowledge and experience gained from glider construction and flight.

But that was not all: Messerschmitt was one of the few of the many former sailplane builders of his day to recognize that any chance of profiting commercially from his work rested solely in the ability to market unusually well-built and inexpensive aircraft with extremely low operating costs.

Thus, it was during those financially lean years, when so many well established companies in so many many fields were crumbling, that an embryonic company under the leadership of a gifted albeit inexperienced Willy Messerschmitt not only established the reputation of its founder, but in many respects began to amaze and astound the aviation world. The name of Messerschmitt soon became synonymous with the concept of extreme lightweight construction while strictly adhering to prescribed structural limitations and embodying a highly-refined aerodynamic form.

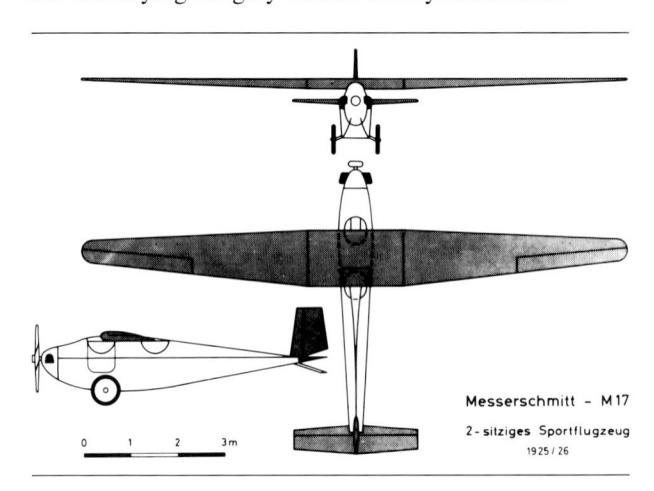

Messerschmitt – M 17
2-sitziges Sportflugzeug
1925/26

M 17 Two-Seater Light Plane (1925/26)

Messerschmitt's first "real" motorized aircraft, the M 17, was a follow-on development of the S 16b two-seat powered glider. In the interim, *Hauptmann* Seywald and other former World War I flyers had formed a civilian sporting flight school from the "worker's union" in Würzburg, and soon after the 1924 Rhön Competition, had ordered an improved version of the "nearly" successful S 16b from Messerschmitt Flugzeugbau, Bamberg. This aircraft then became the M 17, the design which would eventually spawn two further examples, namely: one with an ABC Scorpion engine for *Hauptmann* Seywald, and the other for *Oberleutnant* Noptisch driven by a Bristol Cherub engine. Another interested party was the *Flugsportschule Fürth* founded by Theo Croneiß. This school would be used to train and educate pilots for an airline he intended to form, the Nordbayerische Verkehrsflug GmbH Fürth.

The first version of the M 17, built with the intent of taking part in the "Zugspitzflug" in January 1925 (as contestant number 3), was delivered to Schleißheim with the old 700 cm³ Douglas engine with a chain drive for the propeller, just as the S 16b motorized sailplane had the year before. In fact, the first M 17 may have been the rebuilt S

44

The first M 17, built from remnants of the S 16, was powered by a 700 cubic centimeter chain drive engine. The aircraft was entered in the 1925 Zugspitz Flug competition, but was withdrawn for safety reasons because of its underpowered engine. Here Willy Messerschmitt and pilot Heinz Seywald pose with the aircraft in January 1925 in Oberschleißheim.

16b, for like the S 16b the engine housing did not quite extend back as far as the wing leading edge, leaving the forward cockpit somewhat open, unlike all subsequent M 17s. In any event, the airplane did not take part in the Zugspitz competition. As Schwarz (vaguely) remembers it, the main problem was delays in certifying the plane, and the pilot also failed to show up on time for the start. He claimed that it was due to the variable weather conditions, whereby a lack of upwind combined with a weak engine prevented him from attaining the 3,000 m altitude required to begin. Ultimately, the plane was pulled from the competition at the last minute.

All subsequent M 17s were then fitted with a more powerful engine, either the ABC Scorpion or the Bristol Cherub with a power rating of 18 to 22 kW/24 to 28 hp, and the M 17's subsequent successes confirmed the correctness of such a measure.

According to Professor Messerschmitt and several co-workers from that time, there were 6 to 8 M 17s built. Based on available photographs and copies, this number of aircraft seems to hold true, though with the stipulation that an aircraft simply given a name (for example: "Ello") or competition number could then be given a registration number and photographed with such.

Sequenced according to their known or assumed factory numbers (in parentheses), the following are the seven M 17 aircraft built:

Werk-Nr. (20)	from the S 16b, two seater, with the Douglas engine and competition number 3 at the Zugspitzflug, January 1925
Werk-Nr. (21)	M 17 "Ello" with an ABC Scorpion engine with 17.7 kW/24 hp for the Oberfrankenflug, April 1924
Werk-Nr. (22)	M 17 for the Jurisch-Pipping Firm, Leipzig, with a "Jurisch/24TT250" engine
Werk-Nr. (23)	M 17 D-613 with a Bristol Cherub engine, with 18.4 kW/25 hp, mid 1925
Werk-Nr. 24	M 17 D-612 for Kober Berlin, with an ABC Scorpion engine with 17.7 kW/24 hp, mid 1925
Werk-Nr. 25	M 17 D-779 with a Bristol Cherub engine with 20.6 kW/28 hp, in 1926 for Theo Croneiß. Now in the Deutsches Museum.
Werk-Nr. 26	M 17 D-887 with a Bristol Cherub engine with 20.6 kW/28 hp, delivered in 1926 to *Kommerzienrat* R. Weyermann, Bamberg for Eberhard von Conta.

Design History and Characteristics

Messerschmitt Flugzeugbau Bamberg was now building the M 17 in the old Murmann Brewery. Despite having rented more space in the basement, the facilities were too cramped and the young firm soon moved into the hangars of the former munitions factory (MUNA) in Hauptmoorswald near Bamberg, within the Bamberg industrial area.

The M 17 was a two-seat, cantilever high-wing monoplane built of wood, with a propeller, engine and fuel tank housed inside the forward fuselage, behind which was the passenger seat located at the center of gravity, while behind this was the pilot's seat, set underneath an upward hinged wing canopy. The airplane fell into the light plane category, A-1 class.

The fuselage was designed as an enclosed hull with a trapezoidal cross section created by longerons attached to formers covered with plywood skinning. The fuselage ended in a horizontal cut. Both cockpits had open side cutouts to provide visibility.

The wing surfaces were of a fully cantilever layout, attaching to the fuselage with four bolts. They tapered off at the end both chordwise and in thickness. The wing profile was the same that Messerschmitt had developed for the S 12, this being Göttingen No. 535. It was designed as a single spar (box spar) with a plywood leading edge and fabric covering the surface behind the spar. It incorporated large, unbalanced ailerons made of fabric covered wood ribbing. The empennage: the horizontal stabilizer consisted of a tailplane which could be adjusted on the ground and a rectangular, unbalanced elevator. The vertical stabilizer comprised a tall, narrow tailfin and a rectangular-shaped, unbalanced rudder. Control input was via a simple single stick dual control system. Elevator and rudder control were by means of cables, with even the ailerons driven by cables in the outer wing sections; only in the center wing were two rods used to relay the control inputs.

The undercarriage assembly consisted of a sprung axle housed in a box inside the fuselage; axle ends and wheels protruded from either side of the fuselage. An ashwood tail skid with coil spool spring was fitted at the extreme aft end of the fuselage.

The engine was attached to a welded steel support, which was in turn anchored to the forward bulkhead with four bolts. The multiple pieced engine cowling was hammered out of thin aluminum plates. It was attached using reinforced perforations with tubular rivets in the plates, and was placed over small rivet bolts and secured by cotter pins. Other features included a 26 liter gravity-fed fuel tank behind the firewall, an oil reservoir above the motor, and the propeller at the end of the engine shaft.

M 17's Contests and Races

The disappointments with participating in the Zugspitzflug in early 1925 showed that successful participation in competitions was not entirely hinged upon the qualities of an aircraft design. Timely assembly and registration of an aircraft, an experienced pilot, a design tailored to the competition's requirements, functioning maintenance and information structure, and a healthy dose of luck were all needed in order to beat seemingly superior competitors who had a solid foundation of experience coupled with long years of proven abilities.

From today's perspective, the fact that Messerschmitt succeeded in this effort, and that this success was achieved with his very first motorized aircraft, seems to be more amazing than it probably did to those peers and competitors of his day and age. The following enumeration of some of the events in which the M 17 participated, in one form or another, gives an indication of just how hard the path to success must have been.

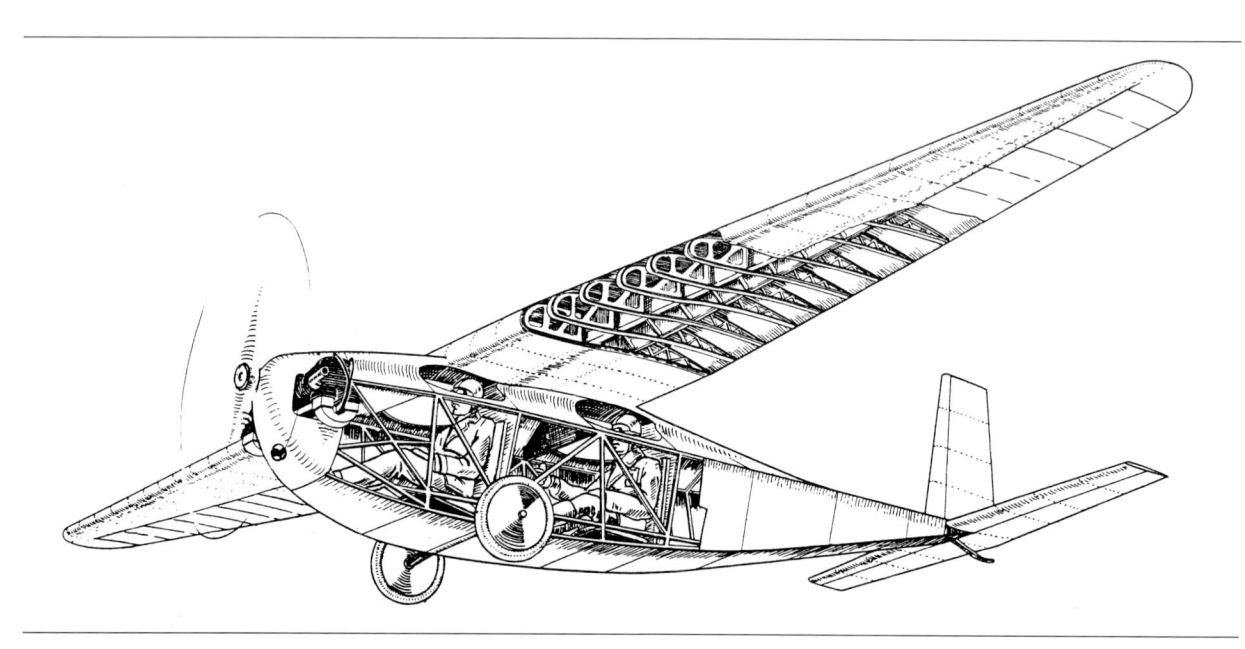

The M 17 was made entirely of wood. Messerschmitt marketed the aircraft for 8,500 marks, with a ten-week waiting period for delivery.

This production example, built in 1926, resides today in the Deutsches Museum as the oldest Messerschmitt airplane in existence. Theo Croneiß is in the aft cockpit, while Willy Messerschmitt sits behind the propeller.

Obenfankenflug from 2 to 4 May, 1925, in Bamberg: M 17 "Ello" with a 17.6 kW/24 hp ABC Scorpion engine, piloted by Seywald won first prize in the altitude and speed variable competitions, plus took second place at the cross-country competition, during which the pilot "wandered" off course.

At the Transportation Exhibition in Munich (15 July to 11 October, 1925), Messerschmitt displayed his M 17 D-612, and at this event's "International Flying Competition" in Schleißheim from 12 to 14 September 1925 it was Carl Croneiß who took first prize in this tough competition in both the altitude and speed competitions (with winnings totaling 6,000 marks), and together with the youthful pilot Maier also placed in the relay race (5th place), taking home another 750 marks.

Süddeutschlandflug 1926. It was the only competition for sporting aircraft for the entire year, running from 31 May until 6 June. The two participating M 17s were only able to take 10th place, with the D-779, piloted by Maier, and 13th place with the D-887, piloted by E. von Conta (both pilots with little competition flying experience), but in the technical performance evaluations, von Conta brought home the bacon when he won 1st prize and 4,500 marks with the D-887. Von Conta's name made the headlines again with his M 17 D-887, when he and the well-known sport pilot and aviation journalist Dr. Werner von Langsdorff made the flight over the Alps from Bamberg to Rome from 20 to 29 September, 1926. This was the first crossing of the Central Alps in a lightplane; the true flight time for the 1,620 kilometers totaled 14 hours and 20 minutes. Trade journals at the time considered this flight to have been one of the major accomplishments of the crew and for Germany's aviation industry as a whole.

The DVL Berlin-Adlershof's assertion that the M 17 was the best airframe designed since the war's end is clear evidence of the esteem Messerschmitt's designs had won in the interim.

The Messerschmitt squadron, consisting of M 17 D-779, the first M 18 and M 17 D-887, seen during a stopover in Berlin, 1926. D-887 was used by E. v. Conta and W. v. Langsdorff to cross the central Alps.

M 18 Feeder Service Commercial Plane (1925-1930)

One of the most astounding design and organizational feats young Messerschmitt achieved occurred during this period (1925/26) when he totally switched over from the purely wooden design hitherto found in his sailplanes, motorized sailplanes, and the M 17 light plane to the all-metal design of the M 18. It is easy for someone familiar with the materials to see how working for decades in the medium of wood, mainly plywood covered ribbing (or occasionally fabric), would lead to extreme lightweight designs. But the fact that, within the space of a year, a young designer was able to switch himself, his team, and his operations over to purely metal construction and develop an all-metal plane suitable for commercial use was something unparalleled in the annals of aircraft manufacturing. In this, Messerschmitt showed himself not only to be a gifted and inspiring designer; he proved himself a talented organizer, blessed with a clear talent for simplification of design and construction.

Situation in the mid-'Twenties

Following World War I there sprang up throughout Europe, especially in Germany (despite the restrictions imposed by the Versailles Treaty), a series of airlines of differing sizes and with different goals, most of which catered to the passenger market.

Since aircraft during the mid-'Twenties were generally larger in size, carrying at least ten passengers, the poor economic situation left many of these seats unfilled; it was impossible to maintain either a national or an international airline operation without considerable financial support from the government. Aircraft manufacturers such as Dornier, Junkers, and Rohrbach developed planes—partly using government subsidies—suitable for cross-country and transatlantic travel, but practically speaking the goal of financial independence remained an elusive one.

Without a doubt, the greatest legacy of Theo Croneiß, founder of the Nordbayerische Verkehrsflug GmbH Fürth, was to have recognized that only by using a small passenger plane designed especially for the feeder air service would it be possible to operate a self-sustaining passenger service, and by doing so, to streamline commercial aviation as a whole.

Croneiß was already acquainted with Messerschmitt from several sport flying competitions. When Croneiß's brother Carl won the Munich International Flying Competition (12 to 14 September, 1925) in a M 17, Theo Croneiß not only obtained this aircraft for his sport flying school, but he also contracted with Messerschmitt for the development of a feeder service plane. Based on the M 17, it was to be a four-seat cruising limousine with metal fuselage, the minimum empty weight possible with the largest possible load capacity designed and built in strict compliance with structural guidelines set forth for commercial aircraft, and was to have good take-off power and normal cruising speed capabilities with a throttled low-horsepower engine. Purchase price was not to exceed 25,000 Reichmarks; at the time, this corresponded to one third the price of competing aircraft. Messerschmitt later explained that Croneiß himself had ordered the all-metal fuselage, but found no problems with using wooden wings. He—Messerschmitt—worked through this idea and informed Croneiß a few days later that he could and would build an all-metal plane which would fully meet all requirements.

The M 17's marketing success, the competition prizes and, most importantly, the trust Croneiß put in the abilities of the young designer, from whom he now intended to order an initial batch of four aircraft, may have all contributed to Messerschmitt daring to broach this task, one which was pioneering both in terms of its design as well as its technology. It was to create an aircraft that would keep pace with the demands of the feeder service industry and, while ensuring that the plane was as robust as required, to imbue it with those qualities which would make it economically superior to those commercial airliners already on the market.

M 18a four-seat commercial plane (1925/26)

Messerschmitt began working out the measurements and design in the fall of 1925. A wooden mock-up of the fuselage satisfactorily resolved the matter of dimensions, and work started on two all-metal airframes immediately thereafter, Werk-Nr.s 27 and 28. They were designed for a flying weight of about 1,000 kilograms and had, aside from the pilot seat, room for three passengers. The engine of choice was an air-cooled seven-cylinder Siemens-Halske Sh 11 rated at 59 kW/80 hp. The wings of both the first and second aircraft comprised a spar divided into several segments, and the cockpit was exposed to the elements. The cabin windows were fitted with glass. Unlike traditional arrangements, the cabin entry door was located on the right side.

Theo Croneiß himself took the M 18a up on its maiden flight on the morning of June 15, 1926, "in rain and wind; it lasted 20 minutes and resulted in a very satisfying flying performance." With the completion of flight testing, the plane was registered as D-947, and starting on July 26 "with

Messerschmitt's first all-metal airplane was the four-seat M 18a.

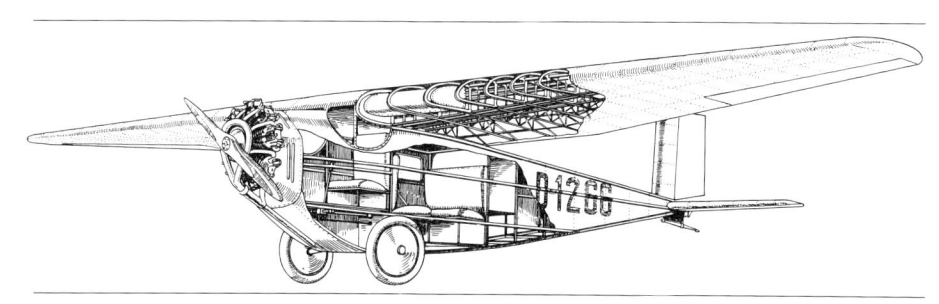

When space became limited at his parents' home on Lange Straße, Messerschmitt and his small team moved to the former Murrmann brewery in Bamberg. Here, work on parts and major components of the M 18a took place.

Fuselage of the first M 18a with Sh 11 engine prior to final assembly, photographed at the Bamberg brewery in the spring of 1926.

The first M 18a, built in June 1926, was still flying in 1934. Here the aircraft is seen following a major overhaul at the Augsburg plant.

director Croneiß as pilot, took off with its first group of passengers for the Nordbayerische Verkehrsflug GmbH."

The second plane was probably ready to take to the air a few weeks later. Initially it was unknown why the aircraft was not immediately put into service. It is assumed that Messerschmitt thought he was to build a four-seat aircraft—the fourth seat being the pilot's seat—while Croneiß had envisioned a plane with four passengers (plus the pilot). In any case, the second plane was redesigned to include the additional passenger seat, and was fitted with the more powerful Sh 12 engine rated at 73.6 kW/100 hp and all associated reinforcements. Following these prescribed modifications and flight tests, the plane was registered as D-1118 and designated the M 18b. At the beginning of the 1927 flying season, this M 18b, Werk-Nr. 28, coded D-1118, was delivered to its new owner, the Luftverkehr Thüringen, who then chartered the aircraft to the North Bavarian Transport Company GmbH.

In the same year work began on two more airframes which incorporated all the "b" variant's design modifications. These were Werk-Nr.s 29 and 30, and they first apppeared at the start of the Nordbayerische Verkehrsflug GmbH's 1927 summer flight schedule with the registration codes D-1133 and D-1177.

M 18b five-seat commercial plane (1927-1930)

The M 18's development and assembly took place parallel to the the company's organizational and dimensional evolution. While the company may have started as the "Flugzeugbau Messerschmitt Bamberg," a change of title was required in order to become a "legal business" (due to its contract with the Nordbayerische Verkehrsflug GmbH) in the business registry at the Bamberg District Court. This happened on April 28, 1926, under the company's new name of "Messerschmitt Flugzeugbau GmbH Bamberg."

As mentioned earlier, construction of the M 18b continued apace. In addition to the previously mentioned modifications (which resulted in a corresponding weight increase to 1,200 kg), a fundamental change to the wing was in order: the box spar design employed up to this point, with four ribs in the inner side, was now combined into a single, individual spar, which formed a rigid tubular construction by a metal skin extending to the wing leading edge. This construction concept, used thus far only in wooden sailplanes, was now being applied to commercial aircraft and initiated heated debate throughout the aviation world. That it finally found acceptance goes to show just how far-sighted Messerschmitt the designer was.

For a time, the M 18b exemplified the standard model of an economical feeder service, or small passenger plane, an aircraft which only required 18.4 kW/25 hp per passenger. This enabled the Nordbayerische Verkehrsflug GmbH to sell its air kilometers for only 60 to 80 pfennigs per km, compared to 2 Marks or more from other airlines, and therefore, with the books always in the black, was able to leave its competitors far behind.

Entering the Messerschmitt Flugzeugbau GmbH Bamberg into the business registry at the Bamberg district court resulted in Theo Croneiß and his Nordbayerische Verkehrsflug GmbH becoming shareholders in Messerschmitt on the one hand, while on the other the delivery of this first plane and subsequent aircraft made Messerschmitt a shareholder in the airline until these airplanes had been paid for. Each craft delivered came to at least 20,000 Reichmarks, and Messerschmitt was obligated to make his shares (up to 1,000 RM each) available for buyback; indeed, he was compelled to do this if for no other reason than to have enough operating capital available to build more airplanes and retain enough creditt. It wasn't just the commercial factors, though, which were of con-

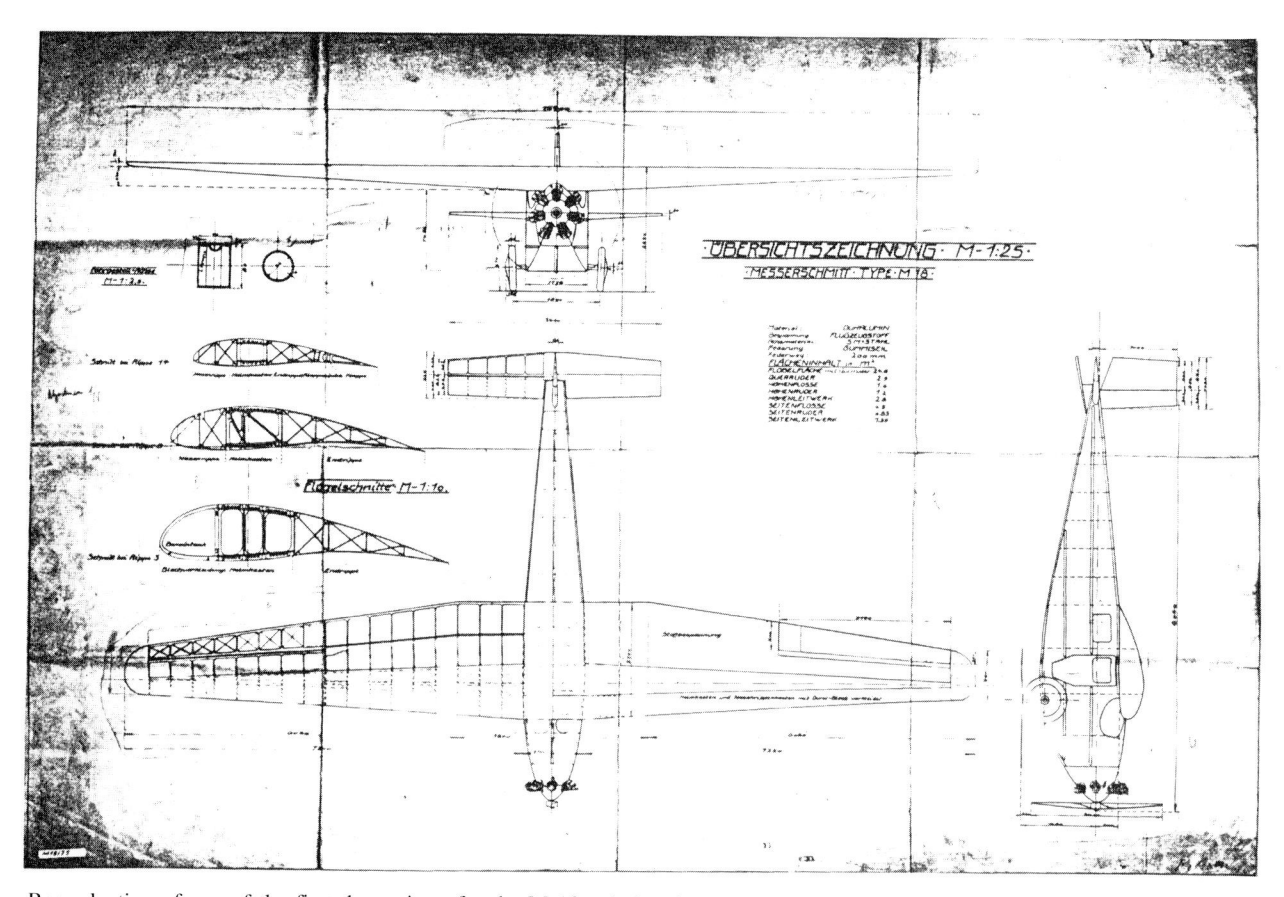

Reproduction of one of the first three-views for the M 18a, dating from 1925. The open wing spar construction is evident here, although Messerschmitt abandoned this approach completely a short time later.

The M 18, commuter airliner for the Nordbayerische Verkehrsflug GmbH (based in Nuremberg-Fürth) provided daily service to a number of cities in southern and central Germany. In so doing, the airline offered connection to Luft Hansa's long distance routes.

A view into the open cockpit of the M 18b.

cern. Space constraints, a shortage of technical personnel experienced in metal work, in qualitiy control, and certification and engine installation, as well as the young company's overall lack of experience in mass production all contributed significantly to miscalculations, production setbacks, delays, and cost overruns. Despite these problems, the spring of 1927 saw the delivery and operational debut of three more M 18bs. In total, the Bamberg firm delivered the following:

    1926, M 18a, Werk-Nr. 27, D-947 Habicht
    1927, M 18b, Werk-Nr. 28, D-1118
    1927, M 18b, Werk-Nr. 29, D-1133, Franken
    1927, M 18b, Werk-Nr. 30, D-1177.

Bamberg also started on Werk-Nr.s 32, 33, and 34, but because of work on the M 19 competition aircraft (Werk-Nr.s 31 and 35), these were not finished until after the company's move to Augsburg under Augsburg's Werk-Nr.s 365, 366, and 367. Beginning in the fall of 1927, the Bayerische Flugzeugwerke AG Augsburg took over continuation of all assembly work as part of its side of the joint interest agreement worked out with Messerschmitt and his company.

M 19 Competition and Sportplane (1927)

The Sachsenflug 1927, a competition tailored to lighter and less costly sportplanes, gave Messerschmitt the opportunity with his M 19 to show just how far the limits of light construction technology could be stretched. The M 19 was Messerschmitt's first low-wing monoplane, and it was the first powered aircraft to have such a high load factor, whose additional load was greater than its empty equipped weight. He entered the competition with two of these planes and, by virtue of the design's excellent empty weight to load ratio of <1, was given a rating of "*unendlich*" (unlimited, infinite) in the technical performance evaluation, and therefore unbeatable, a rating which proved to be true in the subsequent flying portion of the competition.

In January 1927 the details of the competition were made public, and "Flugsport" magazine lost no time in prophesying:

"that the competitors will have a hard time recouping their costs, with the exception of one type which has entered the competition with a good chance of success."

Messerschmitt had studied the DVL's evaluation criteria very closely. He assumed that he was not the only one who had recognized the empty weight to load ratio would be the deciding factor in determining the winner of the technical evaluation. He had also already demonstrated with his first powered aircraft, the M 17, that it was possible to bring this ratio down to a factor of less than 1, and therefore assumed that his competitors would attempt to beat this figure. He even went so far as to call his brother-in-law, Georg Madelung, then at the DVL, and request that he advise Professor Hermann Blenk of this fact, as Blenk was responsible for the evaluation criteria. Professor Georg Madelung reported back a few days later that Professor Blenk had changed the criteria.

Looking back, Messerschmitt felt that in studying the new criteria he saw there was still the possibility of achieving an "*unendlich*" ("unlimited") rating. This he would succeed in doing with the M 19.

The basic shortcoming with the specifications for this competition, which "Flugsport" bluntly called an "unlimited formula gone wrong," had stirred up harsh criticism and dissatisification within the ranks of those competing in the event. Since Theo Croneiß and von Conta had taken

The M 19, a wooden design, was Messerschmitt's first low-wing plane and had an all-through wing spar.

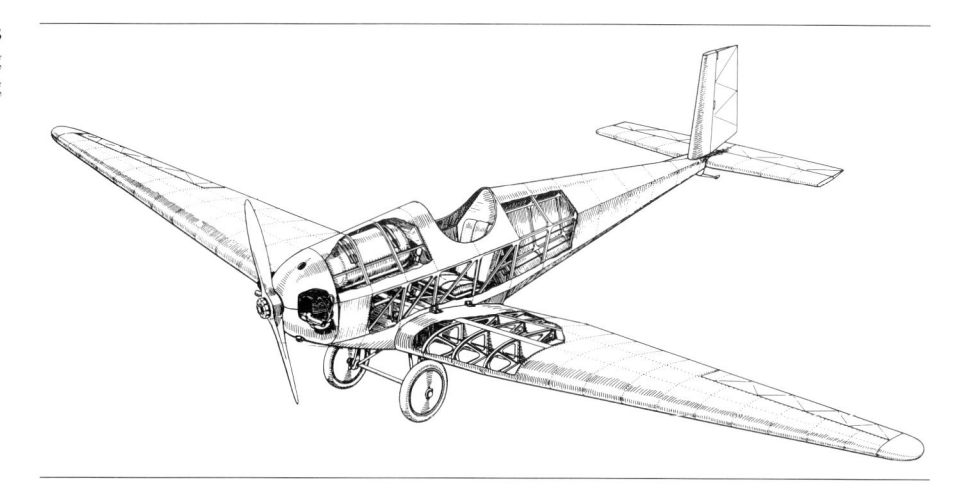

home the entire winnings, the competition management felt compelled to put up additional funding in order to satisfy the other winners.

In any case, Messerschmitt had designed a plane as extremely light and as precisely in compliance with structural requirements that the competition's provisions would allow. The latter specified, among other things, that aircraft with empty weights less than 300 kg would be awarded significantly more points, that takeoff runs must be less than 200 meters, and that the freely determined flying weight, consisting of empty weight plus the load, would have to be the same in all the competition's performance evaluation flights (climb, speed runs).

Since even the safety component, as evidenced by the load factor, was included in the evaluation, the conditions for a standard assessment of all competitors had been clearly marked out. At the beginning of 1927, Messerschmitt began working out the dimensions, the blueprints, and the construction of his version of the competition plane. The workshop then tackled the first aircraft (Werk-Nr. 31). That it was to be designed as a low-wing monoplane was understandable, because of the smaller size of the craft and that

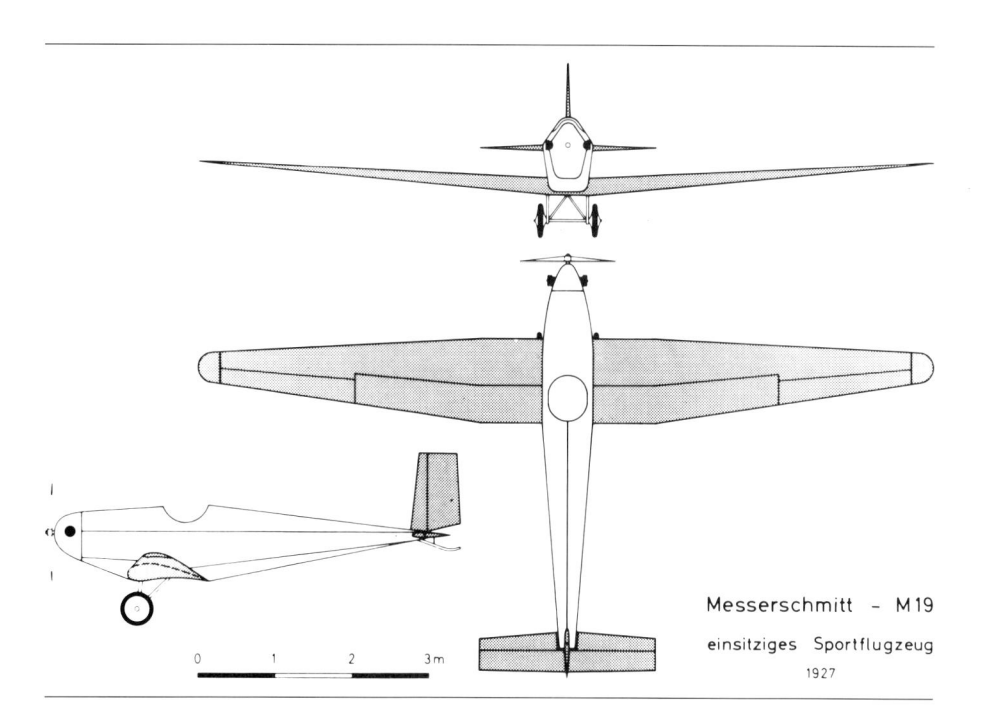

Three-view of the M 19.

Assembling the winning machine at the 1927 Sachsenflug competition using a four-man team. The figure supporting the tail section is Theo Croneiß.

the problem of constantly climbing in and out was more difficult to resolve on a high-wing design. A low-wing design also enabled the fuselage to be set onto a one-piece wing which already incorporated an undercarriage, with a cutout in the fuselage where it mated to the wing, making the low-wing approach easier to design and build. Control inputs were more easily observable, the structural soundness estimates more accurate and the designers could inch closer to the authorized limits specified in the supplemental safety requirements.

If one assumes that aircraft designers such as Messerschmitt favored the low-wing design's aerodynamic advantages based on gut instinct alone (bearing in mind that wind tunnel testing was not yet available), it would seem that Messerschmitt's shift to the low-wing monoplane concept was a purely speculative venture, but one which would later prove to be the correct course of action.

About three months later, a second M 19 aircraft (Werk-Nr. 35) was laid out. The reason: at the end of May von Conta had totally destroyed his M 17, D-887, and as he'd wanted to take part in the Sachsenflug, he ordered a M 19 from Messerschmitt. *Geheimrat* Weyermann was his sponsor, supplying him with the funding, and had hoped to be repaid through the expected winnings. Both aircraft were completed in time for the signup deadline (June 30, 1927). Theo Croneiß had already test flown and broken in the first to be finished, D-1206, at the beginning of August. E. von Conta accepted it in Bamberg on the 25th of August, carried out two test flights lasting ten minutes each, and flew to the competition in Leipzig on the 29th of August. He was a successful participant in the competition, receiving a rating of "unlimited" in the technical performance evaluations. But he lost his plane, completely demolishing it on the last day (4 September) on the last leg of the flight near

Second place winner Eberhard von Conta taking off during the Sachsenflug, starting point Leipzig.

The Sachsenflug winner came to grief on 8 August 1928 when pilot and aviation journalist Werner von Langsdorff went into the trees at the edge of Augsburg's airfield as a result of being given incorrect information on the design's descent angle and landing speed by the factory; Langsdorff's injuries required hospitalization.

Bautzen after covering about 100 kilometers. Despite this, his 2nd place finish at the competition brought with it a prize amounting to 13,074 Reichmarks.

The other M 19, D-1221, which Croneiß had flown and won the competition with, survived the competition's entire 450 kilometer course. Since Croneiß had also received the "unlimited" rating in the technical performance evaluations, and as he had completed the entire race circuit, he was awarded the correspondingly larger portion of the total prize, namely 46,926 Reichmarks. There was no further production of the M 19, a single-seat low-wing monoplane manufactured from wood, with a tractor propeller driven by an air-cooled, two-cylinder Bristol Cherub engine rated at 20-26 kW/28-36 hp. But the design did serve as a springboard for a two-seat development of this same concept, the M 23.

From Bamberg to Augsburg (1927)

The Bamberg Messerschmitt team celebrating the completion of the first M 18b at the Murmann brewery in Bamberg-Hauptmoorswald in the summer of 1926. The move to Augsburg—here humorously reenacted for the photographer—took place starting in September 1927. The "draft horse" in the photo is Herr Heinz, who worked for many years as a foreman in the company.

At just about the same time as Messerschmitt's first M 18a feeder line plane was entering service with the Nordbayerische Verkehrsflug GmbH, the Bayerische Flugzeugwerke (BFW) was founded in Munich (under its old name).[2] One of the major reasons for the establishment of this company was to acquire the assets and liabilities of the now-bankrupt Udet Flugzeugbau GmbH in Munich-Ramersdorf. In conjunction with the establishment of this new company a change of location was in order, and BFW negotiated with the Eisenwerk Gebr. Frisch in Augsburg for the purchase of the former Rumpler Werke AG in Augsburg.

The company was established on July 30, 1926, and was founded by:

- Deutsches Reich, represented by the Reich Minister of Transportation with interests amounting to: 250,000 RM
- Freistaat Bayern, represented by the Bavarian Ministry of Trade with interests amounting to: 100,000 RM
- The bank of Merc, Finck, & Co., Munich, with interests amounting to: 50,000 RM

The BFW's capital thus totaled: 400,000 RM

Incorporation of the new business brought with it changes in leadership. The development manager of the

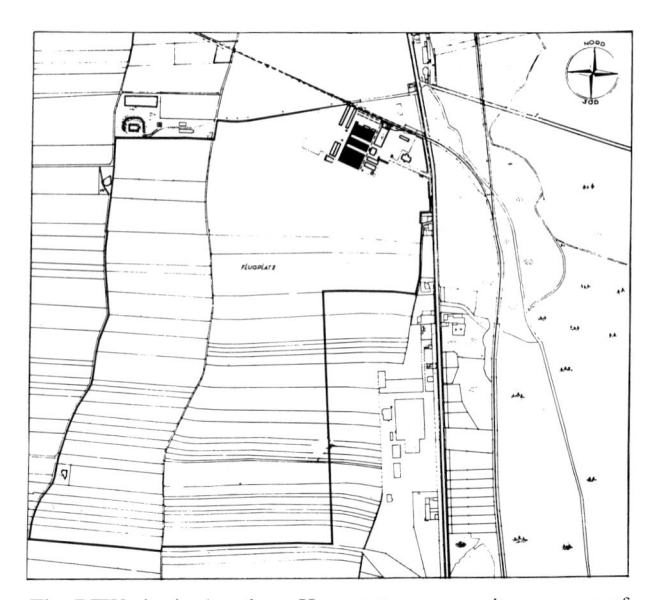

The BFW site in Augsburg-Haunstetten, occupying an area of approximately 870,000 m^2 in 1927, was leased from the city together with the airfield for 99 years.

[2] see page 24.

Udet Flugzeugwerke GmbH up to this point had been Hans Herrmann, and the chief stress engineer was government architect Julius Krauß. Although these two were retained, executive director Hermann Pohl was considered responsible for the problems at Udet and was not welcomed aboard. He had previously nudged out his partners Ernst Udet and Erich Scheuermann.

The move of the Udet people and all the aircraft and machinery, tools, materiel, half-built aircraft and engines all in different stages of completion began on August 2, 1926. Dr. A. Schrüffrer assumed managerial responsibility for the BFW AG Augsburg. By the end of the year the BFW company had incurred considerable losses. The company was kept afloat only by regular infusions of additional financial assistance from the governmental partners. One stipulation, however, involved the Bank of Merck, Finck, & Co. in Munich, transferring its shares amounting to 50,000 Reichmarks over to the *Reichsverkehrsministerium*.

The Bayerische Flugzeugwerke AG in Augsburg had planned on mass production of the Udet-developed U-12a and U-12b trainer, the Flamingo. The *Reichsverkehrsministerium* had contracted for the design. Furthermore, it was expected that this wooden trainer aircraft would undergo further development using mixed wood/metal/fabric construction materials. One of the major changes would have involved building the fuselage of welded steel tubes.

In any event, there were substantial differences of opinion between the managing engineer Hermann and senior management, and Hermann was fired. Therefore, the development section of the Bayerische Flugzeugwerke in Augsburg initially found istelf without a manager.

Messerschmitt exploited the considerable attention his M 17 and M 18 developments attracted from the aviation world and associated governmental figures to negotiate funding in the form of subsidies. However, the unhealthy economic situation at the time prompted the *Reichswirtschaftsministerium* to refuse the state of Bavaria's subsidization of two aviation companies at the same time, namely the Bayerische Flugzeugwerke AG in Augsburg and the Messerschmitt Flugzeugbau GmbH in Bamberg. The *Verkehrsministerium* took the same view, and both ministries made attempts to combine the two firms. Messerschmitt's initial opposition to such a move was centered around preserving his independence, at least in the area of development. Finally, after many months of negotiations, a "joint interest" agreement was reached whereby both firms would merge under the following conditions:

Bayerische Flugzeugwerke AG Augsburg confined itself exclusively to aircraft manufacturing, mainly Messerschmitt designs. Messerschmitt Flugzeugbau GmbH dispensed with building airplanes, moving their location to

The BFW works, formerly the Bayerische Rumpler-Werke, seen from a Flamingo in 1927 looking northwest. The aircraft in front of the test hangar are production Flamingos.

The Udet U 12 Flamingo was born in Munich-Ramersdorf, where approximately 35 of these nimble trainers were built. BFW Augsburg produced 115 of the type from 1926 to 1929 and licenses were sold to Latvia, Austria, and Hungary.

The 100th Flamingo was completed in November 1928 and delivered to the DVS in December. The justifiably proud assembly team poses here for the photographer. The apprentice in the center is holding a wreath with the inscription: "Als Hundertste zur Höh ich geh' zum Ruhm und Ehr der BFW (I'm the hundredth to take to the skies for the fame and honor of BFW)."

Ernst Udet with his famous red Flamingo, D 822. Regarding the Flamingo, Julius Krauß stated that it was "the best trainer of the 1920s."

the BFW AG Augsburg spaces and concentrating fully on aircraft development.

Both firms were located in Augsburg and to the outsider appeared as one single company.

The Bayerische Flugzeugwerke AG Augsburg acquired the following from Messerschmitt Flugzeugbau GmbH Bamberg, of 92,000 Reichmarks against a transfer fee:

- complete rights and all documents for the M 17, M 18a, and M 18b
- all development work for these types
- all necessary working engineering designs for both types and all parts lists
- all personal and business correspondence regarding the types in question, and all instructions and inquiries available
- all devices and means of production to assemble the aforementioned types
- all operational equipment, aircraft, and tools

The company's debts, in excess of 66,000 Reichmarks, were paid by the BFW AG, being debited from the 92,000 Reichmarks paid out as compensation; the BFW would pay off residual claims of 26,000 Reichmarks by June 30, 1928, at an interest rate of 5 percent.

By the same token, the (maximum) amount of 50,000 Reichmarks as compensation for the materiel taken by the BFW was accepted and was to change hands by June 30, 1928.

The agreement was signed on September 8, 1927, with the move to Augsburg beginning immediately thereafter. Messerschmitt himself and those employees brought over from the Bayerische Flugzeugwerke moved to Augsburg,

New or improved designs were to have replaced the all-wood Flamingo. BFW proposed and built two alternatives:
1. BFW 3 Marabu, based on the Flamingo but with tubular steel fuselage (in the foreground) or
2. BFW 1 Sperber, a complete steel-tube construction with fabric covering (second plane, followed by seven Flamingos)

The sole Sperber built went to the pilot Alexander von Bismarck in 1927, who utilized the airplane for American-style aerobatic and barnstorming demonstrations.

The BFW 1's tubular steel fuselage.

Aircraft construction in 1927 was still done by hand: here the Flamingo's aileron is being covered with fabric at BFW.

as well. Messerschmitt naturally took over the technical leadership of the BFW. He immediately arranged for continued assembly of the three unfinished M 18bs brought from Bamberg, and also pressed forward with U 12b production and the completion of two designs he'd not been involved with, the BFW 1 Sperber, a trainer of mixed construction powered by a Siemens Sh 12 engine, and the BFW Marabu, whose fuselage was also a steel tube design. In addition, a further lot of twelve M 18bs was contracted for.

BFW M 20 All-Metal Commercial Plane (1927-1931)

While still in Bamberg Messerschmitt had begun work on the development of a larger, all-metal commercial plane capable of accommodating ten passengers and two pilots. The reason for this development was the growing awareness that engine reliability attained up to that point could also be advanced further still in order to carry 8 to 10 passengers, or approximately 1,000 kilograms of goods for a round-trip of 800 kilometers with a single engine aircraft, if this plane was also built as simply as possible. Naturally,

With the M 18, the company of Theo Croneiß could claim to have the most economical airplane in the world.

the pursuit of this goal would also result in air travel becoming more economical.

It is probable that the many years' worth of experience gained in designing and building the small M 18 commuter plane were put to good use in meeting Messerschmitt's latest challenge. After all, the M 18 had carried four passengers with a mere 100 PS engine, and over several years of

By the end of the '20s, the Nordbayerische Verkehrsflug GmbH in Nuremberg-Fürth operated Messerschmitt types almost exclusively, ranging from the M 17 and M 23 (left) to the M 18 and M 24 (right).

The ten-seat all-metal M 20a commercial airliner.

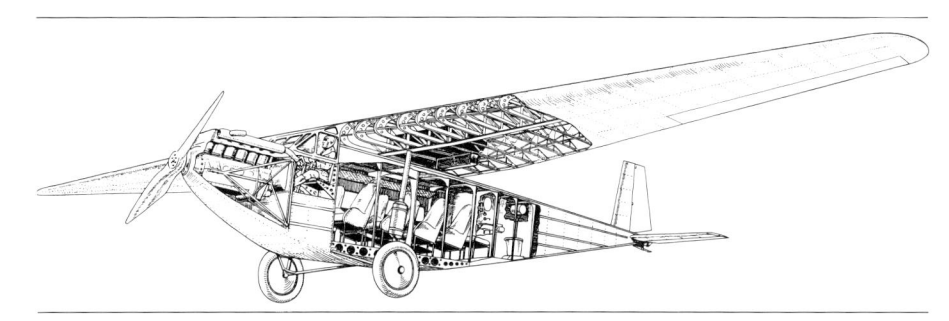

The euphoria surrounding the maiden flight of the M 20a (Werknr. 371) was shattered on 26 February 1928 with the crash and death of pilot Hans Hackmack, a catastrophe for BFW.

operations had proved itself to be an excellent design, both from a flying as well as a technical standpoint. The choice of a relatively powerful engine had also enabled the plane to cruise along at a sharply reduced throttle setting, thus resulting in a relatively long operational life. Accordingly, the M 20 was designed using the well-known practical cantilever high-wing appraoch and constructed of light metal.

At the direction of the *Reichsverkehrsministerium*, the Deutsche Luft Hansa Berlin initially contracted for two experimental aircraft.

Work on these experimental prototypes at BFW therefore began immediately following the Bamberg team's move to Augsburg. Just five months later, on February 26, 1928, the first M 20 was declared ready for flight.

Hans Hackmack, who five years earlier had emerged as the most successful competitor from the Rhön Competion while flying Messerschmitt's last sailplane, the S 14, had since become responsible for the flight portion of prototype testing at the Deutsche Versuchsanstalt für Luftfahrt at Berlin-Adlershof. He welcomed Messerschmitt's invitation to fly the M 20 on its maiden flight, but was killed and the airplane totally destroyed. The fabric covering on the trailing edge of the wing had separated, and this experienced pilot must have mistakenly felt this to be a major problem. He apparently panicked and attempted to bale out even though he was only at an altitude of 80 meters at the time. Certainly one factor in Hackmack's misjudgment in this situation was the fact that shortly beforehand he had been in a fire while testing a Heinkel airplane, thus causing him to react overcautiously. Experienced test pilots came to the conclusion that, had Hackmack stayed at the controls, he could have probably avoided the mishap.

Luft Hansa annulled their contract as a result of the accident. Nevertheless, BFW immediately built a second prototype, and just six months later, on August 3, 1928, the aircraft successfully completed its maiden flight. Theo Croneiß himself carried out the flight and offered his services for the advertising campaign. The plane proved to be quite a success, so much so, in fact, that the Deutsche Luft Hansa demonstrated renewed interest. Following certain changes, the Luft Hansa renewed their order for two BFW M 20a aircraft to be powered by the BMW VI, 5.5 Z, 368-515 kW/500-700 hp engines. After successful test flights, two more similarly modified aircraft were produced for Lufthansa. These improvements led to increased payload capacity and better flight handling characteristics.

Thus was the M 20b created, with the wing surfaces now embodied inside the fuselage and given a slight dihedral and sweep, thereby improving flight characteristics. Reports in the press called the M 20 "the most modern German aircraft."

Theo Croneiß (in the pilot's seat) carried out the successful maiden flight of the second M 20 on 3 August 1928. Below the engine are Willy Messerschmitt and Erich Scheuermann from the DVL. This particular aircraft flew with Lufthansa until 1941.

The second M 20 on its first flight. The single spar wing extending over 25 meters is quite evident here. Compare this photo with a similar one for the Me 323 (q.v.).

Fuselage formers for the M 20a on the building cradle.

The M 20a's forward engine assembly with the 500 hp BMW VI powerplant.

The 25 meter plus wing of the M 20 under construction. The majority of the wing surface was covered in duraluminium, or dural, while the trailing edge and ailerons were fabric covered. A defect in this area led to the crash of the first M 20.

The ten-seat cabin looking forward (above right).

The spacious cockpit of the M 20a.

The type drew considerable attention in the aviation world at the ILA in Berlin in October, 1928, where the M 20a was still being exhibited, not only as a result of its monospar wing concept hitherto absent on such large designs, but also because of its empty equipped weight to capacity ratio (2,800 : 1,800 = 1.55), something which had up to then had been considered impossible for commercial aircraft.

The BMW engine and its 368 kW/500 hp gave excellent take-off and climbing power. Speeds were greater than comparable types at that time (v_{max} = 205 km/h, v_{cruise} = 170 km/h). With a range of 1,000 kilometers it was more than adequate for inner-German flights.

It wasn't until the following year (1929) that the Deutsche Luft Hansa (DLH) accepted the first two M 20a planes, these being:

on June 22, 1929, Werk-Nr. 392 D-1480 "Franken" and on July 20, 1929, Werk-Nr. 421 D-1676 "Schwaben."

A year later, two of the M 20b variants were delivered:

on July 20, 1930, Werk-Nr. 442 D-1928 "Rheinpfalz," and on September 20, 1930, Werk-Nr. 443 D-1930.

M 20b D-1928 crashed on April 4, 1931, during a special flight from Muskau to Görlitz; the pilot, Schirmer, and radioman, Bischoff, were killed. There were eight passengers aboard (Reichswehr officers), four of whom were slightly injured. And just six months prior, on October 6, 1930, the other "b" aircraft had crashed during a flight from Berlin to Vienna shortly before a stopover in Dresden. Cause of the crash was assumed to be a strong downdraft. This accident resulted in eight deaths.

This second accident resulted in a halt in production deliveries of the M 20b$_2$, just getting underway, and a grounding of the type until the reason for the accident and its cure could be found. Investigations showed that the maximum safe load stated in the general specifications for

63

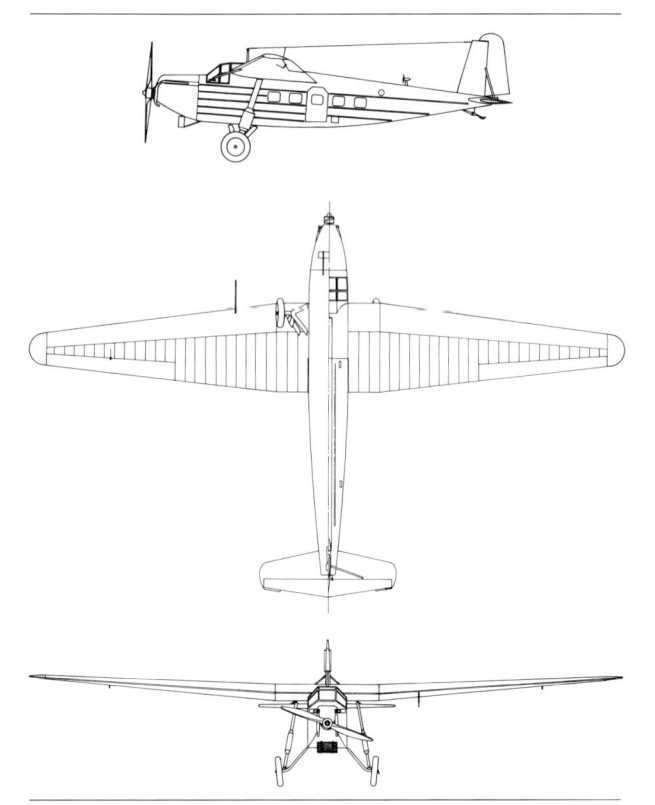

Following two unusual weather-related accidents involving the M 20 in 1930 and 1931 Luft Hansa refused to accept a further ten aircraft on order. These M 20s (which had been built by their deadline) sat on the BFW factory airfield for months. Eventually accepted in 1932, Luft Hansa operated the majority of these machines until well into the 1940s.

This M 20b$_2$ from 1931 was operated by Luft Hansa until 1943.

Three-view of the M 20b$_2$ with BMW VIu engine.

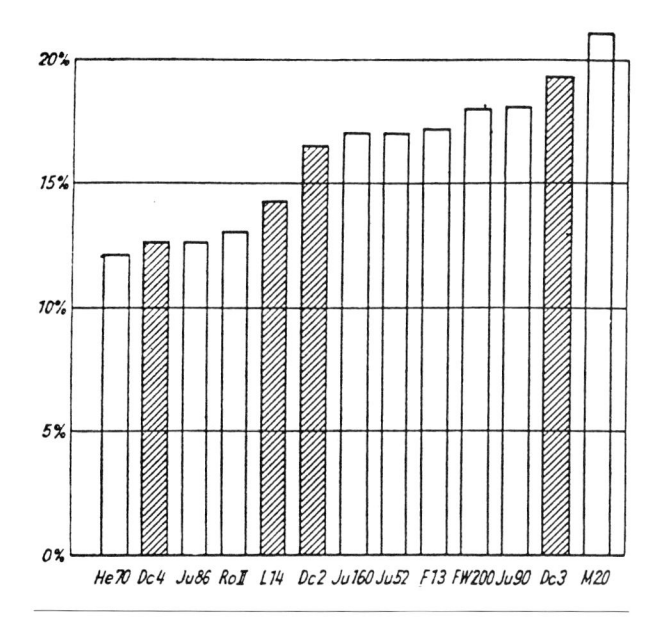

Messerschmitt aircraft were virtually unbeatable with regard to economy, as evidenced by this BFW advertisement from October 1928 and a 1944 DLH economy chart showing the payload of various commercial aircraft as percentage of flying weight.

the aircraft was inadequate. Messerschmitt had adhered to these official specifications and was therefore blameless. It was at this point that the relationship between Willy Messerschmitt and Luft Hansa's director Erhard Milch became strained. This would continue when Milch later became *Staatssekretär* within the *Reichsluftfahrtministerium*. The Luft Hansa, however, took this opportunity to annul the order for another ten M 20b$_2$s and demand their money back. This was the final straw for the BFW, already in financial trouble, and on June 1, 1931, the BFW filed for bankruptcy in the Augsburg district court. The forced settlement reached in the following year (1932) was an agreement that, together with a reapproved certification of the type, led to the airline's acceptance of the ten planes, which acquitted themselves quite reliably and proved to be most economical to operate.

BFW Augsburg was reprivatized in 1928. The reason behind this was the fact that the *Reichsverkehrsministerium* was under repeated attack in the Reichstag for Messerschmitt's ownership of the Augsburger Flugzeugwerke, resulting in the decision to sell the company to private shareholders. Bavaria's Ministry of Trade was also involved in this decision. But the major aircraft companies (Henikel and Albatros, among others) did not have the means since the Ministry of Transportattion in Berlin and the Bavarian Ministry of Trade wanted to sell their shares at 100 percent cost. Messerschmitt feared losing his independence if the shares were acquired by an unknown buyer, and he therefore sought out buyers with whom he was acquainted. He succeeded in persuading his friends in Bamberg, the Stromeyer-Raulino family, to buy the aforementioned shares. On the 1st of July, 1928, the entrepreneurial Stromeyer-Raulino family took over shares at a value of 330,000 Reichmarks; Messerschmitt received the remaining 70,000 Reichmarks to compensate for the assets from the 1927 dissolution of the BFW AG. At the same time, the board of directors and company management was reorganized.

When the M 20 production run stopped in 1931 after a total of 15 machines had been built, Willy Messerschmitt continued with design work on the project. A general aerodynamic reworking of the single engine airframe followed project designs of up to four engines with a corresponding increase in capacity. Here is a wooden model of the twin-engined variant.

65

Effective 1 July 1928, the composition of the new board of directors was as follows:

Otto Stromeyer, Bamberg, Chairman
Prof. Dr. Paul Rieppel, Munich
Dr. V. Scancony, a lawyer from Munich
Dr. Hellmann, state councilor from Munich
Dr. Alexander Schrüffer, the sole chief executive officer up to this point, stepped down, and his place was taken by Fritz Hille and Willy Messerschmitt as CEOs.

The contract between the two firms was also invalidated; however, it was never formally declared null and void since Messerschmitt Flugzeugbau GmbH maintained ownership of all Messerschmitt patents. On the other hand, the company's entire development department was swallowed by the Bayerische Flugzeugwerke AG Augsburg.

In addition to the previously mentioned M 20b and M 20b$_2$, 1928 also saw the development of the following designs:

- the M 21, a two-seat B1-class trainer aircraft, designed as a biplane using mixed construction
- the M 22, which BFW deceptively called a mailplane, designed as a twin engine biplane
- the M 23a and b, two-seat, low-wing monoplane, sportplane and trainer constructed in wood
- the M 24a, commercial plane for eight passengers, constructed of metal.

M 21 Trainer

The *Reichsverkehrsministerium* ordered the M 21 in two configurations. The first, designated the M 21a, was powered by an air-cooled, seven-cylinder Siemens Sh 11 engine rated at 59 kW/80 hp. The more powerful M 21b incorporated the nine-cylinder, Siemens Sh 12 engine with 74 kW/100 hp. This plane was intended as the Flamingo's successor in the transport pilot training schools. In place of plywood, it was to have a welded steel-tubed fuselage with fabric covering, be simpler to assemble, and lighter in weight, as well. Messerschmitt, who only built it as a biplane because of the express wishes of the customer, focused his efforts on the steel-tubed fuselage, while giving his chief designer Wenz free reign in all other areas. The first two examples flew in August and September of 1928 with Theo Croneiß at the controls. Having similar characteristics and performance parameters as their predecessors, they nevertheless enjoyed better take-off and climb capability in view of their weight, which was approximately 10 percent lighter. In any event, they never went into mass production.

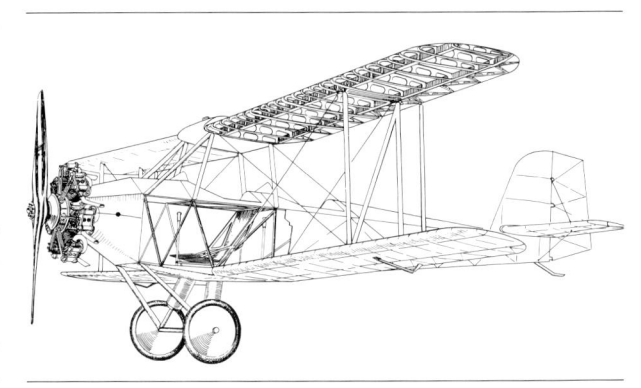

The M 21 trainer, a welded steel-tube design with fabric covering.

On 17 August 1928, just two weeks after the successful first flight of the second M 20, the M 21 took to the air on its maiden flight with Theo Croneiß again at the controls. Among the congratulatory crowd are Julius Krauß (with his hand resting on the cockpit sill) and Willy Messerschmitt (second from right). BFW hoped to garner a follow-on contract to the Flamingo with the M 21.

At the time, much value was laid upon ease of rail transport, ergo the folding wing feature practically demonstrated here while transporting the prototype to the 1928 ILA show in Berlin.

Even in winter, the M 21's versatility was ensured by simply swapping the wheeled undercarriage for skis. The two M 21s which were built went to the DVS.

The biplane layout was a requirement of the contract issuer, the Reichswehrministerium, and was in complete violation of Willy Messerschmitt's principles of creating aircraft shapes with as little drag inducing qualities as possible. The sole M 22 built was flown for the first time in April 1930 by factory test pilot Franz Sido.

M 22 Twin-Engined Biplane (1930)

The M 22 twin-engine biplane was a design which the company had originally conceived of as a bomber aircraft. In actual fact, however, the customer (the RWM) saw it as a pure night fighter and experimental night fighter—designating it the Bf 22 with the code name "Najaku." Messerschmitt was unable to dissuade the customer from the idea of a "twin-engined biplane," and Wenz, the chief designer at the time, was fully responsible for constructing the plane. Although up until this time Messerschmitt designs had been subjected to very little stress analysis during the building phase, in this case Julius Krauß carried out a large number of tests, mainly on individual components. Whether or not this was required of Messerschmitt or the customers themselves is unknown.

What is known is that the stress figures for the biplane were substantially more vague than for his monoplane types; even Messerschmitt didn't trust the findings, not least of which was due to the fact that this was his first twin-engine

Looking somewhat sinister in its dark camouflage scheme, the M 22 night fighter/reconnaissance plane is seen here prior to its planned acceptance by the RWM. On 14 October, RWM pilot Eberhard Mohnike crashed the plane in the Siebentisch Forest near the airfield during one of the acceptance flights. Mohnike was fatally injured. According to testimony by BFW pilot Gerald Klein and prototype construction supervisor Hubert Bauer during the accident investigation, Mohnike had flown a loop, which was not authorized as part of the program. The excessive rpm load on the engine caused one of the three-bladed propellers to break, and Mohnike was not able to bring the aircraft back to controlled flight in time.

The M 22, a steel-tubed and wooden construction with fabric covering, bore the marks of the Udet team at BFW and vaguely resembled a superdimensional Flamingo.

Three-view of the M 22 with two Siemens-Jupiter engines.

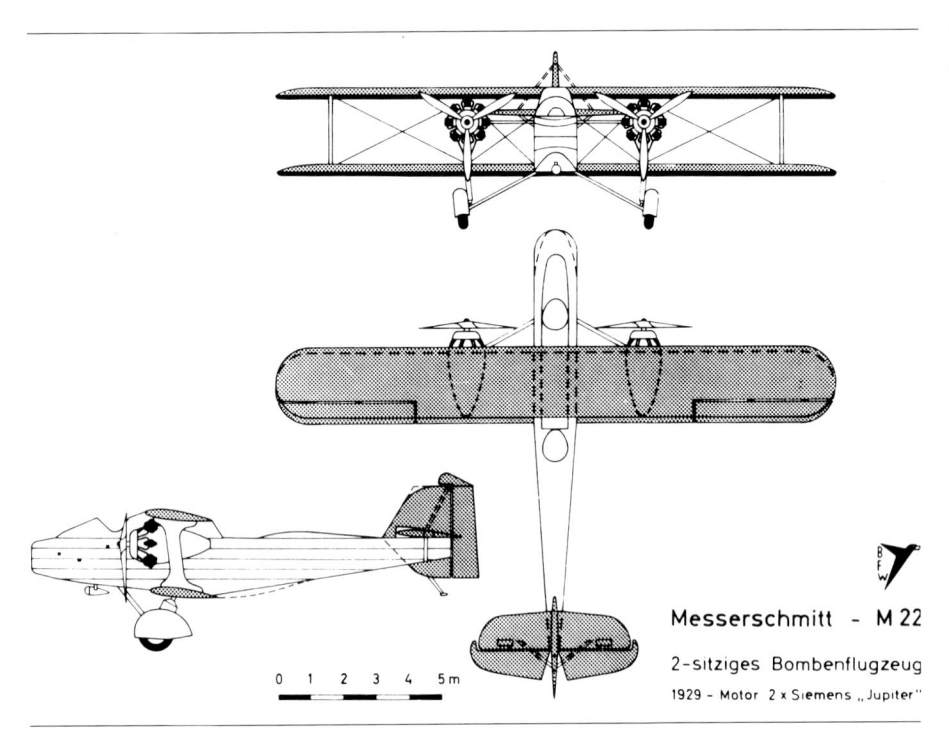

Messerschmitt - M 22
2-sitziges Bombenflugzeug
1929 - Motor 2 x Siemens „Jupiter"

plane. His fears were borne out when a test flight by Eberhard Mohnike (a test pilot with WaPrüf 6 F under *Hauptmann* Student) on the 6th of May, 1930, ended in a crash landing, in which the right undercarriage, the engine assembly, and the upper and lower wing surfaces were ripped apart.

The aircraft was not only repaired, but had the struts strengthened in the areas of the landing gear and engine nacelles between the upper and lower wings. Additionally, it was fitted with an aerodynamic counterbalance to reduce the control forces and mass-balanced controls.

After several flights in Augsburg by the DVL's Joachim von Köppen, the DVL being responsible for official certification of new aircraft designs, the RWM employed pilot Eberhard Mohnike (from the *E-Stelle Rechlin*) to carry out a half-hour acceptance flight on 14 October 1930. The flight ended tragically when a propeller shattered at low altitude, with Mohnike being killed in the ensuing crash. The type was not developed further.

M 23 Trainer and Sportplane (1928-1931)

As was already mentioned, the BFW was working on successors to the Udet-Flamingo, then in mass production series.

Messerschmitt naturally was aware of the Flamingo's excellent flight characteristics, but was of the opinion that its layout, performance, and costs no longer corresponded to the technology of the day. He considered it to be his personal mission to provide a successor on par with this plane, a design which had earned high marks from flight training schools and among the older sport pilots, and one which would be expected to lead to a whole new generation of sportplanes.

The first approach was simply a somewhat larger two-seat M 19, built on exactly the same design, i.e. with fuselage resting on a slender, all-through wing; under the wings came the undercarriage, now split and braced individually. The fuselage had a pentagonal cross-section, with the wings being tapered and ending in semicircular tips with standard ailerons. The tailfin and tailplane each consisted of a tapered fin and control surface lacking any counterbalance and without any rounded edges. The engine was an air-cooled Mercedes-Benz F 7502 engine delivering 18 kW/25 hp. Designated the M 23, early photos show the first aircraft to be painted in glossy dark red over all wooden parts and bright white on the fabric-covered sections of the wing, vertical and horizontal stabilizers. The workers at BFW nicknamed it the Mahogony Coffin.

In the meantime (shortly after the initial press release) Messerschmitt had learned from the patent department at Junkers that the layout used in the design had already been patented under Junkers Patent Number 310610. However,

The first M 23, an enlarged M 19 with a single-spar wing passing beneath the fuselage. This "mahogany coffin" prototype from 1928 was never flown.

The second M 23a, Werk-Nr. 432, was fitted with the French, nine-cylinder Salmson AD9 engine, AD9, with 29 kW/40 hp. The third M 23a, Werk-Nr. 436, received the five-cylinder Genet engine from Armstrong-Siddeley rated at 44 kW/60 hp continuous power; they went directly to the DVL at Berlin-Adlershof following completion, and there underwent a technical performance evaluation prior to their participation in the "Ostpreußenflug 1928/29."

These three were subjected to a comprehensive, laborious evaluation program to ensure elimination of all problems, during which all aircraft not meeting certain minimum standards were removed from the competition; this the DVL carried out between September and December 1928 through their technical and flying departments.

The race was supposed to be a sporting competition under harsh winter conditions, flown by those planes meeting the requirements of the technical performance evaluation. At least ten planes were expected to take part in the competition, but in fact only half as many made it, the Genet-powered M 23a being among them; it was officially registered as D-1571.

he had already come to the conclusion that this idea of an all-through wing supporting a fuselage was not advantageous for small aircraft, and he decided not to pursue acquisition of a license. For this reason the aircraft was never flown, instead being used for exhibition purposes only, as at the Paris Salon (June 29 to 15 July 1928). Critics maintained that the aircraft wouldn't be returned from the show because of failure to pay for the display site.

It soon became obvious to Messerschmitt and his BFW sales department that this underpowered initial M 23 would never be successful in the marketplace. Commercial aircraft demanded an even greater safety factor than competition aircraft. Therefore, Messerschmitt worked on fitting a substantially more powerful engine, something for which the M 23 had been designed from the outset.

Between March and June of 1928 the reworked blueprints spawned three more M 23 airframes, all three being tailored around various high-powered foreign engines. They were given the designator of M 23a.

The first M 23a, Werk-Nr. 418, was powered by the air-cooled, two-cylinder ABC Scorpion outputting 25 kW/34 hp. The aircraft was exhibited on the BFW Augsburg platform at the ILA Berlin (7-22 October, 1928).

Following the ILA in Berlin, in December 1928 Theo Croneiß was the first pilot to fly the M 23a. He was the victor in the East Prussian long-distance race during the winter of 1928/29 in one of the first three M 23a planes built, Werknr. 436 D-1571.

The race itself, flown from the 3rd to the 5th of March, 1929, began at Königsberg. It consisted of three daily "hops" of approximately 480 kilometers each, and only three aircraft completed the race. Theo Croneiß, with the support from his spotter, Fitzeck, was the undisputed winner thanks to his faster plane and his superior flying skills.

This was the breakthrough Messerschmitt had been expecting and the BFW had been hoping for with the M 23, the sportplane and trainer of the future. With a M 23 completed ahead of schedule, the 1928 ILA was followed by an intensive advertising campaign carried out in German sport flying clubs, as well as in Austria, Switzerland, and France, and in French clubs. These efforts paid off with an initial small production run of nine M 23as. The last of these was delivered in May 1930.

Wing and fuselage construction of the M 23b.

The three-point connection of the M 23b's folding wing mechanism.

The instrument panel of the Cirrus-powered M 23b. Of the twelve instruments, the air temperature gauge was a simple off-the-shelf home thermometer.

70

The M 23b light sport trainer was a wood design based on the experience gained with the M 19 from 1927. The selling price of a M 23b with Sh-13 engine was around 13,000 RM.

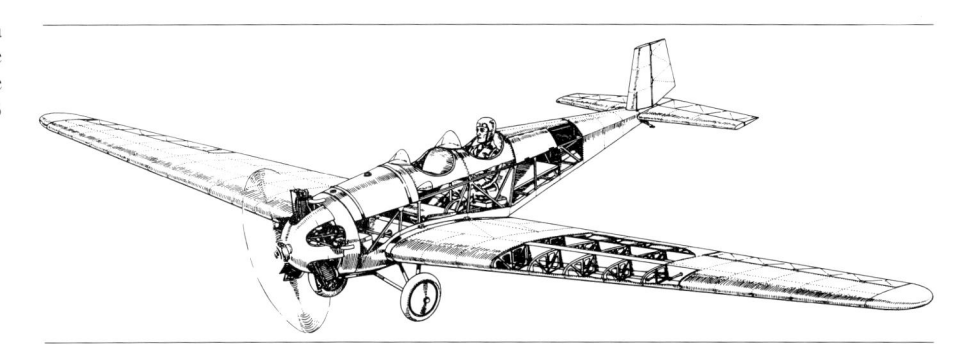

The M 23b

Immediately following the ILA in Berlin Messerschmitt began working on making technical improvements and refinements to the M 23. Aside from the interest he seems to have generated at the ILA from friends and competitors alike, it was also the BFW's multifaceted manufacturing options combined with their first-rate technical department and their relatively advanced talents which inspired Messerschmitt to gradually move from the box-like angular fuselage shape to the more elegant, aerodynamic shape.

Thus was born the M 23b, whose most conspicuous external feature in comparison with its predecessor was the rounded shape of the aft fuselage.

Additional distinctive features of the M 23b included a reinforced, yet finely arranged fuselage assembly, and for the undercarriage, one-piece wheels (i.e. electron cast rims) and, in the later b-series aircraft, an enlarged tail assembly with rounded end caps. A strengthening of the airframe was necessitated by the use of the more powerful engine, one which demanded more fuel and oil and whose speeds generated more dynamic stress. This coupled with an increase in comfort features unavoidably led to an increase in the aircraft's empty weight.

Construction of the b-variant began in September, 1928, with Werk-Nr. 449. It was fitted with the 65/70 kW-88/95 hp, four-cylinder Cirrus III engine manufactured by ADC Aircraft Ltd., London, and had a strengthened undercarriage and rounded tail assembly. With this variant (as with a few other M 23bs) the tailplane and tailfin were braced against each other. In late March, 1929, the aircraft was completed, test flown and evaluated. In July, following the test flight, it received the coding of D-1711 for sport pilot A. von Bismarck, who then unsuccessfully participated with it (entrant number A9) in the 1929 Challenge de Tourisme Internationale.

On the other hand, Fritz Morzik from the Deutsche Verkehrsfliegerschule Berlin was successful. He had used one of the nine M 23bs (from the initial production series)

Relaxed atmosphere during the summer of 1929: Elly Beinhorn accepts her M 23b, Werknr. 466, D-1674. Willi Stör (in the forward cockpit) will be giving her the familiarization flight. Willy Messerschmitt (foreground) looks over Elly Beinhorn's shoulder while the latter adjusts to the cockpit layout.

Willi Stör, who earlier had taken second place behind Gerhard Fieseler in the M 23 during the Deutsche Kunstflugmeisterschaft aerobatic competition, had this to say regarding the type: "During aerobatics, the M 23 performed absolutely amazingly. Easy controllability and great maneuverability ensured that each aerobatic maneuver could be made smoothly and elegantly with power to spare."

In 1930 Ernst Udet was given this As-8 powered M 23b, Werknr. 511, D-1970. After being used successfully in his Africa expedition from November 1930 to April 1931, the machine met a "scripted fate" as a floatplane in Greenland's ice floes in the Dr. Fanck film "SOS - Eisberg."

registered at the 1929 Challenge de Tourisme. This was Werk-Nr. 465, coded D-1673, entrant number A4, and was among the first nine to be powered by the Sh 13 air-cooled, five-cylinder radial engine. The Siemens-Halske engine, rated at 50/59 kW-68/80 hp at 1500/1700 rpm, was comparable in performance weight, lubricant consumption, and reliability to foreign engines. Facing tough international competition, Morzik, with his assistant Schiel, emerged the victors and for the first time earned Germany the French-sponsored wandering trophy. It goes without saying that this victory was also the greatest success that Messerschmitt and the Bayerische Flugzeugwerke had enjoyed up to this point. Despite the world financial crisis, the number of M 23b aircraft sold both within Germany and to foreign countries grew to 70 by the end of 1931.

The M23b's greatest success came at the hands of Fritz Morik, the winner of the 1929 *Challenge de Tourisme Internationale* in the Sh 13 powered D-1673, Werk-Nr. 465.

Romanian M 23b planes at an airfield open house in Bucharest. In addition to eight M 23b machines purchased directly from BFW in 1930/31, the Romanian aircraft company ICAR purchased the license for the type and produced no less than 15 aircraft in 1932/33. On 2 December 1932 a Romanian pilot, Captain Pantazi (accompanied by Grozea), flew one of these machines (with a Sh-13 engine and floats) to set the world's endurance record for a class-C floatplane with a flight of 12 hrs 3 mins.

M 23c Sportplane and the 1930 Challenge de Tourisme Internationale

Just in time for the 1930 Challenge de Tourisme International, the Bayerische Flugzeugwerke Augsburg built eleven M 23c aircraft, test flying and certifying Werk-Nr.s 517 through 526 and 528.

The starting and finishing point of the race, which had been organized by the Deutsche Aero Club, was Berlin, and it began on July 20th. The competition encompassed a flight route of approximately 7,500 kilometers through the countries of Germany, France, England, Spain, Switzerland, Poland, and Czechoslovakia. There were 101 aircraft registered, of which 47 were German competitors, and of these 47, ten were M 23cs—this making the M 23 the most represented for the Challenge de Tourisme.

Of the 35 aircraft successfully completing the flight, seven were M 23cs (and one a M 23b), and once again it was Fritz Morzik who was the overall winner—this time with a M 23c specifically developed by the BFW for this competition.

Developed by Messerschmitt, the M 23c sport plane was based on design specifications of the previous year's winning plane and was tailored specifically to the 1930 Challenge de Tourisme. The aircraft was a low-wing monoplane embodying the typically distincitive Messerschmitt construction concepts. Retaining the tapered monospar wings, the wing surfaces and the rear fuselage section were constructed of wood with diagonal plywood skinning, while the forward fuselage section was of duralumin construction. The control surfaces, housed in ball bearing races, were of a triangular braced wood design with fabric covering. According to the design specifications, the pilot and copilot were to be protected against weather conditions, and the cockpit area was therefore enclosed. The undercarriage was fitted with brakes with an eye towards improving

M 23c production line for the 1930 Challenge de Tourisme Internationale. This variant was the first to be fitted with an enclosed cockpit; the type was powered by the As 8 or Sh 13.

"...Climbs like a modern fighter, holds its own in even the steepest banks without the slightest tendency to slip and, moreover, pilots are able to perform aerobatics quite easily." This assessment of the M 23 by Swiss pilot Walter Mittelholzer is a particularly fitting caption for this photo of a M 23c performing with the BFW buildings as a backdrop.

Measuring the dimensions with wings folded was one point of the technical evaluation for the Challenge de Tourisme Internationale. Again, it was Fritz Morzik who was the victor, flying an As-8 powered M 23c, D-1883 Werknr. 558. In 1971, on the occasion of Fritz Moritz's 80th birthday, Willy Messerschmitt wrote him: "Through your successful flights you have done so much for my work, for my colleagues and, not least, for me personally, that I will be forever grateful to you."

ground maneuverability and decreasing landing rollouts. Weighing in empty at 322 kilograms, the maximum take-off weight amounted to 600 kilograms.

For the first time in the race's history all competing aircraft were powered by German engines: these being either the Siemens Sh 13 air-cooled radial engine with 59/70 kW-80/95 hp, or with the As 8 Argus inline engine with 59/73.6 kW-80/100 hp. The winning aircraft flown by Morzik was the BFW M 23c, Werk-Nr. 518, D-1883, powered by the Argus As 8 engine.

M 23b Floatplane (1930/31)

In 1930, a single M 23b (Werk-Nr. 496, D-1836) powered by an air-cooled, five-cylinder Siemens Sh 13 engine with 50/60 kW-68/82 hp, was kitted out with floats for evaluation purposes and used for feasibility studies on the Ammersee. Erich Scheuermann, a former manager of the Udet Flugzeugbau and later an instructor with the DVL, owned a weekend cottage on the Ammersee that had a spacious area large enough to accommodate repairs and maintenance of the floatplane, and from where it could also be launched.

Three-view of the M 23b (Sh 13) with floats.

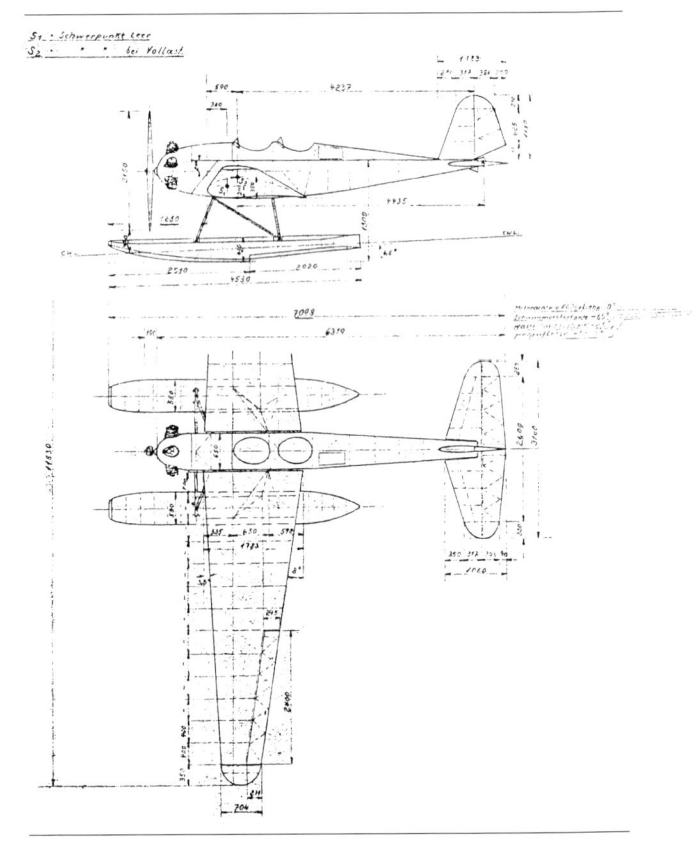

The M 24, an all-metal design for an eight-passenger airliner or cargo plane as well as a specialized type, was planned as the link between the M 18b and M 20.

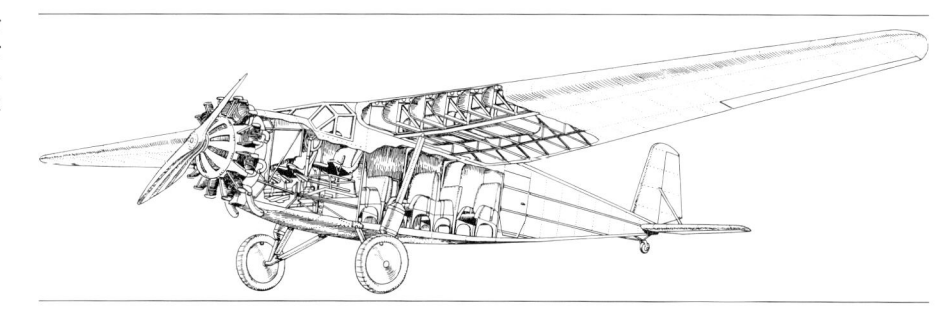

M 24 Eight-Seat Commercial Plane (1929)

The M 24 was numbered among those new designs undertaken in 1928. Messerschmitt drew up the plane, designated the M 24a, as a commercial airliner for two pilots and eight passengers, or as a transport with a load capability of one metric ton and a 500 kilometer operating radius.

The first aircraft was completed (Werk-Nr. 445) in early 1929. The design was built around the water-cooled, six-cylinder Junkers L 5G engine with 250-276 kW/340-375 hp. Like all of Messerschmitt's passenger planes, the all-metal aircraft was built out of duralumin. The plane fell into the class of cantilever shoulder-wing monoplanes, as the continuous wing fitted into the fuselage behind the main spar as was first employed with the M 20b. In the meantime, the Nordbayerische Verkehrsflug GmbH had decided to purchase the plane, with Theo Croneiß himself making the maiden flight on July 8, 1929. Once all the shortcomings pointed out by company pilot Sido had been sorted out, the DVL carried out the acceptance flights and assessment of production types by the DVL in early October, 1929. A few months later, the aircraft was certified under the official coding D-1767, and on 25 February 1930 it was delivered to the Nordbayerische Verkehrsflug GmbH.

Construction of a second M 24a, Werk-Nr. 446, then followed in 1929. Germany's Verkehrsministerium (Ministry of Transportation) in Berlin issued the contract, the recipient being the Deutsche Versuchsanstalt in Berlin-Adlershof (DVL).

This aircraft was first fitted with the water-cooled, six-cylinder BMW Va engine rated at 213-235 kW/280-320 hp. Later the plane received the somewhat more powerful Junkers L 5G. The DVL took delivery of it on May 16, 1930, where it was put to work as a test platform for evaluating new radio-telegraphy systems, among other things.

It is not clear why two more aircraft with more powerful engines were built in 1930. Early hopes for more demand of an eight-passenger commercial airliner proved to

The first M 24a with Jumo L-5 engine was test flown on 8 July 1929, with Franz Sido and Theo Croneiß at the controls, and delivered to the Nordbayerische Verkehrsflug GmbH in early 1930.

The M 24a, D-1767, Werknr. 445, was in service until 1934 and is seen here in Chemnitz on a scheduled flight route.

Of the two M 24b planes built with BMW Hornet engines, one was temporarily fitted with floats and completed its certification check in Travemünde in 1932.

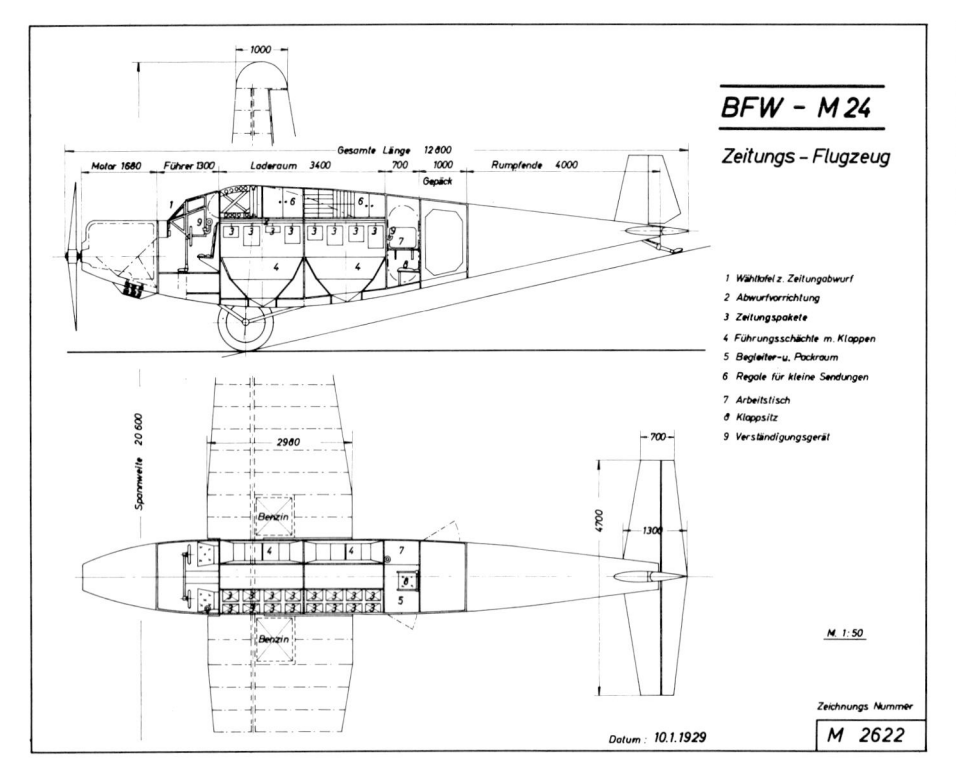

The M 24 was also offered for specialized tasks; this is a "newspaper plane" for dropping newspapers/periodicals and mail in regions such as the North Sea and Baltic Sea islands or in resort areas. (Based on original drawings from 10 January 1929.)

be overly optimistic as the economic conditions rapidly worsened. In addition to the more powerful engine, both of the M 24bs (Werk-Nr.s 515 and 516) were subjected to a series of improvements, including rounded, somewhat larger control surfaces in order to make the aircraft more suited to various other roles. The BFW offered the following uses for the plane (in addition to a commercial airliner and land-based inter-airfield cargo transport):

- Seaplane for passengers and cargo
- Newspaper delivery aircraft with dropping devices
- Photo aircraft for land surveying
- Air ambulance (accomodations for four stretchers or cots)
- Special uses, such as pest control, physics, and meteorological testflights, postal delivery aircraft, etc.

Even before the DVL had completed its comprehensive evaluation program with the first M 24b, the aircraft (Werk-Nr. 515) was refitted with the air-cooled, nine-cylinder Bristol Jupiter radial engine rated at 294-320 kW/400-436 hp; also, floats were to be developed and fitted. In an unexpeceted move, the *Reichsverkehrsministerium* even issued a contract to cover this. In any event, it wasn not until 1932 that the plane was delivered to the *Erprobungsstelle Travemünde*. Werk-Nr. 516 received the air-cooled, nine-cylinder BMW Hornet radial engine rated at 368 kW/500 hp. After additional prototype tests by the DVL, the manufacturer was unable to sell the plane. So it remained at Augsburg until 1934, when the newly established *Reichsluftfahrtministerium* (RLM) acquired it; the M 24b was then registered as D-UHAM and assigned to the Stettin pilot training school.

M 18 Follow-On Developments (1929)

By the end of 1927 Messerschmitt—now in Augsburg—was pondering the idea of follow-on developments to his small transport aircraft, the M 18b, to include increasing the number of passenger seats to six, and a commensurate increase in engine power with its associated increased dimensions and structural reinforcements. All of course with an eye towards expanding the type's roles. The good slow-speed handling and the high-wing monoplane's good downward visibility made the aircraft especially suitable as a photo plane, similar to the earlier special version of the M 18b delivered to Romania.

With an enlarged airframe, the all-metal M 18d could carry six passengers and was available in both land- and seaplane variants.

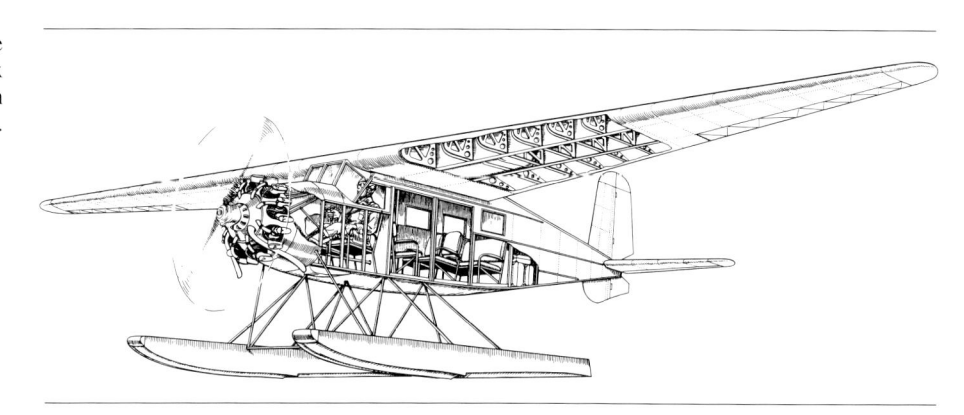

M 18c (1929/30)

Discussions with the Eidgenössische Militärdepartement in Bern (Switzerland) over the delivery of a surveyor plane led to a contract being issued in late 1928.

The Lynx engine built by the British company of Armstrong-Siddeley was requested and anticipated. This was a seven-cylinder, air-cooled radial engine which, as with all British engines, rotated counterclockwise and had to be fitted with an appropriate propeller. Its maximum power rating amounted to 158 kW/215 hp.

The wings, as usual, employed the monospar concept, and were of a pure tapered design with nominally greater span and surface area. Similar to the design of the M 20b and M 24, the wing was buried in the fuselage from the main spar back. Since the engine was set somewhat deeper than that of the M 18b, the pilot's area could be glazed with flat panels, a feature which greatly improved his visibility, particularly to the front. The fuselage, with the door on the right, and the tail assembly remained unaltered.

The undercarriage sat higher to accommodate the larger propeller diameter and was given a wider wheel track, and both axle struts were now set on the outside, at about the fuselage mid-point, with an angled support brace keeping them aligned. The undercarriage main struts were fitted with a strong shock absorber, and angled upward and inward inside the fuselage to join at the wing connection mounts immediately beneath the wing spar.

The first M 18c, Werk-Nr. 447, was announced as finished in September, 1929, but the customer did not immediately accept delivery because the landing rollout was too long. A cure for this was found through subsequent fitting of brake wheels, and the aircraft, now designated CH-144, was delivered to Switzerland. The plane served successfully as a surveyor aircraft with the *Eidgenössisches Grundbuchamt* (Federal Land Registry Office) in Bern. It was later said to have been flown by the well known Swiss pilot Mittelholzer as an expedition plane.

The M 18c from 1929 had an enlarged tapered wing, the more powerful Armstrong-Siddely Lynx engine (158 kW/215 hp) and was available in two versions on the international market; this, the second machine to be built, was delivered to Portugal in 1930.

The second M 18c, Werk-Nr. 448, remained in Augsburg as a demonstration plane. An air-cooled Siemens Sh 12 radial engine, rated at 79.5/92 kW/108/125 hp, powered No. 448, which was certified in June of 1930 as the company plane under the registration code D-1860. Shortly thereafter, it was sold to a foreigner and, at the customer's request, was refitted with the Lynx engine and delivered with the registration CS-AAH in April 1931.

M 18d commercial plane (1929/30)

There was a distinct gap of an airplane between the sizes of the M 18b and M 24 in the feeder line market, a plane designed to accommodate six passengers. Improvements to the M 18c's airframe were primarily influenced by the constructive and functional advancements which had been learned from the development of the two larger designs, the M 20 and M 24.

In addition, demands for comfort in air transportation were growing and would need to be addressed: and ultimately one of the goals was to improve the flexibility of greater operating ranges for feeder planes, or increasing the

Of the eleven M 18d planes built with increased six-seat cabin space and new wing, six went to the Nordbayerische Verkehrsflug GmbH.

BAYERISCHE FLUGZEUGWERKE A.G. AUGSBURG

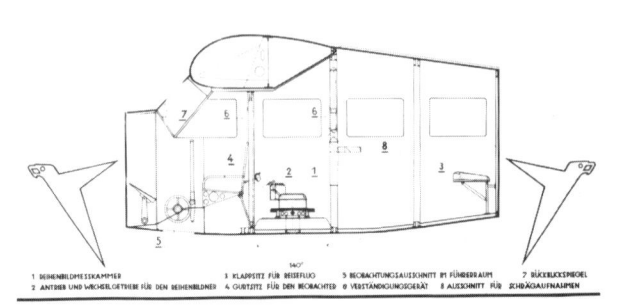

MESSERSCHMITT LUFTBILDFLUGZEUG BFW M 18d

Several M 18d machines were flown as photomapping planes and delivered to other countries, including this Werknr. 470 with Wright Whirlwind engine which was sent to China in the summer of 1930.

number of passengers it could carry. Thus was born the development of the M 18d airframe, a plane designed to accommodate engines up to 240 kW/325 hp depending on the wishes of the customers.

Lengthening the fuselage center portion by 0.7 meters brought not only an improvement in the rudder effectiveness, but also brought with it an increased height and a widening of the cabin by 1.2 meters, plus larger viewports for each of the six passengers, all of which added to the comfort of the aircraft. The entrance was also moved further to the rear, thus making the step lower and easier, as well. In addition, the passenger entrance was moved to the standard left side of the fuselage. The undercarriage, taken from the "c" version, was modified by having the main strut now positioned vertically upward attached to the second wing rib. The elevator and rudder were reinforced, and the design called for a yoke instead of a joystick. The new wing of the "c" version was borrowed without modification. The first three M 18ds were equipped and sold as photo aircraft. Two of these planes received the air-cooled, nine-cylinder Wright Whirlwind engine rated at 239 kW/325 hp.

Werk-Nr. 469, D-1812, was acquired in April 1930 by the Nordbayerische Verkehrsflug GmbH, and later, in September 1931, chartered to the Münchner Fotogrammetry GmbH. It crashed that same month while carrying out a surveyor flight in central Sweden, and though the cause of the crash was unknown it is reasonable to assume it was due to icing.

Werk-Nr. 470 was exported on July 27, 1930, by a German photo company to Hangchow, Province of Chekiang, China, for photo surveyor duties.

The third aircraft, Werk-Nr. 476, was powered by the Armstrong-Siddeley Lynx engine rated at 158 kW/215 hp and delivered in April 1930 to the Adastra Swissair company with the Swiss registration CH 191-HB-IME.

The next four M 18d planes, Werk-Nr.s 477 through 480, were fitted with the Sh 12 engine with 92 kW/125 hp and delivered between May and September 1930 to the Nordbayerische Verkehrsflug GmbH. Just a short time later

The eight year production history of the M 18 ended with the conversion of the last M 18d airframe to the Argus As 10C engine in the winter of 1933/34; all told, about 30 examples of the M 18a through d were built.

it was discovered that the engine power was insufficient at full throttle, resulting in the four aircraft being refitted that same year with the air-cooled, nine-cylinder Mars engine manufactured by the Czechoslovakian firm of Walter, the power rating being 110 kW/150 hp.

Werk-Nr. 481 was fitted with a Wright Whirlwind engine with 239 kW/325 hp and floats. It would end up, just as the last M 18d—Werk-Nr. 482—with no buyer. This aircraft, with registration D-ORIZ, was refitted in 1933/34 with an Argus As 10c with 176 kW/240 hp. Both aircraft were then taken over by the *Reichsluftfahrtministerium* and handed over to land-based groups of the Deutscher Luftsportverband for training purposes.

M 25 Aerobatic Plane (1929)

The M 25 was an aerobatic plane project for the renowned Ernst Udet. During this period he often visited Messerschmitt, who tasked Richard Bauer, design director at the time, to draw up such a project for Udet.

Between February and August 1929 the program incurred costs amounting to 3,618.46 RM. Discussion focused on two designs, one for a single-seat biplane similar to the M 21 and the other for a single-seat mid-wing layout with a wingspan of about 9 m and powered by a BMW X engine (Messerschmitt made use of the latter design as an example of a sample assignment during his later years of teaching).

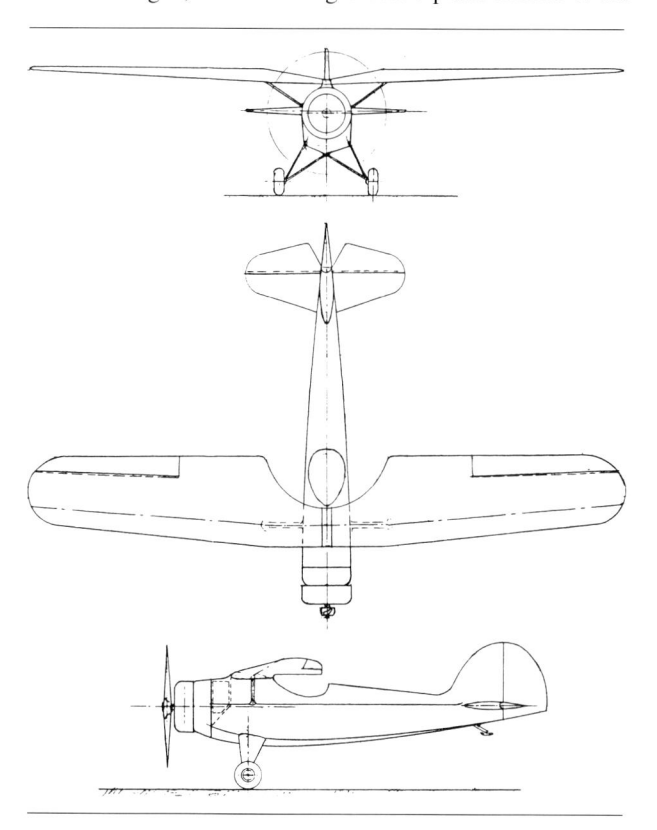

Once Udet flew the Curtiss dive bomber in America and acquired two of these for himself he lost all interest in the M 25, and the project, which had not yet gone beyond the design stage, was canceled.

M 26 High-Speed Commuter Plane (1929/30)

In late 1928 the Bayerische Flugzeugwerke AG Augsburg concluded a licensing contract with the American company of Eastern Aircraft Corporation which stipulated that the U.S. firm would assume sole production of the M 18d and a three-seat high-speed aircraft in America.

The aircraft was designed to accommodate a pilot and two passengers and was designated the M 26. Company employees immediately dubbed it the "Wedding Coach."

Design work started in early 1929, with the airplane being layed out as an all-metal high-wing monoplane with enclosed cockpit. For its engine, the type was fitted with the seven-cylinder Siemens radial engine, the Sh 14, having a power rating of 68/81 kW-91/110 hp.

Only one of the two prototypes ordered was ever completed, this being flown by company pilot Sido in early October of 1929. The aircraft proved to be virtually problem-free from the outset. In the meantime, the world economic situation was rapidly declining. When "Black Friday" hit the New York Stock Exchange, Eastern Aircraft Corporation (EAC) soon ran into financial difficulties and was forced to declare bankruptcy. BFW AG shortly found itself in the same dire straits, and since the EAC couldn't pay its costs, the licensing contract was annulled without substantial losses being incurred.

The prototype, Werk-Nr. 483, D-2085, remained in Augsburg as a factory aircraft and would later (1934) be

In his capacity as an instructor of aviation design at the Technische Hochschule in Munich, Willy Messerschmitt would often use his own or other BFW designs as assignments for students. This was the assignment for the 1933 summer semester and probably represented the M 25 for Ernst Udet (a variation of the design showed a biplane layout). In addition to the three-view, the students were provided a Göttingen 655 wing profile, performance of the BMW Xa engine with N = 60 hp at 2,050 rpm, engine weight of 73 kg, load capacity of 140 kg and landing and maximum speeds of 70 and 180 km/h. The students were required to determine the overall dimensions, weight distribution, proof of performance figures, major cross sections, and design of the individual components.

The M 26, a three-seat all-metal design.

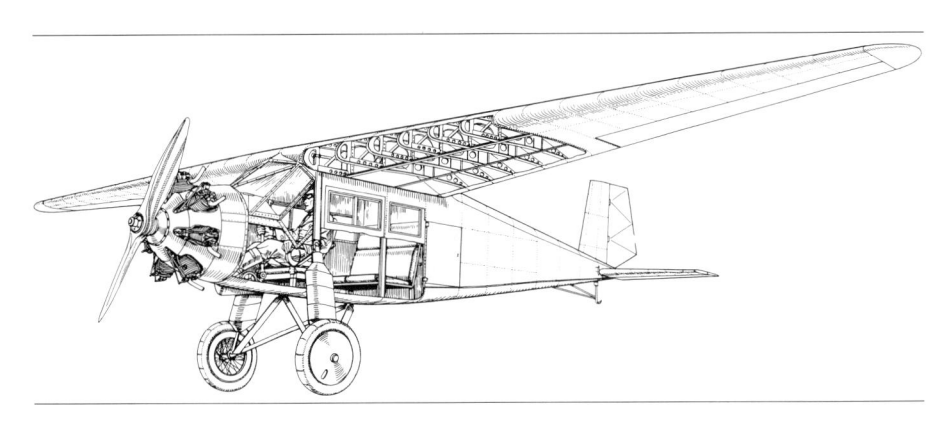

Above, Right: Fuselage of the M 26 under construction.

During the summer of 1930 the BFW hangar displayed the entire production spectrum at the time: the unsold M 26 (left foreground), several M 23s undergoing maintenance (left and right foreground), one M 18c and d each, the second M 20, D-1480 (background) and the M 22 (left in the hangar entrance).

taken over by the RLM. The second aircraft was never assembled, only a few individual components being built before the contract for the aircraft was finally canceled.

M 27 Trainer and Sportplane (1930/31)

With this particular aircraft, Messerschmitt came out with a more powerful and sturdier two-seater tailored to the interests of flying schools. It was a more robust version of the M 23 using mixed construction, designed for engines up to around the 120 kw class. The superior performance of its predecessors was combined with the certain advantages of a fabric covered, steel tubed fuselage design. The prototype M 27a, Werk-Nr. 539, was fitted with the air-cooled, four-cylinder Argus A8 inline engine rated at 74-88 kW/100-120 hp. The plane featured a narrow slat, approximately 10 cm in depth and 1.5 m long, designed to hold the airflow longer in the area of the aileron in order to ensure aileron effectiveness at even greater angles of attack. In effect it was a raised leading edge slat, basically an attempt to circumvent the Handley-Page patent for leading-edge flaps.

The Bayerische Flugzeugwerke built the first plane during the first six months of 1931; Theo Croneiß was able to test fly it in July, and then used it to take part in the Deutschlandflug on August 11, 1931. Lady Luck abandoned him, however, and he was forced to make an emergency landing at Böblingen due to engine damage. It wasn't until the following year that the aircraft received its certification and could be sold.

Two other M 27s, Werk-Nr.s 609 and 610, were totally destroyed in March 1933 when they went into flat spins during spin testing carried out by Willi Stör, the new factory pilot at the BFW and former flight instructor at the *Verkehrsfliegerschule* in Oberschleißheim. In both instances the pilot was able to take to his parachute.

The M 27 was a two-seat sportplane of mixed construction and a fabric covering; the tubular-steel fuselage and Argus As-8 engine is seen here.

The two-piece dural engine shroud made for easy access to the As-8 engine.

This, the first M 27 (Werknr. 539), was used by Theo Croneiß in the Deutschlandflug in August 1931. However, due to engine problems he was forced to drop out.

One of the twelve M 27s built. This tail view shows that it couldn't get much narrower than this.

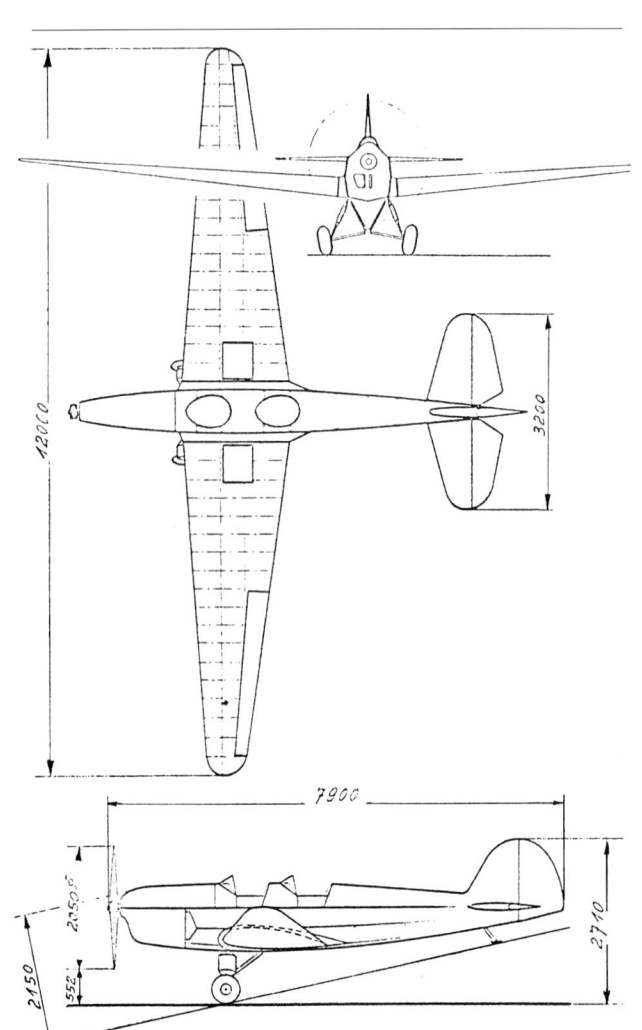

A further nine aircraft (Werk-Nr.s 611 through 619) were registered and marketed as the M 27b from June 1933 until February 1934, some of these being fitted with the more powerful As 8R engine.

M 28 Mailplane (1931/32)

The increasing importance of quick postal delivery between capital cities in Europe gave rise to Lufthansa's decision in 1929 to fly these routes under contract of the German Postal Service, and to subscribe to a special, high-speed aircraft for postal transport.

The Bayerische Flugzeugwerke was invited to participate, as well, and Messerschmitt initially designed a high-wing monoplane based on the idea of his passenger planes. Performance estimates soon revealed the impracticality of this design, and the new prototype, designated the M 28, was built as a low-wing design, with flight trials commencing in January of 1931.

The only M 28 built, Werknr. 527, D-2059, with a BMW Hornet engine, was first flown by Franz Sido in February 1931. However, it wasn't until 1932 that the contractor, Lufthansa, took delivery. In 1935 the airplane was sent to Berlin for inclusion in the German aviation collection.

M 27 three-view.

82

The M 28, a high-speed mail and cargo plane of all-metal construction.

Despite the particularly difficult stipulations mandated for the bid, the M 28 met and exceeded all requirements outlined in the proposal, with the exception of the landing speed which, given the unusually high wing loading of 100 kg/m² at 100 kilometers/hour, was slightly over 100 km/h.

The M 28 was powered by the air-cooled BMW Hornet, a nine-cylinder radial engine with 386 kW/525 hp, had an enclosed cockpit with pilot seats side-by-side, and was designed in duralumin—the exception being the fabric covered areas behind the wing spar. The tapered wings with their V-shape and a streamlined undercarriage gave the aircraft an appealing form despite the overly large appearance of the fuselage.

Since Luft Hansa had recognized during the flight trials of the M 28 and its competitors that the requirements in the request for tender were not optimal, the prototype was delivered to the DVL in Berlin-Adlershof where it was subjected to extensive stress testing.

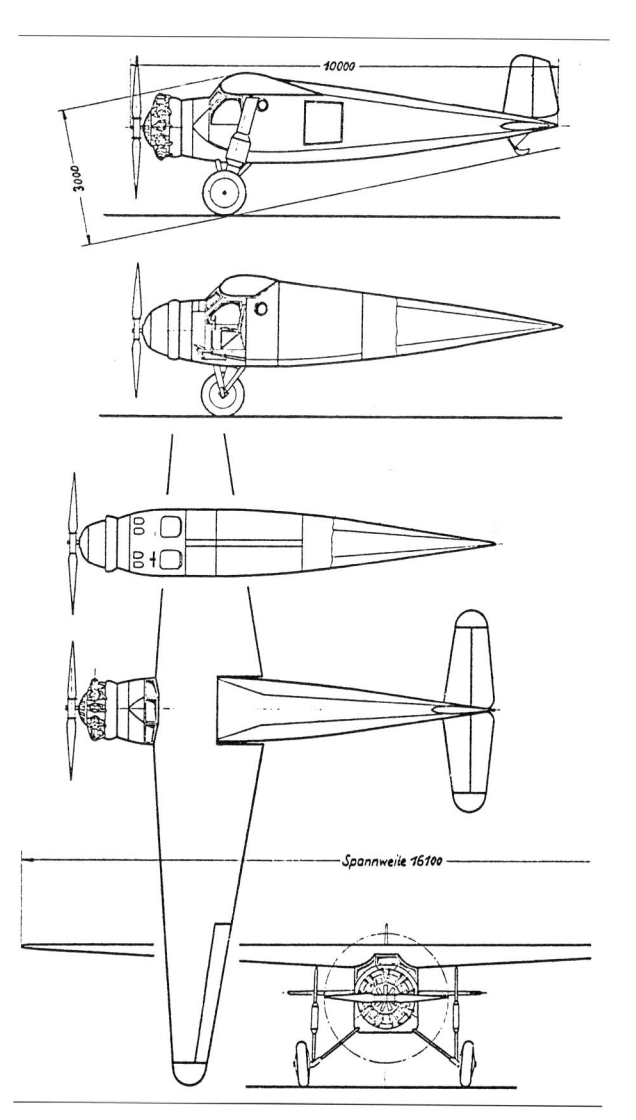

The first M 28 designs revealed high-wing layouts similar to Messerschmitt's M 18, M 20, and M 24. Kurt Tank, who at the time was employed in the project department, subsequently redrew the M 28 as a low-wing design.

Organization of the Developmental Team

Initially, the design department was simultaneously the nucleus of all Messerschmitt's aircraft designs. With the ever-growing spectrum of aircraft types and their variants it became necessary to form a project department. The section concerned itself with designing variants of planes already completed or under production, as well as entirely new aircraft designs. This is how variants of the M 17, M 18, and M 20 were conceived from 1927 onward, most of these being multi-engine designs.

In 1930, Messerschmitt's project department was run by Kurt Tank, later to become the architect of the Focke-Wulf 190. A list of the tasking at that time for the project department would prove how multi-faceted Messerschmitt's developmental plans were:

It is difficult to determine what is troubling Messerschmitt more—worries about BFW's contractual and financial state in 1929 or pre-test jitters prior to getting his pilot's license. In any case, he doesn't appear too happy in this photo of him standing in front of his M 23b. His instructor for his A-class license was Otto Brindlinger, who later became operations director at BFW.

Willy Messerschmitt with Theo Croneiß, his friend, promoter, contractor, and pilot for Messerschmitt aircraft (summer 1928).

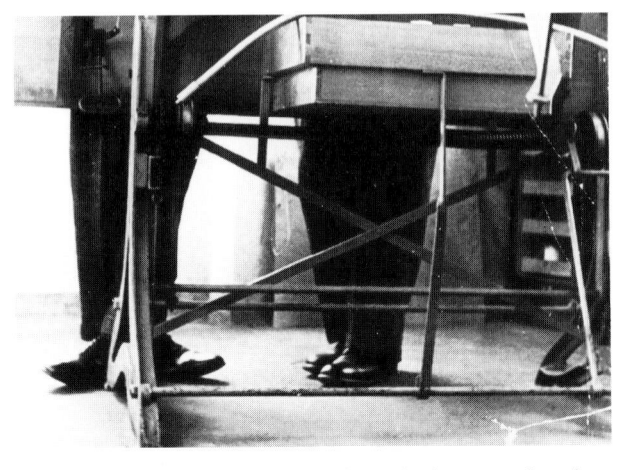

When at the drawing table, Willy Messerschmitt assumed a unique stance which brought smiles to the faces of his colleagues. In addition, it was easy to spot the boss making his rounds through the drafting offices, examining the blueprints. Richard Bauer, manager of the construction department, secretly took this photograph and preserved it over the years.

- M 18d with three engines
- Trainer aircraft
- Postal aircraft based on the Wasp engine
- M 24, three-engine aircraft
- Cargo plane
- Development of an amphibious plane
- Cargo plane with two BMW Hornet engines
- High-speed postal aircraft, Canada Project
- Four-seat airplane (based on M 26)
- Long-range aircraft
- M 24 with three Argus As 10 engines
- Development of a welded fuselage joint
- M 20 with four engines
- M 28 as cargo plane

Approximately 17,850 Reichmarks were spent in 1930 just for the above-named projects, but since none of these would lead to a contract the costs could not be recouped directly.

RVM Subsidies

Even then, it was virtually impossible to offset developmental costs for the wide array of aircraft by the small production runs being sold almost at cost. As mentioned, the development contracts awarded by the *Reichsluftfahrtministerium* (RVM) often covered only a portion of the actual develoment costs, and the RVM generally only offered subsidies to companies after the fact. In the crisis year of 1929, all the major aircraft firms—Junkers, Heinkel, Dornier, Arado, and BFW—received a one-time subsidy in order to cover their losses incurred due to development costs.

In late 1930 the BFW's balance sheets showed still more losses, this time in the amount of approximately 600,000 Reichmarks. This would be supposedly offset using the remainder of 113,000 Reichmarks left over from the previous year's subsidies, but primarily through an infusion of additional capital from the Stromeyer-Raulino Group amounting to 250,000 Reichmarks, as well as through activation of all hidden assets the firm held. In a later report from the *Deutsche Revisions- und Treuhand AG* addressing the interim statement submitted by Fritz Hille on August 31, 1930, it was established that the "development shares" were overestimated by 578,000 Reichmarks and that the BFW AG was completely overindebted and on the verge of bankruptcy.

In this hopeless financial state, business director Fritz Hille unexpectedly announced his resignation effective September 30, 1930. In an all-encompassing 38-page letter to the director of the supervisory board, Otto Stromeyer, he detailed his reasons for such a step, laying total blame for the BFW's financial problems at the feet of its technical director, Willy Messerschmitt. His employment with the Heinkel company in Warnemünde a mere 14 days later easily confirms the suspicion that Mr. Hill had been working toward this end for quite some time, and he was apparently correct in assuming that Heinkel would gladly accept a prominent employee from the competition.

World famous sailplane pilot Wolf Hirth reported that he was particularly pleased that Willy Messerschmitt himself was learning to fly.[3] When he landed in Augsburg one day in 1930 for leisurely Sunday flight operations, Willy Messerschmitt happily informed him that he was already quite comfortable flying himself. It is not generally known that Messerschmitt had acquired his A2 and B1 class pilot's licenses.[4] What Wolf Hirth, an old schoolmate and test pilot at Harth-Messerschmitt, considered a substantial victory for aircraft designer Willy Messerschmitt was felt by sales director Hille as Messerschmitt's irresponsible distraction from his duties as developmental head of the Bayerische Flugzeugwerke and one of the "reasons for his betrayal" of Messerschmitt.

Bankruptcy (1931)

Due to the serious financial difficulties of the BFW AG, in early 1931 the RVM recommended entering into merger negotiations either with the Heinkel GmbH in Warnemünde or with Dornier Metallbau GmbH in Friedrichshafen. The conditions of these firms were so poor, however, that such an action would have meant the complete dissolution of the BFW.

As a last means of exerting pressure on the BFW, Deutsche Luft Hansa annulled its contract with the BFW for building ten M 20 planes while at the same time demanding reimbursement of its downpayment. This demand could not be met, since by this time production of these aircraft was at an advanced stage. Given the situation, the Bayerische Flugzeugwerke AG had no other recourse than to file for bankruptcy in the Augsburg courts, an act which occurred on 1 June 1931. For Willy Messerschmitt personally it was at first impossible to continue working under the guise of the BFW AG. But he then mobilized his old company, the Messerschmitt-Flugzeugbau GmbH, which had lain dormant for some time yet owned several of the patents not affected by the bankruptcy proceedings. Some startup capital was also scraped together, and thus a handful of personnel laid off by the BFW AG could be rehired (albeit at a sharply reduced salary) by the MTT-GmbH (as Messerschmitt's new bureau was abbreviated to). Richard Bauer became the director, but Willy Messerschmitt con-

[3] Rolf Italiaander, Wegbereiter Deutscher Luftgeltung. Berlin 1941.
[4] Dt. Akademie der Luftfahrtforschung (pub.), Wer ist Wo? Berlin 1939.

tinued providing the design work himself—often in minute detail.

The bankruptcy attorney, Konrad Merkel, understood the need for safeguarding and retaining as much of the company's warehouse inventory as possible in order to provide a foundation for any new beginning. Through Merkel, the BFW company managed to work out a deal with Deutsche Luft Hansa whereby the latter would reissue its contract for further construction of the ten M 20b planes once the empennage had been redesigned.

M 29 Competition and Sportplane (1931/32)

In October 1931 the *Reichsvehrkehrsministerium* issued a request for tender to several design companies, including the Messerschmitt-Flugzeugbau GmbH Augsburg, for six aircraft with an eye towards standing up a team for the 1932 Challenge de Tourisme Internationale. In addition to the company's chief test pilots, only a select few others were taken from DVS flying schools to be competitors with the Messerschmitt planes. One of the natural choices was Fritz Morzik from the RVM, the winner in 1929 and 1930. Messerschmitt hoped to avoid disappointing those who expected that he would field the winning plane this time as well by coming up with a completely new design for the competition. By incorporating a refined aerodynamic shape with futuristic new concepts he succeeded in creating an aircraft which not only had sleek lines but also most impressive performance figures. Some of the features worth mentioning are Messerschmitt's first use of a cantilever strut fitted with teardrop shaped spats to cut down on drag. For reducing landing speeds the landing flaps and ailerons were designed as split flaps (Handley-Page flaps). The tailplane did not have a separate elevator, being was instead designed as an all-moving surface. The two seats in tandem were covered by side-opening dural-framed cellophane canopies which could be jettisoned by "emergency cables."

Breaking with tradition, in preparation for the competition the M 29 underwent a systematic three-month evaluation period which was so thorough that nothing was expected to go wrong. Company test pilot Erwin Aichele flew the plane on its maiden flight on April 13, 1932, and at a flying display in Schleißheim he suffered a mishap (nose-over). The plane was only slightly damaged and was soon repaired. All the other aircraft remained mishap-free during their test flights and evaluations. Several aircraft logged up to 30 flight hours during this period, flown by various pilots, and the prototype itself clocked up more than 100 flight hours after its repairs.

Thus, when all six competition aircraft stood ready in early August, nobody doubted their impending success. However, any hope of success was dashed when the type was immediately pulled from the competition following two mishaps[5] involving the M 29 on 8 and 9 August which resulted in fatal consequences, without clear explanation of either cause. Despite this setback—the competition was won by a Pole—the technical world still considered this design as a brilliant achievement, providing valuable experience for the development of future high-speed aircraft. This not only applied to Messerschmitt's Me 108 *Taifun* "air limousine" from 1934, or to the Me 109 light fighter from 1935, but it also extended beyond Germany's borders to reverberate in aviation design throughout the world.

Description

The M 29 was a two-seat, low-wing monoplane with cantilever wings of mixed construction.

Fuselage: four-spar steel tube construction; the forward section was designed as an aerodynamic shell for the engine, with the aft tapering off into the vertical fin. Long-

---

[5] At least one cause of the mishaps may have been vibrations affecting the elevators (all-moving tailplane), which sooner or later would have led to a crack in the tailplane itself.

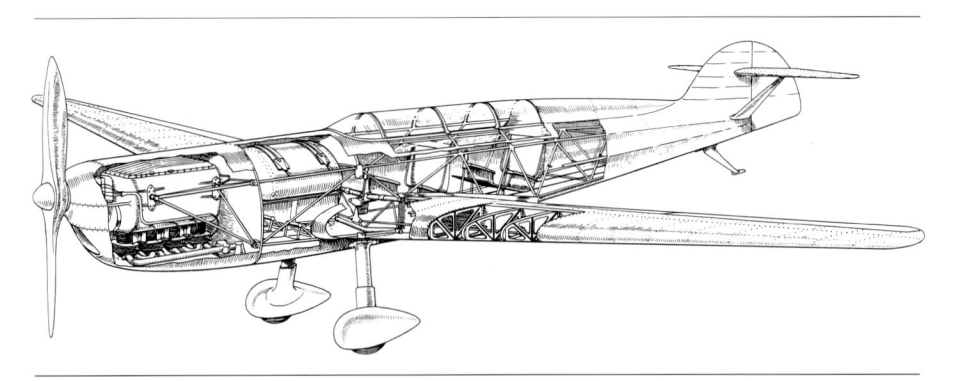

The M 29, a two-seater of mixed construction, introduced a new era in sportplane design with its cantilever undercarriage and landing flaps. Ludwig Bölkow once labeled the design as a "superplane." Built as a contest winner, misfortune prevented it from fulfilling its role.

The first of six M 29s took to the skies on its maiden flight on 13 April 1932 with Erwin Aichele at the controls. Seen here on final approach, all the new features on this design are quite apparent—the single-strut landing gear (a patent by Willy Messerschmitt), the first use of landing flaps, and the all-moving tailplane.

erons and formers gave it an octagonal cross-section, which for the most part was covered in fabric. The wing-fuselage join sections and the point where the undercarriage struts attached to the fuselage made use of specially reinforced crossties. The cockpit aerodynamically tapered off to the rear. The upper part of the cockpits were englassed in cellophane and opened to the side; for both seats the canopy slid to the rear. As was already mentioned, in the event of an emergency the entire caonopy could be jettisoned to the side using a lever. The cockpits were roomy, and the armrests comfortable. Back-pack parachutes could be used. Behind the copilot seat was a small storage compartment. Both seats were adjustable heightwise by an easy-to-reach lever, and could also be adjusted forward or rearward. The rudder pedals could be adjusted according to the leg-length of both the pilot and copilot. The wheel brakes were activated by tapping the control pedals.

Wings: these were designed in two parts and were of a tapered layout. Both halves were attached to the fuselage at three points, and by means of a hand lever could be quickly folded alongside the fuselage. A landing flap ran the entire length of the trailing edge right up to the aileron, enabling greater lateral control and countering of the air flow through flap adjustment.

Tail assembly and control surfaces: all rudders were of wood construction with fabric covering. The rudder and tailfin together formed a semi-circle, which was clipped off at the bottom and ran directly into the aft fuselage. The elevator, set high on the tail assembly, pivoted at a point cut out of the tailfin and was braced against the fuselage.

Undercarriage: this was the first use of the so-called single-strut undercarriage invented by Messerschmitt. It differed from all other types of landing gear primarily in that the wheels, attached to the axles (which protruded directly from the struts), were connected to the fuselage solely by the legs and not by means of any separate bracing. The legs only extended from the fuselage far enough to allow for sufficient propeller clearance when compressed, and to avoid nosing over on landing. The wheels and legs were covered with streamlined shrouds, minimizing as much as possible the negative drag effects of the undercarriage. One

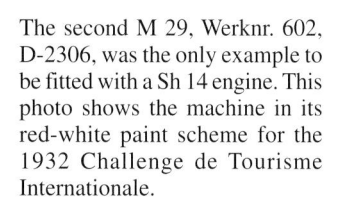

The second M 29, Werknr. 602, D-2306, was the only example to be fitted with a Sh 14 engine. This photo shows the machine in its red-white paint scheme for the 1932 Challenge de Tourisme Internationale.

The third M 29, Werknr. 603, D-2307, was delivered to Fritz Morzik on 8 August 1932 for this previous winner's participation in the 1932 Challenge de Tourisme Internationale. However, two crashes on 8 and 9 August by other potential participants in the Challenge de Tourisme Internationale resulted in the type being banned from the competition. Information later released by the accident investigation committee showed that the tailplane, designed to be an all-moving surface, was subject to vibrations during flight which sooner or later led to structural fatigue and the eventual failure of the control surfaces.

particularly novel and effective approach exemplified on the M 29 was having the upper part of the legs braced into the airframe above the wheels, so that when attached, they formed the sides of the main fuselage former. The pneumatic cushioning effect of the legs was softer than that of a pure oleo strut; it allowed dropping the aircraft at a rate of 3.5 m/s. The metal skid was cushioned with rubber webbing and could swivel.

The engine: five M 29 aircraft were powered by the air-cooled, four-cylinder Argus As 8R series engine with 92-110 kW/125-150 hp, and one plane had the roughly similarly rated, air-cooled, seven-cylinder Siemens radial engine, Sh 14a. The engine frame assembly was of welded steel construction held in place inside the fuselage with rubber buffers. A 120 liter fuel tank was located behind the firewall in the fuselage upperside so that there was sufficient drop for the fuel into the carburetor. An air-cooled oil reservoir was situated on the underside of the fuselage. The drag incurred by the Siemens radial was offset by burying it in a NACA cowling. This engine required the oil tank to be installed in front of the bulkhead on the fuselage

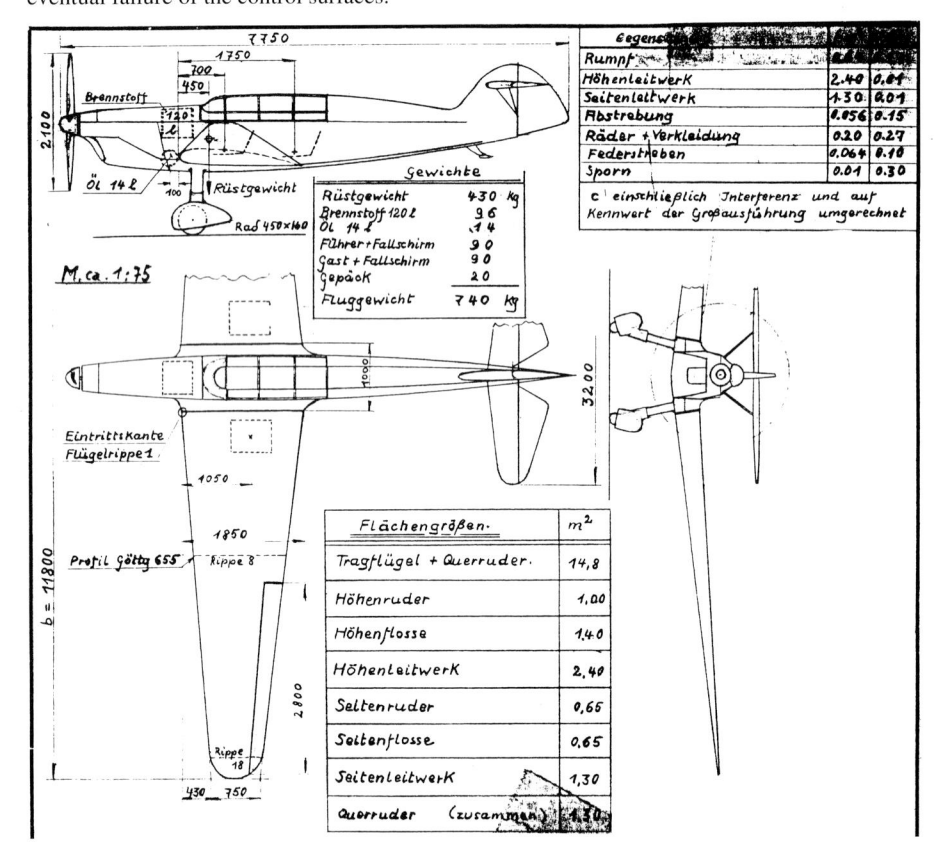

In the summer semester of 1936 the type sheet for the M 29 served as the basis for one of Messerschmitt's assignments to his students at the TH Munich.

upperside and shielded from the engine's heat. The instruments on the instrument panel in the cockpit were quite adequate for the time, consisting of airspeed indicator, altimeter, turn and bank indicator, compass and onboard clock for navigation; for monitoring the engine there was a rpm gauge, an ignition cut-out and circuit breaker switch, oil pressure indicator and oil thermometer, as well as a fuel gauge. The instruments were organized according to their functions.

M 30, the Electron M 26 (1929)

In 1929 the *Reichsverkehrsministerium* issued contracts to several airplane manufacturing firms for exploring the possibility of using electron in aircraft construction. Electron was a new, light metal alloy with a magnesium base produced by the IG Farbenindustrie AG in Bitterfeld, and had a specific weight of 1.8 (duralumin 2.7). The Bayerische Flugzeugwerke were to build the M 26 from electron for comparison studies. It soon became quite evident that, because of the more expensive processing method involved and the metal's lower sturdiness, it was not simply a matter of transferring the M 26 blueprints over. In the majority of cases, new designs were needed and new blueprints would have to be prepared. Furthermore, in several instances it would not have been possible to work within the space of the M 26 design and larger external dimensions would be required. Consequently, work on the project very quickly came to a halt, and electron was only used for a few parts not subjected to wear and tear, e.g. as a coating material.

M 31 Sportplane and Trainer (1932)

The former trend to build sportplanes with more power reserves, i.e. with more powerful engines, while simultaneously equipping them with better aerodynamics—see the M 17, the M 23, and M 27—was followed by a desire to lower purchase and operating costs as the economic situation worsened. Since relatively less expensive German engines under 80 hp were available, Messerschmitt hoped to meet this demand with an inexpensive, robust, two-seat trainer and sportplane. From this idea arose the M 31, a two-seat low-wing monoplane of composite design.

The tapered wooden wing was designed in the customary Messerschmitt monospar construction method, with a torsion-resistant leading edge and end ribs connected to the support spar with plywood skinning. Also, the wing was designed with a slight negative outside twist to improve lateral stability and reduce the risk of stalling.

The fuselage consisted of a series of triangular shaped frames made of welded chromium molybdate steel tubes and covered in fabric. The tapered cross section had an arched back terminating in the engine cowling at the front and the vertical fin at the rear. The open cockpits were arranged in tandem and had a windscreen, with a small storage area just behind the aft cockpit. The vertical stabilizer, protruding from the steel tubed fuselage end, served as the anchor point for the wooden, fabric covered rudder. The horizontal stabilizer, adjustable in flight, was braced against the vertical fin and made of wood and plywood covering. The elevator was separate and constructed like the rudder. All control surfaces rotated on ball bearings.

The only example of the two-seat M 31 sportplane was put on display at the 1932 DELA in Berlin. The low-wing design of mixed construction embodied some elements from the M 25 project of 1929, despite the latter being a high-wing layout (q.v.).

The controls for the aircraft were designed for ease of both maintenance and adjustment. A joystick could be fitted in the passenger cockpit for training purposes, which could be rendered inoperable by simply turning the grip on the control stick in the front cockpit.

The undercarriage assembly had separate axles and made use of oversized wheels with a nod towards the plane's intended trainer duties. Low pressure tires were used to help cushion landings. The freely swiveling skid was also built for strength. Of the engines available at that time, the one selected was the air-cooled, five-cylinder BMW Xa radial engine, with a power rating of 29-44 kW/40-60 hp, and hidden beneath a NACA cowling for drag reduction.

The aircraft, Werk-Nr. 607, was completed in early August of 1932, and test flown by company pilot Erwin Aichele. Following several months of testing the aircraft it finally reached its expected attributes and performance, but no official certification was applied for. Instead, in the spring of 1933 the plane was refitted with the more powerful air-cooled, four-cylinder Hirth HM 60 rated at 47-52 kW/65-70 hp, and then was registered as D-2623. It wasn't until the following year that the RLM acquired the sole example of the M 31. The expected interest never materialized even though the M 31 was exhibited at the 1932 DELA.

M 32 Trainer Project

For quite some time there was a project for a biplane trainer on Messerschmitt's "to do" list, something which Udet pressed for and Messerschmitt actually had little desire to build. It was to be a successor to the "Flamingo."

Design sketches for such a plane had been around for quite awhile. When in late 1930 the BFW reviewed its production plans for the following year, it was discovered that the "mixed construction" manufacturing department would not be fully utilized. It was therefore decided to construct a batch of five planes "out of pocket," and work began immediately on the blueprints. The thought was that such small, relatively inexpensive aircraft would be the most likely to sell, plus there was sufficient materiel already on hand for the project.

Six months later the company declared bankruptcy, and all work on contracts not covered was halted. A large percentage of the work force, including the entire design department, was laid off, with the influential architect of the

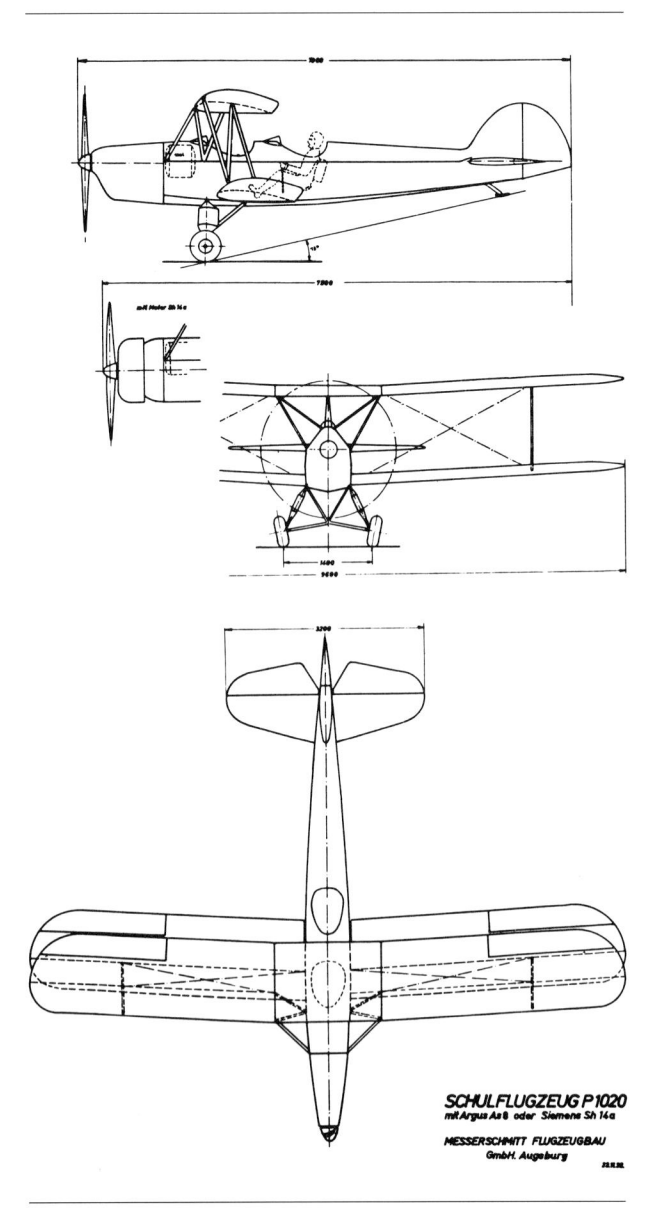

A biplane trainer project in the tradition of the Flamingo was born in 1932 as a competitor to projects being developed by other companies, such as the Fw 44 Stieglitz. Although work started on the P 1020 project at BFW at the instigation of Ernst Udet, further construction was brought to a halt due to the company's bankruptcy. Designer Paul J. Hall, who went to Heinkel after the mass layoffs at BFW, was able to complete the project at Heinkel. Thus, Messerschmitt's blueprints from 23 November 1932 bear a remarkable resemblance to the Heinkel He 72 Kadett.
Optional As 8 or Sh 14a engine, 7.50 meter wingspan and a length of 7.90 meters were just some of the design's characteristics.

M 32, Paul J. Hall, transferring to the Heinkel company in Warnemünde. It was under his direction (and based on the M 32) that the He 72 Kadett trainer was born. According to Hall, he was given a BMW automobile as a present from Heinkel "as recognition for this work"!

The five M 32 planes cost a total of 28,000 Reichmarks and ended up being sold for scrap metal as part of BFW's assets.

M 33 Kitplane (1932)

At the 1932 German Aero-Sport Exhibition (DELA) in Berlin Willy Messerschmitt and the Messerschmitt Flugzeugbau GmbH unveiled a full-scale mockup of his M 33 kitplane with the following specifications: 9.26 m wingspan; 5.65 m length, 1.9 m height, and a two-stroke Type P DKW engine with an 11 kW/15 hp power rating. The M 33 was to be ideally suited for sport flying, with part of its attraction being the ability to build the aircraft oneself and thereby fulfill "the long-awaited dream of the younger generation of flying enthusiasts," as the advertisement claimed. An instruction sheet, accompanied by diagrams, showed how to put the aircraft together and included the undercarriage, tail assembly, and fuselage. Another sheet and series of diagrams broke down the individual parts for the engine, the fittings for the wings and tail assembly, undercarriage, controls, and fuselage fittings with engine connections and the smaller pieces.

A large-scale mockup of a "weekend amphibian," based on a "Kleinhenz patent," was shown at the BFW stand at DELA 1932. The DELA management board had only admitted BFW's mockup as evidence that prototype development was in full swing. Willy Messerschmitt distanced himself from the project, a fact which must have played a role in the design not being assigned a BFW project number or type designation.

The entire idea seemed quite plausible, and one could imagine that the *Flug- und Arbeitsgemeinschaftsgruppen* (Flying and Workers Unions-FAG) within the aero-sport clubs would have enthusiastically pursued the kitplane idea and powered flight just as they had the construction of trainer gliders. In any event, nothing was mentioned about the M 33's characteristics or expected performance, nor its delivery times or costs. And since in times of need few young people even had pocket money, it was to be expected that the FAG groups were hard pressed to come up with the funds for the construction material or, especially, the prebuilt components. In actual fact, it is not known whether or not a single M 33 was ever built or even if one was ever flown.

The M 33 Volksflugzeug (people's plane) was offered as an inexpensive kitplane in modular form. Although exhibited on the BFW stand at the 1932 DELA in Berlin, it was only built in mockup form.

M 34 Long-Range or Antipode Plane (1932)

At the 1932 DELA Messerschmitt Flugzeugbau GmbH exhibited a model for a long-range aircraft, the M 34, a design conceived by Messerschmitt together with his brother-in-law Professor Georg Madelung (creator of the world-record holding sailplane from 1922, the Vampyr). The M 29's ingenious aerodynamic refinements, whose drag was less than half that of its M 23 predecessor, as well as having a landing speed lower than that of the M 23, led Messerschmit to conclude that incorporating these concepts into new designs would offer considerable advantages. Assuming that new aircraft engines designed with improved shape and dimensions—diesel engines—were now available, and that there was no headwind, the type was expected to have an estimated operating range of 20,000 kilometers. This meant that the airplane could go from one spot on the earth to any other without any outside aid. The cruising speed for such a venture was to be at least 200 km/h, and the plane could even carry a payload. Since any distance hardly measures more than 12,000 to 15,000 kilometers, the large operating range provided a safety buffer. Powerplant might consist of a single engine with 600 kW/800 hp or two engines with 300 kW/400 hp each. With two engines, one could be shut down after takeoff and used only as a reserve engine. By switching the propeller to "free jet" mode, the operating efficiency was expected to jump from approximately 65 percent for standard configurations up to 85 percent. The slim fuselage, the wing's aspect ratio similar to that of the sailplane, possibly locating the radiators in

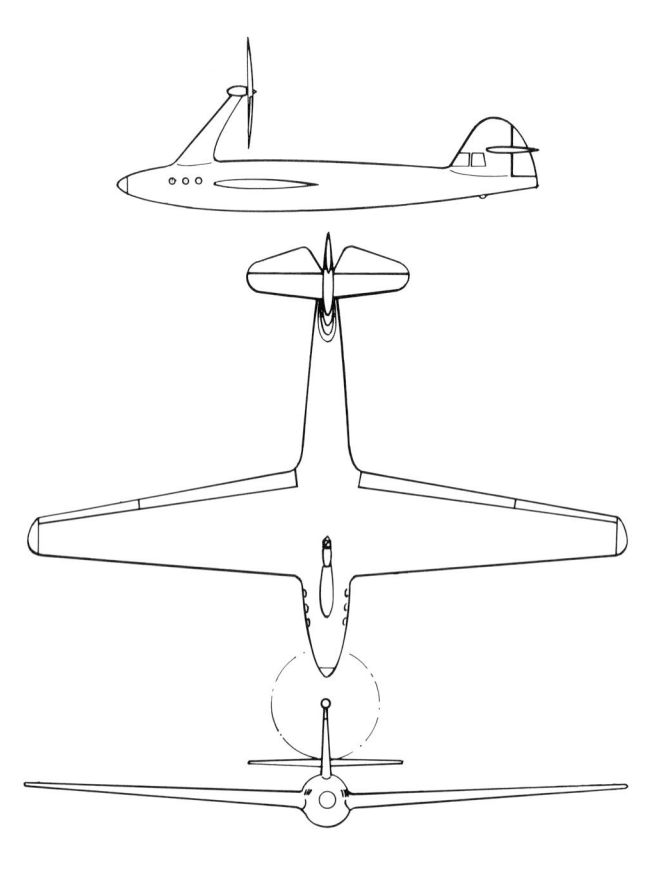

This reconstructed three-view of the M 34 should only be considered a basic plan.

A model for a long-range commercial airliner was introduced at DELA 1932 under the designation M 34. This was an idea that Messerschmitt could never fully get out of his head and was later embodied (at least partially) in the Me 261 and 264. All project data for the M 34 have been lost.
This is a photograph of the model together with one of a M 20. Assuming that both models were built to the same scale, we can make a rough estimate of the dimensions. Based on this, the M 34 would have had a wingspan of about 34 meters and a length of 14 meters.

A model of a high-speed mail-plane was also shown at the DELA. Given the project number P 1012, it evidenced all the elements of the current trends in aviation design, such as a refined aerodynamic shape and retractable undercarriage—features which Messerschmitt first introduced at BFW in 1934 with the Bf 108.

the wings, moving the cockpit to the vertical stabilizer at the extreme aft end of the fuselage, and a retractable landing gear were all options which, although not yet within reach at that time, could someday be accomplished given sufficient means. The long-range aircraft was not intended to be any sort of utopian concept, but was rather an attempt at showing just how far aircraft design could go.

Forced Settlement (1933)

In late 1932 the sales department of BFW AG, still struggling under the cloud of bankruptcy, hired Rakan Kokothaki who, working with the receiver, made a concerted effort to bring about an end to the bankruptcy. This was finally achieved when the majority of the creditors signed a settlement agreement. In a preliminary meeting in December of 1932 all creditors then gave their tentative approval to a forced settlement, or compulsory composition. This led to ratification of the settlement on April 27, 1933, at the Augsburg District Court. The Bayrische Flugzeugwerke AG Augsburg could resume work; the company reopened its doors on May 1, 1933, with a work force totaling 82 people. The Messerschsmitt Flugzeugbau GmbH and the Messerschmitt design department were gradually absorbed into the Bayrische Flugzeugwerke AG over the following months.

M 35 Trainer and Aerobatic Plane (1932/33)

Messerschmitt, together with his small team, had in the interim weathered the storm with the construction of the M 29, the design and construction studies for the M 30 (M 26 made of electron), and the construction of the M 31.

Now he once again turned to composite construction principles to develop an aircraft similar to the M 31, but substantially more robust and designed with twice the power; the M 35. This was to be his last sportplane.

Both of the first aircraft, Werk-Nr.s 620 and 621, were completed in the first half of 1933 under the designation M 35a/b. They were fitted with the air-cooled, seven-cylinder Siemens Sh 14 engine rated at 110 kW/150 hp and were also test flown by company pilot Aichele. One of these airplanes was publicly displayed for the first time at the International Aero-Sport Exhibition in Geneva from April 27 to 6 May 1934. The third aircraft, Werk-Nr. 622, was powered by the air-cooled, six-cylinder four-stroke Argus As 17A engine with a 147 kW/210 hp power rating and an electron propeller. First flown by Willi Stör on May 14, 1934, the propeller fractured and the aircraft suffered minor damage. This and all remaining twelve aircraft, which were (probably) all powered by the Sh 14a engine, were assigned Werk-Nr.s 623 to 631 and 644 to 646 and designated the M 35b. They were delivered between March of 1934 and September of 1935, with one each going to Spain and Romania and the remainder to Germany. The last M 35b, Werk-Nr. 646, D-EQAN, was given to Willi Stör, and it was in this aircraft that he became Germany's *Kunstflugmeister* (aerobatic champion) in May 1935 at Stuttgart and again in July 1936 at Munich-Oberwiesenfeld.

The aircraft was a two-seat low-wing monoplane built of wood/fabric and metal construction with cantilever wings, designed for training and aerobatic flight. The two-piece tapered wing with rounded tips was fastened to both sides of the fuselage at three points; the upper and lower bolts could be pulled out of the attachment points using a lever. The third bolt ran into a swivel mount in the leading edge, allowing the wings to fold up against the fuselage. Naturally, the design employed Messerschmitt's monospar wing concept with a torsionally rigid leading edge, and the wing covering went right up to the aileron assembly where a sup-

The mixed construction M 35 trainer and aerobatic plane.

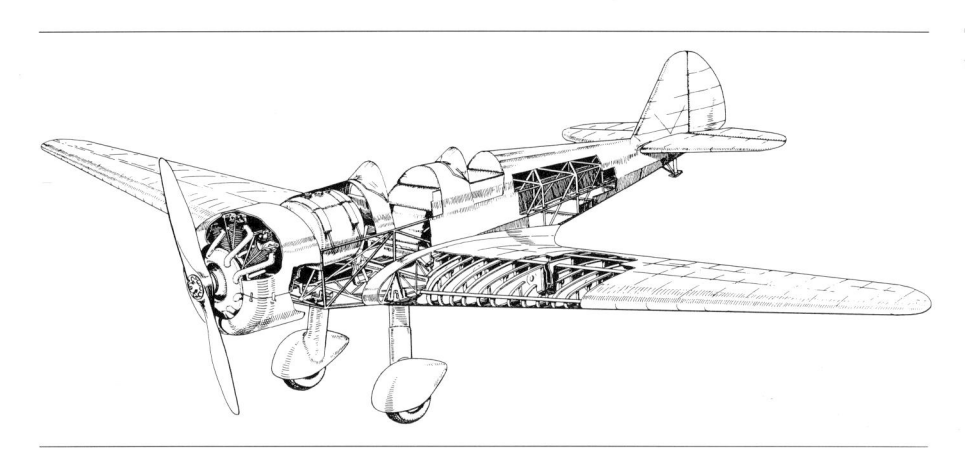

Much admired for its perfect harmony of design, the M 35 was displayed at the International Sportplane Exhibition in Geneva from 27 April to 6 May 1934.

The second M 35, Werknr. 621 from 1933, seen over the towers of Augsburg.

port spar capped it off nicely, forming a rigid box structure behind the main spar, as well. The wing areas behind the support spar were covered in fabric, as were the the ailerons. The spaces inside the wing where it joined the fuselage were designed to hold luggage. The fuselge consisted of a welded, diagonally braced steel tubed structure covered in fabric. The tandem pilot and co-pilot seats were designed to accommodate seat parachutes. A hinged flap in both cockpits made climbing in and out quite easy. The fuselage gradually tapered to the plywood-covered vertical stabilizer via a series of steel tubed triangles. The horizontal stabilizer was designed in a tapered form with rounded edges. It was made of wood with plywood skinning, and the elevator and rudder were covered in fabric. For trimming, the horizontal fin was adjustable, manipulated by a hand wheel from the pilot's position. A joystick was used for controlling the plane. A dual joystick and second instrument panel could easily be installed when needed for training purposes.

The undercarriage was "cantilever" like that of the M 29. Cushioning was provided by rigid oleo struts. The especially large travel stroke took into account the aircraft's training role and the anticipated hard landings. The braked wheels were made of cast electron with internal expanding brake and low pressure tires. Braking was activated by a hand lever in the cockpit. Struts and wheels were covered in aerodynamic fairings and spats. The tailskid was free-moving. Panels in the aft fuselage gave easy access to the skid spring coil. Engine mounts were made of welded steel. The engine cowling could be easily opened by lifting two lever catches. The air-cooled Siemens radial engine was housed beneath a special NACA cowling designed to reduce drag. The M 35 had what was considered especially good flight handling characteristics. The plane's spin characteristics were risk-free; by letting up on or releasing just one of the three control inputs the plane would recover from

The support mount for the cantilever strut inside the wing of the M 35.

Willi Stör posing in front of his baby blue M 35, D-EQAN, Werknr. 646, in which he won the 1935 and 1936 German aerobatic championships in Stuttgart and Munich, respectively.

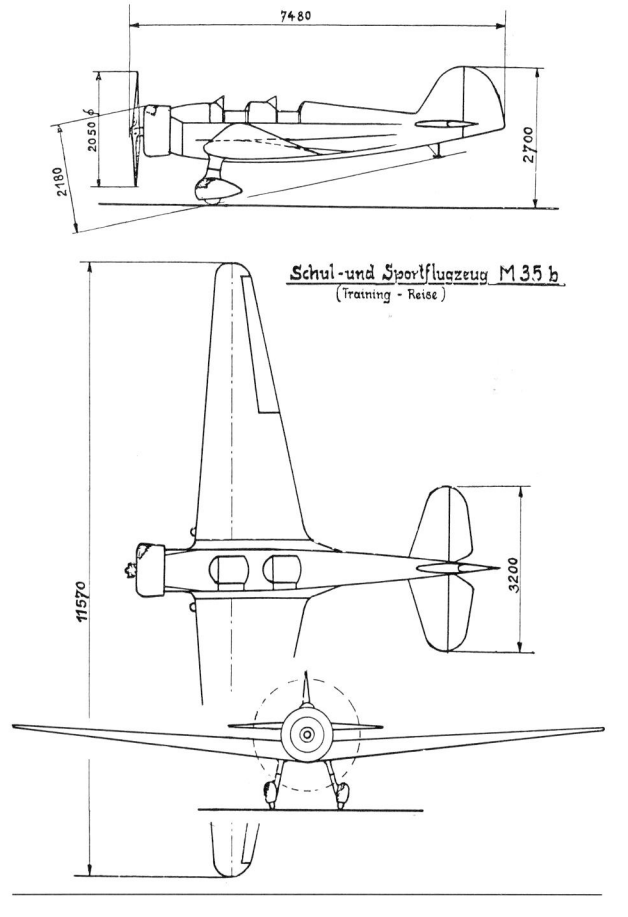

M 35b three-view from 24 September 1934.

the spin itself. It therefore could only be put into a controlled spin, a feature extremely beneficial during training.

Any fresh start for the Bayerische Flugzeugwerke after 1 May 1933 was hampered by the fact that the *LuftVerkehrsamt* had become subordinated to *Staatsekretär* Milch following the rise to power of the National Socialists. Milch's aversion to Messerschmitt was so great that not only did he deliberately overlook Messerschmitt when issuing contracts to every other aircraft manufacturer for expanding commercial aviation, he even demanded that BFW only build aircraft under the license from other aircraft firms. Since the BFW's senior management wanted to avoid further problems and the company was clearly struggling to remain a viable player, for the short term at least it was forced to comply with these demands.

The whole situation became even more distressing when Milch forced the BFW AG to build the He 45c biplane under license from Heinkel. This was after Heinkel, ten years Messerschmitt's senior, had found Messerschmitt to be his toughest opponent during the flying competitions in the 1920s and had often attempted to call into question the reputation of the young upstart.

M 36 Commercial Plane for Romania (1934)

Since the Reichsluftfahrtministerium was no longer issuing contracts to Messerschmitt's small construction department, BFW's senior managers were compelled to seek them out abroad. The result was negotiations with Romania, where good business connections between BFW and the ICAR (Intrepindere Pentru Constructii Aeronautice Romane, Bukarest VI) aircraft company had already been established with the 1931 delivery and transfer of license for the M 23b, and where Messerschmitt enjoyed a solid reputation as a designer. The fruits of these negotiations was a construction contract for the M 36 high-wing passenger plane of mixed construction. The aircraft was to be designed to accommodate six passengers and two pilots. Initially, a Gnôme et Rhône engine, the K 14, with 335 kW/ 456 hp was to be installed, but then ICAR decided to use the air-cooled, ten-cylinder Armstrong-Siddeley Serval engine with reduction gear and 265 kW/360 hp. Even with this markedly less powerful engine, the aircraft successfully completed its initial test flights (in late 1934). The following details apply to the aircraft's construction:

Fuselage made of chromium molybdate welded steel tubing, covered in fabric, monospar wing manufactured from Romanian pine, covered with thin plywood skinning, tubular steel engine block, engine beneath NACA cowling, four fuel tanks—two in each wing, trimming tank in the aft fuselage. Tailplane was trimmable during flight, with a counterbalance fitted on the vertical stabilizer. Landing flaps and aileron deflector were available during landing.

The cockpit area had two seats side-by-side with dual controls, and a roomy passenger compartment with six seats facing in the direction of flight. Each passenger had his own individually-controlled ventilation system, with a large luggage compartment aft. Undercarriage, tailwheel, and braking system were all products of Messier.

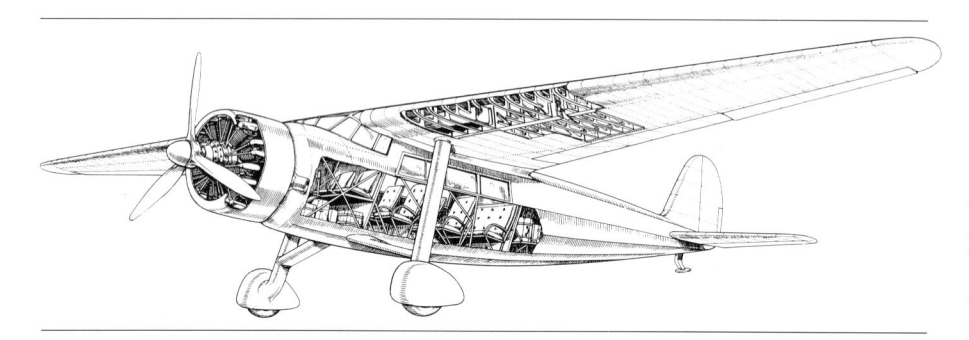

The six-seat M 36 passenger plane of wood and metal construction, powered by an Armstrong-Siddeley Serval engine.

The only machine constructed was designed in Augsburg and assembled at ICAR in Bucharest in 1934. It flew with the Romanian airline LARES on scheduled routes from 1936 to 1938.

Collapse and Rebirth (1935)

Restructuring the Aviation Industry

With the coming to power of the National Socialists in 1933, the rearmament program for all branches of the military broke the bounds of the 100,000 man strong army the Weimar Republic had been limited to. Added to this was also a program for building up an air force which, after Germany's 1 March 1935 proclamation to maintain its own military, officially became the third branch and was initially called the Reichs*Luftwaffe*. (The old Fliegertruppe, founded in 1912, was dissolved in 1920 after the Versailles Treaty went into effect).

At first, the foundations for equipping the new air force were simply not there because none of the available aircraft construction firms were equipped for mass production. The civilian airliners, trainers, and sportplanes permitted up to this point had been built in such small numbers that they were virtually considered one-off projects. The only companies which could come close to fulfilling the demand for mass production were Junkers, Heinkel, Dornier, and Arado. These were undoubtedly the largest German airframe manufacturers in the 1920s and in the early 1930s. And they were to supply the *Luftwaffe* with its first airplanes. But these companies and their limited production resources could not hope to meet the demand for the number of aircraft required, which climbed to the thousands, and production licenses had to be awarded to subcontractors.

The organizer for all of this was the new *Reichsluftfahrtministerium* (Reich Ministry for Aviation (RLM)) under the direction of the newly created office of Reich Minister of Aviation, taken over by Hermann Göring (1893-1946), who had also become the commander-in-chief of the *Luftwaffe* in May of 1935.

The second in charge of the RLM was the perennial enemy of Willy Messerschmitt, the Secretary of State in the RLM Erhard Milch (1892-1972), who in 1940 was promoted to the rank of *Generalfeldmarschall* and took over Udet's post following the latter's death.

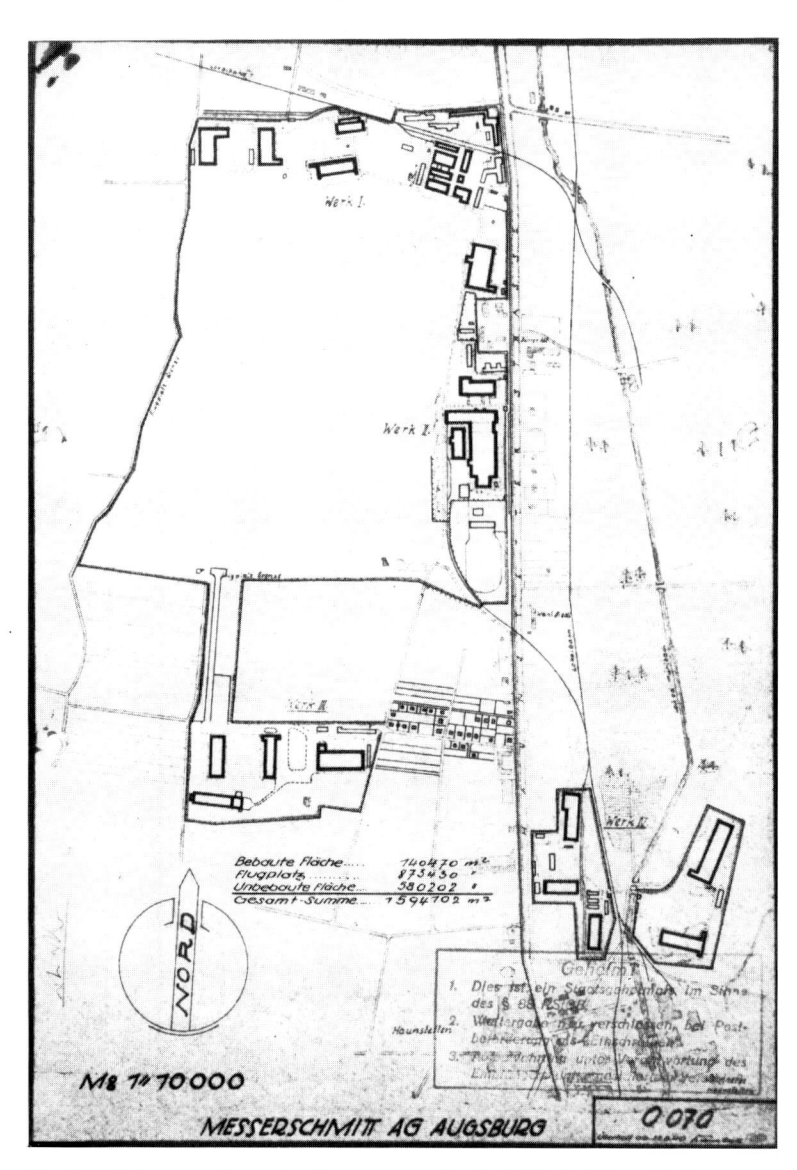

The RLM's large-scale production contracts for the Bf 108, 109, and 110 forced the continuous expansion of the Bayerische Flugzeugwerke. Between 1934 and 1937 the factory grew around the airfield to four complexes (Werk I through IV) and covered an area of over 140,000 m².

There was incessant building work going on along the Haunstetter Straße during the years of growth from 1934 to 1937. This picture shows a view looking southeast toward the airfield. Biplanes from the first license-built contract from the RLM are seen parked in front of the new test hangar.

Messerschmitt's friend, Ernst Udet (1896-1941), former co-founder of Udet Flugzeugbau which BFW had taken over, and the most popular flyer in Germany, became an *Oberst* and chief of the RLM's *Technisches Amt* (Technical Office) in 1936, and then the *Generalluftzeugmeister* in 1938.

The RLM, which issued development-, production-, and licensing contracts, was now required to include the small Bayerische Flugzeugwerke (BFW), whose attorney Konrad Merkel had narrowly avoided bankruptcy by transferring the settlement among the companies being issued contracts.

In one of the first large aircraft acquisition programs during the 1935-1937 planning period, established by the RLM's *Technisches Amt* LC II department on November 1, 1935, there were 17 airframe firms named with a total of 20 (later 23) factories as contractors[1]:

AGO Flugzeugwerke GmbH, Oschersleben*
Arado Flugzeugwerke GmbH, Potsdam*
Arado Flugzeugwerke GmbH, Warnemünde*
ATG-Maschinenbau GmbH, Leipzig

Bücker Flugzeugwerke GmbH, Rangsdorf
Bayerische Flugzeugwerke AG, Augsburg
 (Messerschmitt AG from 1938 onward)
Dornier-Werke GmbH, Friderichshafen*
Dornier-Werke, Wismar
Erla-Machinenwerk GmbH, Leipzig*
Gerhard Fieseler Werke GmbH, Kassel*
Focke-Wulf Flugzeugbau GmbH, Bremen*
Gothaer Waggonfabrik AG, Gotha*
Hamburger Flugzeugbau GmbH, Hamburg*
 (Blohm & Voss, Abt. Flugzeugbau from 1938 onward)
Ernst Heinkel Flugzeugwerke, Rostock
Henschel Flugzeugwerke AG, Schönefeld*
Junkers Flugzeug- und Motorenwerke AG, Dessau*
Hanns Klemm Flugzeugbau, Böblingen*
 (beginning in 1936 Siebel Flugzeugwerke Halle GmbH)
MIAG Mühlenbau- und Insudstrie AG, Braunschweig*
Weser Flugzeugbau GmbH, Bremen*

And just three years later, three additional new firms and factories were mass producing aircraft:

[1] The aircraft acquisition program 1935-1937 from November 1, 1935. RLM, LC II/13271/35.

Luftwaffe leaders and RLM staff members often paid visits to the factory, particularly in 1941. This photo shows just such a group gathered to witness a flight demonstration (left to right): Willy Messerschmitt, Generalfeldmarschall Hermann Göring, Director Fritz H. Hentzen, General Ernst Udet, and the Regensburg chief, Theo Croneiß.

RLM boss Erhard Milch, Willy Messerschmitt, Fritz H. Hentzen, Ernst Udet and Hubert Bauer (left to right) are captured on film discussing the latest plans.

Messerschmit GmbH, Regensburg
Wiener Neustädter Flugzeugwerke GmbH, Wiener-Neustadt*
Heinkel-Werke GmbH, Oranienburg

(Those firms designated with an asterisk were license holders between 1937 and 1945 for various Messerschmitt aircraft.)

In accordance with the program, among other designs the BFW was given the license-building contract for critically needed trainers. These included:

70 He 45s
90 Ar 66s
115 Go 145s
35 He 50s
50 Ju 87s (though not assembled at BFW, instead substituting:)
30 Do 11s

The BFW's own developments which were named and contracted for were:

7 pre-production Bf 108s trainers	(in competition with the Kl 36)
45 production BF 108s	
10 pre-production Bf 109 fighter aircraft	(in competition with the He 112, Ar 80, and Fw 159)
8 pre-production Bf 110 fighter aircraft	(in competition with the Fw 57 and Hs 124)

(Not included in this listing are prototypes of the Bf 108 and Bf 109 already flown at this point in time, and the Bf 110, then under construction.)

There are four more BFW designs named in other RLM planning documents for aircraft development from 1936/37[2,3]:

Bf 161 reconnaissance aircraft	
Bf 162 high-speed bomber aircraft	(in competition with the Ju 88, the He 119, the Hs 127, and the Do 17M)
Bf 163 liaison aircraft	(in competition with the Fi 156 and the Klemm Fh 201, later the Si 201)
Bf 164 aircraft for round-the-world flight	(in competition with the He 119 and Ju 88)

In addition there was the Bf 165 long-range bomber project under consideration for a short time.

These contracts seriously taxed the abilities and resources of the project department (*Probü*) under Robert Lusser, who had arrived at BFW in 1933, and the design department (*Kobü*) under the direction of Richard Bauer, who had been with the BFW since 1929. New personnel had to be constantly brought in. Yet in spite of the increased manpower they were unable to accomplish everything. The partially drawn up Bf 163 project was subcontracted to the

[2] Aircraft prototype table of LC II, 1, as of October 1, 1936. LC II no.7920/36 1 from October 14, 1936.
[3] Aircraft development program (as of October, 1937). LC II no. 1588/37, 1 zbV, Secret, Berlin, December 22, 1937.

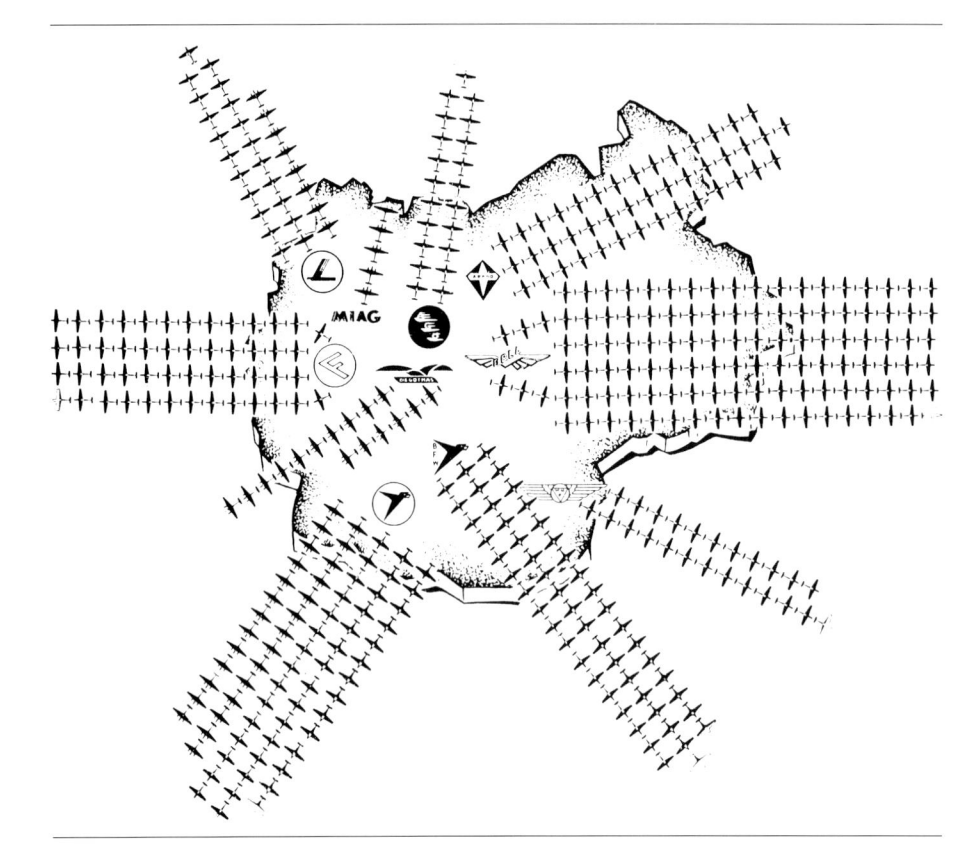

Production orders for Messerschmitt aircraft were so great that other companies were brought in to license build the types. By early 1940 ten different companies in the Reich were producing Messerschmitt planes, making Willy Messerschmitt the largest employer in the aviation industry. By this time, these companies had built approximately 4,300 Messerschmitt aircraft, although only about 700 of these were actually built at the parent works in Augsburg.

Weser Flugzeugbau. The *Amt* canceled other developments, such as the Bf 164 and Bf 165.

Prototype construction under Hubert Bauer, with the BFW since 1929, faced a similar dilemma when it was burdened with constructing the V-prototypes for the Bf 108, 109, and 110 and their pre-production series.

This license-construction program, although bringing in much money, in no way matched the concepts of Willy Messerschmitt. Nearly all of them were of mixed wooden and fabric construction (other than the Do 11), and Messerschmitt, with his all-metal monocoque method used almost exclusively from the Bf 108 onward, was advancing into a new era of aircraft construction. The license construction program, running from 1934 to 1937, nevertheless had the advantage of building up a skilled core of technicians for the BFW, which built about 400 aircraft during this period. And these workers would be urgently needed for the ensuing production series.

The number of new Messerschmitt Bf 108, Bf 109, and Bf 110 models the RLM now called for made it necessary to curtail the license construction program. In a reversal of roles, Willy Messerschmitt was now forced to seek out license-builders for his aircraft. The situation grew so dramatically during the war that approximately two-thirds of

[1] Elly Beinhorn, Alleinflug. Mein Leben, Munich 1977, pp. 210-211

the German aircraft manufacturers named in earlier paragraphs were producing Messerschmitt aircraft. Messerschmitt therefore became, directly or indirectly, the largest employer in the aviation industry, although the Messerschmitt group itself, with its 25,600 employees in early 1943 was in fourth place for personnel strength, behind Junkers (41,600), Heinkel (31,000), and Arado (26,800). One year later, Messerschmitt jumped to third place in front of Arado with over 30,000 employees.[4]

(Not included in these figures is the continuously fluctuating number of foreign workers, war criminals, military prisoners and, in the latter half of the war, concentration camp detainees assigned as laborers.)

Bf 108 Taifun Four-Seat Commuter Plane

Probably the most famous of German commuter planes was the Messerschmitt Taifun, developed for the 1934 Challenge de Tourisme Internationale. At the time it caused quite a stir, being an all-metal, four-seat aircraft with retracting undercarriage, wing slats, and slotted landing flaps—as well as proving to be the fastest aircraft in the competition. It

The Bf 108A V-6 with Hirth HM-8U powerplant, one of the competition aircraft in the 1934 Challenge de Tourisme Internationale, seen here after being fitted with the production-standard control system consisting of leading edge slats, ailerons and landing flaps.

soon became one of the most renowned airplanes in the world—as the ideal commuter, it was a milestone in light aircraft construction.

Development of the Bf 108

The *Reichsluftfahrtministerium* (RLM) in Berlin decided quite late in the game to take part in the Challenge de Tourisme Internationale of 1934. Poland, land of the winning pilot from 1932, was to establish the conditions and lay out the course. The conditions were advertised in June 1933, and the competition was to take place in the summer of 1934 with Warsaw as the starting point.

In September 1933, the RLM awarded a development contract to the Bayerische Flugzeugwerke (BFW) for a four-seat touring plane, of which six prototypes were to be built. Parallel contracts went to Fieseler for the Fi 97 and Klemm for the Kl 36. The Messerschmitt touring plane, which was designated internally as the M 37, was given the type des-

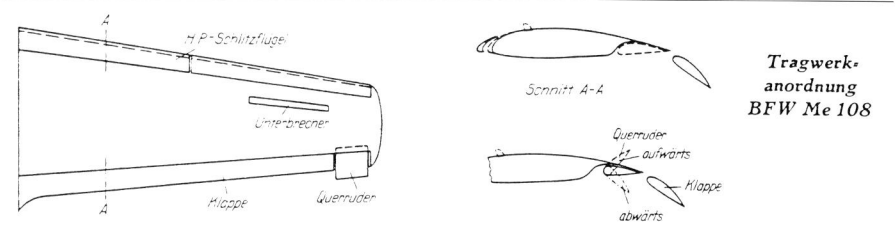

The original control surfaces of the Bf 108 competition planes with two-piece leading edge slats, spoiler, smaller aileron, and landing flap.

Schlitzflügel (nach Handley Page - Lachmann) längs Spannweite durchlaufend, unterteilt. Äußerer Teil (Schlitzendflügel) selbsttätig wirkend; innerer Teil mit Landeklappe gekuppelt, vom Führersitz aus mechanisch gesteuert. Schmales behelfsmäßiges Querruder ist mit Unterbrecherklappe auf Flügeloberseite gekuppelt derart, daß Unterbrecher nur bei Querruderausschlag aufwärts hochklappt. Unterbrecher ist auch bei geschlossenem Schlitzflügel wirksam. Die Landeklappe ist ein längs fast der ganzen Spannweite nach rückwärts herausschiebbarer Hilfsflügel (Flugelfläche in Richtung Flügeltiefe wird um 8 vH vergrößert), der gleichzeitig maximal 31° abwärts geschränkt wird (Düsenschlitz an der Hinterkante des Hauptflügels).

ignator of Bf 108 from the RLM and had to be developed, designed, and built in nine months. Final assembly of the V-1 began in April 1934, with the maiden flight taking place on June 13, 1934, in Augsburg at the hands of Carl Francke. Technical evaluations started in Warsaw on August 29.

In contrast to the previous Challenge de Tourisme Internationales in 1929, 1930, and 1932, the technical evaluations counted to a much larger degree, comprising almost 80% of the total points possible. In earlier competitions it was more the overall consistency and speed over the route which had mattered, whereas in 1934 the evaluation conditions were more suited to the design requirements for a modern touring plane. The technical aspects, assessed by an evaluation committee made up of technical experts, comprised approximately one-third of the total number of points possible. For example, the committee looked for visibility from the pilot's and passenger's seats, the readability and positioning of the gauges, adjustability of the seats, safety features, means of adjusting load distribution, and so forth. Specific categories also included minimum speeds, takeoff and landing runs before and after an 8 meter barrier, respectively, fuel consumption over 600 km, startup time, and setup and teardown to fit through a gate 4.5 m wide and 3.50 m high. During the race, speeds over 210 km/h did not earn additional points, but in the high-speed category every kilometer per hour over 210 km/h did earn one point. There were other aspects in the evaluation table, such as if the wing folded through one axis 12 points were awarded, through two axes earned 6 points, regardless of the complexity of the mechanism. The airplane designers were forced to tailor their designs to the point structure and come up with solutions which would ensure the maximum number of points in the overall evaluation.

The Solution: Monocoque Construction and Technical Characteristics of the Bf 108

In creating the Bf 108 as the "ideal aircraft," Willy Messerschmitt and his rapidly growing team (with Robert Lusser as the project director and Hubert Bauer as the prototype construction manager) extended the limits of new design principles to the point where the type not only served as an example for sportplane construction, but for profitable mainstream aircraft production, as well.

The single-spar all-metal stressed-skin wing—fitted with leading edge slats and large Fowler flaps—could easily be folded up against the fuselage after loosening the connector bolts in the center section. In accordance with manufacturing procedures, the oval fuselage behind the cabin comprised a right and left fuselage half, which in turn consisted of sequentially layered monocoque sections with one edge flanged to the inside. The unusually light and stiff shell was then further reinforced by riveted stringers. The use of flush rivets resulted in an extremely fine aerodynamic quality to the surface of the plane. The upper portion of the fully glazed canopy could be popped off in an emergency. The landing gear was fully retractable, with each leg consisting of a cantilever oleo shock strut. Retraction took place mechanically by hand using a simple worm gear, with the wheels folding up into the wells in the wings. Audio and visual signals indicated when the landing gear had not extended prior to touchdown when the engine was

Introduction of the landing flap meant a significantly shorter landing. This new technology was only reluctantly accepted by the pilots at the time. The Messerschmitt company therefore issued special leaflets showing the proper landing techniques and the advantages of using the flap system.

The Bf 108's four-seat cabin provided the same type of comforts found in any luxury automobile. The aircraft had dual controls. The lever between the seats is the retraction ratchet for the landing gear.

throttled back. The wheels were fitted with pneumatic brakes and were activated via a lever on the control stick. The tailwheel did not retract. The Bf 108 was the third German airplane with retractable undercarriage, after the He 70 and Ju 60.

Since the competition rules stipulated that minimum and maximum speeds would play a major role in point tallies, Willy Messerschmitt and his developmental team took great pains to ensure that the plane had both good slow-flight handling as well as a high cruising speed. To this end, the group developed an unusual wing design using an efficient high-lift system. In order to use the Handley-Page leading edge slats on the Bf 108, Messerschmitt traded the license in exchange for his patent rights in England for the single-spar construction method.

During the competition the Bf 108 attained the speed range of 4.64, i.e. the minimum speed registered at 62.7 km/h and the maximum speed at 291 km/h. The rather extensive lifting aids, consisting of flaperons, spoilers, and

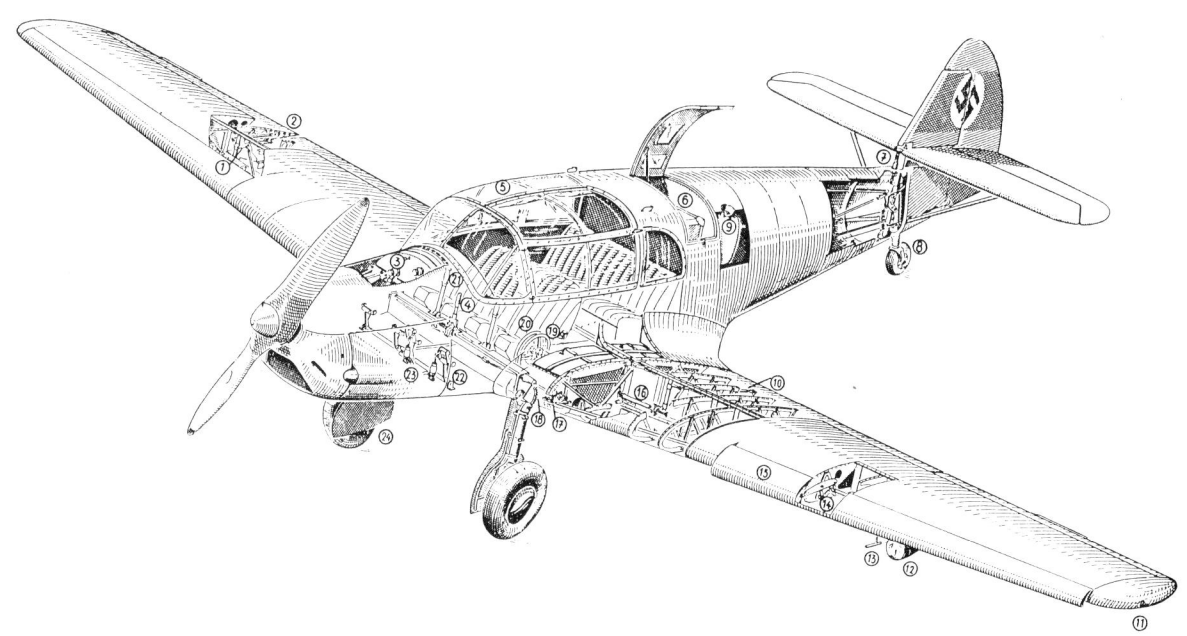

1. Steuersäule für Quersteuerung
2. Inneres Querruderlager mit Querruderantrieb
3. Schmierstoffbehälter
4. Einziehratsche für Fahrwerk
5. Kabine (Plexiglas)
6. Verschließbarer Gepäckraum
7. Höhenflossenverstellspindel
8. Luftbereiftes Spornrad mit selbsttätiger Geradführung
9. Einfüllstutzen für Tank
10. Stoßstange für Klappenantrieb
11. Kennlicht links
12. Landescheinwerfer
13. Staurohr für Fahrtmesser
14. Getriebe des selbsttätigen Vorflügels
15. Durch Luftkraft gesteuerter Vorflügel
16. Verstellspindel für Landeklappe links
17. Verbindungshebel mit Druckrollen für Querruderantrieb
18. Zahnsegment für Fahrwerkseinziehung
19. Handrad für Klappenverstellung
20. Handrad für Höhenflossenverstellung
21. Doppelsteuer
22. Seitensteuerpedal
23. Bremspumpen für Laufräder
24. Laufrad mit Abdeckung

Parts listing for the Bf 108

dual slats were not all incorporated into the Bf 108B's design.

The first Bf 108A V-1 prototype, Werk-Nr. 695, D-IBUM, was the only one to have a wooden wing and was followed by five more all-metal Bf 108As, with the last one taking to the air for the first time on 28 July. With only one exception, all Bf 108A planes made use of the Hirth HM 8U engine with reduction gear (3:2) and a three-bladed wooden airscrew having a diameter of 2.20 m. Only the V-4 had the somewhat less powerful Argus As 17A. In late July the V-1 crashed during slow-flight testing when it struck a tree branch, killing RLM pilot *Freiherr* Wolf von Dungern. It was only through great effort that all Bf 108s were prevented from being grounded.

The Bf 108 in the 1934 Challenge de Tourisme Internationale

Four of the five Bf 108s registered for the competition actually participated:

No.	Type	Werk-Nr.	Registration	Engine	Pilot
12	Bf 108A V-3	697	D-IZAN	HM 8U	Otto Brindlinger
15	Bf 108A V-4	698	D-IGAK	As 17A	Carl Francke
14	Bf 108A V-5	699	D-IMUT	HM 8U	Theo Osterkamp
16	Bf 108A V-6	700	D-IJES	HM 8U	Werner Junck

Elly Beinhorn following her spectacular one-day flight from Gleiwitz to Istanbul and back to Berlin, a distance of 3,570 km which she covered in 13 1/2 hours in Bf 108 D-IJES on 13 August 1935. This was the first Bf 108 to bear the name "Taifun" (Typhoon), christened by Elly Beinhorn, and this name was later adopted for the entire production series.

Theo Osterkamp came in fifth place in the overall competition, with Werner Junck in sixth place and Carl Francke in sixteenth place. Otto Brindlinger was forced to drop out when he strayed off course. The Bf 108 gave quite a good showing in the technical portion of the competition. The fastest plane was D-IMUT, piloted by Theo Osterkamp, which registered the highest speed of 291 km/h over a 300 km stretch. During the long-distance portion of the competition, which covered some 9,530 km, speeds of more than 210 km/h were not rewarded with extra points. The Bf 108 pilots were also disadvantaged by not having enough time for proper training due to the fact that the competition planes were completed at such a late stage.

An unusually high amount of favorable interest in the Bf 108 was shown in trade publications between 1934 and 1936, particularly with regard both to its design construction as well as its aerodynamic shape and the excellent flight performance spawned by both.

How the Bf 108 became the "Taifun"

The Taifun (Typhoon) moniker stemmed from aviatrix Elly Beinhorn, who described the "christening" in her book "Alleinflug"[1]:

"There she stood, the much talked-about Me 108, tucked away in a far corner of the *Bayerische Flugzeugwerke's* hangar, all covered in dust and looking quite forlorn. It was love at first sight between her and me. And this love stood fast until the day my last 'Me' (I'd gone through seven or eight, each being the latest model) was drafted for the war effort. I made a few circuits in the Messerschmitt, and my love grew even stronger. What were the pilots complaining about this fantastic machine for? To be fair, the planes used in the Challenge de Tourisme had special ailerons, which may not have been everyone's cup of tea. Nevertheless, any pilot must have noticed what a racy machine he had in his hands.

Without a moment's hesitation I headed in her direction and asked to borrow the Me 108 for awhile. My request was granted. It's important to remember that the plane's 300 km/h maximum speed made it the fastest in its class in the world at the time. Its highly refined aerodynamic qualities gave it the ability to carry enormous amounts of fuel without being detrimental to its takeoff and landing characteristics.

For an entire day I puzzled over my giant atlas, trying to plot the maximum distance I could cover in a single day safely—insofar as this word had meaning in the world of sport aviation. I considered the whole adventure more than

---

[1] Elly Beinhorn, Alleinflug. Mein Leben. Munich 1977, pp 210-211.

just propaganda for the machine, but the thought was in the back of my mind that I could use this success to garner financial support for my future flights by demonstrating the Me 108. But this was no title for an airplane which was to become a world-beater! After thinking about it for awhile I named my 'Me' 'Taifun.'

This was an appellation which would be understood in English- and French-speaking countries as well, and conjured up images of speed and thunderous, roaring winds. The name was later applied to the whole series."

Move to the B-series

The Bf 108 was so persuasive from a technical and performance standpoint that the RLM ordered a production batch for the *Luftwaffe* to be used as a liaison plane and trainer. After a thorough reworking of the design, the Bf 108B went into production in 1936, rolling off the assembly lines in the fall of that year. Production continued at Augsburg until early 1938 before being transferred to the Regensburg plant.

The main differences between the B-version and the A-model used in the Challenge de Tourisme were not only to be found in the powerplant (Argus As 10C vice Hirth HM 8U) and the potential for variable pitch props (Messerschmitt Me P7), but also in the modified wings using standard ailerons, simple automatic slats extending along two-thirds of the leading edge, and an increase in span from 10.31 m to 10.62 m. The empty equipped weight for the four-seater rose to 860 kg (A-model: 560 kg), and the maximum takeoff weight was now 1,400 kg (A-model: 1,050 kg). The larger 220 l capacity of the fuel tank (A-model: 150 l) gave it a range of 1,000 km. The speed range was 3.53, with a 300 km/h maximum speed and an 85 km/h landing speed. The interchangeable variable-pitch propeller reduced the takeoff run from 270 m to 195 m and cut back on the time to climb to 3,000 meters from 16.7 minutes to 12.6 minutes.

As the degree of awareness of the Bf 108 grew, so did the number of export contracts, including 21 planes for Japan in 1938 and the Manchurian Airlines operating under Japanese sovereignty, which received 15 to be used as a small commuter airliner for three passengers. For military purposes as trainers and liaison planes Spain received 22,

During his visit to Messerschmitt AG Augsburg in October 1938, Charles Lindbergh flew Bf 108B-1, D-IBFW, and gave the type high marks for its outstanding handling characteristics. He is seen here with Dr. Hermann Wurster (left), who is familiarizing Lindbergh with the controls prior to the flight.

The Taifun was one of the most successfully exported German aircraft. From 1936 to 1940 over 100 of the type were delivered to various countries. Japan received 21, of which 15 machines were employed as small commercial passenger planes with the Manchurian Airlines. This advertisement from 1938 shows the routes flown in Manchuria.

Yugoslavia 12, Bulgaria and Hungary each 6, and Switzerland 13, the latter as trainers for its Bf 109Ds delivered in 1938.

One of the Bf 108B-1s refitted with an Argus variable-pitch propeller, operated by the Luftwaffe during the war.

A production Bf 108B-1 at the Messerschmitt GmbH factory airfield in Regensburg-Prüfening during the winter of 1939/40. Hundreds of this Bf 108 variant were delivered to the Luftwaffe, where they served in the courier and liaison roles.

However, Messerschmitt's main customer remained the *Luftwaffe*. Even during the years prior to the war the RLM only authorized a limited number of planes for civilian and export purposes.

During the war there were hardly any machines for civilian use. Even those planes delivered to private customers or companies in Germany were later requisitioned for the *Luftwaffe*.

The *Luftwaffe* employed the type right up to the end of the Second World War, primarily in the role of a courier plane, commuter plane for commanders, headquarters, flight readiness and aviation officials, and as a trainer. During this period the Taifun was a kind of status symbol.

The first "operational" use of the 108 came in 1937, when a handful of machines served the Condor Legion in Spain in the liaison role.

Enthusiasm for the Taifun

Elly Beinhorn was given D-IMXA from the first batch of Taifuns delivered in 1936. It was in this plane that she carried out many spectacular long distance flights, some alone and some with her husband Bernd Rosemeyer. In her book "Alleinflug" she wrote:

"We both were enamored of this plane that we could fly with our little finger. During our flights we never ceased to be thrilled by our plane's high cruising speed, its well-behaved and relatively low landing speeds, and last but not least its low speed handling in bad weather or when flying over a train station to get our exact bearings. During those days such measures were commonplace, and we often did it—for of course we had no radio equipment on board."

On 19 October 1938 the American pilot Charles Lindbergh had the opportunity to make two flights in the Me 108 while a guest of Willy Messerschmitt during a visit to the Augsburg factory. Lindbergh recorded his impressions of the flights in his journal:[2]

"The 108 is by far the best plane in its class that I've ever flown. It has excellent control handling—stick and rudder are quite light..."

The British magazine "Aeroplane" from 28 June 1939 wrote:

"Structurally the Taifun is interesting because it is about the first fully stressed-skin all-metal aeroplane of its size yet to be built. The striking thing about it is its marvellous finish with flush riveting everywhere and a very fine cellulose covering."

And "Flight" from 29 June 1939:

"But the Taifun's really outstanding qualities are those of control and stability. Certainly no other civil machine...has aileron control which is so absolutely positive and light ..."

Nearly 50 years later, in the fall of 1988, well-known test pilot Dieter Schmitt had the opportunity to fly and assess a Messerschmitt Bf 108D-1 Taifun built in 1943 (a MBB "traditions plane"). The following paragraphs are excerpts from his report published in the 2 January 1989 edition of "Aerokurier":[3]

"All in all, the cockpit layout was surprisingly modern even by today's standards, even though landing gear and flaps had to be manually activated.

[2] Charles A. Lindbergh, Kriegstagebuch 1938-1945, Vienna-Munich 1976, p. 71
[3] Dieter Schmitt, pilot-report: Messerschmitt Me 108D-1 Taifun. In: Aerokurier no. 2/1989, pp. 34-36

106

The closer one looks at the Me 108, the more pressing becomes the question: has there really been anything new since then in aircraft construction in this class? Of course much has changed since that time. But what ultimately matters is a design's performance and how a pilot feels when he's at the controls of a plane. In both cases the Me 108 receives the highest marks. Today the designers from 1933/34 can proudly claim to have created an aircraft back then which can be considered a role model for an entire generation of touring planes.

They convincingly succeeded in melding the Me 108's flight and control handling, unique even today, with excellent construction and application qualities into a harmonious synthesis of design and function. And just as remarkable is the fact that this plane's performance and character were not restricted to a single prototype or a handful of aircraft, but applied to the entire spectrum of production."

Bf 108C record-setting variant

The Bf 108 design was originally considered for a number of indigenous and foreign-built powerplants available at the time. Aside from the few examples using HM 8U, As 17, and Sh 14 engines, the As 10C remained the standard motor for the Bf 108. But at least two planes were fitted with the somewhat more powerful HM 508C (200 kW/270 hp) by Hirth at Stuttgart-Zuffenhausen. A few modifications to the airframe necessitated by the engine resulted in this variant being given the new designation Bf 108C. Chief test pilot for Hirth Motorenwerke, Hermann Illg, used one of these machines on 8 July 1939 to climb to an altitude of 9,075 m and thereby set a new Class C record.

On 7 July 1939 chief test pilot of the Hirth-Motoren GmbH, Hermann Illg, flew Werknr. 1078, coded D-IAXC and powered by a Hirth HM-508-C engine (270 hp), to an altitude of 9,075 m and set a new record for category 1 light aircraft. Only two of these Bf 108C types were built.

Right: Three view of the Bf 108D-1, built for the Luftwaffe in Regensburg from the summer of 1941 onward and later at SNCAN in France.

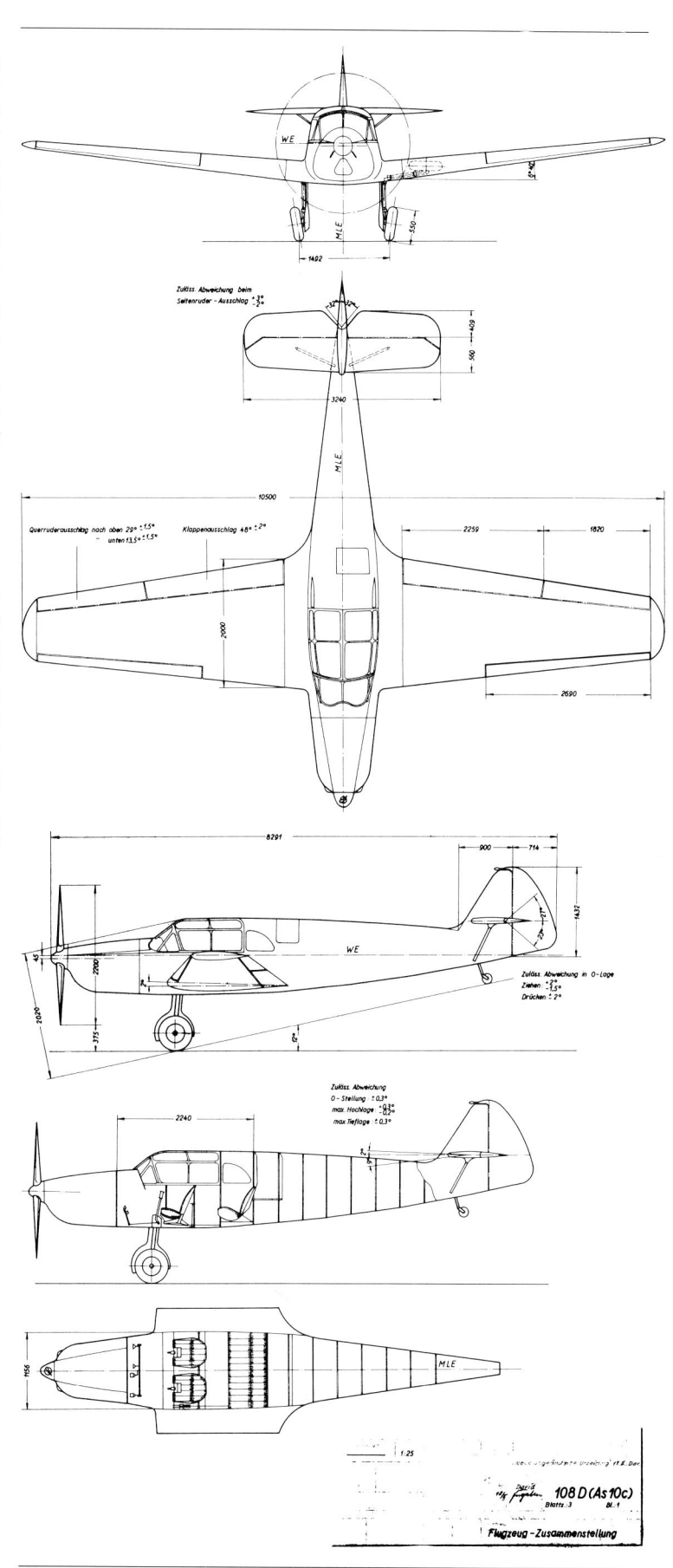

D-series for the *Luftwaffe*

In 1940/41 the Bf 108 design was reworked yet again for its main customer, the *Luftwaffe*. Some of the features of the new variant included a rudder using a horn balance, as well as a 24 volt electrical system and a production variable-pitch propeller—either the Messerschmitt P7 or an Argus design which made use of a rotating cowling with angled ribs ahead of the propeller hub that provided adjustments in flight.

The 108 was tropicalized for operations in North Africa, incorporating special cockpit ventilation, air filters with sand scrubbers, increased capacity oil reservoir, supplemental fuel tanks, and a more powerful generator. These modifications reduced the 108's capacity to just three persons.

Production of the D-series began in the fall of 1941 at Messerschmitt's Regensburg works. When Bf 109 and Me 210 production tempo increased, 108 production shifted to France. Production resumed in February 1942 once all assembly jigs, materials, and production blueprints had been transferred by rail to Messerschmitt's daughter company at Les Muraux near Paris, the Société Nationale de Construction Aéronautique (SNCAN). The RLM initially contracted for 234 aircraft of this type. The vast majority of the Argus As 10C engines were also built by a non-German firm, this being the Walter company in Czechoslovakia.

Me 208 - Bf 108 nosewheel variant

At the instigation of the RLM project studies were undertaken at Augsburg in 1941 for a follow-on development of the Taifun with a retractable nose gear. At the same time range would be increased to 1,300 km through the installation of larger fuel tanks (260 l), and a larger luggage compartment would provide more storage room. A new side opening door would make cabin access easier, as well. Wingspan rose from 10.62 m to 11.50 m, length from 8.29 m to 8.85 m, and takeoff weight to 1,580 kg. The powerplant was to alternatively be the Argus As 10C or the Hirth HM 508, both easily interchangeable. The initiator of this tricycle version was Gerhard Caroli, supervisor of Messerschmitt's flight testing program. Preliminary tests were carried out on a Bf 108 suitably modified by SNCAN at Les Mureaux. An extract from the flight report on 15 October 1942 reads as follows:[4]

"Takeoff with control stick hands-off was easily accomplished...as were three-point landings at ca_{max}...

The nosewheel was the tailwheel from a Potez 63. It is flutter-free without the need for a friction brake and is drawn in using a simple mechanical retraction method. In the event that flutter were to set in while taxiing on concrete surfaces, it could easily be eliminated by a friction brake.

[4] Messerschmitt AG Augsburg, flight report no. 838/11, Me 108 with nosewheel, 10/15/42

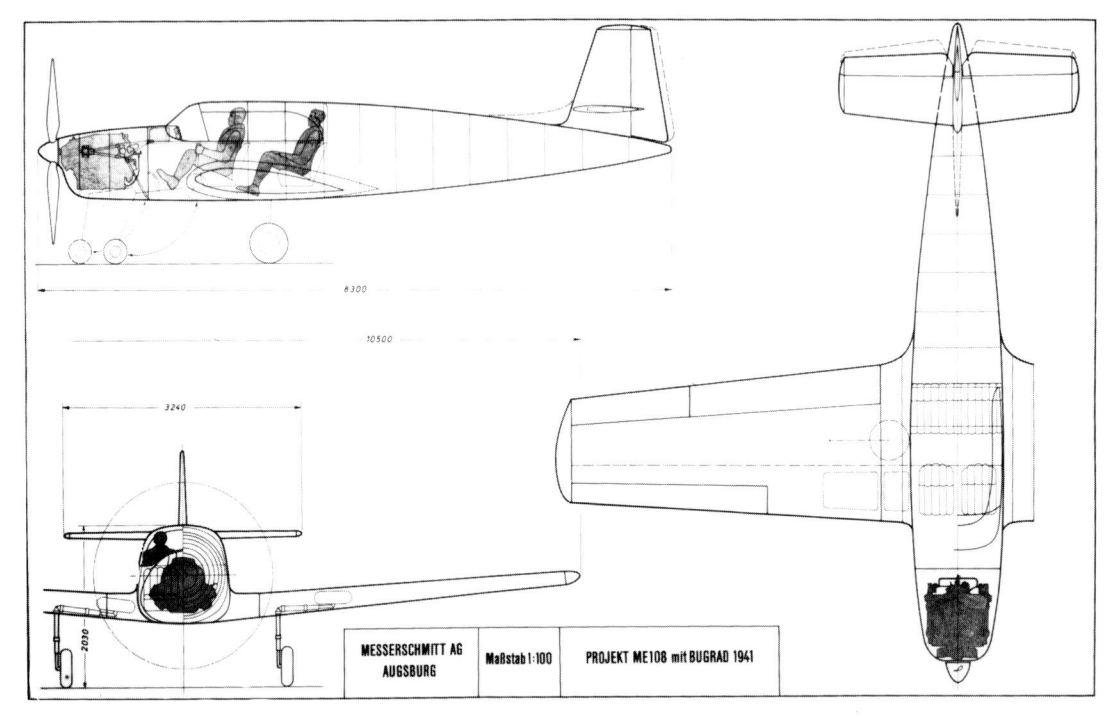

The first Me 208 built in 1944/45 under French supervision at SNCAN (later Nord Aviation) in Les Mureaux. It was produced for the French air force with a Renault engine as the Nord 1101 Noralpha. 200 of the type were built. The Nord 1101 design dispensed with the leading edge slats.

For ease of construction it was decided to dispense with the leading edge slats and replace them with fixed slats in the outer wings. To this end the leading edge slats on the test plane were reduced to 1.20 m span. As a result, however, wing stall handling was shown to have markedly worsened. The landing speed also seemed to be somewhat higher, since apparently the airflow in the center part of the wing not covered by the slats separates too early. More precise research will be undertaken at Les Mureaux with the clear-cut objective of ensuring that the 108's virtually idiot-proof stall handling is preserved."

Following completion of the project work in 1941, development of the Me 208 was handed over to SNCAN in France. Construction of the five prototypes contracted for by the RLM, of which the first was to have flown in the summer of 1942, suffered serious delays. Maiden flight of the Me 208V-1 didn't occur until July 1943. American bombing raids crippled the factory at Les Mureaux so badly that there seemed to be little sense in continuing work on the project. The termination of the Me 208 announced that same year corresponded to the *Luftwaffe*'s aircraft commonality program just going into effect.

Opposite Page: A 1941 project drawing of an enlarged Bf 108 with nose wheel. The type was developed in France as the Me 208 and flew there for the first time in July of 1943.

Continued production of Taifun in France

Shortly after the German troops pulled out of Paris in August 1944 French officials made a concerted effort to restart those aircraft and aero-engine factories which had formerly been under German administration with the goal of rapidly equipping the Arméee de l'Air with trainers and liasion/courier planes. As a result, three German types saw continued production virtually unchanged, including the Messerschmitt Bf 108D as the Nord 1001 Pinguin with a Renault 6Q engine. There followed other versions with the Potez 6D engine.

Even the tricycle-gear Me 208 prototypes begun in 1943 were finished as far as was possible and modified to accept the six-cylinder inline SNECMA-Renault 6Q powerplant with an output of 176 kW/240 hp. Over 200 of these machines were built for the French air force as the Noralpha and Ramier. A single Nord 1110 acted as a flying testbed in 1959 for the Turboméca Astazou piston engine.

The Messerschmitt Taifun also served as the impetus behind an all-French design for a small four-seat touring plane, the Nord 1201 Norécrin, powered by a 108 kW/147 hp SNECMA-Regnier engine, which saw large-scale production orders.

Prof. Willy Messerschmitt was given this Bf 108B-1, Werknr. 1545, in October 1937 with his own unique registration coding of D-IMTT

Willy Messerschmitt and his Taifun

Due to his relationship to Ernst Udet, Kurt Schnittke (born in 1920) got his pilot's wings early on, and between 1937 and 1939 was apprenticed and trained as an aircraft construction engineer at the Augsburg works. He related the following anecdote.

When Prof. Messerschmitt operated his Me 108 D-IMTT he always flew with another pilot—even though Messerschmitt had a license himself. One day, when Messerschmitt wanted to fly his plane—this time from Augsburg to Nuremberg —Kurt Schnittke got the machine ready and charted out the route. As Messerschmitt climbed in, he told the young man: "Schnittke, today I'm not going to be the co-pilot, I'd like to do everything myself." Schnittke obliged him, and Messerschmitt took off. But he neglected to retract the landing gear, an operation that had to be done by hand. When they landed in Nuremberg, Messerschmitt pulled his small slide rule from his jacket pocket, made a few calculations and said: "Schnittke, we spent way too much time in the air for this journey. I thought my 108 was faster than that." To which Schnittke replied: "Indeed it is, but you forgot to retract the landing gear— you wanted to do everything yourself." The absent-minded professor was compelled to laugh.

Prof. Willy Messerschmitt often flew his Bf 108 himself during the peaceful interwar years, although he was always accompanied by another pilot (such as Fritz Wendel seen here). Even this machine, however, had to be given up to the Luftwaffe during the latter part of the war.

Taifun accomplishments

Aug/Sep 1934	1934 Challenge de Tourisme International, 5th, 6th, and 16th place Fastest plane in competition
Aug 1935	Elly Beinhorn Single-day flight, Gleiwitz-Istanbul-Berlin, 3,570 km in 13.5 hrs.
Aug 1936	Freiherr Speck von Sternburg Victor of Olympia Star Race to Berlin
Aug 1936	Elly Rosemeyer-Beinhorn Three continents in a single day, Damascus-Cairo-Athens-Berlin, 3,450 km
Dec 1936/ Jan 1937	Africa Flight Berlin-Cape Town-Windhoek-Berlin, 25,550 km
Feb 1937	Otto Robert Thomsen Second place in International Oasis Race
Feb 1937	Wolfgang von Gronau Fourth place in International Oasis Race and subsequent Africa Race, 30835 km
Mar/May 1937	Theo Blaich Africa Flight to Cameroon, 26,000 km
May 1937	Hans Seidemann Victor in Isle of Man Race
Aug 1937	Otto Brindlinger European Circuit, 6,500 km in 32 hrs.
Jul 1938	Otto robert Thomsen Victor in Queen Astrid Race, Belgium
Jul 1938	Raduno del Littoria (Italian Circuit) Second thru sixth place
Aug 1938	Hannes Gentzen Victor in Star Flight to Dinard
Jan/Aug 1938	Otto Brindlinger, Inge Stölting, Horst v. Salomon South, Central, and North American Flight, 40,000 km
Apr/Jun 1939	Elly Rosemeyer-Beinhorn Greater India Flight, 34,000 km
Jul 1939	Raduno del Littoria (Italian Circuit) Overall winner was Christian Dietrich, who also took third and fourth place
Jul 1939	Hermann Illg Class C altitude record at 9,075 m

Bf 108 Taifun variants

Bf 108A	V-1 thru V-6 Competition touring plane, developed for the 1934 Challenge de Tourisme Internationale Engines: Hirth HM 8U, 175 kW/240 hp, or Argus As 17A, 165 kW/225 hp
Bf 108B-0	B-1 pre-production series Engines: Argus As 10C, 176 kW/240 hp, Hirth HM 8U, 176 kW/240 hp, or Siemens Sh 14A, 110 kW/150 hp
Bf 108B-1	Main Taifun series Motor: Argus As 10C, 176 kW. 240 hp
Bf 108B-1s	B-series aircraft retro-fitted with variable-pitch propeller (Me P7 or Argus)
Bf 108C	Only two built, altitude record for Class C Engine: Hirth HM 508, 200 kW/270 hp
Bf 108D-1	*Luftwaffe* production model with series-built variable-pitch propeller, 24 volt electrical system, modified rudder, some tropicalized
Me 108F	Project from 1975 for a follow-on six-seat version
Me 208	Follow-on development with nose gear, only two completed in France in 1943/44 Engine: Argus As 10P, 176 kW/240 hp
Nord 1000	same as Bf 108D-1
Nord 1001	Pinguin I, engine: Renault 6Q 10
Nord 1002	Pinguin II, engine: Renault 6Q 11
Nord 1003	Engine: Potez 6D
Nord 1004	Engine: Potez 6D
Nord 1100	same as Me 208V-1
Nord 1101	Noralpha, engine: Renault 6Q 10
Nord 1104	Engine: Potez 6D
Nord 1110	with Turboméca Astazou (first flight 1959)

Me 108 - Postwar Projects in Germany

In the late 1940s, Messerschmitt was able to reestablish his business operations in Munich, and it wasn't long before queries about the unforgettable Taifun began pouring in from abroad—especially from Scandinavia, and South and North America. These letters were requests for replacement parts, design details, for old and new Me 108s, and for production licenses.

But since all documentation pertaining to Messerschmitt planes up to 1945 had been confiscated by the Allies, Messerschmitt's technical department had nothing in hand to deal with these requests.

However, since interest in the Taifun appeared to be world-wide, in mid-1951 Willy Messerschmitt, Wolfgang Degel, and Hans Hornung began thinking about a new version of the Me 108 with a view to the day when Germany would regain its air sovereignty. Those involved in the project understood that the new 108 would have to be modernized. Willy Messerschmitt soon began designing the aircraft (designated Me 108W, the "W" being for *Weiterentwicklung*, or follow-on development) from scratch in his own unique style. He generated sheet after sheet with dozens of his sketches and calculations. Hans Hornung took care of the aerodynamic data, calculated the performance, and gave the project the number P 1118. This indicated that it was one of Messerschmitt's first postwar aviation projects, continuing from the old project number list which had ended in 1945 with the project 1112.

This work proceeded at a leisurely pace until 1955. In the eventuality of an issuance of a production license for the new 108, Wolfgang Degel summarized the concept in mid-1954 as follows:[5]

"The Me 108, modernized and brought up to today's standards may appear to the casual observer to be generally similar to the prewar Me 108B model.

In addition to a nosewheel, this follow-on development utilizes a somewhat modified wing outline which obviates the need for leading edge slats and a redesigned form to the fuselage center and aft sections brought about for ease of manufacturing. Priority has been given to an oval cross section in this area. Use of a butterfly tail has also been considered as an option.

In order to meet range requirements, internal fuel tanks have been complemented by wingtip mounted tanks, thereby increasing the range to a maximum of 1,500 km.

The equipment, particularly the radio system, are of necessity up to today's standards. The same applies to sound insulation, landing gear operation, starter system and other similar areas.

The new design also must include an enlarged baggage area.

A trainer aircraft built with the assistance of Prof. Messerschmitt in Spain was laid down from the outset in such a manner that the experience gained, along with entire sub-assemblies, can be applied to this Me 108 development."

Fundamentally, this was basically the same concept behind the Me 208 developed in France during the war and now being license built for the Armée de l'Air as the Nord 1101. An agreement at the time between Messerschmitt and Nord Aviation for using this plane for the Germans might

---

[5] Messerschmitt Technical Dept., Munich. Consideration of License Issuance for Me 108 (draft, undated, around May 1954)

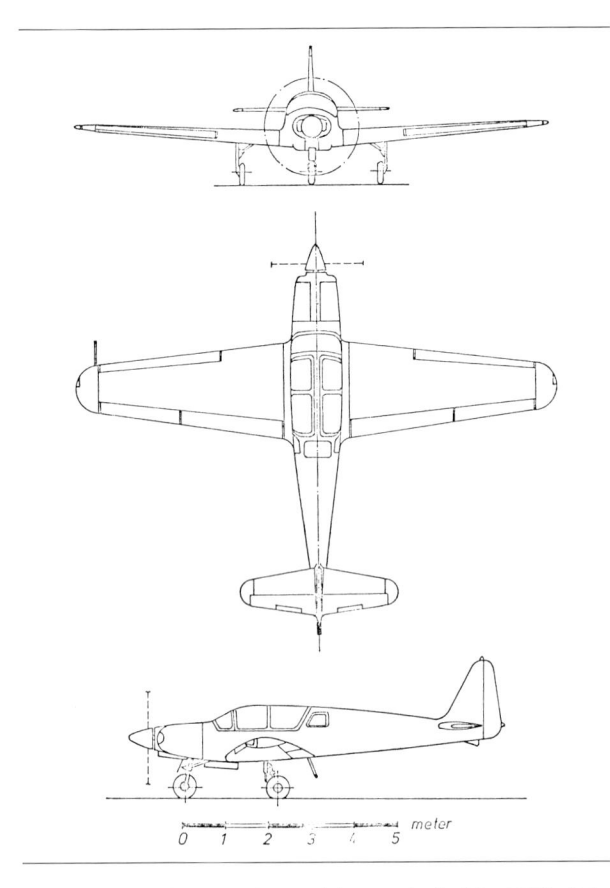

Project study of a new six-seat Messerschmitt Me 108F of the "Taifun-Flugzeugbau GmbH" from 1975. With a 300 hp Lycoming IO-540 engine, the plane was expected to reach a speed of 315 km/h. Empty weight was 822 kg, with the takeoff weight being 1,550 kg. Wingspan was 11 m and length 8.60 m.

possibly have led to the first Franco-German joint venture program in the aviation industry. But the early 1950s political atmosphere and later negotiations for license-building the Magister made the topic so sensitive that it was never even brought up.

Interest in the new 108 rapidly tapered off, however, when the technical department at Messerschmitt estimated that construction and testing of a prototype would cost DM650,000. This figure did not even take into account production spin-up costs and production blueprints, and only a rough estimate of the operations and maintenance network was provided. There was simply not yet enough money for such investments. In addition, by this time the company was heavily involved in preparations for the first license-built planes of the Bundeswehr.

In the meantime, the American competition had become so well established in the sport- and touring plane market that it became virtually impossible for Germany to compete effectively.

Yet the venerable Taifun came up repeatedly in conversation. Oldtimer fans throughout the world lovingly guarded the few remaining examples. Even the Messerschmitt AG acquired an old version with the registration D-EFFI. In 1966 Willy Messerschmitt invited Elly Beinhorn to fly this plane and visit him in Estepona. She subsequently wrote an enthusiastic description:[6]

"...after thirty years my reunion with the Me 108, lovingly christened 'Miss Evergreen' by the press, was not a disappointing one. At the time, my opinion that this plane was ten years ahead of its time was something of an understatement. Where can you find today a sportplane that can make circles at 300 kilometers per hour as easily and safely as at just over 100 km/h?"

This was most likely not the only song of praise which prompted Mano Ziegler, at the time press secretary for the Messerschmitt AG and also a pilot, to undertake another initiative for developing a new version of the Taifun. Yet again, several years passed until the establishment of the Taifun Flugzeugbau GmbH in December of 1973. It was the intention of this company to build a modernized, six-seat Me 108F Taifun at a facility to be built in Weiden, Oberpfaffenhofen.

Numbered among the enthusiastic supporters of this initiative were Willy Messerschmitt, Kurt Tank, Ludwig Bölkow, and Gero Madelung. Reinhold Ficht was to have been the managing director, who brought with him plans for the Me 108F which had been reworked by Willy Messerschmitt. However, a lack of a suitable site for the factory and a shaky financial situation led to the financial supporters distancing themselves from the project even before actual work began, and in 1975/76 the company ceased to exist.

Thus, the Taifun lives on only in those examples which have survived from the few produced in Germany and France, whose owners not only tenderly care for them, but also occasionally take to the air in Messerschmitt's speedy touring plane.

---

[6] Alte Liebe rostet nicht - Elly Beinhorn flog zum Professor nach Spanien. In: Messerschmitt - Die Zeitschrit der Messerschmitt-Gruppe, vol. 1, January 1967, pp. 10-11.

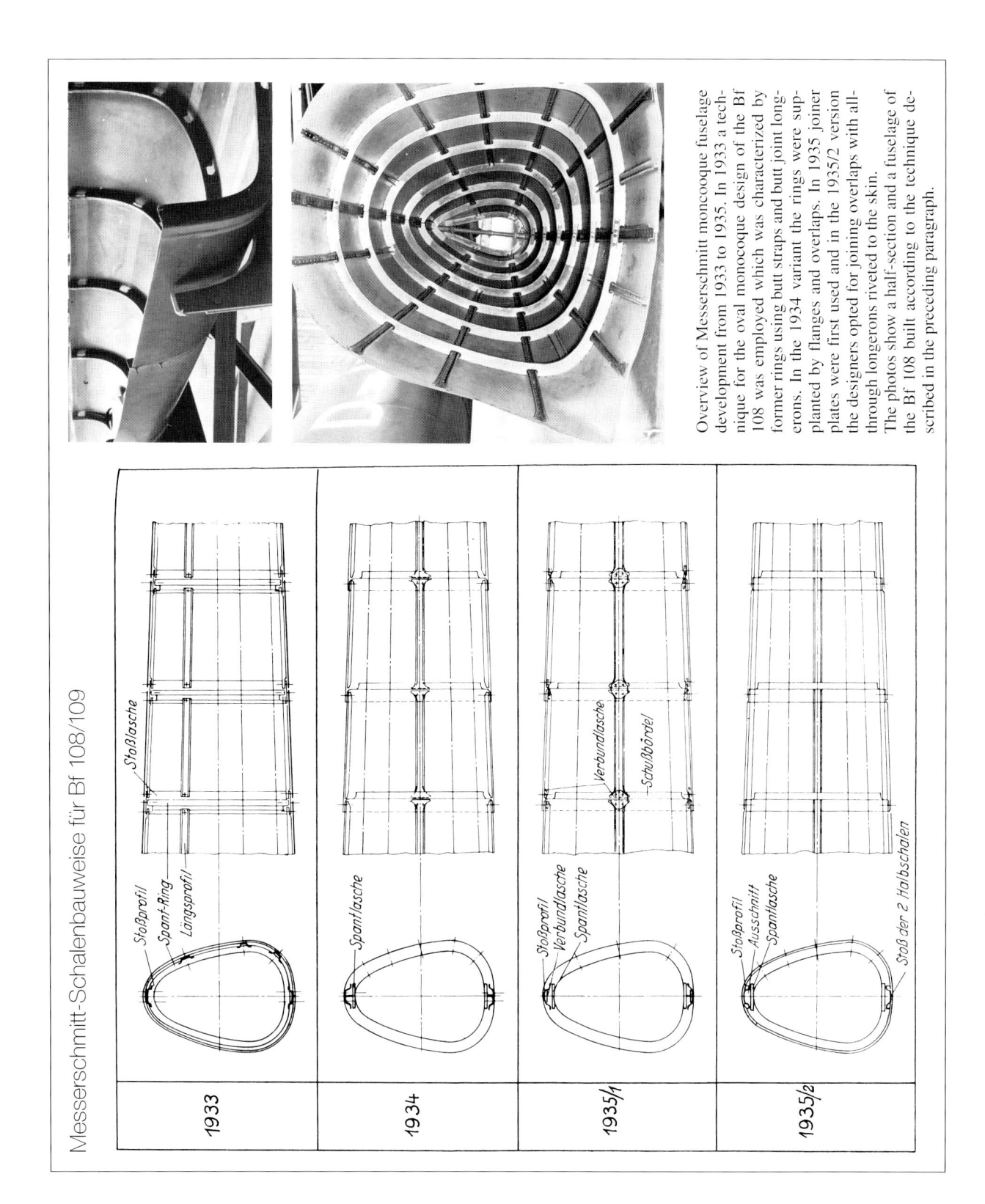

Overview of Messerschmitt moncooque fuselage development from 1933 to 1935. In 1933 a technique for the oval monocoque design of the Bf 108 was employed which was characterized by former rings using butt straps and butt joint longerons. In the 1934 variant the rings were supplanted by flanges and overlaps. In 1935 joiner plates were first used and in the 1935/2 version the designers opted for joining overlaps with all-through longerons riveted to the skin.
The photos show a half-section and a fuselage of the Bf 108 built according to the technique described in the preceding paragraph.

Messerschmitt-Schalenbauweise für Bf 108/109

| 1933 | 1934 | 1935/1 | 1935/2 |

Bf 109 - Messerschmitt's Entry into Military Aviation

The foundations for a new Germain air arm were laid by the *Reichswehr* during the years of prohibition in the '20s and early '30s. At the *Reichswehr's* prompting, various aeronautical firms drew up, built, and flew "disguised" prototypes of fighters, bombers, reconnaissance aircraft, and trainers. Several production contracts were issued subsequent to the National Socialists coming into power. Numbered among the fighters were the Arado Ar 65 and Ar 68, and the Heinkel He 51, all biplanes of modern construction, which began equipping the new *Luftwaffe* in the spring of 1935.

Up to this point, the Bayerische Flugzeugwerke (BFW) contributions to the new military aviation program had been limited to its unsuccessful M 22 reconnaissance prototype, a biplane design not of Willy Messerschmitt's personal doing. At first, the BFW was only given license-production contracts for various trainers and recce planes from Heinkel, Gothaer Waggonfabrik, and Dornier.

But by the autumn of 1933 Theo Croneiß, who in October that year had become the BFW's chairman of the board while simultaneously becoming responsible for the expansion of Bavaria's aviation industry, was able to inform Willy Messerschmitt that there were new contracts for military aircraft on the horizon. Couched in the terms "commercial airliner" and "high-speed courier plane" in Göring's letter to Croneiß on 20 October 1933 were hints at developmental contracts for a bomber and a fighter:[1]

"Dear Croneiß,

This is an extremely confidential matter, but I would like to let you know that I consider your new role of organizing the aviation industry in Bavaria to be an extremely important one. I would expect you to devote your energies without delay to the establishment of a company which hopefully will soon be able to produce a first-rate commercial airliner!!! Just as important, however, is the development of a high-speed courier plane, which doesn't need to be anything more than a single-seater. In the event that you are not yet involved in the development of such airplanes, we can discuss the matter of assigning the company license-production contracts so that when the time comes, it will already have a large work force in place to deal with the work and also work toward developing new types based on the experience gained. Ergo, I would ask you to review these matters and give me your views as soon as possible. It is fully within our interests to have a large factory established in Augsburg."

Initially, only the companies of Arado, Focke-Wulf, and Heinkel were given developmental contracts for a "light fighter." Their prototypes, the Ar 80, Fw 159, and He 112

[1] Confidential letter from Hermann Göring to Theo Croneiß dated 20 October 1933.

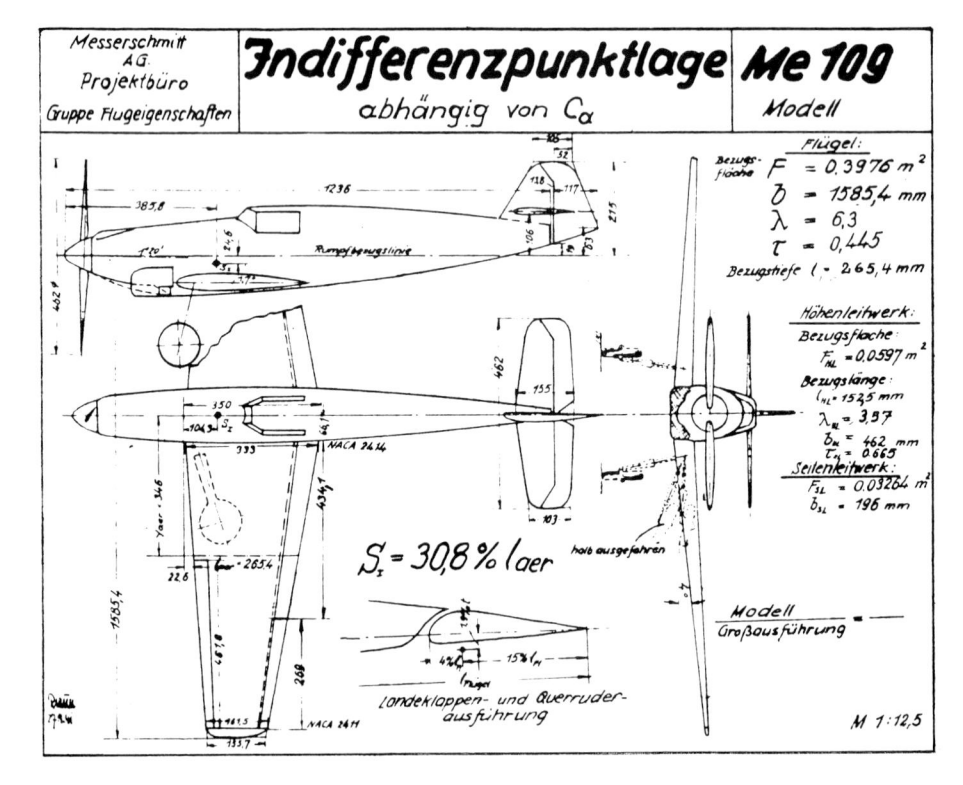

Three-view of a 1934 wind tunnel model showing the original shape of the Bf 109. Even at this early stage, the landing gear design is shown to good effect. Also clearly recognizable is the offset rudder profile, designed to counter the propeller torque and retained throughout all production variants.

made use of the Junkers Jumo 210A aircraft engine, with a rating of 450 kW/610 hp. The request for tender from the new RLM for a single-seat monoplane fighter stemmed from December 1933, with the request mandating a speed in excess of 400 km/h at an altitude of 6,000 m and good maneuverability coupled with good spin handling.

Willy Messerschmitt, who after the war spoke relatively little of his earlier designs, related the beginnings of the Bf 109 in a Spiegel interview in January 1964[2] which addressed the management of modern aircraft projects:

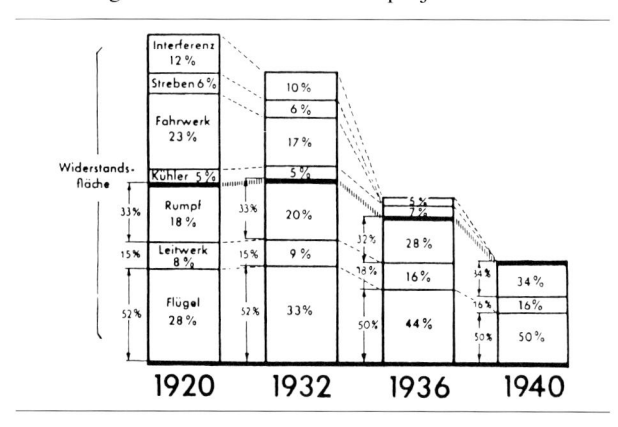

Distribution of drag for fighter aircraft from 1920 to 1940. Drag was reduced by two-thirds over this twenty year period. An amazing development in which Willy Messerschmitt played a significant role with his Bf 109/209 designs.

"Look at how I developed the Me 109. At the time, in 1933, I'd built an airplane for a Romanian airline company, and then the Chief of the *Technisches Amt*[4] sent me a letter with not altogether friendly undertones. Asking if I wasn't aware of the times, and how I could be having my staff of 25 or 30 people working on foreign projects...

I was asked to come to Berlin and was told: we've already awarded three companies the contract for developing a modern fighter, but we would be willing to give you such a contract, too. I studied the requirements in detail and went back to Berlin, where I told the gentlemen that the conditions weren't to my liking. Such a fighter would never be able to bring down a bomber embodying the latest technological advances, and these would most assuredly be appearing shortly. What would then be the point of the fighter in the first place...?

Generalstabschef Wever, a very intelligent man, saw to it that I got the contract without any preconditions attached and that I was able to build an airplane as I imagined a fighter to be...

When the 109 was just about ready to fly for the first time, my friend Udet came to me...and said: 'You're building a fighter, I hear. I'd like to take a look at it, if possible.' I showed it to him, and his face got quite a comical expression on it. A mechanic opened the canopy and he sat down inside it, the canopy closing over him. Udet looked out, opened the canopy and climbed out again. Slapping me on the back, he said 'Messerschmitt, that will never be a fighter...

...a fighter pilot must feel the speed,' he said, 'and the one wing you've got there needs another one above it with struts and wires between—then it will be a fighter'...After flight testing his mind quickly changed."

In early March 1934 project department chief Robert Lusser began discussing numerous detail issues with the RLM about the Bf 109 pursuit fighter (VJ, or *Verfolgungsjäger*), as it was now designated. These discussions focused on the powerplant to be used, the armament, and the equipment. Willy Messerschmitt also remembers that another point was the folding-wing requirement. This seems to have been an important one, and led to the landing gear being attached to the fuselage, a solution which gave the design a narrow track and later on resulted in considerable criticism.

Two engines in the 20-liter class were envisioned, both either in development or testing at the time. These were the Jumo 210A (450 kW/610 hp) and the BMW 116 (550 kW/750 hp), both of which were liquid-cooled twelve-cylinder overhead cam engines.

There were several different options when it came to armament configuration:
a) 1x MG C30 (20 mm) firing through the hollow propeller shaft
b) 2x MG 17 (7.9 mm) firing through the propeller arc
c) 2x MG 17 (7.9 mm) and 1x MG FF (20 mm) engine mounted cannon

The original requirement also called for a light bomb payload to be carried internally (behind the pilot's seat).

Like the Bf 108, whose maiden flight occurred on 13 June 1934, Rober Lusser and Willy Messerschmitt designed the Bf 109 as an all-metal type of monocoque design. The Bf 109, however, was of a much more rugged construction. Willy Messerschmitt once addressed this light construction method in 1955:[5]

"...You are aware that I've been an advocate of light construction for thirty years now. The technical competitions in the 'twenties and 'thirties gave me the opportunity to cultivate an expert light construction organization within our company. Year after year we were given responsibility for developing light aircraft, whereby light construction wasn't the means to the end, but the end itself. In 1933 our

---

[2] *Flugzeugbauer* Willy Messerschmitt, interview in Der Spiegel, No. 3 from 15 Janaury 1964, pp. 34-37.
[3] see M 36 section.
[4] *Major* Wimmer

Original Bf 109 section, the airframe used for structural testing (Werknr. 757) from 1934. Note the non-standard wing-fuselage join and the wrought iron undercarriage support mount.

efforts finally paid off. We succeeded in developing the four-seat Me 108 Taifun within the international weight limitations for lightplanes, and this even with a retractable landing gear. This was the fruit of our many years of labor in the field of light construction. One year later this focus on light construction brought us our greatest success—the Me 109 fighter. It was so small and therefore so light, and so light and therefore so small, that it beat all the other competitors. The secret of its success wasn't so much an extremely refined design, something which would have led to costly manufacturing and reduced tolerance, but rather a true application of light construction principles: minimal dimensions, minimal surface area, lightest materials, equal distribution of stress loads, and above all the combination of multiple functions within the same component. The consequential application of these principles not only resulted in an extremely lightweight design, it also led to simplification of construction. In spite of its extremely light weight, we were able to keep construction time down...with the Me 109 requiring less man-hours per kg of weight to build than any contemporary design."

In his book "Die Entwicklung der deutschen Jagdflugzeuge," Rüdiger Kosin writes:[6]

"The construction of the BFW Me 108 and the experience gained certainly played a factor in accelerating the work on the Bf 109 fighter. A considerable number of fundamental questions, such as retracting the undercarriage ahead of the wing's main spar and diverting the airflow around the (necessary) wells in the wings, fitting the control surfaces and many others had already been thought through and resolved from a design standpoint. The Me 108 can almost be considered a prototype of the Bf 109, or vice-versa: the 109 was an adaptation of the 109 concept into a fighter."

[5] Letter from Willy Messerschmitt to Prof. Dr.-Ing. Paul Brenner, Vereinigte Leichtmetallwerke GmbH, Bonn, from 3/13/55.

Design work began in August 1934 under Richard Bauer, and in December of that year assembly work started on the V-1. In the meantime, it was becoming increasingly apparent that the planned engines would not be ready in time for the first flight. The RLM, which at this time still was trying to cultivate good relations with the Royal Air Force (a Heinkel He 70 had been sold to the Rolls-Royce engine company) acquired a few Rolls-Royce Kestrel engines for use, two of which were passed on to the BFW and the remainder distributed to Arado and Heinkel.

Thus, the Bf 109V-1 was initially powered by a foreign engine, a twelve-cylinder, 510 kW/695 hp inline motor, which changed the aircraft's external appearance considerably. On 28 May 1935 (15 months after the contract was awarded) Hans-Dietrich Knoetzsch, a new test pilot at Messerschmitt, took the V-1 up on its maiden flight. Other

Engine view of the Bf 109V-1 with the Rolls-Royce Kestrel II powerplant. Cooling was in part provided by evaporative skinning on the wings. The bumps on the wings were needed on the V-1, otherwise the landing gear would not retract flush due to its (non-standard) larger tires.

[6] Rüdiger Kosin, Die Entwicklung der deutschen Jagdflugzeuge, Die deutsche Luftfahrt, vol. 4, Koblenz 1983, p. 108.

Prototype for all Bf 109s, the V-1, Werknr. 758, D-IABI, was first flown on 28 May 1935 by Hans-Dietrich Knoetzsch.

than mechanical problems with the retraction of the undercarriage, the flight went smoothly. However, Knoetzsch crashed the V-1 during the summer as it was being readied for testing at Rechlin, and the machine was out of commission for several months. Joachim von Köppen of the *Deutsche Versuchsanstalt für Luftfahrt* (DVL) and Willi Stör of the DVS subsequently assumed responsibility for managing the flight test program. On 12 December V-2, Werk-Nr. 759, D-IILU, was flown for the first time. Beginning in January, Dr. Hermann Wurster from the DVL began running the test program. He proved to be an extremely knowledgeable and involved Messerschmitt test pilot, whose skilled flight demonstration undoubtedly contributed to the RLM's decision to select the Bf 109 as the *Luftwaffe*'s new standard fighter.

A postwar debate on the decision between the Bf 109 or the He 112/113 prompted him in 1966 to write:[7]

"The Reichsluftfahrtministerium's developmental contract for a single-seat fighter (monoplane) was founded in the *Luftwaffe* High Command's idea of a single-seat fighter with superior flight performance, outstanding turn and roll capabilities, the ability to hold together in any type of dive, and flawless spin handling.

At the time, there was no such thing as a pressurized cockpit, and the airflow using the Dräger oxygen mask had to be slow and steady, meaning that at the altitudes pilots sought (above 10000 m) to gain an advantage in combat there was the constant danger of succumbing to aeroembolism.

In such situations, where these fighters operated at their service ceilings virtually on the edge of stalling, there was the serious risk of involuntarily going into a spin and dropping several thousand meters. After regaining consciousness in the denser air, it would then be possible to recover from the spin. It was this thinking which led to the stringent demands for spin safety. Fighters had to be able to be spun at least ten times to the left and right (due to the unidirectional motion of the propeller's rotation), since it would take three to five revolutions before the pilot was out of the danger zone for flat spinning, a situation from which it was impossible to recover in many cases.

During the acceptance flight at the *E-Stelle Travemünde*, in addition to the standard aerobatics I flew the Me 109 at its most tail-heavy setting through 21 revolutions to the right and 17 to the left and made a vertical dive from 7500 m to nearly ground level. At that time Dr. Jodlbauer was still insisting on the '*Männchen*' (lit. "little

Willi Stör, Ernst Udet and Willy Messerschmitt in January 1936 next to the Bf 109V-2. Udet flew the Bf 109 for the first time here, and as a result was able to revise his valued opinion on biplane fighters compared to the latest aircraft, an opinion which stemmed from his experiences during the First World War.

[7] Me 109 or He 113? (discussion) by Dr.-Ing. Hermann Wurster in: Jägerblatt, Offizielles Organ der Gemeinschaft der Jagdflieger e.V., No. 6, 1966 pp. 5-8.

man"), because when in a tailslide, with the air flowing over the elevators and rudder from behind, the airplane supposedly underwent considerable stress when it suddenly nosed over into a normal dive.

The demonstration made a good impression on the well-represented military and civilian commission, and the Me 109 gained their trust. One contributing factor was the news that chief test pilot Nitschke had been forced to bale out of the He 112 during spin testing a short time previously when he apparently was unable to recover from the spin...

At the test center, we test pilots had the opportunity to swap out planes at leisure for comparison flights. Before I climbed into the He 112, the center's director of fighter testing, Dipl.-Ing. Francke, cautioned me: 'Herr Wurster, with the He 112 you mustn't do the the things you did with your Me 109...'

I got the impression that the He 112 was not as easy on the controls during aerobatics as the Me 109, particularly with regard to its roll rate, which is after all an important criterion for both aerobatic flying as well as combat...In addition, when making a turn with a change of control functions the He 112 had a greater tendency to stall toward the inside of the turn than the Me 109, thanks to the latter's leading edge slats. When gently pulling the nose up into a stall, the He 112 would behave admirably and nose over into the stall. A sharp, sudden pullup and the 112 would bank over more so than the Me 109, again thanks to the Messerschmitt's leading edge slats.

The Me 109's better handling in aerobatic flight was also confirmed by all the numerous military and civilian pilots who flew it. Not least of whom was *General* Udet, one of our most successful pilots from the First World War and just as talented a civilian aerobat, a figure who virtually no one would contradict when it came to his expert opinion regarding the suitability of an aircraft design for aerial combat. The same also applied to the *General der Jagdflieger* Ritter von Greim, who expressed virtually the same sentiments. Both Udet and Greim not only flew fighter combat missions for their own sakes, but also used the time to good effect in order to train under tough combat conditions—something I can confirm after having flown against them in mock dogfights...

The Me 109 had somewhat of a smaller surface area (than the He 112) and therefore generated less parasite drag. Additionally, despite its enormous ruggedness proven during flight testing its static advantages made it somewhat lighter, and in horizontal flight it had a bit lower lift coefficient. This meant that in all horizontal speed categories the Me 109 was somewhat faster. Both aircraft were roughly on par when it came to climb rate.

The Me 109's landing gear was not problematic, and the slightly angled-out appearance was fully suited to poor landing conditions. The Me 109 could land on one leg without serious damage or flipping over, a situation often encountered when the undercarriage malfunctioned or the plane had seen combat and its cables had been shot out.

To be sure, I've never made a single-strut landing with the He 112, but its wider track undercarriage would have meant that it would not have stayed upright for as long during rollout and would have dropped over onto one wing at greater speeds, leading to ground looping. Which in turn would increase the danger of flipping over. Aircraft with wider tracks also had tendencies to ground loop if there were problems with the wheel brakes...

By attaching the undercarriage to the Me 109's fuselage it was possible to swap out the wing without having to resort to a makeshift undercarriage or lifting jacks. This was an indisputable advantage for both the manufacturing process on the assembly line, as well as in the repair shop at the airfield. This approach also had advantages when it came to stress, as the landing shock was absorbed directly by the wing supports, i.e. in the static center, and not carried from the wing to the fuselage as with the He 112...

And ultimately, the Me 109 had advantages when it came to production, as well. Its construction was simpler and it was cheaper to manufacture, among other things because of its easy-to-manage straight tapered wings and superior fuselage construction, with the fuselage join plate and transverse frame being pressed from a single piece of metal. Compared with the He 112, the Me 109 required significantly fewer rivets and needed less building cradle and jig work.

All in all, it can be said that in comparison with the He 112 the Me 109 had only those advantages which were the most critical and decisive for flying...

The RLM's verdict in favor of the Me 109 was not made at a round table session, nor was it solely up to Messrs. Milch and Lucht, but was accomplished in the main by the concurring voices of the entire flying community. The decision to produce the Me 109 instead of the He 112/He 113 was undoubtedly the correct one, for in fact the Me 109 was the better fighter plane..."

Bf 109 Goes into Production

The flyoffs between the Bf 109 and He 112 at Travemünde between 26 February and 2 March were accomplished with the Bf 109V-2, one of the eight prototypes which had flown by late 1936. All were powered by the Jumo 210B, C and D series engines, whose performance ranged from 500 kW/680 hp to 530 kW/720 hp. Once the decision had been made in favor of the Bf 109 preliminary work on production got underway immediately; plans called for 654 Bf 109B series aircraft to be produced in 1937/38. In December 1936 the first six Bf 109B-1 production aircraft were finished. But even the planned expansion of the Augsburg facility would not have been able to meet these ambitious goals, for production of the Bf 108 and Bf 110 was underway. It was therefore decided to build a brand new factory in Regensburg. The RLM also arranged for the AGO, Arado, Erla, Fieseler, and Focke-Wulf companies to license build the Bf 109. Augsburg drew up the entire license program, including blueprints for the jigs, in such a short time that the Erla and Fieseler companies, along with the parent company, were able to produce a total of 326 Bf 109s in 1937. The Bf 109B-1, with its two forward fixed MG 17s firing through the propeller arc, was delivered to the *Luftwaffe* as a replacement for its He 51s and Ar 68s. A total of 45 Bf 109Bs, including some armed with the MG C 30 (later designated the A-1) were sent in 1936/37 by way of German Baltic ports to Spain. There the Condor Legion was operating on the side of General Franco in the Spanish Civil War. The Bf 109B replaced the first generation of fighters, in particular the Heinkel He 51 biplane, and was able to regain superiority in the air. The operations in Spain proved to be a valuable learning experience, and provided the impetus for follow-on developments of the Bf 109, especially with regard to its armament. That the operation of a new aircraft type in a foreign country was not without its

Right: Beginnings of full-scale production: the first Bf 109B-1s in early 1937 at Augsburg. By 1945 about 34,000 had been manufactured, with a further 1,000 subsequently built in other countries.

Below: The second largest prewar series was the Bf 109D with the Jumo 210-D engine. About 650 were built at Arado, AGO, Erla, Fieseler and Focke-Wulf from 1938 to 1939.

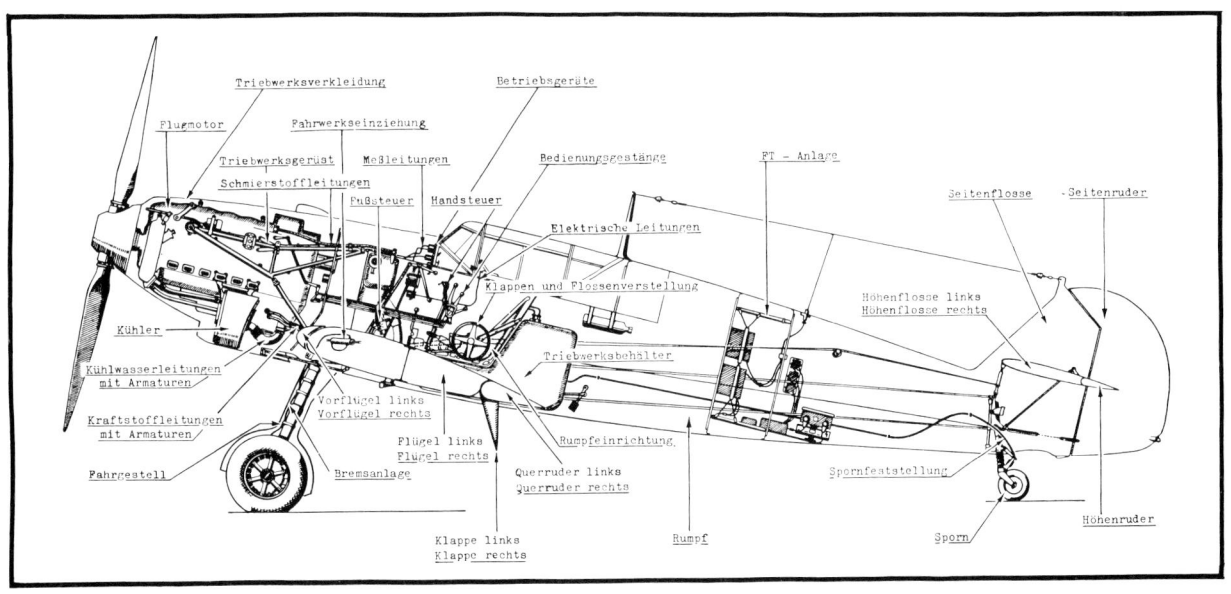

problems is evidenced in several excerpts from Hannes Trautloft's journal from the period of December 1936 to January 1937, mentioned in his book "Als Jagdflieger in Spanien":[8]

"The Bf 109 is finally ready. Takeoff is something I'm not entirely used to, but as soon as I'm in the air I feel at home in the new bird. The machine is great. An Italian Fiat hovers over the airfield at the same time. Up to now the Fiat had been considered the fastest airplane General Franco had. But as I fly under him, my eye caught him for only a moment before leaving him far behind...

I've been in Seville almost two weeks now, for the Bf 109 goes from one teething trouble to the other. Small things, really. The tailwheel has a problem, the water pump, the carburetor, or the gear door...Once my engine gives out while making a circuit and I have to make an emergency landing; at least I didn't crash...

I had to make another emergency landing. There are repeatedly problems with carburetor adjustment and the cooling system due to the incessant heat here. An airplane designed for operations in Spain can't be so precisely built as one for operations in Germany; the differences in climate must be taken into account.

On landing the tailwheel on my Bf 109 refuses to extend; this is the second time it's happened to me. The rudder is crushed... After thinking about it awhile I decide to keep the tailwheel locked in the down position."

For the *Luftwaffe*, the experience in Spain with the Bf 109B brought with it the understanding that the aircraft would be more successful against ground and air targets the more firepower per unit of time it could deliver. This resulted in a call for more guns at the expense of less ammunition per gun. These wishes were turned into reality in short order, and the Bf 109 was fitted with two additional MG 17s installed in the wings. To this end the thin wings had to be redesigned to accommodate the gun mounts and ammunition belts, and the Messerschmitt team once again rose to the occasion. A small production run of these Bf 109C-1s was built at the Augsburg parent factory with the Jumo 210G fuel-injected engine and a fuel tank increased by 100 liters, some being sent for duty to Spain. At the same time, plans got underway for large-scale production of this variant at the license manufacturing companies of AGO, Arado, Erla, Fieseler, and Focke-Wulf, which built about 650 of the now designated Bf 109D-1 variants in 1937 and 1938. Again, these were powered by the Jumo 210D (as with the Bf 109B). A handful from this series also went to Spain in August 1938.

British author Herbert Molloy Mason commented on the experience with the Bf 109 in Spain as follows:[9]

"In December 1938 the Condor Legion...had over 45 Me 109B-2 and C-1 planes available. They were the best fighter planes in the world and—not counting dogfights with the Mosca (I-16 Rata low-wing monoplanes)—were the unchallenged lords of the skies over Spain. The German pilots soon found that the I-16 was a very maneuverable aircraft. To counter this advantage required the employment of new tactics, developed by the *Luftwaffe*'s up-and-coming aces, such as Adolf Galland, Werner Mölders, Herbert Ihlefeld, Walter Oesau and others..."

The D variant was the first Bf 109 to be sold to a foreign country. Switzerland bought 10 examples in 1939, some of which soldiered on until 1949 as trainers for the Bf 109E versions delivered later.

Major Seidemann and Carl Francke, participants at the international flying meeting held at Zurich-Dübendorf in July 1937, pose here in front of a Bf 109B powered by the Jumo 210-G.

---

[8] Hannes Trautloft, Als Jagdflieger in Spanien. Aus dem Tagebuch eines Legionärs, Berlin (no year), pages 167, 182, 186-187.
[9] Herbert Molloy Mason, Die luftwaffe. Aufbau, Aufstieg und Scheitern im Sieg. Vienna/Berlin 1973, p. 222.

Bf 109 Becomes a Sensation

From 23 July to 1 August 1937 Zurich-Dübendorf in Switzerland was the scene of an international flying meet, the first of its kind to anticipate speed, climb, and dive competitions for military aircraft. The *Luftwaffe* showed up with several new aircraft models, including six Bf 109s. Ernst Udet, chief of the RLM's *Technisches Amt* and recently promoted to *Generalmajor*, had great hopes of winning the speed competition. His red-painted Bf 109V-14 (Werk-Nr. 1029, D-ISLU) had one of the first Daimler-Benz DB 601A engines, serial number 161. Udet gave it full throttle on the transfer flight from Augsburg to Zurich, taking just 23 minutes. In the first competition, though, he was forced to prematurely return to Dübendorf after his engine began sputtering. The race circuit of four 50 km laps was won by Carl Francke from the *E-Stelle Rechlin* in the Jumo 210Ga powered Bf 109V-7 with an average speed of 409 km/h. Francke also was the clear winner in the climb and dive competitions. During the next day's Alpine Circuit, Udet's plane lost oil and the engine seized up, necessitating an emergency landing. The airplane was so seriously damaged that it had to be written off. Udet had only light injuries to his arm. Major Seidemann was the victor in this race, completing 367 km in 56 minutes. Three Bf 109s also won the Alpine Circuit for the three-plane team. The Messerschmitt Bf 109's superiority over foreign military aircraft was plain to see and made an impression on the spectators. These successes caused quite a sensation in other countries, especially since it was claimed that the German fighter wings had already been equipped with this speedy fighter. In reality the first Bf 109s were delivered to the flying units in the spring of 1937, where they were chiefly used for pilot conversion training.

Dr.-Ing. Hermann Wurster in the cockpit of a Bf 109V discusses events with Willy Messerschmitt.

Bf 109V-13 Sets World Record

The successes at Dübendorf spurred both Willy Messerschmitt and his team, as well as Daimler-Benz engineers under Fritz Nallinger—with the support of the RLM's *Technisches Amt*, to discover just how fast the Bf 109 could really go. The V-13, one of the players at Dübendorf, underwent several modifications. It was fitted with a slimmer engine cowling, a VDM variable-pitch propeller, a more streamlined canopy, and a higher-performance DB 601 fuel-

After returning from Zurich the V-13 was refitted with a DB-601 engine for a world record attempt. The fuel injection engine had a higher rpm and an output of 1,250 kW/1,700 hp. Aside from its completely new powerplant and spinner the cockpit canopy was rounded off, all gaps were taped over, the pitot tube removed and the machine was buffed to a high-gloss finish. On 11 November 1937 Dr. Wurster was able to grab the landplane speed record for Germany for the first time with an average speed of 611 km/h.

injected engine, which could briefly attain 1,220 kW/1,660 hp at higher rpms and with an alchohol fuel additive. The V-13 was polished to a high glossy sheen, all openings and gaps were spackled, and the pitot tube removed. On 11 November BFW chief test pilot *Dr.-Ing.* Hermann Wurster was able to set a new landplane speed record of 610.95 km/h at an altitude of about 75 m along a three-kilometer stretch marked out above the Augsburg-Buchloe railway line. This record, which the FAI verified almost immediately, was over 100 km/h faster than the old record.

The record-breaking plane was reported to the FAI as a Bf 109 with a DB 600 engine, and was registered as such. The National Socialist propaganda machine was already using this event to its own ends.

The Bf 109V-13's success gave Ernst Heinkel renewed impetus to develop the record-setting He 100, and for Willy Messerschmitt it was cause for a low-drag high-performance airplane, the future Me 209. Daimler-Benz succeeded in increasing the DB 601's performance to 2,035 kW/2,770 hp with the Re V version and thus facilitated the world record flights of the He 100 and Me 209 in March and April 1939, respectively.

Bf 109E with the 30 liter DB 601 Engine

The new DB 601A was available in quantity beginning in 1938. The engine had some 50 percent more power, rated at 805 kW/1,100 hp, and was destined to be installed in the Bf 109. Some prototype engines had already been mated to Bf 109B airframes and tested the previous year. A complete redesign of the entire engine assembly was required for production of the Bf 109E variant. This particularly affected the cooling system. The water-cooled radiators were split, i.e. relocated beneath each wing, and the enlarged oil cooler integrated into the lower engine cowling. The best aerodynamic shape had to be found for the turbocharger's intake scoop, and this involved a comprehensive flight test program using a dozen different scoop designs. Since the takeoff weight of the new Bf 109E compared to the previous D-series jumped from 2,170 kg to 2,610 kg, the landing gear also had to be beefed up, as the heavier weight also led to higher landing speeds. In comparison with the D-series the maximum speed increased from 460 km/h to 570 km/h at an altitude of 6,000 m.

Production began in 1938, at first at Augsburg and then by drawing in the license manufacturers at Messerschmitt GmbH Regensburg and the Wiener Neustädter

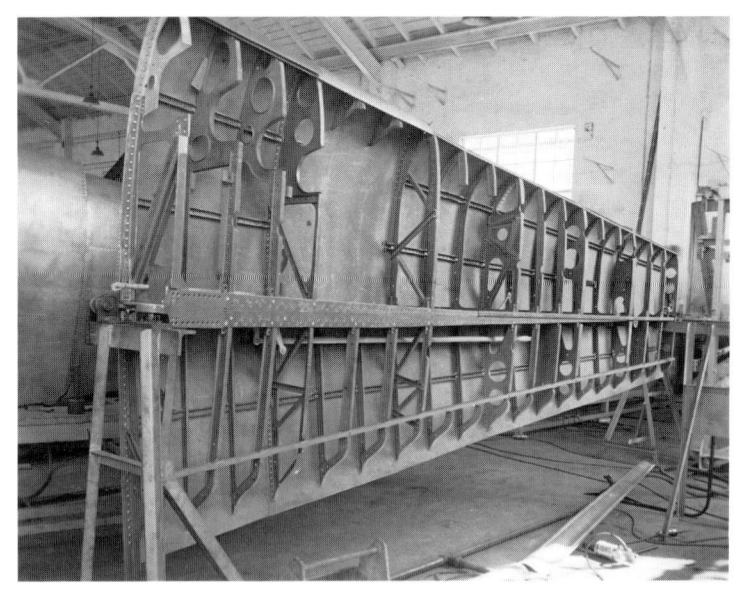

Structure of the single-spar wing of the Bf 109E.

The first Bf 109s fitted with DB-600 and DB-601 engines were built from October 1938 on and were designated E-0. After much experimentation, the turbocharger intake was modified even further for the production version. These machines were delivered as unarmed trainers to the fighter pilots' school in Werneuchen.

Flugzeugwerke in Wiener Neustadt. Both plants began supplying the *Luftwaffe* with Bf 109Es starting in 1939. The Regensburg factory built the most export aircraft, for which many orders had been received following the type's outstanding success at the Zurich flying meet. Thus, Switzerland took delivery of 90 Bf 109E-3s, Spain 40 E-1s and E-3s, Bulgaria 19 E-4s, Hungary 40 E-3s, and Yugoslavia 73 E-3s.

At the start of the Second World War most of the *Luftwaffe*'s fighter wings had been equipped with the E-model, or "Emil," as the troops called it. The Bf 109 proved to be the superior fighter on all fronts, even in the West, where the *Luftwaffe* encountered new generation Dewoitine 520 and Morane 406 fighters. At the time, the Armée de l'Air found itself in the unenviable position of converting over to new aircraft types. Even the first Spitfires the British sent to help proved incapable of changing events.

It wasn't until the Battle of Britain, fought between July and October of 1940, that the *Luftwaffe* ran into an enemy on par and even superior in some cases. Adolf Galland, later *General der Jagdflieger*, addressed this situation in his book "Die Ersten und die Letzten":[10]

"The modern Vickers-Supermarine Spitfire may have been some 20 to 30 km/h slower than us, but was clearly superior when it came to maneuverability. The older Hawker Hurricane, the predominant fighter used by the British, was no match for the Me 109 when it came to speed and climb rate. Our ammunition and guns were also much better. Another advantage proved to be our engines, which, unlike the British, had fuel injectors instead of carburetors. This meant that during critical moments in a dogfight the engine didn't cut out at negative *g*. The British fighters would attempt to shake their pursuers by making a split-S (half-roll) when we got on their tails, whereas we simply pitched down and pressed our attack home with full throttle and eyes bulging out of our heads..."

Soon, however, the faults of the Bf 109E began to show through, primarily with regard to its extremely limited range—after all, it had originally been designed as a point defense interceptor. Its main opponent over England was the Spitfire, which not only enjoyed a "home field" advantage, but also had greater range and better maneuverability. When the Bf 110 proved to be unsuitable as an escort for the German bomber formations, this role had to be filled by the fighter units flying the Bf 109. Yet because of their limited endurance they increasingly found themselves at a disadvantage. Galland explains[11]:

"All units were forced to take the most direct course to London, since the escort fighters had a fuel reserve of only ten minutes. It was thus impossible to make large-scale feints or avoid British flak zones...The disadvantages of the Me 109's limited endurance became ever more pronounced. On one of the *Geschwader's* missions alone twelve planes were lost, not to enemy fire, but simply because the bomber formation they were escorting had not yet made landfall on the return leg after two hours' flying time. Five of them made it to the French coast with their last drop of fuel and belly-landed on the beach, but the other seven had to ditch in the Channel."

When Göring paid a visit to the Channel Front, he inspected the *Geschwader* units there and asked the *Kommodore* of JG 26, Adolf Galland, and the *Kommodore* of JG 51, Werner Mölders (who was also the most successful fighter pilot at the time), what they wanted. Galland half-jokingly asked that his *Jagdgeschwader* be reequipped with Spitfires, while Mölders wanted Bf 109s with more powerful engines.

An entire Bf 109E could be suspended in the wind tunnel at Braunschweig.

---
[10] Adolf Galland, Die Ersten und die Letzten. Die Jagdfligher im zweiten Weltkrieg, Darmstadt 1953, p. 84.
[11] Galland, pp. 100-101.

This Bf 109E-3, Werknr. 1952, was experimentally fitted with four MG 17 and 1 MG C 30 guns. Here Dr. Hermann Wurster poses for company photographer Margarete Thiel on 13 June 1940.

Experience and demands from the front-line units during the war's first two years were immediately passed on by the RLM to the aviation industry, and Messerschmitt engineers quickly turned these into reality. The Bf 109E's operational spectrum grew significantly by incorporating modifications:

- Demand for increased range was met by developing and introducing a 300 liter drop tank.
- Demand for carrying a 250-35- kg bomb load was resolved by fitting a bomb rack which could be swapped out for the drop tank pylon.
- Combat experience resulted in additional armor for the pilot and armored fuel tanks being introduced on the production line and retrofitted to existing airframes.
- For operations in the South, particularly in North Africa, the aircraft were tropicalized, one prominent feature being sand filters for the turbocharger's airscoop.
- Special variants for tactical reconnaissance were fitted with photo equipment.
- Supplemental field conversion weapons sets were tested and made available. Beginning with the E-3 the MG 17s in the wing were replaced by MG FF 20 mm guns.

Nearly all these modifications were tested using the Bf 109E before being incorporated and eventually carried over to the new Bf 109F and G variants.

Mölders' wish was fulfilled in 1940 with the availability of the more powerful DB 601N rated at 860 kW/1175 hp (using C3 fuel), which was installed in the E-4 through E-7 subvariants.

Bf 109T for the Aircraft Carrier Graf Zeppelin

Even before the aircraft carrier Graf Zeppelin was launched on 8 December 1938, the *Luftwaffe* was already well underway testing the airplanes it would use in Travemünde. Some of these planes were based on modified or converted versions of existing types, such as the Junkers Ju 87C carrier dive bomber, the Bf 109T carrier fighter, and the Arado Ar 96T carrier trainer, as well as entirely new designs planned to operate from carriers from the outset. Examples of the latter included the Arado Ar 195 multi-role plane and the Fieseler Fi 167.

Some E-series were configured as fighter-bombers in 1940. This is the first machine, Werknr. 1361, CA+NK, shown with an ETC 500 rack capable of carrying 250 or 500 kg bombs, photographed in June of 1940 in Rechlin.

This Werknummer 1776 was pulled from the D-series, and in the spring of 1938, as the V-17, was fitted with catapult attachment points and landing hook needed for carrier takeoffs and landings. The machine crashed during the delivery flight to the test center at Travemünde in July of 1938 and was replaced by several different planes from the E-series.

In the case of the Bf 109, as early as 1938 a V prototype from the "D" series was kitted out with an arrester hook and catapult fittings. However, this aircraft crashed just prior to delivery to the test center. In 1939 several Bf 109 E-series planes were fitted with aircraft carrier components and evaluated at Travemünde by test center engineers, as well as instructor pilots from the existing *Trägergruppe 186*.

Catapult launches were made using a 20-meter long Heinkel compressed-air catapult (maximum acceleration 2.4 g) mounted on a pontoon, and landings were made using an electro-magnetic braking system on a runway strip laid out to simulate the Graf Zeppelin's flight deck. The braking distance varied between 20 and 30 meters, and the maximum deceleration was 2.6 g. Counting all carrier test aircraft, well over 1,500 brake landings were carried out. This testing continued until 1943, when work on the aircraft carrier itself came to an abrupt halt.

The Bf 109 test results in 1930/40 led to the Bf 109T series, with the Fieseler company in Kassel taking up production, and by the middle of 1941 they had built a total of 70 aircraft. The basis for the T version was the Bf 109E-4/N and E-7/N (with the 860 kW/1,175 PS power rated DB 601N engine) then in production. Unlike the base model, the Bf 109T also embodied the following modifications in addition to the reinforcements to the fuselage and arrester hook area:

- The wing span was increased from 9.90 meters to 11.08 meters, i.e. lengthening each wing by 0.59 meters, and thereby enlarging the wing surfacees from 16.35 m² to 17.50 m².
- Installation of the catapult fittings for the four-point catapult system, whereby the two aft connection points were bent hooks which protruded laterally from the fuselage. Both of the other connection points were installed under the fuselage beneath the cockpit.
- The installation of an arrester hook with a ball-and-socket joint and steel cable leading to a release handle in the cockpit.
- Arm supports and thicker head padding on the rear armor. Supplemental equipment included a Patin remote indicating compass, a lighting system, and a rubber raft.

The wing surfaces could be folded, similar to the arrangement for rail transport, and were unbolted at the wing roots.

However, when the *Luftwaffe* took delivery the airplanes lacked their catapult fittings and arrester hooks. The few aircraft which were already fitted with these features had them removed. The small series was designated as Bf 109T-2, with testing and certification taking place in the fall of 1941.

Bf 109 - Special Developments

Production of the Bf 109 soon smoothed out and developed a momentum of its own, leading to numbers previously unheard of. Constant armament and equipment requests by field units were met in large part by swapping out components directly or through the use of field conversion kits.

Bf 109 - Two-Seat Trainer

Other requirements on the part of the *Luftwaffe* and the RLM meant a continuous reworking of the basic Bf 109 design into new variants, as well as the follow-on Me 209/309. These projects were accomplished by the Messerschmitt project and design departments, as well as the Bf 109 program management team, now under Richard Bauer.

Bf 109G-12 two-seaters at a training wing (1944).

Among several proposals coming directly from the company was one just before the war from Willy Messerschmitt for a two-seat trainer version. For this, the Bf 109 was to be equipped with a less powerful Argus As 410 engine. This proposal was not taken up by the RLM, which instead opted for a new light trainer, the Arado Ar 79. It wasn't until during the course of the war, when the Bf 109 was fitted with more powerful engines—becoming heavier and therefore more difficutl to fly—that the field units renewed their requests for a Bf 109 two-seater with dual controls. Augsburg undertook work on a prototype modification in the middle of 1943, and a conversion contract was issued to Blohm & Voss in Hamburg for a small series using G airframes. Starting in mid 1944 these began making their way as the G-12 to the fighter trainer schools. In all, over 90 two-seat Bf 109s were built.

Bf 109 with Radial Engine

While the Focke-Wulf Fw 190, whose standard powerplant was to be the BMW 801 twin radial engine, was still in its development phase, the Messerschmitt company was awarded a contract for converting a Bf 109 to accept an American Pratt & Whitney Twin Wasp twin radial engine. This aircraft, the Bf 109 V-21, was flown by Dr. Hermann Wurster as early as August of 1938.

Bf 109X, V-21, Werknr. 1770, D-IFKQ, was based on an E-series airframe, but was fitted with an American Twin Wasp radial engine. The wings were modified in addition. First flight took place on 18 August 1938 and showed that the conversion offered good handling characteristics.

The aircraft's flight handling proved acceptable, and it had even better stalling characteristics than the Bf 109E series. But the personnel at Messerschmitt were so set on using the Daimler-Benz engine that the project was not pursued further.

When the Fw 190 and its BMW 801 engine flew in mid 1939, the RLM gave the Messerschmitt company a BMW 801R, a special version of the 801D rated at 1,470 kW/2,000 hp, to be installed in a Bf 109 airframe. On September 2, 1940, Dr. Wurster test flew the Bf 109 X (Werk-Nr. 5608, registration D-ITXP). In his flight report, he addressed both the positive and negative points of this unusual variant:[12]

"Flight characteristics: the stability around the yaw axis is not as good as that of the Me 109E, but it is still sufficient for the combat environment (approaching targets). There seemed to be no changes in the stability of the pitch and roll axes.

Stall characteristics for the aircraft are clearly better than those of the Me 109E. The aircraft wobbles only slightly when flaring out on landing. Its better pitch behavior means that the aircraft isn't heavier to land in spite of its higher approach speed (v_a = 200 km/h); if anything, it feels somewhat lighter than the Me 109E. The view from the cockpit is somewhat better on landing than the Me 109E and F thanks to its different fuselage outline. The improvement in pitch resulting from the installation of the radial engine has already been noted on the Me 109 with the Twin Wasp powerplant...

With the engine running, the moment of rotation around the roll axis is quite noticable. Heavy rudder must be applied to keep the aircraft in the air when flying slow.

Criticisms: The engine does not respond well to the throttle control due to the flawed engine command system. Both when applying throttle as well as reducing throttle the engine response time is irregular with long lag times. This can have serious consequences on final approach or when taking off."

The Bf 109X went to Rechlin in October 1940 where it was flown against the Fw 190. Despite a positive assessment only one prototype was ever built. By this point the lines of production between the Bf 109 and Fw 190 and their two designers, Willy Messerschmitt and Focke-Wulf's Kurt Tank, had already been defined.

Bf 109F - the Perfect One-Oh-Nine

Based on the experience gained in combat in 1940, in December of that year the Messerschmitt team began a thorough reworking of the Bf 109 with a view toward regaining the superiority over the British Spitfire which the Bf 109E had enjoyed at the beginning of the war.

This rework involved redesigning the airframe using the aerodynamic data obtained in wind tunnel testing and thus saving time in the manufacturing process. Specifically, these included:

- new wing layout with rounded tips for improving the plane's flight handling
- new cowling with blended in spinner
- aerodynamically improved radiator housing
- absence of elevator braces
- reinforced armor for pilot and fuel tanks
- fuel tank capacity increased by 100 liters
- armament concentrated in the fuselage, two MG 17s fixed forward firing through the propeller arc, and an MG 151/20 firing through the hub

While the preproduction models were still powered by the Daimler-Benz DB 601A and E, the main series was equipped by the DB 601N rated at 860 kW/1,175 hp. This engine gave the Bf 109F a top speed of 635 km/h, a considerable improvement over the Bf 109E's 570 km/h.

One of the first F-series from early 1941 shows off the completely redesigned "one-oh-nine" in its most perfect form.

[12] Messerschmitt AG, Augsburg, flight report No. 402/159 from 9/2/40.

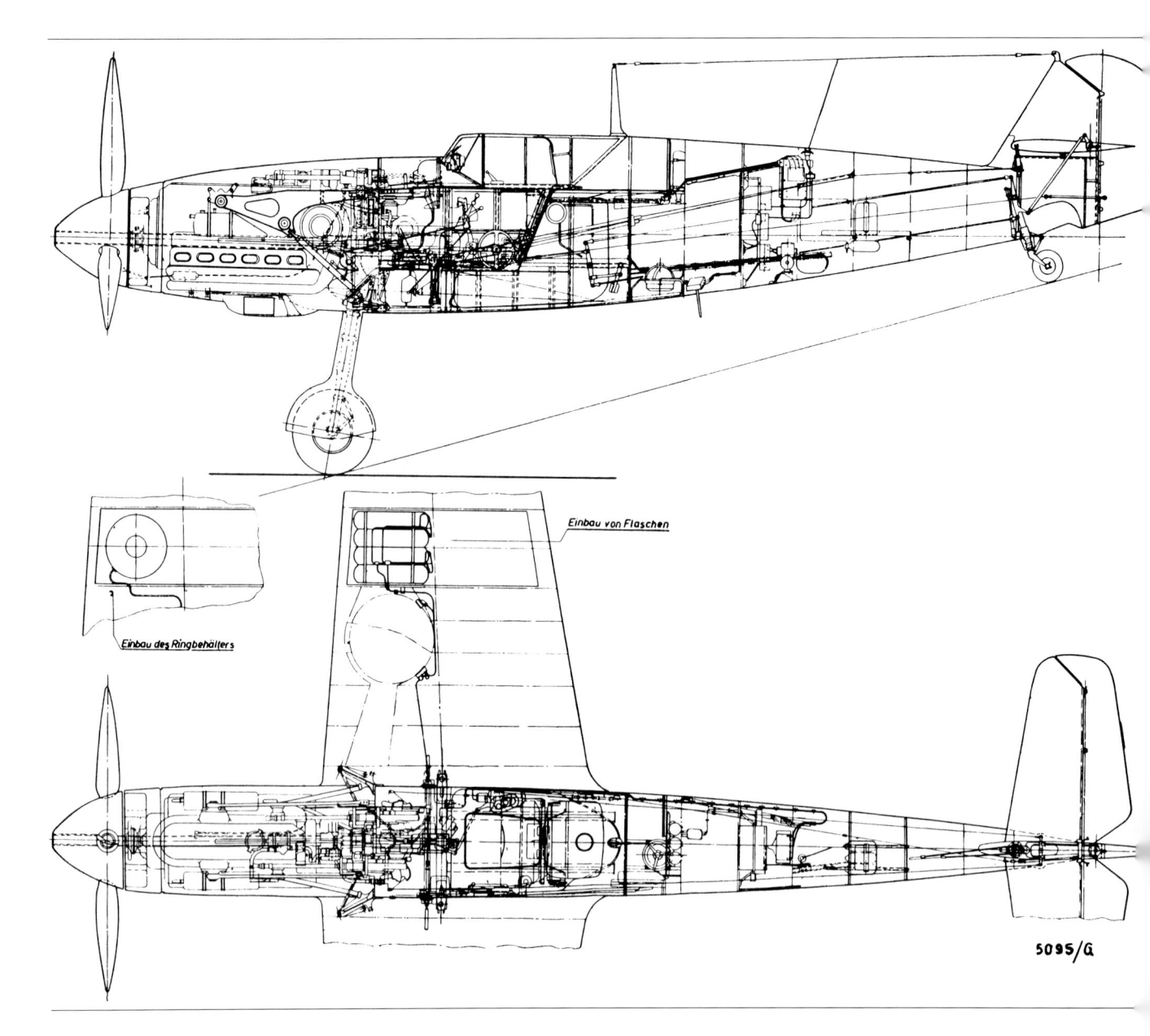

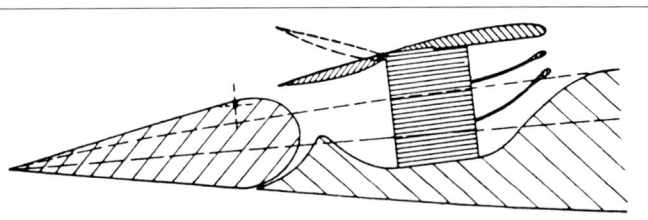

This redesign also had the benefit of reducing production time from about 9,000 man-hours to approximately 6,000. In any event, however, only about 2,400 Bf 109F models were built through 1942, a number limited by the availability of the DB 605 engine; the G-series would follow shortly.

Side and top view of the Bf 109G-2 from the summer of 1943.
Top of Page: Conceptual sketch of the (Junkers) airflow cooler installation in the wing of the Bf 109F.

Bf 109G - the Most-Produced Variant

The constant improvement in the performance of enemy aircraft led to the new, improved Bf 109G, H, and K series in 1942 and 1944. Powered by the DB 605A engine having an output of 1,085 kW/1,475 hp, the Bf 109G-1 reached a maximum speed of 650 km/h at an altitude of 6,000 m. Numerous changes were made to the airframe, as well. There were variants with pressurized cockpits and simplified versions lacking such an amenity. The lines for the GM-1 injection system were incorporated into the design, mounts for drop tanks attached, the oil reservoir capacity increased, the armor strengthened, and an improved canopy introduced. Armament initially consisted of two MG 17s and a single MG 151/20. From the G-5 onward firepower increased to two 20 mm MG 131s and a MG 151 engine cannon, improvements which resulted in the "face" of the Bf 109 changing yet again. The typical bumps on either side of the cowling, a consequence of the MG 131s, as well as bulges in the wings resulting from the reinforced undercarriage led to the troops nicknaming the airplane "the Bump." Takeoff weight had now climbed to 3,150 kg.

Production of the Bf 109 "Gustav" occurred in greater numbers than ever before, taking place at three facilities whose resources were concentrated almost exclusively on the Bf 109: Messerschmitt GmbH Regensburg, Erla-Maschinenwerk GmbH in Leipzig, and Wiener Neustädter Flugzeugwerke GmbH in Austria. By taking exemplary streamlining measures, director Karl Lindner at the Regensburg factory was able to reduce the man-hours in pre-assembly and on the assembly line by up to 40 percent.

Manufacturing time for the fuselage dropped from 203 to 163 hours. Monthly output of the Bf 109G at the Regensburg works rose from about 50 aircraft in 1942 to over 250 in 1943.

A British Test Pilot Assesses the Bf 109G

Eric Brown, a Royal Air Force fighter pilot during the Second World War, became a test pilot with the Royal Aircraft Establishment at Farnborough in 1944. While flying for the RAE, he had the opportunity to evaluate 55 different German aircraft and compare them to their British counterparts. He was able to fly several of Messerschmitt's Bf 109 variants, including the Bf 109G-6/U-2. In his book "Wings of the *Luftwaffe*" (published in Germany as "Berühmte Flugzeuge der *Luftwaffe*") he provides a detailed account of his impressions of the *Gustav* model. What follows are a few excerpts:[13]

"...The Bf 109 was simply a well-conceived, soundly-designed fighter that maintained during maturity the success that attended its infancy; its fundamental concept facilitated the introduction of progressively more powerful

[13] Eric Brown, Wings of the Luftwaffe, Shrewsbury 1987, pp. 149-157.

The first Bf 109G-1s at the Regensburg factory airfield in July 1942, prior to delivery to field units.

Full-scale production at the Regensburg factory: DB 605A engines are readied for fitting to the fuselage.

Full-scale production at the Regensburg factory: the finished fuselage sections are to be rolled over to the middle line in the hangar for wing fitting and final assembly.

armament and engines which enabled it to stay in the forefront of fighters for three-quarters of a decade...

The cockpit was small and narrow, and was enclosed by a cumbersome hood that was difficult to open from the inside and incorporated rather primitive sliding side panels. The windscreen supports were slender and did not produce serious blind spots, but space was so confined that movement of the head was difficult for even a pilot of my limited stature...

The Bf 109G-6 climbed well and at a steep angle, the acutal rate of climb being of the order of 3,800 ft/min (19.3 m/s) at sea level. Stability proved excellent in the longitudinal and lateral planes but was almost neutral directionally. Control harmony was poor for a fighter, the rudder being light, the ailerons moderately light and the elevators extremely heavy... Overapplication of longitudinal control in maneuvers easily induced the slats to open...

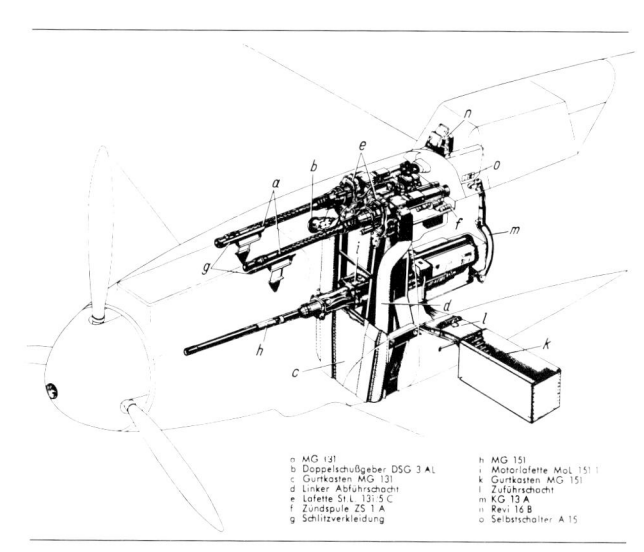

From the Bf 109G-5 on two MG 131s replaced the two MG 17s, a change which necessitated protrusions in the engine cowling called "*Beulen*," or bumps. While the ammunition belts for the MG 131s were behind the motor, the box for the hub-firing MG 151 was in the wing.

At its rather disappointing low-level cruising speed of 240 mph (386 km/h) the *Gustav* was certainly delightful to fly, but the situation changed as speed increased; in a dive at 400 mph (644 km/h) the controls felt as though they had seized! The highest speed that I dived to below 10,000 ft (3,048 m) was 440 mph (708 km/h), and the solidity of control was such that this was the limit in my book. However, things were very different at high altitude, and providing the Gustav was kept where it was meant to be (i.e. above 25,000 ft/7,620 m) it performed efficiently both in dogfighting and as an attacker of bomber formations...I measured 384 mph (618 km/h) in level flight at 23,000 ft (7,010 m)...

A substantial change of altitude was called for on the flare or round-out before touchdown, and even after ground contact the lift did not spill rapidly, and ballooning or bouncing could easily be experienced on rough ground. Once the tailwheel was firmly on the ground the brakes could be applied quite harshly, thus giving a short landing run, but care had to be taken to prevent any swing as the combination of narrow-track undercarriage and minimal forward view could easily result in directional problems."

Eric Brown went on to relate how the Fighter Development Squadron evaluated the Bf 109G-6 tactically against the Spitfire LF IX, XV, and XIV variants, as well as a P-51C Mustang. In climbing and diving the Gustav's relative performance was something of a mixed bag (generally su-

Bf 109G-5 planes from Jagdgeschwader 27 on an escort mission over the Mediterranean (1943).

One of the most successful Finnish fighter pilots, Eino Luukkanen, sits on alert in the cockpit of his Bf 109G-6 (summer of 1944 in Immola).

The improved-visibility canopy for the Bf 109 was available from early 1944 as a conversion kit and was fitted to production G-10s as well as K-series aircraft. The upper half of the previous steel head and back armor was made of armored glass to provide better rearward view. This had been done at the request of General Galland. The canopy was therefore called the Galland hood.

perior at lower altitudes), while the British and American fighters enjoyed a clear advantage when it came to speed, especially when using boost power.

Brown summed up the *Gustav* thusly:

"Allied bomber formations were certainly finding *Gustav* a formidable antagonist for it had heavy firepower, a reasonable overtaking speed and presented a very small target profile to the gunners..."

Performance Boost Using the MW 50 and GM-1

The Bf 109G-6 series was the first to be fitted with the improved Db 605D engine, which was initially rated at 1,140 kW/1,550 hp and had a larger turbocharger, made use of C3 fuel (100 octane), and had a maximum pressure altitude of 6,500 m. For a short term performance boost the Bf 109G also employed the MW 50 (water-methanol) fuel injection system, later incorporated into the K-version, as well. To activate the so-called "emergency boost"—limited to ten minutes—the pilot pushed the throttle beyond the stop point and thereby triggered the water-methanol injection while simultaneously adjusting fuel flow, rpm, and manifold pressure. The MW 50 system, effected by the plane's cooling system, could only be used up to maximum pressure altitude. Above that, the pilot resorted to the GM-1 system, a

A view of the cockpit in the Bf 109G-6. Notice the butt of the hub-firing MG 151 jutting out between the pedals.

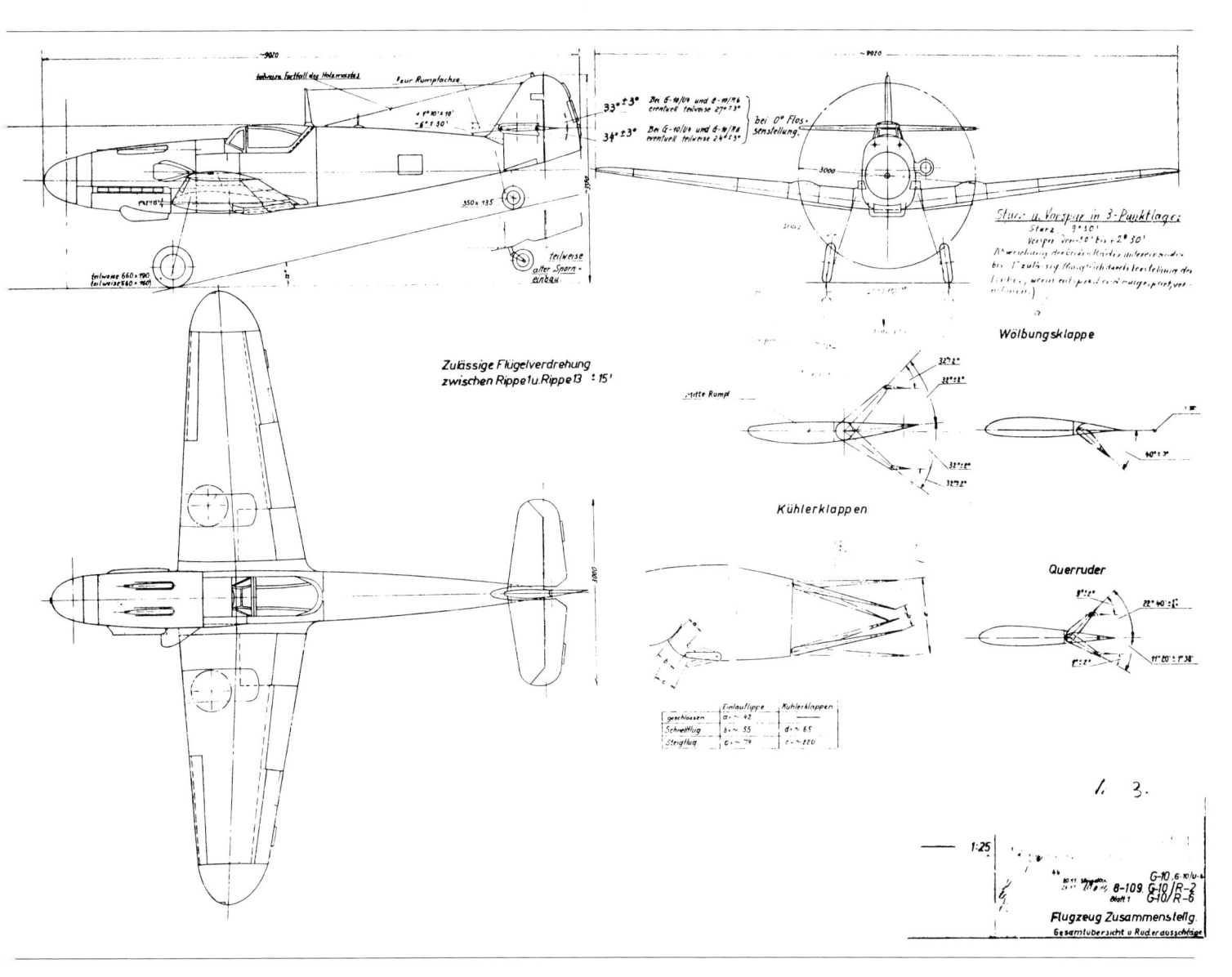

Three-view of the production Bf 109G-10 showing several improvements, such as the canopy, enlarged rudder, retractable tailwheel, and optional antenna mast. The asymmetrical rudder is shown in detail.

method which enhanced the engine's oxygen by injecting liquid nitrous oxide. It gave a brief burst of speed of up to 100 km/h at altitudes over 8,000 m. Both systems were often retrofitted to existing aircraft types, the MW 50 for fighter-bombers and the GM-1 for fighters operating at high altitudes.

Various Types of Special Evaluation

In addition to the ongoing evaluation programs involving improvements, modifications, and field conversion kits affecting the Bf 109 variants, during the years between 1938 and 1944 the type was subjected to numerous special trial programs for follow-on developments and underwent detailed flight testing. Among the most important were:

- In 1938 a Bf 109B was used for stall evaluation. To avoid stalling at high angles of attack Messerschmitt fitted leading edge slats to all his aircraft beginning with the Bf 108. This method had not been perfected, however, and to make it safer the DVL in Berlin under Dr. Wolfgang Liebe carried out tests which Messerschmitt at Augsburg continued with in 1938 using a Bf 109B. Dr. Hermann Wurster flew the tests. Cotton strips were glued to the aircraft's wing surfaces, and the airflow over the wing was observed and photographed. The strips made visible the separation of the spanwise boundary layer flow as it spread outward from the leading edge. Dr. Liebe had metal strips installed on the upper wing surface perpendicular to the leading edge (later called wing fences) which kept this spanwise flow contained. This method, which significantly improved the plane's stall handling, was kept a secret up until the end of the war. After the war, it found its way into aircraft design for such types as the Spanish Me 109, as well as many foreign types ranging from the MiG-15 to the Caravelle.

Bf 109V-48, Werknr. 14003, VJ+WC. A G-series airframe also served as a testbed for a butterfly tail, which was planned as a feature not only for the Bf 109 series but also for other Messerschmitt and Lippisch projects. First flight was on 21 January 1943.

- In 1940/41 a Bf 109E was fitted with fixed skis and tested for operations in the far north. These were never used operationally, however.
- In 1941 an underwing weapons gondola was tested with MG 151s, and this was subsequently offered as a field conversion kit.
- In 1942 a jettisonable heavy-load tailwheel was evaluated on the Bf 109 as a fighter-bomber for takeoffs in overloaded configuration, i.e. with a 250 kg underfuselage bomb and two 300 l drop tanks under the wings.

Bf 109V-31, Werknr. 5642, SG+EK, was an F-series airframe used to test the concept of a wide-track undercarriage in the autumn of 1941. An extendable belly radiator was installed in the fuselage. Both features served as preliminary design tests for the Me 309.

A Bf 109G-6/Trop in April 1943, used for testing various improvements for the K-series. The MG 131 protrusions over the engine cowling have been faired into the nacelle lines here, improving speed by some 6 km/h. Notice also the extra undercarriage doors. The MG 151 wing gondolas and the ETC 50 racks behind them could be fitted to different variants.

- In 1943 the Bf 109V-31 was used to test a wide-track undercarriage and a retracting oil cooler for the Me 309.
- In 1943 the Bf 109V-48 tested a V-tail in connection with several Messerschmitt and Lippisch projects. Following initial favorable results, it was later discovered that the tail suffered from stability problems in all axes and the idea was dropped.
- In 1943 and 1944 several firing trials were carried out using air-to-air rockets, whose launch tubes were affixed beneath the wings of Bf 109E and G testbeds. These trials included the Rheinmetall-Borsig RZ 65 73 mm rocket and the *Nebelwerfer* Br 21 210 mm rocket. The latter was used operationally against Allied bomber formations in the form of a field conversion kit.
- As part of the Bf 109's performance improvement program, in addition to the GM-1 and MW 50 fuel boosters introduced later there were also attempts at short-term speed boosts using rocket propulsion and Schmidt-Argus pulse-jets. Both never advanced beyond the evaluation stage.

The installation of turbojet engines in the Bf 109 was limited to project studies based on the Me 155 carrier fighter project.

Bf 109 High-Altitude Fighter Projects: Me 409, Me 155, P 1095, Me 109H

As construction continued on the Graf Zeppelin aircraft carrier in 1942 and the now obsolescent Bf 109T found itself in need of replacing, the RLM awarded the Messerschmitt company a contract for a "special carrier-based single-seat fighter," shortly followed by another contract for a "special high-altitude fighter." With the Me 210 crisis in full swing at this time and faced with the impossibility of starting two entirely new developments simulta-

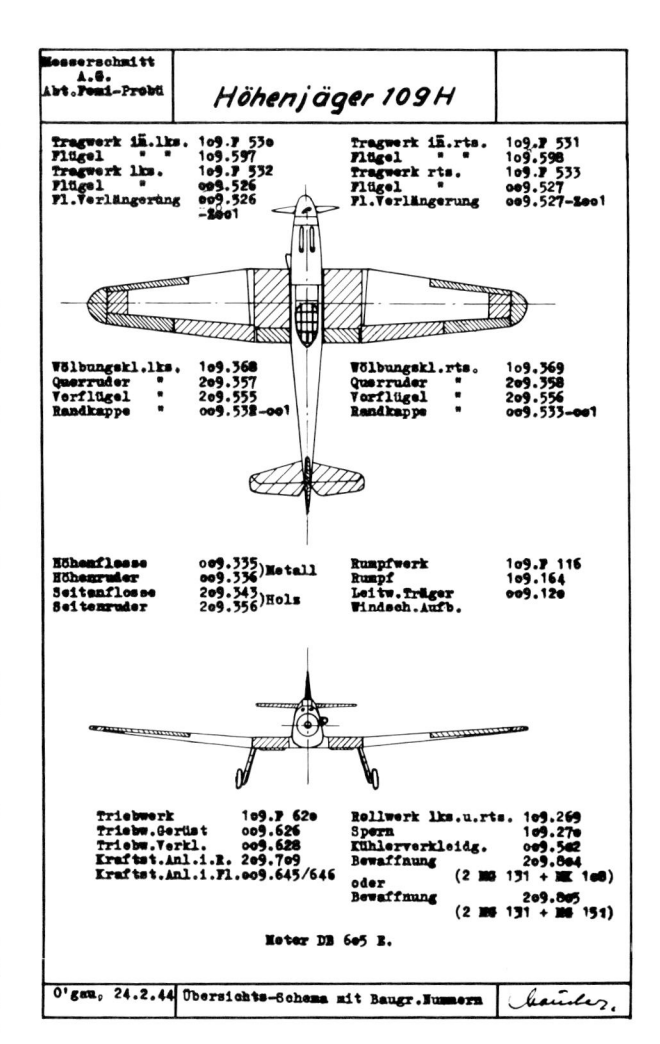

Overview of the Bf 109H components. The shaded areas show new components compared to the original Bf 109G-5.

Conceptual drawing of the Bf 109H with wing-mounted MK 108s.

| Stufe I | Stufe II | Stufe III |

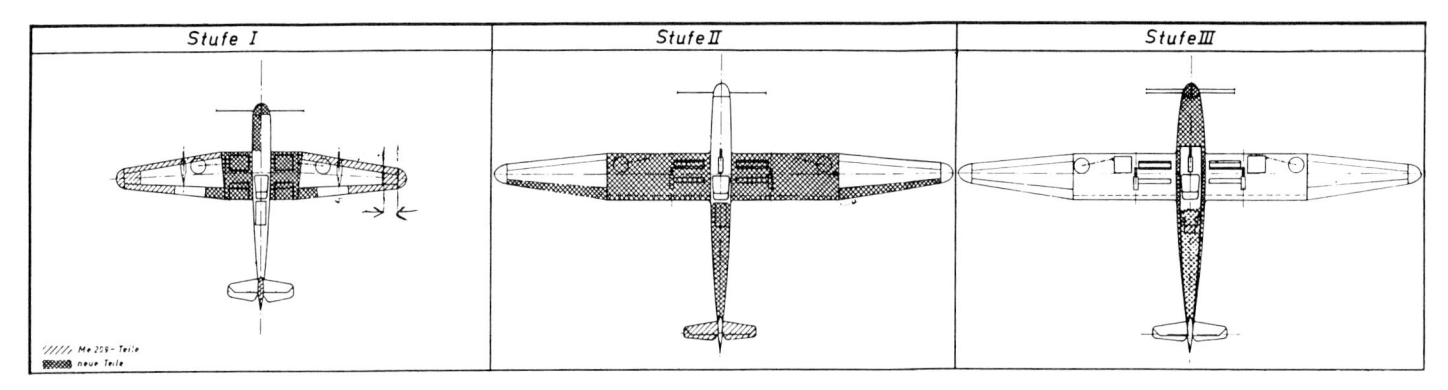

Three-stage plane showing the development of the high-altitude fighter based on the Me 109/209, dated 26 July 1943. The third stage was project no. P 1091, which was turned over to Blohm & Voss as the Me 155.

neously, it was decided to make use of the Bf 109G as the basis for both projects. In both cases the wings would be enlarged, and the high-altitude fighter would utilize the DB 628 high-altitude engine by Daimler-Benz. Both Bf 109 derivatives were briefly carried under the designation Me 409 in the project department before being reclassified as the Me 155.

Due to limited capacity, Willy Messerschmitt gave the work and data for the Me 155 carrier fighter to the project department of the Paris-based SNCAN, a company under the control of Messerschmitt. Here the work proceeded at a sluggish pace, and the blueprints were only about 50 percent complete when the carrier fighter project was canceled in February 1943.

At the same time in Augsburg, work on the Me 155 high-altitude fighter version shifted to a Me 209-based design, which had in the interim replaced the Me 309 in terms of priority. The "Me 209 High-Altitude Fighter with DB 628" proposal was submitted to the *Amt* on 23 April 1943. The performance figures for this project were practically the same as those of the Me 155 high-altitude fighter version. In lieu of a contract, however, the *Amt* voiced its interest in putting more effort into the project to study the possibility of attaining higher service ceilings than those on offer.

Based on previously calculated "optimum data" for an "extremely high-altitude fighter," the P 1091 project was born utilizing either the DB 628 or the DB 602 + TKL 15 turbocharger. Estimated service ceiling was about 17,500 m.

Planning forecasts showed, however, that production would not get underway before late 1944, and Willy Messerschmitt was therefore instructed to come up with a quick solution for a high-altitude fighter using available Bf 109 aircraft with minimal conversion work. The *Amt* expected the type to have a ceiling between 13,000 m and 15,000 m.

Fritz Hügelschäfer and Ludwig Bölkow in the project department worked out a three-stage plan for developing a high-altitude fighter in July of 1943, the first stage involving modification of two BF 109G airframes as prototypes.

The most significant change to the DB 605 powered Bf 109G-5 base model was increasing the wingspan from the G's 9.92 m to 13.26 m for the H model, accomplished by inserting two plugs into the wing center section. The inherent strength dropped to a load factor of 4.5. Two additional V-types were built using the DB 628 high-altitude engine as a basis, which was expected to be the engine of choice for the high-altitude fighter series.

The conversion and repair facility at Guyancourt in France tackled a pre-production batch of six aircraft by converting Bf 109G airframes into the quick solution high-altitude fighter. As part of an information-gathering trip to the Western Front, Fritz Wendel was able to test fly the Bf 109H on 5 April 1944 and discovered structural problems in the wing center sections. On 14 April, when exploring these problems further during high-speed flutter trials in Augsburg with the Bf 109H V-50, the aircraft broke up in flight and Fritz Wendel was forced to take to his parachute. As a result, production of the remaining Bf 109H aircraft at Guyancourt was delayed until the summer of 1944. Those few aircraft completed were generally employed as high-altitude reconnaissance aircraft in the West.

The planned variants within the first stage included:

H-0   with DB 605A and 1x MK 108 firing through the engine and 2x MG 151 in underwing gondolas.
H-2   with DB 605E and 1x MK 108 engine cannon plus 2x MG 131 in the wings.
H-2/R2   high-altitude reconnaissance aircraft with 1x MK 108 engine cannon and an Rb 50/30 or 75/30 camera.
H-3   with DB 605E and 1x MK 108 engine cannon and 2x MK 108 cannons in the wings.

High-Altitude Fighter, Stage II

This called for yet another extension of the wings from stage I to 39 m² and a span of 21 m. In addition, the fuselage was lengthened and the design incorporated the wide-track undercarriage and empennage of the Me 209. The planned engine was the DB 605A with a DB 603 turbocharger and

The Bv 155V-2, not quite finished. The British handed the machine over to the Americans, and now the aircraft rests in a warehouse of the National Air and Space Museum in Washington.

O_2 fuel injection. Armament was expected to be 2x MK 108 engine cannons and 2x MG 131s or 2x MK 108s in the wings.

High-Altitude Fighter, Stage III (P 1091)

Due to the project department being overtaxed as it was, this revamped project based on the Me 155 high-altitude fighter variant was eventually turned over to the Blohm & Voss Flugzeugbau in August 1943 as the "High-Altitude Fighter, Stage III, Me 155B-1." As only limited data was available, chief design engineers Richard Vogt and Hermann Pohlmann had a strong influence on the design from the outset. Even in one of the first project updates submitted to the RLM it was identified as the "Me 155 - Blohm & Voss Design." As work continued, the project increasingly became known as the BV 155. It was, in effect, a new design altogether.

When the first BV 155V-1 took to the air on 8 February 1945 with chief test pilot Helmut Rodig at the controls, it was obvious that virtually nothing from the Bf 109G or Me 209 had carried over.

Bf 109K - Final Large-Scale Production Variant

The Bf 109G had been in large-scale production since the summer of 1942 and embraced some 80 sub-variants. Changes and requests for changes had piled up so much that swapping out parts during repairs was fast becoming a nightmare.

Werner Göttel, at the time in charge of coordinating Bf 109 license production, estimated that the Bf 109 had undergone about a thousand changes up to 1943. It was therefore decided to "clean up" the Bf 109 from a design standpoint, working in all the changes needed and drawing up new blueprints. It was hoped that the new K-series would offer an improvement in performance, especially at high altitudes, and thereby regain a certain superiority over the North American P-51 Mustang escort fighters. To this end, the K-series was to incorporate the Daimler-Benz DB 605L engine with two-stage turbocharger and a maximum pressure altitude of 9,600 m.

In late 1942 the project was assigned to Ludwig Bölkow, who at the time was working in the project department at Augsburg. The "clean up" program took place at the Wiener Neustädter Flugzeugwerke (WNF) at Wiener Neustadt in Austria. With a staff of 140 workers—designers, stress analysts, draftsmen, and mechanics—the program began in early 1943 and had progressed to an advanced stage when the WNF was first attacked by American bombers on 13 August 1943. 500 people were killed in the factory.

Despite serious damage and subsequent raids, the vast majority of the new blueprints were rescued. In any event, though, it wasn't until the autumn of 1944 when the first Bf 109K aircraft were delivered to the field.

Like this Bf 109K-4 here, in May of 1945 both flyable and damaged Messerschmitts could be found scattered throughout Germany awaiting the scrapyard.

137

Externally, the K-series differed from the G-series by an aerodynamically refined engine cowling minus the bumps, a modified airscoop, a more rounded canopy, strengthened wings with covers for the wheel wells, an enlarged rudder made of wood and a taller retractable tailwheel. Bombing raids had also interrupted production of Daimler-Benz engines, and since the planned DB 605L engine was not yet available, the K model initially utilized engines from the DB 605 ASCM and DCM series. Nearly all Bf 109Ks were fitted with the MW 50 injection system, with a few fighter variants also incorporating the GM-1 system for improving performance at higher altitudes. Although a broad spectrum of variants for the K-series had been envisioned, ranging from the K-1 to the K-14 (including the K-12 trainer), ultimately the variant delivered by far in the greatest numbers was the K-4. This sub-type made use of two MG 131s and a MK 108 firing through the propeller hub. One of the field conversion kits included a MG 151/20 under each wing. For the K-6 variant Messerschmitt returned to the "gun wing" last used on the Bf 109E, which saw a MK 108 fitted directly into each wing instead of a field conversion kit.

With GM-1 boost the Bf 109K attained a speed of approximately 700 km/h at an altitude of 7,500 m and therefore was the fastest of the one-oh-nines. Its service ceiling was about 12,500 m.

Contracts seemed to imply that several thousand Bf 109Ks would be in production by mid-1944. In actual fact, however, it is certain that hardly more than a thousand were built.

Erich Hartmann, a successful fighter pilot with 303 kills to his credit when he visited the Messerschmitt AG in September 1944, was given the opportunity to test fly a prototype of the Bf 109K. He took the viewpoint that nothing should be changed with regard to the flight handling in comparison with the Bf 109G. However, with regard to the armament, he recommended—primarily in view of the average fighter pilot's skills—that the type be fitted with many light caliber guns for fighter vs. fighter combat to give a wide field of fire. Naturally, for fighter vs. bomber combat other recommendations would apply (*General* Galland called for more cannons). Up to that point Erich Hartmann had scored virtually all of his victories with the Bf 109F and B models. Ending the war as the most successful fighter pilot in the world, his tally ultimately rose to 352 kills. Hartmann's overall impression of the 109 was conveyed in writing to Willy Messerschmitt on the latter's 75th birthday with the words "My victories wouldn't have been possible without your 109."

The Bf 109 Goes International

Following on the heels of the Bf 109's successes at the International Flying Meet in Zurich in 1937 was an increased interest in this new fighter plane, not only in Europe, but also across the waters. This interest was so strong that the Reich established a quota on exports. Even before the war and well into its first year Switzerland received about 100 Bf 109s, Yugoslavia 73, and the USSR 5. Other interested parties, such as Sweden, Norway, and Turkey, could not be supplied after the war had started.

During the war only those air forces allied with Germany were given the Bf 109. These included Finland (160), Bulgaria (170), Romania (140), Slovakia (15), Hungary (100), and Spain (15 plus the license).

Rights for license production were given to Hungary, as well. There the Hungarian wagon factory in Györ supplied about 320 of the 1,900 Bf 109Gs ordered. The Romanian company of IAR in Brasov could only deliver a few of the 600 Bf 109Gs on order before the war ended.

Stock and replacement parts for the Bf 109 were also manufactured in occupied France, as well as by the Italian aviation industry.

Postwar Production - the Czech 109

The Avia company in Prague in occupied Czechoslovakia was also to have license-built the Bf 109G-14 (as well as the Me 262) for the *Luftwaffe* beginning in 1945.

As production had started before the war's end, it was decided to use the type for building up the new Czech air force and continue with production. But there were few DB 605 engines to be had, and it was decided to utilize the

Three successful Bf 109 pilots: Josef Priller (1915-1961), Fritz Wendel (1915-1975), and the most successful fighter pilot in the world, Erich Hartmann (1922-1993), seen here in 1960 during ceremonies donating the Bf 109E to the Deutsches Museum in Munich.

One of 25 Avia S 199 aircraft supplied to Israel by the CSSR in 1948.

Jumo 211 engine from other production resources. The mating of airframe and engine involved a considerable rework of the entire front of the aircraft, as well as the use of a new propeller. The resulting single-seater was designated the S 199, while the two-seat trainer—equivalent to the Bf 109G-12—was given the designation CS 199.

550 machines were built from 1946 to 1949, with some flying up until 1957. In 1948 the newly established air force of Israel acquired 25 Avia S 199s, where they were used in the first Arab-Israeli War and occasionally saw action against Egyptian Spitfires.

Spanish License Production of the Bf 109G

With every variant of the Bf 109 up to and including the E having seen action with the Condor Legion in Spain between 1937 and 1939, the Spanish were certainly well acquainted with the type, and in 1942 Spain's Ministry of Defense obtained the license for the Bf 109G. However, the war prevented the contract from being filled. Germany was only able to supply 25 aircraft without DB 605 engines. Consequently, the Spanish were forced to install other engines which had not yet been tested in the Bf 109.

The first production batch rolled off the Hispano Aviacion assembly line with the Hispano Suiza 12 Z engines as the HA 1109 K1L (military designation C.4J), with production terminating in 1952. Subsequently, these aircraft, and all others, were fitted with the Rolls-Royce Merlin and were designated the HA 1112 M1L (C.4K). Unlike the DB 605, the Hispano Suiza 12 Z-17 and the Rolls-Royce Merlin were both inline engines, and this necessitated a redesign of the engine mounts and cowling.

It was a query from Spain addressing the static and design problems of this conversion which provided Willy Messerschmitt the opportunity to again become involved in aircraft construction following the war.

"Spanish Me 109s": Ha 1112-M1L fighters with Rolls-Royce Merlin 500/45 engines having an output of 1,610 hp on takeoff. The armament here consisted of two Hispano-Suiza 20 mm cannons as well as four Oerlikon 80 mm air-to-ground rockets.

Production and variant statistics for the "international" Me 109.

139

A total of 240 Spanish Bf 109s were built under license up to 1957, with the last of these retiring from service in 1965. Some of these were sold to a British film company and saw their last "operational" sorties in a film about the Battle of Britain. They were subsequently sold off and can now be found in museums and private collections throughout the world. Even the MBB mascot plane is a "Spanish" Me 109 retrofitted with a DB 605 engine.

Bf 109 Variants

Stemming from the single Bf 109 basic type of 1935 were a total of 10 major variants during the plane's twenty year developmental history (counting foreign programs). However, these variants also branched out into numerous sub-types as a result of conversion, retrofitting, or the use of field conversion kits. Taking these into account, there were over 100 different models of the Bf 109.

The following are the most important of the 10 major variants and sub-variants:

Bf 109A	retroactively assigned designator for some V-types and B airframes with 1x MG C30 (20 mm), saw action in Spain.
Bf 109B-1	first production variant with Jumo 210D, armament 2x MG 17 fixed and a third MG 17 firing through the hub.
Bf 109C-1	small subvariant of B series with Jumo 210G and new "gun wing" for 2 additional MG 17.
Bf 109D-1	production version of C with Jumo 210D, armament 4x MG 17.
Bf 109E-1	major changes through installation of DB 601A engine and three-blade variable-pitch propeller, armament 4x MG 17.
Bf 109E-3	as E-1, armament 2x MG 17 and 2x MG FF (20 mm).
Bf 109E-4/BN	as E-3 with DB 601N engine and ETC 50 or 500 bomb rack, thus the first capable of fighter-bomber operations.
Bf 109E-7	as E-4, first with combined pylon for bombs or jettisonable 300 l drop tank.
Bf 109E-9	as E-7, first variant capable of tactical recce with Rb 50/30 camera in fuselage.
Bf 109T-0 thru T-2	carrier-borne fighter derived from E-7 with increased span, catapult-launched, some with arrestor hook. Armament same as E-3.
Bf 109F-1 thru F-6	aerodynamically refined Bf 109 with DB 601E engine. Armament relocated to fuselage with 2x MG 17 and 1x MG 151 engine cannon. Optional add'l 2x MG 151 underwing field conversion kit. Fighter-bomber and recce equipment optional.
Bf 109G-1 thru G-4	light fighter with DB 605A engine, pressurized cockpit optional. Armament as Bf 109F. Fighter-bomber and recce equipment optional thru field conversion kits.
Bf 109G-5 thru G-14	as G-1, increased armament: 2x MG 131 (bumps on cowling), 1x MG 151. Fighter-bomber and recce equipment optional, as engine boost, using R1 thru R6 and U1 thru U6 conversion kits. G-12: two-seat trainer.
Bf 109H-0 thru H-4	high-altitude fighter based on G-5, wingspan increased to 11.92 m. Only a few built. Several stages projected using various powerplants (DB 605L, DB 628, Jumo 213).
Bf 109K-1 thru K-12	fighter with DB 605D engine, aerodynamically refined cowling, longer tailwheel (retractable), enlarged wooden rudder, Gm-1 or MW 50 systems, armament: 2x MG 131, 1x MG 108 or MG 151, optional conversion kits as for G model, K-12 planned as trainer.
Foreign/CSSR: Avia S 199	copy based on Bf 109G-14 with Jumo 211F engine, armament 2x MG 131, 2x MG 151, CS 199 trainer as G-12.
Foreign/Spain: Hipsano Aviacion HA 1109K1L	copy using HS 12 Z17 for fighter pilot training.
HA 1112M1L	fighter with Rolls-Royce Merlin 500/45 engine. Armament 2x HS 804 (20 mm) cannon in wings, 4 - 8 Oerlikon rockets.

140

Me 209 Record-Breaking Airplane

In mid-1937, Robert Lusser and the project department began looking at a specialized airplane which would seize for Germany the world's absolute speed record for airplanes. Thanks to the success of the first Bf 109B production aircraft, which had won several competitions in Zurich, it was decided to modify Bf 109V-13 (D-IPKY) for the attempt on the record. On 11 November 1937 chief test pilot Dr. Hermann Wurster flew the rail line between Buchloe and Augsburg at an average speed of 610.95 km/h. This was 44 km/h faster than the current speed record for landplanes, set by Howard Hughes in the U.S. on 13 September 1935 with his Special 1B plane.

But Messerschmitt wanted more, for the absolute speed record had since been set at 709 km/h by the Italian Francesco Agello in the Macchi-Castoldi MC 72, a floatplane originally developed for the Schneider Trophy competition. Up to this point, Germany had not participated in the Schneider Trophy for two major reasons: a lack of funds and the absence of a high-powered engine.

With an eye towards advancing aviation development using special experimental designs, in 1937 Willy Messerschmitt turned to the RLM:[1]

"We therefore need experimental planes, purely experimental planes. And when developing them, the designer must not be restricted to available tools and equipment or limited by regulations, nor should he be forced to make concessions to any consumer for the airplane's practical use. The experience gained in developing such an experimental plane would then make it easier to produce a functional aircraft of advanced design.

And just as we should build planes purely for experimental purposes, so should we accelerate research. It is an unhealthy sign when we can now fly faster than our fastest wind tunnels can blow."

These arguments prompted the RLM's research department to approve initial funding, but with the stipulation that the military applicability of any such aircraft would have to be taken into consideration. Internally, the Messerschmitt project was designated P 1059, while officially it was carried on record as the Bf 209. With the renaming of BFW as the Messerschmitt AG in mid-1938, however, it subsequently became known as the Me 209. Design of the P 1059 began in early 1938, with component assembly beginning shortly thereafter. Four V-types were laid down in prototype construction and were built using the same techniques as for the Bf 109. But the 209 was a much smaller design than the 109, with about 30% less fuselage surface area. Drag was reduced by dispensing with the commonly used oil and water radiators. Liquid cooling for the DB 601 was accomplished by utilizing the surface evaporation method (after flowing around the engine, the superheated water was piped out to the tightly-riveted wings, where it was cooled through the principle of heat exchange and pumped back to the engine). The system normally utilized 220 liters of water, but this figure doubled for the record-setting flight. However, the cooling system proved to be less than satisfactory, entailing an unacceptable water loss rate of between 4 and 7 liters per minute. The annular oil cooler was situated behind the spinner.

On 1 August 1938 Dr. Hermann Wurster took off from Haunstetten to make the Me 209V-1's maiden flight. Even from the start, it was obvious that the 209 was not a pilot's dream plane by any stretch of the imagination. There was little in the way of positive results throughout the testing program for the four prototypes. It began with the plane's tendency to ground loop on takeoff and other landing gear problems, continued with heavy control forces and inadequate stability along the pitch and yaw axes, and ended with cooling problems for the special-built DB 601 engine, which had been boosted to an output of 2,090 kW/2,770 hp and only a few of which had been produced thus far. In the case of the Me 209V-2 these deficiencies led to its loss after just a few flights, when a seized engine forced Fritz Wendel to set the plane down hard on the runway. The machine was damaged to the point where it had to be written off. There could be little thought of setting records under such circumstances.

In the meantime, after the Bf 109 had beaten out his previous He 112 design for the *Luftwaffe*'s standard fighter, Messerschmitt's great rival Ernst Heinkel had had his project team develop a new fighter, the He 100. On 5 June 1938, Ernst Udet had "only just" set a 100-km record when he flew the He 100V-2 to a speed of 634.73 km/h, thereby beating Wurster's record from the previous November by 25 km/h. Three months later Heinkel also made an attempt to wrest the absolute speed record from the Italians with the He 100, but the He 100V-3 was lost in a crash. Nonetheless, on 20 March 1939 Hans Dieterle was able to set a new record of 746.606 km/h at Oranienburg in the He 100V-8 (called the He 112U for propaganda purposes).

The Messerschmitt team now put all their resources into beating Heinkel's success. Hubert Bauer, manager of prototype construction, recalls the conditions and mood which prevailed on the day of the (nearly failed) record-breaking flight:[2]

---

[1] Willy Messerschmitt, Problems of High-Speed Flight. Lecture at the Third Scientific Conference of official members on 26 November 1937. In: "Schriften der Deutschen Akademie der Luftfahrtforschung, vol. 31. Munich and Berlin, 1940.

[2] Hubert Bauer, notification of Me 209 record-breaking flight. Undated (for press release from the Messerschmitt AG, Augsburg)

Of the four spectacular speed records set in Germany between 11 November 1937 and 26 April 1939, the first and the last were established by Messerschmitt, while Heinkel took the second and third. Ernst Udet, who was also a gifted artist, recorded the first two record flights as a cartoon, while company photographer Margarete Thiel photographed the smiling victors of 26 April: Fritz Wendel and Willy Messerschmitt.

The brand-new Me 209V-1, still without paint, as flown on its maiden flight by Dr. Hermann Wurster on 1 August 1938.

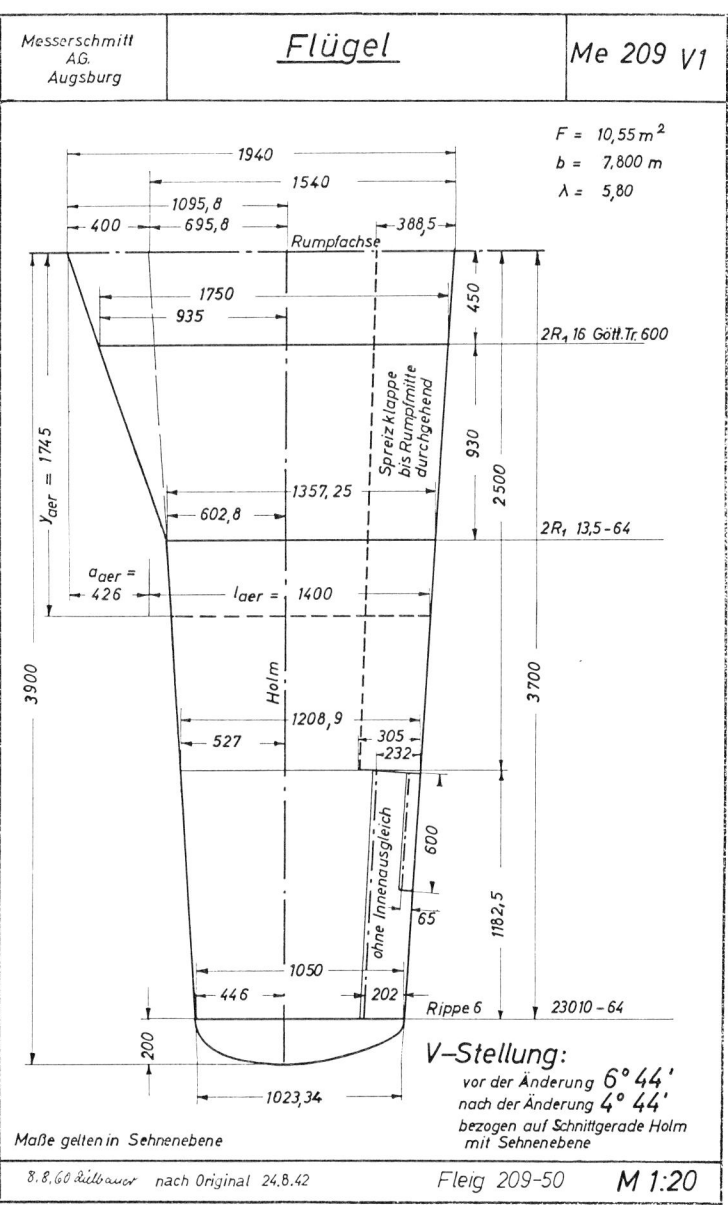

Sketch showing the thin, closely-riveted sealed wing and the coolant circulation for the engine's evaporative cooling system.

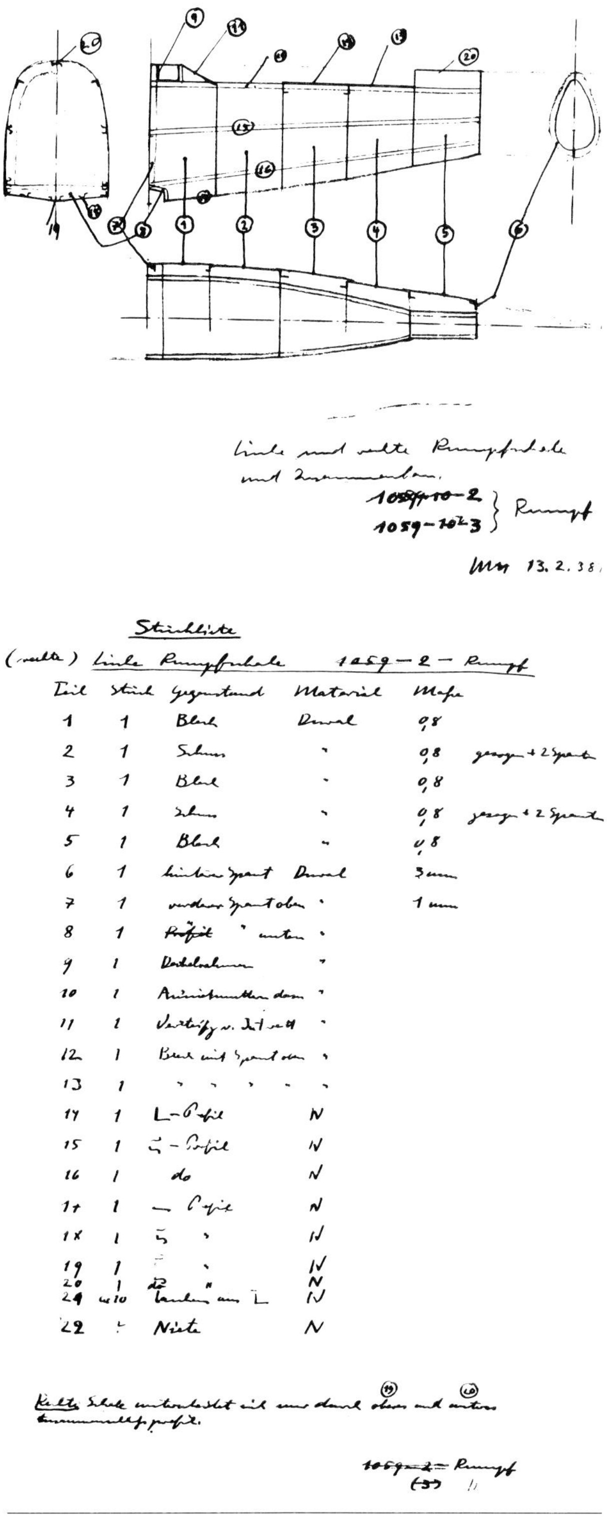

Design sketch by Willy Messerschmitt for the P 1059 (Me 209) project, dated 13 February 1938. The two fuselage sections have been roughly drawn and sectionalized with the parts breakdown listed below.

"In the early morning hours of April 26th, 1939, the Me 209 yet again took off from the company field of the Messerschmitt AG in Augsburg. It was another attempt by Fritz Wendel to break the world's absolute speed record, held at that time by the Italian Agello. The firm's aerodynamics engineers had figured that, with its stubby wings and a highly-tuned Daimler-Benz DB 601 engine, the Me 209 would be able to attain a speed which would break the current record. The fact that the machine was quite difficult to fly, especially when landing on the small airfield, came with the territory. The machine demanded everything a pilot could give it.

But this attempt failed as well—the weather was too uncooperative. When Prof. Messerschmitt arrived there was a situational analysis meeting in his office. General Udet, who had the final say for the RLM, agreed with Prof. Messerschmitt to call off the attempt. No more flights, the course was to be dismantled and the specialists sent home. It was over for the record flight. With heads hung low and many a curse on their tongues, the people involved slowly filed out of the conference room. The mood of the players sank to zero. All the efforts had been for naught, the risks which had been taken were inconsequential, and the ground crew had labored night and day on the plane and its engine for nothing.

Then Peter cast a sympathetic eye on the scene. Around noontime he parted the clouds and the sky turned a spotless

```
                    Z e i t n a h m e p r o t o k o l l .

Betriff: Rekordversuch: Geschwindigkeit über Grundstrecke
         Flugzeugführer: Fritz Wendel
         Flugzeug: Bf 109 R, D - I N J R
         Datum: 26. 4. 1939
         Rekordversuchsstrecke bei Augsburg
         Zeitmeßgerät: Olympia-Zielzeitkamera.

Als Zeitnehmer 1. Klasse und Sportzeugen des Aero-Club von
Deutschland bestätigen wir, daß die Zeiten der 4 Grund-
streckendurchflüge anläßlich des obigen Rekordversuches
entsprechend den FAI.-Vorschriften mit der Olympia-Ziel-
zeitkamera genommen wurden. Die Auswertung der Filmstreifen
wurde nach dem der FAI. bekanntgegebenen Verfahren vorgenom-
men. Sie ergab folgende Durchflugzeiten und-geschwindigkeiten:

                    1. Durchflug
von 13,8049 Sek, entsprechend    782,330 km/h
                    2. Durchflug
von 14,7072 Sek, entsprechend    734,334 km/h
                    3. Durchflug
von 13,9230 Sek, entsprechend    775,694 km/h
                    4. Durchflug
von 14,8312 Sek, entsprechend    728,194 km/h

                                              - 2 -
```

Conceptual drawing showing how Askania cameras were used to measure the record-breaking flight. The three-kilometer long route along the railroad line between Augsburg and Buchloe had to be flown four times at an altitude of 75 meters.

```
                    - 2 -


Das für die Anerkennung des Rekordes maßgebende arithmetische
Mittel der 4 Durchfluggeschwindigkeiten beträgt:

                    755,138 km/h.

Augsburg, den 28. April 1939

                    Rieckmann.  Hild


            Unterschriften der Zeitnehmer 1. Klasse
                des Aero-Club von Deutschland


                  (Dr. Rieckmann, Dr. Hild)
```

Statement by the timekeeper from 26 April 1939 as handed in to the FAI in Paris as evidence of the record-setting flight. The aircraft was given as a "Bf 109R." The highest speed clocked by the Me 209V-1 was over 782 km/h.

The Me 209's military application was researched with this V-4 having increased wing area and a DB-601A engine. When the aircraft proved unsuitable as a fighter, the V-4 was nevertheless given a military paint scheme as a propaganda ploy in order to be able to announce to the world the birth of a new fighter plane.

blue. I couldn't stay in my office any longer; by 3 o'clock that afternoon I knew that such ideal weather had to be exploited. But none of the senior managers were in their offices, and I wasn't able to get ahold of anyone by telephone to reverse the decision of that morning. I was able to get the director of test flying on the line, but that wasn't enough. I made a judgment call and, taking responsibility for the consequences, ordered another flight be made ready—so long as Fritz Wendel would volunteer to fly later on that day. Fritz told me he was willing, and landed around 1800 hrs—everything had gone well. We waited a long time at the Hotel 'Drei Mohren,' until around midnight when the results had been tallied and recognized by the gentlemen from the FAI. The absolute speed record of 755 km/h was ours."

In June there was also an attempt to use the 209V-3 to break Heinkel's 100 kilometer record. In any event, the RLM was no longer interested in this record-setting see-saw between Messerschmitt and Heinkel; Messerschmitt's record with the 209 (called the Me 109R in propaganda literature and for the official FAI listing) was intended to show the world that the Me 109, the *Luftwaffe*'s standard fighter, was also the fastest plane in the world.

Flown by Dr. Wurster for the first time on 12 May 1939, the Me 209V-4 had been planned from the outset for the military role. It was fitted with a standard 810 kW/1100 hp DB 601A engine. The primary focus of the flight test program, which extended into early 1941, was on the plane's flight handling characteristics. Yet these remained poor despite increasing the wingspan by 1.50 m to 9.30 m and thereby giving it an area of 11.7 m². Dr. Wurster ended his 15 September 1940 flight report with the words:

"Takeoff in the plane is quite bad. When using the longer diagonal runway it barely cleared the obstructions at the end, even though we're taking off with 1.42 *ata* boost and running the engine at excess rpm with a takeoff weight 300 kg below maximum. I would like to point out that such a takeoff is unacceptable for a fighter plane."

Heinrich Beauvais and Alexander Thoenes, the official test pilots for the *E-stelle*, also saw no advantages to the Me 209 over the Bf 109F (undergoing flight testing at the same time)—at least when it came to flight handling. Officials in the ministry had held out little hope for the 209's success from the start, as can be seen in a statement from the RLM-LC flight development program from September 1938:[3] "Bf 209, light fighter: no intention of procuring type other than a pre-production series, as continued development of the Bf 109 is more advantageous from a field perspective." Not even the pre-production series mentioned was ever built.

[3] RLM/LC Aircraft Development Program LC No. 243/38, Secret, from 3 Sept. 1938

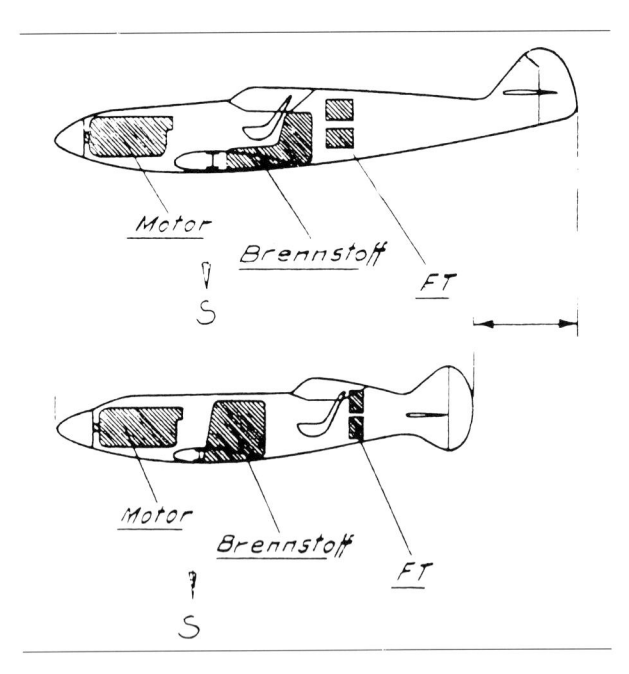

Available space for fixed and variable loads would have been utilized much better in the Me 209, with its fuselage some 21% shorter and having approximately 30% less surface area, than in the Bf 109E. However, it would have suffered from having a higher wing loading and, consequently, higher landing speed and significantly worse visibility for the pilot on landing and takeoff.

Yet another rework of the military Me 209V-4 from November 1939 envisioned a high-speed fighter with DB 601E engine, two MG 17s and one MG FF or MG 151, as well as the cockpit being moved forward. It was hoped that performance would be improved by increasing the wing loading. Empty weight was expected to have been 2,152 kg, with takeoff weight at 2,660 kg. This was the last of the projects based on the record-setting design.

Thus ended this first 209 program. Its record and reputation stood for another thirty years. Even though the war and later periods saw the development of faster propeller-driven planes which could have broken the 209's record, this didn't officially occur until 16 August 1969 when Lockheed test pilot Darryl E. Greenamayer attained 771 km/h in a modified Grumman F8F-2 Bearcat. At the time of this writing, the record stands at 832 km/h, set in a modified North American P-51D Mustang.

Parts of the record-setting Me 209V-1, D-INJR, survive today in the Krakow museum in Poland. The plane had been handed over to the German Aviation Collection in Berlin in September of 1942, then shipped off to the eastern Reich during the war for storage.

The fate of the other V-types is also well documented: V-2 was disassembled and put into storage in a Reich-controlled warehouse; the V-3 was set aside for a planned Messerschmitt museum at the Augsburg works, where it was destroyed in a bombing raid; the V-4 was sent to *Luftzeugamt* Erding for scrapping.

The "New" Me 209

Constant improvements to the Bf 109 and the newly announced Daimler-Benz DB 603 engine (fighter/*Zerstörer* standardized engine) led to the idea in early 1942 of resurrecting the Me 209 type number for a new type of airplane incorporating all these improvements into one. A comparison study made by Richard Bauer's newly established "109/209 program management" section on 4 March 1942 showed that many of the Bf 109G's components (e.g. fuselage, wing sections, tailplane, and cockpit) could be utilized on the one hand, while on the other hand the entire engine assembly, the wide-track undercarriage, the vertical stabilizer (a butterfly-tailed variant was also studied), and the installation of more guns would all have to be designed from scratch.

At the same time, however, development and prototype construction of the Me 309 was in full swing. There were high hopes that the Me 309, which had completed its first flight on 18 July 1942, would be the one to replace the Bf 109, and because of this development of the Me 209 continued throughout all of 1942 with little urgency. Bottlenecks in men and materials became increasingly common as the war progressed, and one of the consequences was a reduction in the number of new developments within the aviation industry. This approach had reached its peak by late 1943. One of the first victims of this trend was the Me 309, whose development was halted by the RLM on 26 January 1943—in favor of the Me 209. Just prior to this, the project department had submitted another comparison study between the Me 209 and the 309 showing that there was

146

little difference between the two types in their planned operational variants. Just the opposite was true, in fact, for the Me 209 stood in a much more favorable light given that roughly 50% of its components and production capacity could be pulled from the Bf 109G production, which at the time was running at full bore.

A high-level official decision was made, and in 1943 planning began for production of the Me 209 to begin at the Regensburg Works. Given the fact that the Messerschmitt Corporation was producing or refining the Bf 108, 109G, 110, Me 163, 210/410/, 262, 323, and 328, plus building the Me 309 and Me 264 prototypes, at the time it was decided to concentrate on only the major aspects of such an undertaking. These involved:

- Completion of the construction documents by Richard Bauer at the *Kobü* in Augsburg, at the same time making the wings and control surfaces out of wood.
- Awarding contracts for the acquisition of raw and semi-finished materials, and finished components such as the undercarriage struts, annular radiator for the engine, armament, DB 603 powerplant (as a fighter/*Zerstörer* standardized engine*)*.
- Jig and tooling construction (for which additional technicians would be needed)
- Construction of five prototypes (three were laid down in Augsburg, two at Regensburg)
- Production of 40-60 aircraft in a pre-production series for operational evaluation (requested by *Generalmajor* Adolf Galland)
- Breakdown of variants for the various production lots.

Given that a quick production spin-up could only be accomplished by using the makeshift tooling and assembly jigs, Regensburg director Karl Linder pointed out the urgent need for technicians to be assigned to jig construction.

The Messerschmitt company officially submitted the project description to the RLM in April of 1943. According to the description, the Me 209 was to have the following roles and features:[1]

"1. Role.

The type is primarily to be employed as a standard fighter. For this role it is to be equipped with five guns (1x MK 109, 2x MG 131 in the fuselage, 2x MG 151 in the wing roots). In addition, with 9 guns it can be used as a heavy fighter/strike fighter (add'l 4x MK 108) in the wings. Furthermore, there are plans to use it as a fighter-bomber (3 guns and 1,000 kg bomb), long-range fighter-bomber

[1] *Reichsminister der Luftfahrt* GL/C-E2/*Festigkeitsprufstelle*, minutes from the Me 209 structural soundness requirements. Conference on 2/23/43 at Augsburg. Berlin-Adlershof 3/5/43

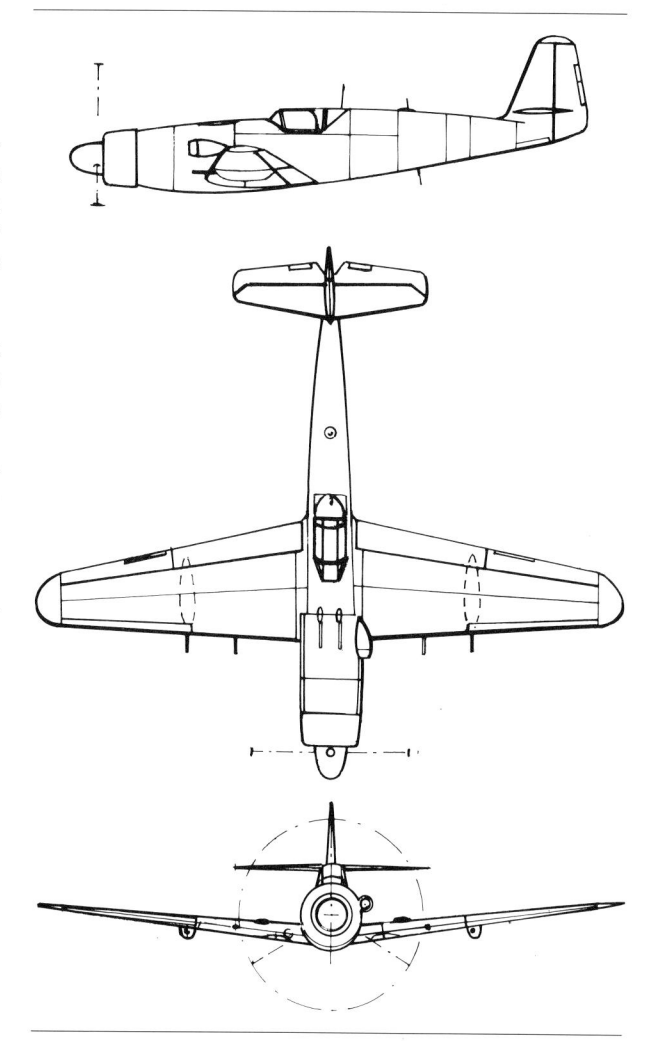

Three-view of the planned production Me 209 with DB 603 engine and six guns (four MK 108s and two MG 151s).

(500 kg bomb and 2x 200 l drop tanks), reconnaissance aircraft, and high-altitude fighter.

2. Dimensions and Performance.

Mtt has supplied the current data as follows: Takeoff weight as fighter G = 4 metric tons, maximum takeoff weight as long-range fighter-bomber G_{max} = 4.88 metric tons, takeoff weight as strike fighter G = 4.74 metric tons (including 275 kg add'l armor). Wing area F = 16.4 m², wingspan b = 10 m. Larger wing area is expected for high-altitude fighter and strike fighter roles, with F = 21.6 m² and b = 13 m. DB 603 engine, with possibility of fitting BMW 801, Jumo 213 and Jumo 222.

Maximum speed in level flight with DB 603
   at sea level v_h = 570 km/h
   at 7000 m v_h = 725 km/h."

In April 1943 a high-altitude fighter variant of the Me 209 with DB 628 engine was researched, drawn up, and offered to the *Amt*.

In November of 1943 another project was offered, this time based on the Junkers Jumo 213E engine then in development. Ludwig Bölkow, who at the time was responsible for the K-variant of the Bf 109G at Wiener Neustadt, estimates that throughout the 209's developmental life over 100 different designs were studied in the project department, most of these having been drawn up by Fritz Hügelschäffer.

The program suffered a setback in the midst of this startup phase (much of the construction documentation had already been sent out and the five prototypes were either under construction or awaiting parts), which virtually heralded the end for the Me 209, even though exchanges between the contractor, the company, and the *Luftwaffe* continued on for some time afterward. It was sparked by General Galland's observations following a Me 262 test flight on 22 May 1943 at Lechfeld. On 25 May Galland wrote a letter to *Generalfeldmarschall* Milch enthusing over the Me 262's qualities and "requesting permission to make the following suggestions" regarding the 209:[2]

"We've got the Fw 190D in development, a design which can be considered on par with the Me 209 in all areas. Neither type will ever be able to surpass their enemy counterparts in terms of performance, especially at higher altitudes. The only areas which can be improved are armament and speed. Therefore:
a) terminate the Me 209 program.
b) focus all fighter production on Fw 190s with BMW 801 and DB 603/Jumo 213 engines.
c) convert freed-up construction and fabrication capacity to building the Me 262 at once."

In a memorandum to the chairman of the board Friedrich-Wilhelm Seiler, operations manager Rakan Kokothaki, and production manager Fritz H. Hentzen, dated 24 May 1943, Willy Messerschmitt spelled out the consequences (which, by the way, would also have sealed the Bf 109's fate) and his arguments for continuing with the 209:[3]

"General Galland's test flight in the 262 has prompted a discussion on accelerating production of this type. A suggestion was made to drop the 209 and start up 262 production in its stead. General Galland has now demanded that fighter numbers not be allowed to drop below the figures set by the increased output program and that, by canceling the 209, the 190 is expected to cover the numbers when Bf 109 production ran out. In order to achieve commonality of fighters within the German *Luftwaffe*, shifting to a *single* piston-engined fighter type is only advantageous when there is no loss in production and when the best performing aircraft is the one selected. Which brings up two points:

1. In my opinion, it is very doubtful whether it will be possible for those factories producing 190s to increase their output when fighter numbers drop with the end of Bf 109 production.
2. Such a decision can only be made when the better performing airplane has been determined through fly-offs. This latter point can be expected to have taken place by the end of the year.

In any case, these two matters must be reviewed with the utmost care to avoid having to reverse any decision once it's made."

Milch, with Göring's approval, dropped the Me 209 from the program on the 25th of May, the same day as Galland's letter.

Messerschmitt took the opportunity of a visit to Hitler's Obersalzburg headquarters on 27 June as part of an emergency conference of the seven leading aircraft designers to renew his defense of the Me 209. He amended his previously stated arguments with the claim that the Me 262's fuel consumption was too high and therefore would jeopardize the current fuel supplies. Willy Messerschmitt was permitted to renew his work on the 209 and, under the personal supervision of *Staatssekretär* Milch, competitors Willy Messerschmitt and Kurt Tank would compare the advantages and disadvantages of their respective designs.

Willy Messerschmitt soon seized upon one of his typical arguments for the Me 209: the Ta 152 was larger and therefore about 400 kg heavier than the 209. This would have an impact upon its flight performance and affect the stockpile of raw materials, already recognized as being in short supply. In a letter to Friedrich-Wilhelm Seiler on 2 June, Willy Messerschmitt expressed his deep concern about renewed talk of canceling the 209 in favor of the 262, this time with another counter-argument that the latter's engines had not yet reached production maturity. He asked Seiler in Berlin to encourage the decision-makers to carefully rethink the cancellation of the 209. Hidden in this request was clearly an unspoken resolve to avoid any repetition of the 13th of March, 1942, the day which saw production halted on the Me 210. Willy Messerschmitt was still reeling from the shock.

On 3 November 1943 Fritz Wendel took off from Augsburg for the first time in the new Me 209 (V-5), followed by the V-6 on 22 December that same year. By this time nearly 85 percent of the blueprints and documentation for production had been completed. However, with other aspects of the program running into delays, particularly in the area of jig construction, it was finally decided to kill the

---
[2] A letter by the *Luftwaffe's* General of the Fighters (Galland) to *Generalfeldmarschall* Milch on 5/25/43 in David Irving's "Die Tragödie der Deutschen Luftwaffe. Aus den Akten und Erinnerungen von Feldmarschall Milch." Frankfurt am Main, Berlin 1970
[3] Messerschmitt AG, Augsburg, Willy Messerschmitt, memoradum from 5/24/43 to Messrs Seiler, Kokothaki, Hentzen

The new Me 209 (V-5, SP+LJ) with DB 603 engine, as flown on its first flight by Fritz Wendel on 3 November 1943.

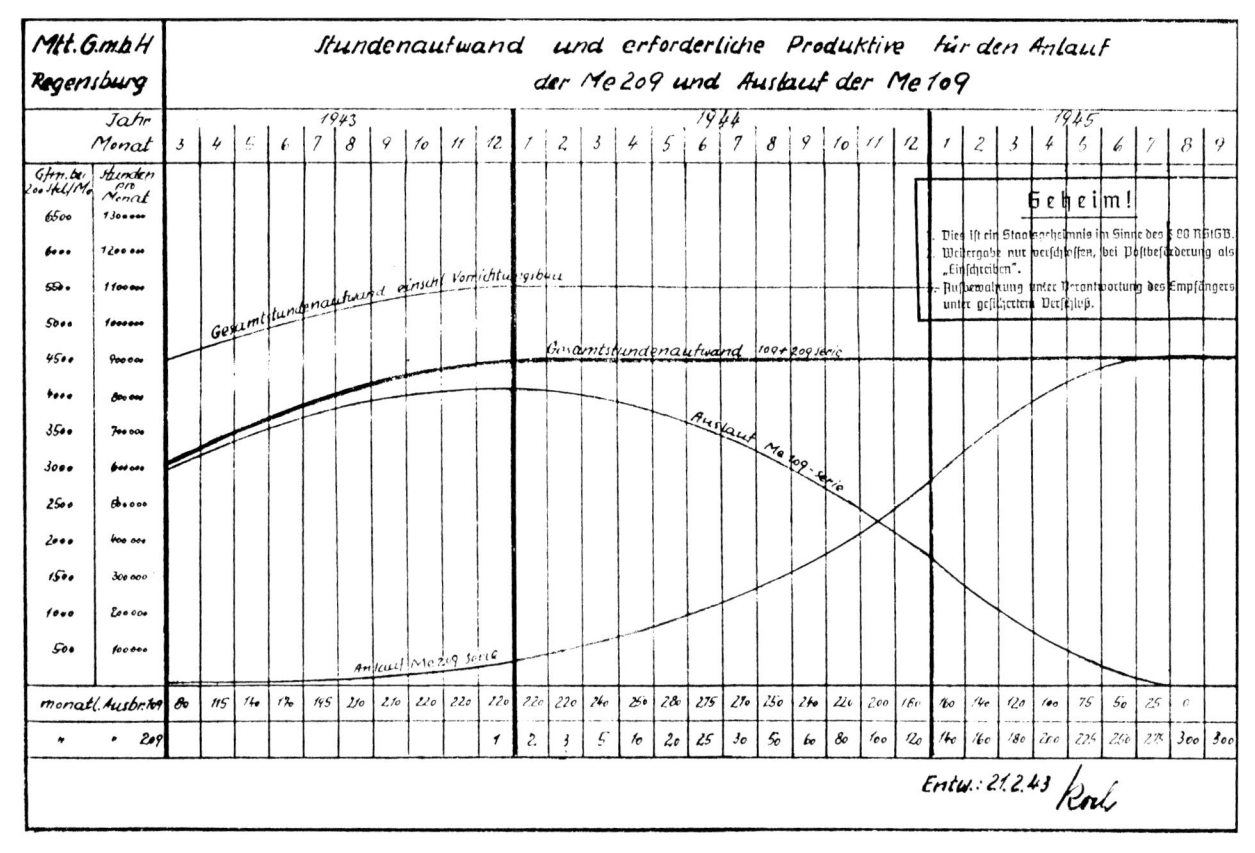

The high risk associated with introducing a new aircraft type during a war can be gleaned from this planning graph prepared by the Regensburg company, dated 21 February 1943, showing initiation of Me 209 production and dropoff of Bf 109 output. Notice the two year period from initial to peak production of 300 aircraft per month for the Me 209. This was not taking into account the as-yet unidentified problems which field units and combat operations would undoubtedly reveal for the new type. Willy Messerschmitt had on several occasions warned of the much greater difficulties which were to be encountered with the introduction of the Me 262, also under discussion at the same time.

149

209 for good on 24 November and clear it for export to Japan (in place of the Me 309) at the same time. Messerschmitt's minutes from 14 December 1943 portray the overall picture thusly:[4]

"As Mtt.AG is already aware, the *Herr Reichsmarschall* has ordered that further development of the Me 209 is to cease immediately.

The first test flights with the 209 have shown that, excepting the tailplane, the aircraft's handling is in order. Since most of the production documentation has been completed for the type, the RLM intends to offer the 209 for export in place of the 309.

Mtt.AG. is tasked with taking one of the two completed prototypes, V-5 or V-6, and conducting remaining stability evaluations using the larger 2.8 m² tailplane prior to carrying out a review of the type's general handling characteristics. This work (basically flight testing) is to be organized so that it does not interfere with the high-priority Mtt. development program."

The Japanese military attaché in Berlin, Major General Otani, had been negotiating with the RLM and the Messerschmitt AG for months in an effort to obtain a Me 309 prototype and its associated documentation. During one of these discussions, on 14 January 1944, he was informed that the 209 had been approved for export. Otani was told that the 209 enjoyed the following advantages over the 309:

- 85 percent of documentation already completed.
- Satisfactory flight handling even after just the initial evaluation flights, plus better turning radius than the Bf 109G.
- 10-15 percent higher speeds than the Me 309.

The response to Otani's query as to why the 209 and 309 had not gone into large-scale production for the *Luftwaffe* was that the 309 had been dropped from the program because it had been deemed too much of a financial risk to start up production, as well as because of the nose gear (which the Americans had also had problems with on their Airacobra).

Otani learned that the reason for canceling the 209 was because Germany was involved in full-scale production of the Fw 190. However, in no way did this imply that the former was an inferior machine, since the *Technisches Amt* had been just as pleased with the 209 as it was with the Fw 190. Selection of the Fw 190 had been made with a view toward commonality of production and in light of the Fw 190's ability to accept three different engine types, the BMW 801, the Jumo 213 and the DB 603.

[4] Messerschmitt AG, Augsburg, minutes (Me 209 authorization for export) from 12/14/43

In any event, Japan initially leaned toward the purchase of the 309 (see chapter). Thus ends the story of Messerschmitt's last two propeller-driven fighter planes falling between the Bf 109 and the Me 262. Had everything gone according to plan, production would have been well underway by this point (monthly output for the Me 209 had been projected at 100 by November 1944, jumping to 300 by August of 1945). Within the company, however, feelings of animosity between Willy Messerschmitt and Rakan Kokothaki smoldered for a long time afterward because of Messerschmitt's active campaigning for the 209.

Me 209/309 Performance Table
compiled by the *Probü* on 1/15/1943 for comparison purposes[5]

Aircraft:	**Me 209**	**Me 309**
Weight:	3935 kg	4085 kg
Area:	17 m²	16.5 m²
Aspect Ratio:	6.3	7.6
Engine:	DB 603A(603G)	DB 603A(603G)

Maximum Speeds:
Climbing and Full Military Power

Altitude (m)	Speed (km/h)	Speed (km/h)
0	571 (571)	574 (574)
3,000	653 (653)	657 (657)
7,200 (8,500) V_h	724 (742)	730 (748)

(maximum speeds of the Me 209 can be increased by approx. 9 km/h with fitting of Me 309 wing)

Climb Rates and Times: Climbing and Full Military Power
Me 209

Altitude (m)	Climb Rate (m/sec)	Time to Climb (min)
0	17.8 (17.8)	- (-)
6,100 (7,400)	13.0 (11.6)	6.2 (8.2)
8000	8.7 (10.0)	9.2 (9.0)

Me 309

Altitude (m)	Climb Rate (m/sec)	Time to Climb (min)
0	17.3 (17.3)	- (-)
6,100 (7,400)	12.8 (11.4)	6.3 (8.3)
8,000	8.7 (10.0)	9.3 (9.1)

Ceilings (0.5 m/sec climb rate):

Altitude (m)	11.6 (12.2)	11.7 (12.3)

Landing Speed:

G_L	3,468 kg	3,600 kg
v_L	149 km/h	161 km/h

(with 309 wing landing speed increases approx. 10 km/h.)

Boosting the DB 603G engine's climbing and full military power by 200 hp (possibility according to *Oberst-Stabsing.* Mann from 1/20/43) will provide a corresponding improvement for both types.

Weights for both types have been provided with comparable configuration (2x MG 131, 2x MG 151, 1x MK 108)

[5] Messerschmitt AG, Augsburg (project dept.), Me 209-309 comparison study from 1/15/43

Me 309 showing its planned armament of seven machine guns and cannons in the fuselage and wings.

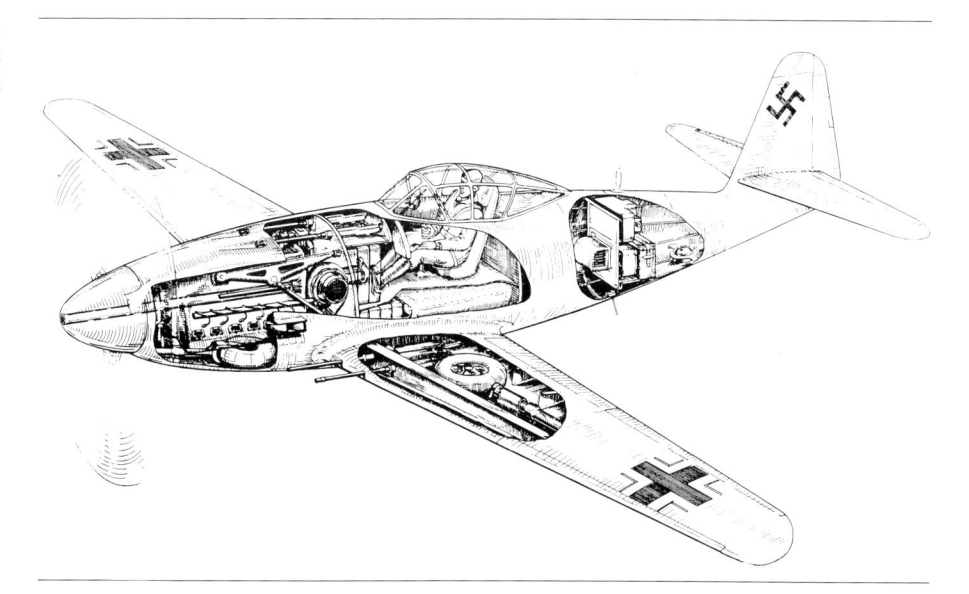

Me 309 - Successor to the 109

With thousands of examples built in the seven years since the Bf 109 had left the drawing boards, in early March of 1941 Woldemar Voigt in *Probü* and Richard Bauer, the 109/209 project director from the *Kobü,* began working on a completely new fighter which would make use of all the innovative technology uncovered up to that point. These innovations included:

- Installation of the new Daimler-Benz DB 603 engine, just announced, which was expected to develop 1,100 kW/ 1,475 hp and therefore be more powerful than the DB 605 then in production. (In December 1941 talks began with Junkers on fitting its Jumo 213, also just announced).
- Significant increase in firepower with up to seven machine guns/cannons.
- Improving takeoff and landing characteristics by utilizing a wide-track undercarriage and a nosewheel—a first for Messerschmitt.
- Aerodynamic refinements, such as the incorporation of a laminar flow wing profile (later also a swept wing), and oil and water radiators capable of retracting into the fuselage.
- Construction of the airplane using modern manufacturing techniques, with special emphasis on the Messerschmitt method of monocoque construction. (Fuselage divided into ammunition section, cockpit, and tubing, the latter joined together using screw-type formers.)
- Other improvements, such as pressurized cockpit, ejection seat, and a Me P 6 variable/reverse pitch propeller.

Daimler-Benz had told Messerschmitt that he could expect delivery of the first DB 603 engines for his Me 309, Me 310, and lastly, the Me 209, by August of 1941. Willy

Me 309V-1, not yet armed, as flown on its maiden flight by Karl Baur on 18 July 1942.

Next to the He 280, the Me 309 was the second German aircraft with a nose gear.

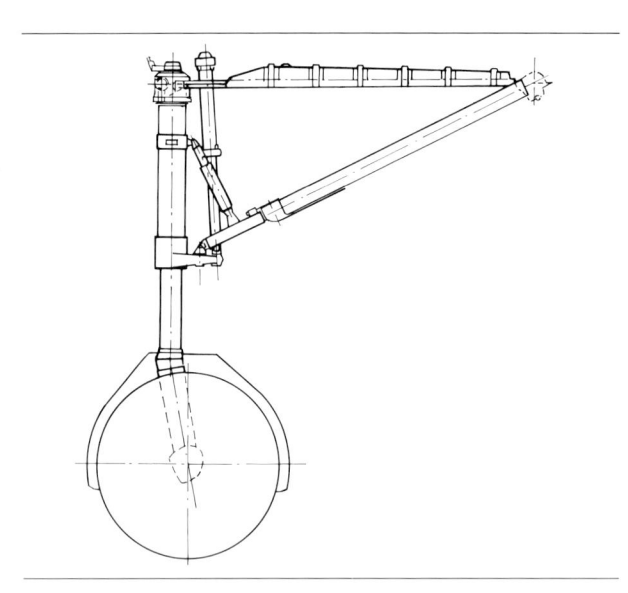

Detail of the nose gear, designed in part by Willy Messerschmitt himself.

Messerschmitt therefore optimistically estimated (and with the Me 309's special prioritization status in mind) that the first 309 prototype and Me 310, also in development, would be finished by the late summer of 1941. But it was soon apparent that Daimler-Benz would not be able to meet its promised date. The first pre-production engine didn't ar-

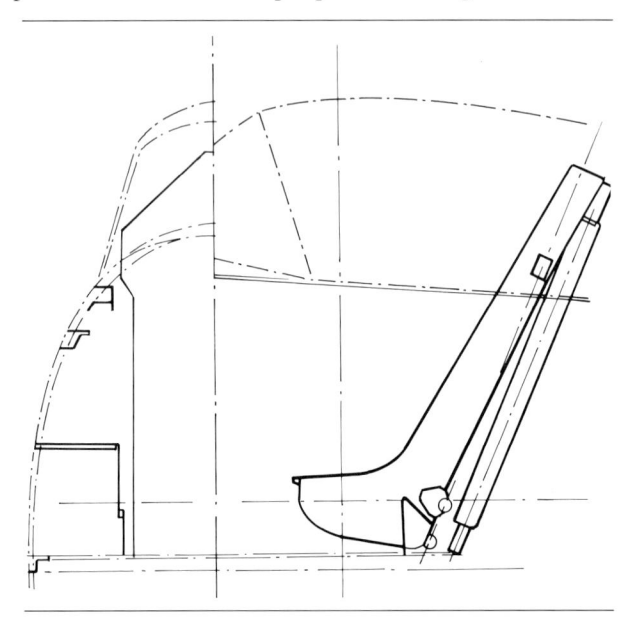

Conceptual drawing of the ejection seat installation in the cockpit of the Me 309V-1.

rive until six months later. Similar circumstances occurred with the Jumo 213, which didn't arrive in pre-production form until November 1942.

These delays precluded Hubert Bauer in prototype construction from laying down the first of the RLM's ten Me 309 V-types until October of 1941. Nevertheless, by this time many of the innovations to be applied to the Me 309 had already been tested on Bf 109V-series, such as the pressurized cockpit, the retractable radiators, the wide-track undercarriage, and the nose gear. Even the Me P 6 variable pitch propeller, designed and built in a special department under the supervision of Robert Prause, was also tested beforehand on the Bf 109.

As always, and especially with new designs, Willy Messerschmitt was personally involved in the design details. Such as the retractable nose gear, whose attachment point was a mounting flange on the engine's reduction gear housing, an idea probably employed for the first time and for which Messerschmitt applied for a patent. Subsequent testing, however, revealed that this approach placed a greater stress on the engine and made it susceptible to problems.

The first flight of the Me 309V-1 took place on 18 July 1942 at Augsburg with Karl Baur piloting the airplane. That same day, just a few kilometers away at Leipheim, Fritz Wendel flew the Me 262V-3 under jet power for the first time. It is somewhat ironic to know that Messerschmitt's two most modern airplanes had closely linked problems with their undercarriage. The most modern propeller-driven plane had a nosewheel (after the He 280, the second in Germany) which caused problems in the ensuing evaluation phase, something which the designers had anticipated. And the most modern airplane of the period, the Me 262, had a tailwheel which increased the difficulties in taking off. Thus, beginning with the V-5 the Me 262 was fitted with the nosewheel intended for the production version. For the time being the Me 262's designers wanted to avoid the problems which the Me 309 was having with its nose gear, as they expected to have their hands full with the 262's new engines in any case.

As would be discovered over the coming weeks and months, the Me 309's flight testing kept a pilot much busier in the cockpit than the trusty Bf 109G. There was a price to pay for the new technology, from problems with the nose gear (flutter, failure to extend or retract, hydraulics) and the new, immature engine and its new oil and water radiators (loss of oil and water, overheating), to the propeller, which could be set to reverse pitch as a braking aid on landing.

Nor were the flight handling characteristics satisfactory, especially the roll rate and turning radius, which were no improvement over its predecessors. Heinrich Beauvais noted that "the Bf 109 enjoyed quite an advantage in a turn" following one of the first test flights and a fly-off between the 309V-1 and a Bf 109G by the *E-Stelle* in November 1942. At this time the Me 309 test program took top priority. Yet nearly all flights took place with the V-1 only, for the V-2 did not arrive for evaluation until 29 November, and then was out of commission for even longer when its nose strut broke while taxiing the same day. The V-3's first flight followed on 28 December 1942. Assembly of the V-4 and V-6 through V-10 was delayed because of holdups in powerplant deliveries.

Production installation documentation for the DB 603 had still not been made available by October 1942 (in the meantime, the engine was also needed for the Me 410), and Willy Messerschmitt informed the *Amt* that without this documentation he could not meet the Me 410's early 1943 and the Me 309's late 1943 deadlines. As the type had been given top priority, it meant that some serious planning decisions now had to be made. This affected the production run, the breakdown of the variants, and any projected follow-on developments.

The following sub-types had been planned:

- Me 309 with DB 603 as a light fighter with 2x MG 131 and 1x MK 103 or 108 in the fuselage.
- Heavy fighter with 2x MG 131 and 1x MK 103 or 108 in the fuselage and 2x MG 131, 151 in each of the wing roots.
- Dive bomber/fighter-bomber with 2x MG 131 and 1x MK 103 or 108 in the fuselage and 2x 250 kg bombs under the wing.
- Bomber with wingspan increased by 75 cm and 2x MG 151 in the wing roots and 1x 1,000 kg bomb (SC 1000 or

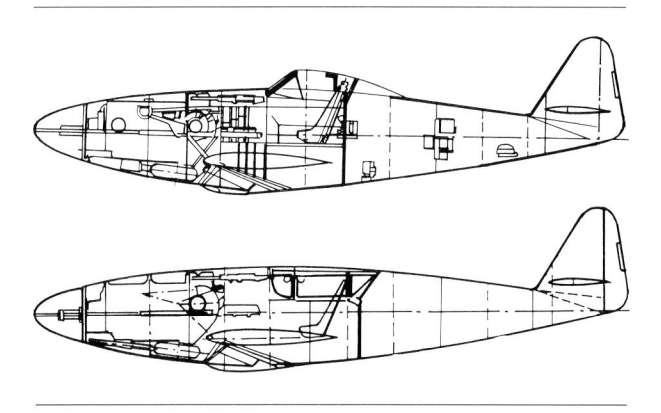

The Me 309, showing standard layout and a version proposed by Friedrich Schwarz with sunken cockpit (periscope plane), providing better protection for the pilot.

PC 1000) on a starboard mounted wing rack, plus a 1,300 l fuel tank on the port side close to the fuselage.

Follow-on projects included:

- High-altitude fighter using various engine types.
- Penetration reconnaissance aircraft with camera.
- Periscope plane with the cockpit completely enclosed within the fuselage, giving the pilot increased protection. This was a design by Friedrich Schwarz, director of the armaments division and one of Messerschmitt's colleagues from his Bamberg days.
- Project for a Me 309 with swept wings, expected to make it 30% faster. Caudron in France was to have taken over this high-priority project, including associated wind tunnel testing, in late 1942.
- Twin variants as a high-speed bomber using many of the Me 309's components.

The fighter-*Zerstörer* standardized Jumo 213 engine was to have been considered as an alternative to the DB 603 when designing many of the above-named variants.

Development of the Me 309 was officially stopped and the type certified for export on 26 January 1943. This action came as no real surprise, for by the end of 1942 the aviation industry had reached its capacity and bottlenecks in materials were becoming more prevalent. Plus, the Me 209 (based on the Bf 109G), also in development, was showing as good or better promise with regard to performance and material consumption. Only minor follow-up work and the maiden flight of the Me 309V-4 on 21 April 1943 were undertaken after January.

In the summer of 1942 a Japanese delegation under the direction of Berlin's military attaché, Major General

Otani, had visited the Messerschmitt Works and witnessed flight demonstrations of the Bf 109G, Me 210, Me 309V-1 and the *Giganten*, the Me 321 and 323. As a result, Japan bought two Bf 109Gs and four Me 210s, which had been cleared for export. Not approved for export, however, were the Me 309 and the Me 323. Then, in 1943, trade negotiations with Japan picked up again. General Otani agreed to Messerschmitt's offer of the Me 309V-3 plus 2,000 associated drawings (albeit not all complete) and 1,000 parts lists for the sum of 1.2 million Reichsmark. Otani ordered the airplane through a Berlin export company on 22 January 1944. (In the meantime, Otani had also heard that the Me 209 had been approved for export, see chapter.)

Then, with the Me 309V-3 ready for shipment from Augsburg, the event which everyone had long feared finally happened—the first large-scale Allied bombing attack on Augsburg and its armament industry. For two days, on the 25th and 26th of February 1944, American and British bombers sowed their seeds of destruction. Over 1,200 people died in the city and at the Messerschmitt Works. Another loss was the Me 309V-3, thus ending the story of the Me 309.

Some preventive measures against the expected bombing raid had already been taken at Messerschmitt, in that the technical developmental team under Willy Messerschmitt, the project department and the design/fabrication department were all relocated to Oberammergau in October/November 1943 to set up shop in the mountain infantry barracks there. Flight testing had already been transferred to Leipheim and Lechfeld at an earlier date.

For the aviation industry's production facilities, the pace was now stepped up for dispersing them to aboveground and underground sites (bunkering). It had not been completed by the time the war ended.

Bf 110 - A Heavy Fighter Defined

In the 'thirties, the air war theories of the Italian General Guilio Douhet and the Frenchman Camille Rougeron were increasingly finding their way into the growth, planning, and expansion of the world's air forces. These theories stated that bomber fleets would play a dominant role in any future war. In Germany, this viewpoint was championed by General Wever during the *Luftwaffe*'s early growth years, and he had accordingly organized the planning of the Do 19, Ju 89, and Bf 165 long-range bombers. Although his ideas were not initially pursued after his death, as the *Luftwaffe*'s commander-in-chief Göring became more and more interested in the concept of the high-speed medium bomber, the main examples in this category being the He 111 and Ju 88. It wasn't until 1939 that the long-range He 177 arrived on the scene, even though this latter type never really proved satisfactory from a technological standpoint.

The need for protecting one's own bomber fleets and at the same time fending off potential attacking bombers necessitated an entirely new type of aircraft, and this type had to first be defined and created. While an interceptor was expected to have "high speed, good climbing rate, and good maneuverability" (Bf 109), the new type must also feature "high speed, heavy firepower, outstanding maneuverability, plus long range." These "heavy fighters" were expected to engage and destroy attacking bomber formations long before the enemy had reached the country's borders. Thus was born another definition for this class of aircraft—the "destroyer" (*Zerstörer*).

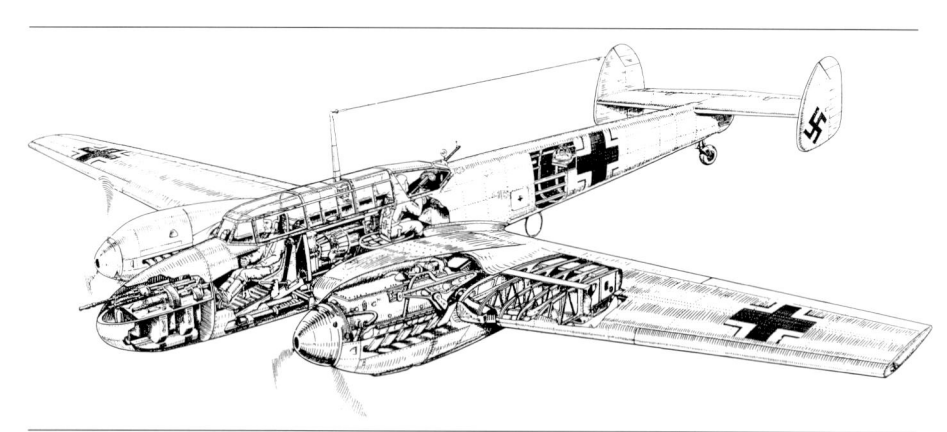

Bf 110C with DB 601 engines.

Project section chief Robert Lusser played a significant, if not defining, role in the development of all Messerschmitt designs from the Bf 108, 109, and 110 to the early stages of the Me 262.

Below: The original shape of the Bf 110, as tested in January 1936 by the AVA in Göttingen in model form, was an enlarged and twin-engined Bf 109 which offered a twin-tail design as an alternative.

Development of the Bf 110

These were some of the theoretical considerations which played a role as Willy Messerschmitt and project chief Robert Lusser carried out ongoing discussions on the topic with the RLM from April to June of 1934. From these talks, the general role and shape of the Bf 110 took form. When the discussions concluded, Messerschmitt provided the RLM with a memorandum for the P 1035 (as the design was known in the company), in which he offered his solution for the *Zerstörer* role (which later became the Bf 110), with the recommendation that the basic type could also be built as a high-altitude strategic reconnaissance plane (the later Bf 161) and as a bomber (the later Bf 162). Different roles could be met by making various modifications to the design, mostly in the area of the fuselage. This was the first time that Messerschmitt had proposed something along the

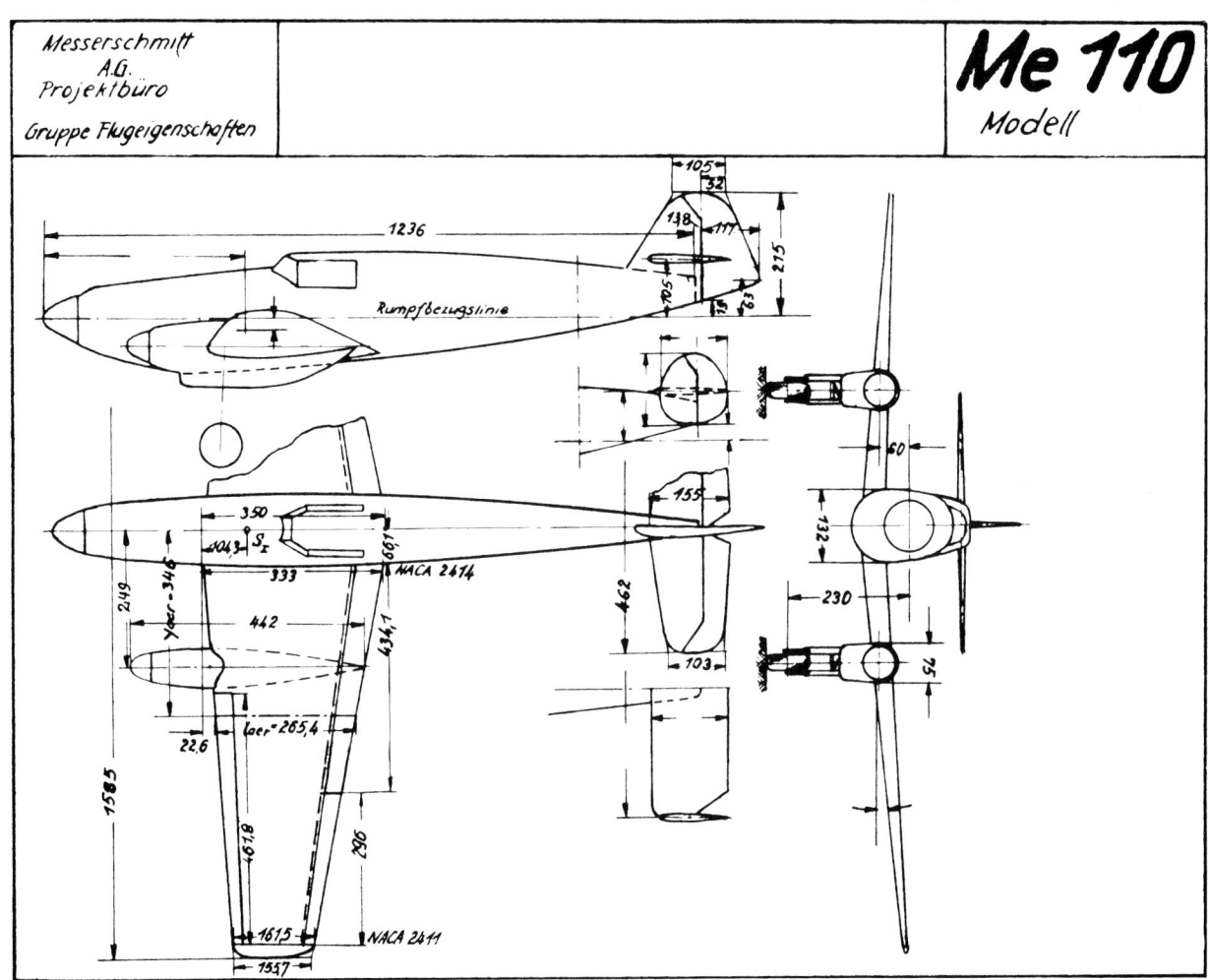

The Bf 110V-1, Werknr. 868, in its original form with Jumo 210 engines and squared-off rudder edges, as flown by Dr. Wurster on 12 May 1936 on its maiden flight. Carrying the registration D-AHOA, it was delivered to the Travemünde test center in October 1937.

lines of a modular construction method, an idea which he subsequently employed with increasing frequency, e.g. for the Me 410, P 1090 and even the Me 262. The basic premise behind this concept was to save on developmental resources, materials, assembly jigs and tooling, and therefore produce the design more economically and smoothly.

The RLM embraced the idea, and within just 15 months awarded contracts for all three models. The ministry gave the go-ahead for development of the *Zerstörer*/heavy fighter and assigned it the official designation of Bf 110 in early August of 1934. At the same time it contracted with the BFW for construction of three prototypes and an airframe for stress analysis. A pre-production contract for a batch of seven planes followed in January 1935.

The *Probü* looked at several different layouts in its search for the optimal design. At least two options were evaluated in the AVA Göttingen's wind tunnel. The one variant was a twin-engined, enlarged Bf 109, the prototype for which was still being built. The other type was an aircraft with new, tapered wings and a long, slender fuselage with a multi-seat cockpit under a protruding canopy and twin rudders (thereby providing a better field of fire for the rear machine gun). This was the design which was selected. There followed an extremely long design phase, lasting two years, before the first prototype took to the air. This had been brought on by an unusually high number of changes requested by the contractor, although most of these were able to be made on the drawing board. Further delays in building the prototypes were caused by bottlenecks in the delivery of materials and equipment, which were plaguing the *Luftwaffe* even at this early stage in its expansion. With no prior examples or experience in building an aircraft with such a layout to fall back on, the question of the plane's armament, e.g. number and caliber of machine guns and cannons and their location, became a major point of concern.

Eventually the *Kobü* pulled off a brilliant feat of engineering design by bundling four MG and two MG FF guns in the nose, despite the fact that Robert Lusser wasn't overjoyed with the resultant fattened and unattractive nose. It wasn't until just before delivery of the pre-production aircraft on 1 July 1938 that the matter of armament was finally decided.

Powered by two Jumo 210 engines, the maiden flight of Bf 110V-1 (Werk-Nr. 868, D-AHOA) took place on 12 May 1936 with Dr. Hermann Wurster at the controls. It was the first Messerschmitt prototype to be flown by BFW's new chief pilot. Dr. Wurster had been working for the company since coming from *E-Stelle* Travemünde in January 1936 and had replaced Hans-Dietrich Knoetzsch (who had flown the Bf 109V-1 on its first flight the previous year).

The V-2 (Werk-Nr. 869, D-AQYE), also with Jumo 210 engines, followed on 24 October 1936, and Dr. Wurster was able to fly the V-3 (Werk-Nr. 870, D-ATII) on Halloween the same year. This was the first to be powered by Daimler-Benz DB 601 engines.

As the early DB 601s were prone to breaking down, it seemed too risky a proposition to equip the production Bf 110s with this powerplant. The aircraft therefore entered production in March of 1938 with the less-powerful, but more reliable, Jumo 210G. Companies such as Focke-Wulf, MIAG, and Gotha were also brought into the program as license-manufacturers for the type. The first machines went to *Zerstörer-Lehrgeschwader 1* at Barth.

Like the Bf 108 and Bf 109, the Bf 110 was of an all-metal duralumin construction (designed under the direction of Richard Bauer). The fuselage consisted of two half-sections, which in turn were comprised of multiple segments. These segments had their edges bent at right angles to form the bulkhead formers. These and the longitudinal members gave the airframe its stiffness and prevented twisting.

The single-spar wings were attached to the fuselage at four points. The number of ribs was kept to a minimum by using stronger outer panelling sections. The wings incorporated leading edge slats, and the ailerons and trailing edge slotted flaps, plus the liquid coolers on their undersides for the Daimler-Benz engines. The four self-sealing wing tanks were located in the area between the engine nacelles and the fuselage, and held a total of 1,270 l of fuel. The undercarriage retracted hydraulically (or with compressed air in

The first Bf 110B-0 production models, photographed in the spring of 1938 at the Augsburg factory airfield. These machines were not yet armed due to the fact that type and number of guns had not yet been established. In the summer of that year the aircraft were transferred to the heavy fighter group in Barth, while Bf 110V-4 (D-AISY) went to Rechlin.

an emergency) into the engine nacelles. An inflatable rubber dinghy could be carried in the extreme tail end of the fuselage.

Situated beneath a plexiglas canopy, the crew compartment was so roomy that a third seat was installed for the night-fighter variant. The radio operator/gunner's compartment also housed all radio communications equipment, as well as RDF and blind flying systems, plus an MG 15 for providing defensive fire.

When the Americans were given a captured British Bf 110 for evaluation in 1941, they were able to see for themselves what Willy Messerschmitt had tried to achieve with his aircraft designs: simplified construction in order to be able to economically integrate complicated components into production. The American engineers were amazed at the heavy offensive armament, the ease of access for maintenance, the good view from the cockpit, and the well furbished cockpit instrumentation for the pilot. They were critical of the weak rear defensive armament.

The Bf 110B and its two Jumo 210G engines, plus four MG 17s, two MG FF and a defensive MG 15 had a takeoff weight of 5.6 metric tons and could reach speeds of approximately 400 km/h with a range of 1,800 km. Even though the range fit the original description, its speed was far below that of the announced high-speed bomber's at 500 km/h. Nevertheless, this early Bf 110 was an excellent machine. Willi Stör, chief of the production acceptance flight

A Bf 110B-1 with Jumo 210 engines and rounded wing and tail surface edges.

testing department and German aerobatics champion, proved this claim time and time again in mock dogfights with the Bf 109 which the company staged for ever more frequent visits from foreign delegations and purchasers. Stör considered the Bf 110 to be the best twin-engined aerobatic plane in the world.

With the numbers of DB 601 increasing in their availability by 1939, the Bf 110 was fitted with the new powerplant and, with a few other changes, went into production as the C-model in January. For *E-Stelle Rechlin*,

One of the first Bf 110C-1 types with DB 610 engines and modified wings, circa late 1939.

this change of engines was of limited value as evidenced in this report on the modification:[1]

"In no way does the propulsion system of the Bf 110C fulfill the requirements expected of a new-generation engine. Taking an aircraft designed for 2x 20 liter engines and retro-fitting it with 2x 30 liter engines without making any change to the airframe was a step backwards in development. Once the deficiencies listed on the opposite page have been corrected, only then can the propulsion system be considered ready for operations."

This conversion increased the takeoff weight by 600 kg to 6.26 metric tons, but speed also increased by about 100 km/h to over 500 km/h. Range dropped, however, to about 1,000 km. Given its original mission, the results were yet again something of a mixed bag.

Fritz H. Hentzen, the man responsible for all large-scale production contracts within production circle F 2, also flew all of his products from the Bf 108 to the Me 210; here he is seen in the cockpit of a Bf 110.

[1] E-Stelle Rechlin E3b2, Mustererprobung Bf 110C, Triebwerkerprobung, Report No. 5 from 15 Aug 1939.
[2] T. Osterkamp/F. Bachér, Tragödie der Luftwaffe, p. 128

Bf 110 in Combat

With its much heavier firepower compared with the single-engined Bf 109, the Bf 110 could be used in a wider variety of roles. Even as early as the Polish campaign, a lack of suitable targets in its original capacity as bomber destroyer led to the Bf 110 being employed in the ground support role. This role was repeated in the offensives in northern Europe and the West. There the Bf 110 was given another important task. It wasn't until the Battle of Britain, when the Bf 110 was used as an escort fighter for bomber formations, that the Bf110's limits became painfully obvious. Theo Osterkamp, at the time a fighter pilot on the Channel, expressed his opinion of the type in a postwar study:[2]

"The *Zerstörer* concept was one of fighter command's problem children. Neither those types developed in Germany, nor the British equivalent, the Defiant, were able to fully meet expectations as defined by the idea of a 'heavy day fighter' in terms of their flight handling characteristics. Based on a good showing in Poland and France, senior officials had expressed high hopes for the type, hopes which were dashed to pieces. In combat, the heavy-handed controls required such physical exertion that they became unbearable. The airplane proved to be so inferior to the British fighters—especially the Spitfire—in a dogfight, that it was useless as an escort for the bomber formations; indeed, the type required its own protection in the face of an energetically led attack from enemy fighter units. In such cases, pilots had no recourse other than to go into a so-called "defensive circle" for protection. When they eventually had to break this circle, it was generally accomplished with serious losses if there were no friendly fighters to cover them. In some cases British fighters were able to rack up kills 'like a string of pearl beads.'

Only in cases where the target area lay outside the range of the Me 109 were they pressed into service as bomber escorts. This almost invariably resulted in extremely high losses for our bombers and *Zerstörer*s."

This 1,050 liter fuselage tank, commonly known as a "Dackelbauch" (dachshund belly), was fitted to D-series Bf 110s in an attempt to improve the range. However, the loss in speed was so great that it was decided to use wing tanks instead.

Like any airplane, the Bf 110 had its proponents as well as its enemies. Pilots often varied widely in their opinions. Johannes Kaufmann's account of his training at Oberschleißheim in 1941 fluctuated from neutral to positive (he went on to fly the 110 on the Eastern Front):[3]

"After completing the first three weeks of instruction, we began training on the Bf 110. The airplane was considered to be one of the so-called well-behaved types and, from a purely aeronautical perspective, offered no challenges. Takeoff was normal, with no particular tendency to ground loop. Landing was just as easy. Maneuverability was acceptable, given the size of the airplane. Acceleration and climbing weren't particularly noteworthy. Nevertheless, one has to agree that for its time this aircraft type fit the bill in the so-called *Zerstörer* role.

Air-to-ground gunnery training against targets using the Bf 110 was not demanding. The plane was relatively stable in the air and sighting could easily be corrected in flight. As on other airplane types, there was a problem with pulling out at the right time."

The opinion of Walter Horten, a technical officer in *Jagdgeschwader 26*, has more of a negative overtone. He was able to test fly the Bf 110 at *Zerstörergeschwader 26* before the Western Offensive and subsequently outmaneuver the type in his Bf 109E. He warned against using the 110 against England:[4]

"I test flew this 'mule' and found it quite well behaved and pleasant, no comparison with the sluggish Do 23 from bygone days in Giebelstadt. However, in key areas it was lacking—it was simply not maneuverable enough to be considered a fighter aircraft. Control forces were too heavy at somewhat higher speeds, and with its 16 meter wingspan and high weight was just too clumsy—pulling up out of a dive required the strength of an ox."

And:

"For me it was inconceivable how an escort fighter for bombers could ever have gone into production as such, despite testing at Rechlin and in the field! For in this role it is clearly inferior to the more maneuverable, single-engined fighters it must clash with."

Over the channel six months later, Horten witnessed the catastrophe that befell ZG 26 when one day 24 of the 28 Bf 110s it sent out failed to return:

"A catastrophe on several levels; one was the loss on this mission, another was the awareness that four *Zerstörer* wings were tactically ineffectual as fighters on this front! And that through military aviation planning errors another eight fighter wings were missing from the Battle of Britain (two Bf 109s could be built for every Bf 110). We fighters now had to protect these 'kites' if they were to remain in action over the Channel."

In its role as a long-range bomber, the Bf 110E could carry bombs or drop tanks (or a combination of both) under its wings and fuselage.

[3] J. Kaufmann, Meine Flugberichte 1935-1945, p. 89
[4] Jägerblatt No. 5/1986, p. 16

159

Another debate was brewing among the night fighters: what type, how many, and where should weapons go in the Bf 110? The idea of an upward firing weapon came directly from the field; this would allow the night-fighter to move directly underneath the bomber, out of the way of its rearward firing guns. Here, too, it came down to details: should the guns be angled at 70 or 80 degrees, should they be angled forward or back, or point upward vertically?

For the nose armament, there was the question of fitting the MG 151/20 and/or the MK 108. As firing the MK 108 often led to one's own Bf 110 being damaged by pieces from the bomber, some of the pilots were given a choice on what guns they wanted installed in their aircraft.

This role, so different from ground support and anti-shipping missions, highlighted the need for a fast, maneuverable airplane whose firepower lay somewhere between that of a pure fighter and a dive bomber. The Bf 110 looked to be a good candidate. Its mission portfolio was expanded even further to include reconnaissance, night-fighter, and bomb-carrying versions. Drop tanks were developed for increasing range to accommodate the ever expanding theater of operations. The Bf 110 was tropicalized for operations in the Mediterranean and North African theaters.

In late 1942 the Bf 110 went into production as the G-series with more powerful DB 605 engines. This step had been made necessary by the installation of more complicated and heavier nightfighting equipment and the fitting of extremely large caliber guns. The takeoff weight of this version reached over 9 metric tons.

Bf 110G-2/R1 with a BK 3.7 (Flak 18) gun as part of a field conversion kit in the autumn of 1942. Although the gun combination was initially planned as an anti-bomber weapon, it subsequently was used in the tank-busting role.

However, even this measure was not enough to help it through the day war beginning in 1943, especially in its original capacity as a bomber destroyer—now fighting against the Allied bombers streaming across the skies of the Reich. Like in the Battle of Britain, the Bf 110 was hopelessly outclassed by the Allied Thunderbolt and Mustang escort fighters. The same was true for its successor, the Me 410, a type which was delivered and operated in limited numbers.

The Bf 110G-2/R3, with a supplemental underfuselage MG 151 pack and small wing tanks, was employed as a heavy fighter in the defense of the Reich, as seen here in the winter of 1943/44.

Bf 110 Achieves Success as a Nightfighter

At the time, the planners in the RLM had not intended to use the Bf 110 nor any other type for this new aspect of aerial conflict. The need for a night-fighter was not felt until after British airplanes began striking at northern German cities in 1939/1940, and it was obvious from the beginning that the Bf 110 was tailor made for the job. An *E-Stelle* Rechlin report on suitable types for the night fighting role from 1939 stated:[5]

"In view of the fact that in the future the Bf 110 is the only type to be used for night-fighting, the evaluation results from flying at night can be summarized as follows: the aircraft is well-suited for night operations."

And:

"During the course of prototype trials, the Bf 110 was flown at night over Rechlin and elsewhere. These flights took place over a long period of time and were made in conjunction with a wide variety of evaluation requirements. These conditions included clear moonlit skies, as well as dark and gloomy overcast skies with low-hanging clouds, at altitudes up to 4,000 m and carrying out all maneuvers normally encountered in combat. The following details were noted:

1. Flight and Landing

With the exception of heavy control forces (which poses no more problems at night than during the day), the flight handling characteristics of the Bf 110 can be considered good; the aircraft's behavior while flying inverted is especially problem-free. There is no tendency to stall, so that landing poses less of a problem for any pilot than the Bf 109, for example. Here, however, should be mentioned the relatively high landing speed and correspondingly long roll out of 145 km/h/450 m. In view of this, night flying operations from small airfields should be avoided.

An early Bf 110 night-fighting scene dating from the early '40s, painted from memory by combat artist Alfred Hierl.
Above Right: A Bf 110G-4 nightfighter with FuG 220 Lichtenstein SN-2 antenna and belly pack housing two MG 151 guns.

[5] E-Stelle Rechlin E2b3, Bf 110 Eignung für Nachtjagd, Report No. 1 from 20 Jan 1940

2. Instrumentation and RT Equipment

Unlike the single-seat types originally planned for nightfighting, the Bf 110 is fitted out with all equipment needed for blind flying, even the RT sytem (FuG X) meets all requirements in this area."

With another lease on life as a bomber destroyer, the Bf 110 would operate in this role under the control of *General der Nachtjäger* Josef Kammhuber until the end of hostilities, and achieve its greatest success. The first attempts began on a moonlit night when, with the aid of ground-based searchlight batteries, *Oberleutnant Werner Streib* from *Nachtjagdgeschwader 1* made the first night kill of the Second World War when he brought down a RAF Whitley bomber on the night of 20 July 1944. Many highly decorated nightfighter aces scored their kills in the Bf 110, including *Major* Wolfgang Falck, who initiated the concept of night fighting, *Major* Heinz-Wolfgang Schnaufer, *Oberst* Helmut Lent, and *Oberleutnant* Wilhelm Johnen.

Even though other types were either converted for night operations (Ju 88, Do 217, Me 262) or purpose built (He 219, Ta 154), the 110 remained the standard night-fighter through 1945. The type may have started out without any specialized equipment, suited to the role only because of its blind flying capability, but as the war progressed its armament kept up with demands and radar technology was developed. *Oberleutnant* Wilhelm Johnen points out the advantages of his Bf 110G-4 in early 1944 (shortly before his forced landing in Switzerland—see box):[6]

"One of its high tech devices is the Naxos system, quite an advanced piece of equipment. If a British night-fighter were to creep up from behind in the dead of night, the Naxos would light up when he got to within 500 meters and you'd know he's there. A quick half-roll and the pursuing Tommy had better watch out. Just like the electric altimeter; what a great invention. The search radar is also top-notch. It can pick up a bomber from 6,000 meters out dead-on and lead the hunter to his quarry on 'Ariadne's thread,' as it were. And finally, there's the weaponry: two cannons and four machine guns sticking out front, and the 'jazz music' upward firing cannons, an idea that my crew came up with and have used successfully; now these guns are being fitted to production models at the Messerschmitt Works. I'm really quite proud of my C9+ES..."

With its heavy caliber weapons (MG 151, MK 108), the new oblique-firing cannons, the radar equipment and its antennas, a third seat for the radar operator, and its increased armor, the Bf 110 had obviously become a much heavier machine; it now weighed up to 9.4 metric tons on takeoff, nearly double the weight of the Bf 110B from 1938.

Bf 110 Production

Despite a rising wave of criticism against the Bf 110—brought on by the superior performance of the enemy's day fighters—the type was in production almost to the war's end. The 232 machines originally planned to meet the acquisition program of 1938/1939 swelled to over 6,000 aircraft in the end. This made the Bf 110 one of the most produced German types ever, in fifth place behind the Bf 109 (about 35,000), the Fw 190 (20,000), the Ju 88 (15,000), and the He 111 (7,300).

As with the Bf 109, license manufacturers became involved in the program at an early stage. By 1941 there were

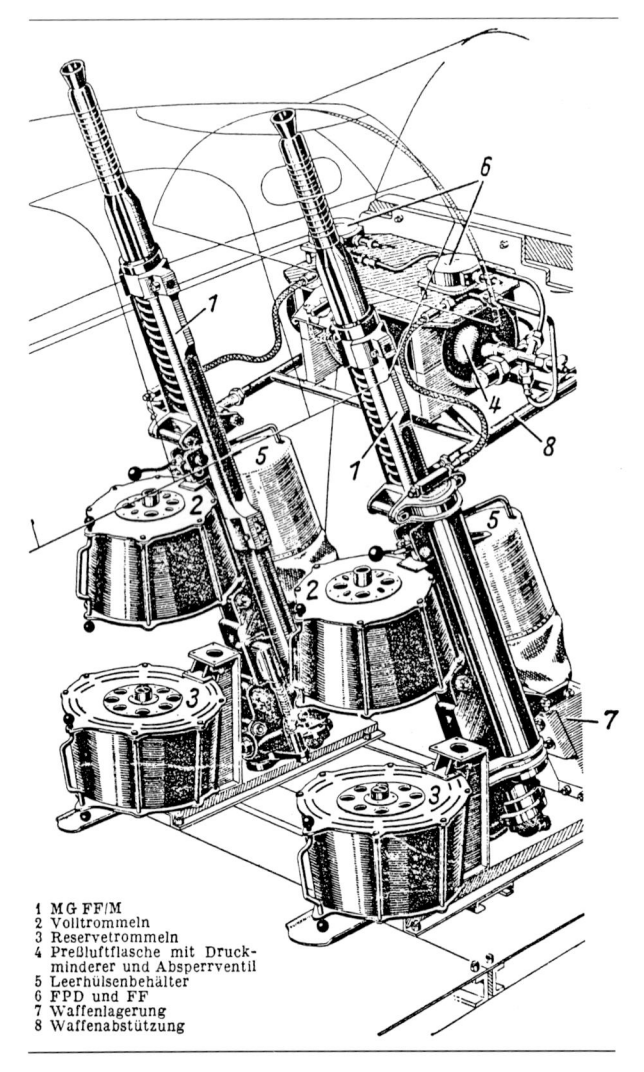

1 MG FF/M
2 Volltrommeln
3 Reservetrommeln
4 Preßluftflasche mit Druckminderer und Absperrventil
5 Leerhülsenbehälter
6 FPD und FF
7 Waffenlagerung
8 Waffenabstützung

Upward firing twin MG FF-M guns, each carrying 120 rounds, as fitted in the aft section of the Bf 110G-4 cockpit.

[6] W. Johnen, *Nachtjäger gegen Bomberpulks*, p. 126

Full-scale Bf 110 production—here the fuselage sections—in Augsburg. These fuselage shells would be completed to include the fitting of equipment. Wings and empennage would be assembled in another hangar and fitted to the fuselage during final assembly.

four companies producing the type at a rate totaling 50 to 60 per month: Messerschmitt/Augsburg, Focke-Wulf/Bremen, Gothaer Waggonfabrik/Gotha(GWF), and MIAG/Braunschweig (from 1942 its branch Lutherwerke Braunschweig, LWB).

While Willy Messerschmitt may have been praised in writing by the *Luftwaffe*'s commander-in-chief, Hermann Göring, in May of 1940 for the Augsburg Works having exceeded its monthly quota of Bf 110s, it wasn't more than a year before he had a crisis on his hands. Its Me 210 successor was not being delivered because of technical problems in the design. Even the above named license manufacturers had—in addition to winding down their 110 production—set up assembly lines for the Me 210 or were in the process of converting their resources over when the Me 210 was terminated in March of 1942.

It was extremely difficult to resume production of the Bf 110 at this point (in some cases the assembly jigs had

Full-scale production of the Bf 110 ran simultaneously in Augsburg, at Focke-Wulf in Bremen, at MIAG in Braunschweig, and at the Gothaer Waggonfabrik in Gotha.

already been sent to the scrapyard). Production was now concentrated at two sites, GWF and LWB. (Focke-Wulf had its own problems making enough room to accommodate its Fw 190, and the Mtt AG was concentrating on fixing its mess with the Me 210.)

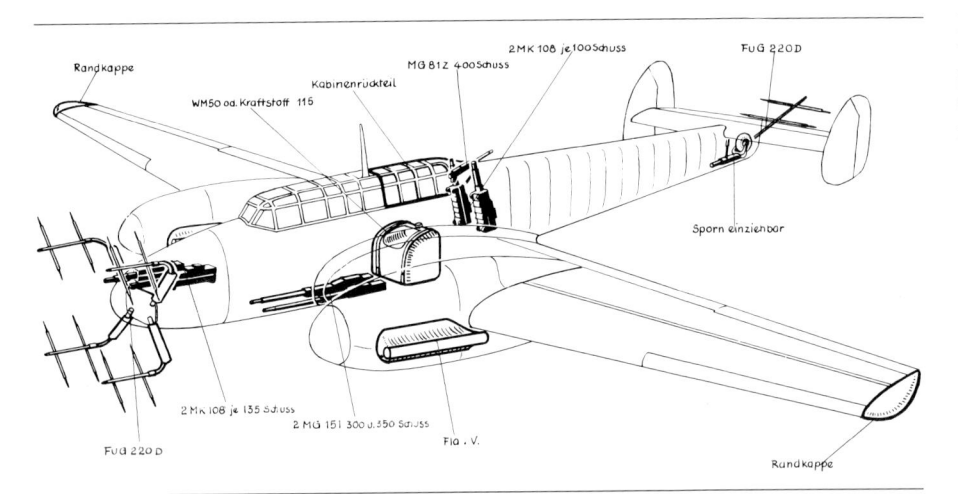

Bf 110H: as the final production variant, the "H" model was expected to have gone into production at the end of 1944 with numerous improvements.

The GWF and LWB were now building more Bf 110s than ever before. The RLM had lifted production quotas from 50 per month to 150 to 250. For a time, monthly production actually reached 150. In 1943/1944 there were just as dramatic interruptions in production caused by bombing raids on the facilities. In those two years alone, it was estimated that over 2,000 Bf 110s were lost to lack of production capacity.

In 1942 the GWF was given responsibility for the Bf 110 program under the direction of Karl Maiershofer, meaning that all improvements to the new G-series and its DB 605 engines would take place at Gotha.

The bulk of aircraft being manufactured at this point were night-fighters (all told, about 3,000 built). *Zerstörer*s took second place as before, although these were now fitted out with the heavy caliber MG 151, MK 108, and Flak 18 guns. Included in the total numbers of Bf 110 built are also about 500 reconnaissance-configured variants.

It was decided to produce the H-series beginning in the latter half of 1944 after reviewing improvements and simplifications in the manufacturing process and with a view toward introducing such refinements as increased wingspan, lengthened fuselage, and more armor. However, these plans were canceled, along with production of the Bf 110 altogether, in favor of the turbojet-powered airplane.

Five Episodes with the "Political" Bf 110

The Bf 110 was involved in several "affairs" throughout its production life, most of which were known only to a small circle of participants for security reasons.

1. General Vuillemin and the Bf 110

When France's General Vuillemin and a delegation visited German *Luftwaffe* and aircraft production centers in August 1938, he became the first foreigner to be shown the still-secret Bf 110B. At the Augsburg Works he was allowed to observe ground trials of

The French delegation under General Vuillemin (center right) was shown an impressive program, including an aerobatic demonstration by a Bf 110 flown by Willi Stör. The delegation was hosted by W. Messerschmitt, T. Croneiß, and General Milch (left to right) in August 1938 in Augsburg.

one of the first Bf 110B-1s in the firing stand prior to its delivery to the *Luftwaffe*. According to Hans Kaiser, who was also present, the firepower of the four MG 17s and two MG FFs made a visible impression on him. In addition, Willi Stör gave a masterful flight demonstration in the 110.

2. *The Stolen Bf 110C*
On 10 May 1939 Franz Oettil, a former *Luftwaffe* pilot, absconded with a Bf 110C (Werk-Nr. 979) which had just been test flown by Willi Stör. Oettil was a colleague of Stör's at the Augsburg flight acceptance testing center. Subsequent investigation of the incident revealed that Oettil refueled somewhere outside the Augsburg Works and took onboard his brother Johann. He then set off in a westerly direction. The aircraft crashed the same day near Levier in the Vogesen Mountains and both brothers were killed. Whatever the motive behind the theft of the plane, the French authorities were able to profit even from the wreckage and learn from its armament and new DB 601 engines. This was confirmed following the Battle of France when a RLM team came across an analysis of the DB 601 from the crash in a French safe.

3. *Rudolf Heß and "his" Bf 110*
What a spectacular and politically charged affair it became when Rudolf Heß failed to return from a long-range training flight in a Bf 110E.

Born in 1894, Rudolf Heß had been a pilot in the First World War, then became one of the first NSDAP members in the 'twenties, was Hitler's secretary and from 1933 onward was "deputy to the Führer." In April 1929 Heß got his private pilot's license, class A, and became active on the airfield circuit. In July 1930 he was known to have owned a Messerschmitt M 23b, and in the following year two further M 23s were entered alongside his name in the aircraft registry. It is possible that these sportplanes were flown as advertising for the Party.

In 1934 Heß was the victor of the Zugspitz Race in the city of Nuremberg's own M 35. He once even flew aerobatics in a Bf 108, even though this type was not cleared for such stunts. Heß had contact with Messerschmitt on this and other occasions. He was a welcome guest at the plant on his official visits. In 1940 his visits to Augsburg increased in frequency. Helmut Kaden, a pilot under Willi Stör at the production flight testing center, remembers:

"In the fall of 1940, shortly after our successful campaign in the West, Rudolf Heß came to Willi Stör and asked him if he might be able to fly a Bf 109E. For his reason he stated that he wanted to be available to his Führer as a combat pilot, as well. Although Stör knew of the aerobatic flying skills of Heß, he refused to take responsibility for the latter flying a military plane (a polite way of declining). When Heß then asked Messerschmitt, Prof. Messerschmitt also supported Stör's position. Heß did not initially seem to be bothered by these rejections. Kaden later found out that Heß made the same request during visits to the Wiener Neustädter Flugzeugwerke, Arado, and Fieseler (all of which were building the 109 under license), and everywhere he went, this request was declined.

Sometime around October 1940 Heß came to Augsburg again, this time asking Willi Stör if he could fly a Bf 110. Since the Bf 110 was a two-seater and therefore had the potential as a trainer, after consulting with Messerschmitt, Stör agreed to the Nazi leader's request. After only a few flights under Willi Stör's tutelege, Heß was allowed to fly the machine solo in the vicinity of Augsburg.

Helmut Kader took Stör's place when Stör left in October to go to the Balkans on a 109 advertising campaign and then, in early 1941, began making preparations for an extended trip to Japan. He now took over responsibility for Heß's flight training. With his visits becoming more frequent and Heß making special requests, it was decided to reserve for him a Bf 110 from the pool of aircraft awaiting acceptance testing prior to their delivery. It was Bf 110E-1/N, Werk-Nr. 3869, coded VJ+OQ. Heß expressed his desire to make long-distance flights—alone—and from Messerschmitt he was able to wrest several

Of all the high-ranking National Socialist politicians, Rudolf Heß was the most frequent guest at Messerschmitt in Augsburg up until his flight to England on 10 May 1941 in a Bf 110E. Here from left to right are T. Croneiß, H. Bauer, R. Heß, and W. Messerschmitt.

modifications to his plane. These included moving the radio control switches and aft oxygen supply switch from the radio operator's seat to the front and making a few changes to the heating controls. When Heß saw the new drop tanks (900 l) from Junkers and discovered that these could be fitted to the 110 beginning with the D series, he immediately wanted to have them fitted to his machine, a request which was also fulfilled.

On Saturday, the 21st of December 1940, Heß wanted to make the first (probably) long-distance checkout flight to check radio calibration. Augsburg had 40 cm of snow on the ground and the runway had to be cleared. He then took off with 3,000 l of fuel (1,200 l internal, plus 2x 900 l drop tanks).

After about three hours Heß returned and made a landing with his vertical stabilizer jammed. He confessed to having accidentally pulled the flare pistol from its holster, and it had become caught in the control cables inside the fuselage.

On 18 January 1941 he made a similar flight, from which Heß also returned after about three hours. He stated that he couldn't figure out how to work the radio direction finding set and hadn't been able to locate the transmitter at Kalundborg. It was explained to him that that wasn't possible with the FuG 16 and advised him to use the Reich transmitting station at Munich for determining his location.

On 30 April Heß was ready for another flight when his adjutant, Karlheinz Pintsch, came up to the waiting plane and informed Heß that the Führer would not be able to make the ceremony on 1 May recognizing the Messerschmitt Works as a *Nationalsozialistischer Musterbetrieb*, or National Socialist Model Operation. Heß stood in for him.

Heß's next and final flight occurred on 10 May. After the weather station reported good conditions over northern Germany, he took off at around 1800 hrs in a flight suit which Kaden had loaned to him, and for which Heß had conscientiously signed for. When after four hours Heß had not yet returned, his adjutant Pintsch had been instructed to open an envelope, which he did, and found that it included letters to be passed on to Hitler, Heß's wife, Ilse, Willy Messerschmitt, and Helmut Kaden (in the latter, Heß confirmed that the flight suit had indeed been loaned to him by Kaden. It was found among his belongings when Heß died at Spandau Prison in 1987.)

With this information, those present understood that Heß would not be coming back. As is subsequently known, at around 2300 hrs he bailed out near Glasgow, Scotland, after a carefully planned and disguised flight. He had hoped to establish political connections in England. The Bf 110 crashed and broke up, but the wreckage is today under the care of the Imperial War Museum at Duxford.

Willy Messerschmitt was ordered to report to Göring on 12 May. The *Reichsmarschall* demanded to know why he let a maniac have access to one of his planes. According to an account he gave a newspaper after the war, Messerschmitt retorted dryly: 'How am I supposed to believe that a lunatic can hold such a high office in the Third Reich. You should have made him resign, *Herr Reichsmarschall*!' Göring laughed out loud at this reply. 'You're incorrigible, Messerschmitt! Go back and keep on building your airplanes. I promise you that I'll get you out of any trouble if the *Reichsführer* tries to give you grief over this affair.'

It is worth noting that Ernst Udet, in his capacity as *Generalluftzeugmeister*, also had a Bf 110E at his disposal for urgent duties during his last year in office, which he enjoyed flying. The machine was based at Berlin-Tempelhof. According to Kurt Schnittke, one of the destinations they flew to with this plane was Salzburg, where Udet had been summoned by Hitler to provide a report after the Heß flight.

4. Bf 110 Export to the Soviet Union

In April 1940 five Bf 110C-2 planes were delivered to the Soviet Union by air as part of a Russo-German trade agreement. Included in the deal were also five Bf 109Es and two Bf 108Bs. The package had a total value of 1.123 million RM. Hans Kaiser, who worked in the *Probü* at the time, recalls that despite being used to the export successes of the Bf 108 and 109, it still gnawed at the Augsburg team to have to supply the Soviet Union, of all countries, with the Bf 110, a design of which they were justifiably proud. The Soviets' interest in the radio equipment was noteworthy.

5. Night-fighter Secrets and State Security

On the night of 28 April 1944 Oberleutnant Wilhelm Johnen of NJG 6 accidentally strayed over the Swiss border following a successful encounter with British Lancaster bombers. An engine fire caused him to shut down one of his two DB 605 motors, and then his aircraft was caught in searchlight beams and he was forced to land at Dübendorf. Even though the three-man crew was permitted to return to Germany after days of negotiations over their release, the valuable

The Bf 110G-4/R7 in which Wilhelm Johnen was forced to land at Dübendorf on 28 April 1944.

Bf 110 remained behind. (According to Swiss records, it had a combination of FuG 8, 10, 16, 25, 202 and 220 radio systems on board.)

To prevent the secret radio equipment from falling into the hands of enemy agents, the German State Security considered risking a commando raid to destroy the craft. But then it was decided to make use of official contacts with the Swiss authorities, and in return for blowing up the plane the *Luftwaffe* offered to sell Switzerland the modern Bf 109G (at the time Switzerland was operating the obsolescent Bf 109D and E models purchased prior to the war). The Swiss agreed, and the Bf 110G-4 was blown up in the presence of witnesses on 19 May. There then followed the transfer of 12 Bf 109G-6 airplanes from Regensburg to Dübendorf. Cost per Bf 109G-6 with Db 605 engine, a 20mm MG 151/20 firing through the propeller and two MG 131/13 machine guns was 500,000 Swiss francs.

It was soon discovered, however, that these machines had numerous problems, and the engines in particular required considerable maintenance. The Messerschmitt and Daimler-Benz companies continued to provide the Swiss with warranty service even after the war was over.

Bf 110 Variants

The Bf 110 had a similar developmental cycle to the Bf 109. Three different engines were fitted to six basic variants: the Jumo 210 for the B-series, the DB 601 for the C- through F-series, and the DB 605 for the G-series.

All told, over fifty variants were built, although some of these were through factory and field conversion kits. The most important of these are:

Bf 110B-1	*Zerstörer* with Jumo 210G engines. Armament: 4x MG 17, 2x MG FF, 1x MG 15, first production variant with *Luftwaffe* officially taking delivery in July 1939.
Bf 110C-1 thru C-4	*Zerstörer* with DB 601A engines. Armament:: 4x MG 17, 2 MG FF or FF-M, 1x MG 15.
Bf 110C-5	Recce version with RB 50/30 camera. Armament: 4x MG 17 and 1x MG 15.
Bf 110C-7	Bomber, as C-4 but with add'l pylons for 2x 500 kg bombs under the fuselage.
Bf 110D-0 thru D-4	*Zerstörer* with increased range through 1050 l belly tank (*Dackelbauch*) and 2x 900 l wing tanks. In some instances, belly tank could be swapped out for bomb payload.
Bf 110E-1	Bomber with increased range, improved equipment, including K 4ü autopilot, add'l pylons for 2x 50 kg bombs on each wing.
Bf 110F-2	*Zerstörer* or bomber with DB 601F engines, as E, able to carry 2 MG 151 under fuselage.
Bf 110F-4	Night-fighter with specialized equipment. Armament: 4x MG 17, 2x MG FF-M, 2x MG 151 as field conversion kit under fuselage, 1x MG 81Z.
Bf 110G-2	*Zerstörer* with DB 605 engines. Armament: 4x MG 17, 2x MG 151 as field conversion kit, 1x MG 81Z.
Bf 110G-2/R1, R2, R5	*Zerstörer* as G-2, but 1x Flak 18 conversion kit and 72 rounds in place of 2x MG 151 kit.
Bf 110G-4	Night-fighter (3-seat) with specialized radio equipment. Armament 4x MG 17, 2x MK 108, 1x MG 18Z and 2x MG 151 conversion kit or 2x MG FF or 2x MK 108 upward firing guns in fuselage (R8, R9).

Me 210A with twin Daimler-Benz DB 601F engines.

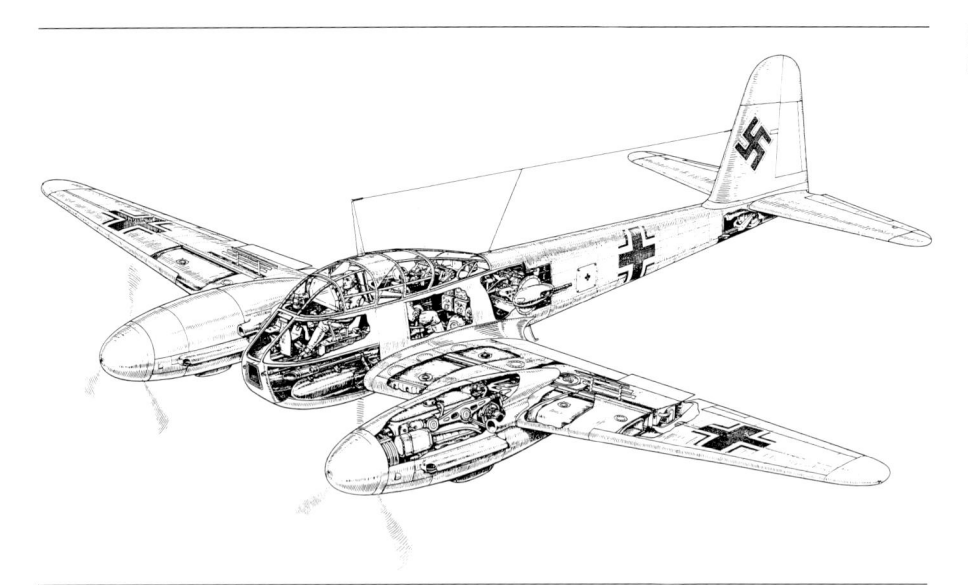

Me 210 - A Major Setback

Drawn up in 1937/1938 under the designation of Project P 1060, the Me 210 was a program which followed on the heels of the successful Bf 108, 109, and 110. The contractor had demanded that the Me 210 be built in the shortest amount of time in the largest numbers possible as a dive-bombing *Zerstörer*, or heavy fighter. Both this general feeling of self-confidence and the rapid mass production demanded would have such a negative backlash that it would lead to a major setback for both the contractor and for the Messerschmitt AG.

A certain optimism must have prevailed at Messerschmitt, evidenced by the absence of critical management during the aircraft's design gestation period. Walter Rethel had transferred from Arado to become the design department's senior manager in early 1938, and as the lines of authority had not been clearly defined with the former director and current manager of single-engined fighter types, Richard Bauer, this may have had a negative effect on the entire development team. In any case, grave errors in the design of the wing profile and several construction mistakes were the result.

The scapegoat for this misfortune was, of all people, Messerschmitt's friend and *Generalluftzeugmeister* Ernst Udet. It was unfortunate that the three new types destined for the *Luftwaffe*—the Me 210, the He 177, and the Fw 190, which all flew for the first time in 1939—were still in the testing stages as a result of technical difficulties and were not available in the numbers planned in 1941. A situation which bordered on the disastrous with the launch of the invasion of Russia (22 June 1941) and Hitler's internally advised declaration of war against the United States (December 1941).

Shouldering the burden of a stagnating armament program and facing war with the United States, something which Udet and his friend, Friedrich Wilhelm Siebel, had strongly advised against, Udet suffered a nervous breakdown and shot himself on 17 November 1941 in Berlin. The warning was not heeded. The new top man in the RLM now became Erhard Milch, who made no secret of his animosity toward Messerschmitt.

As early as 1937—the Bf 110 was only in prototype form at the time—the RLM and Messerschmitt began thinking about a follow-on development to the *Zerstörer*. The contractor (read RLM) also wanted the Bf 110's successor to be able to carry bombs and deliver them from a dive. Willy Messerschmitt and the project department under Robert Lusser felt that, rather than simply improving the Bf 110, the solution would be to design an entirely new airplane.

The *Probü*, sometimes working in two parallel groups, drew up about 70 proposals in 1937/1938, and by late 1938 the project solidified as the P 1060 and was passed on to the design department. The Bf 110's basic dimensions were kept, but the designers were forced to accept a higher wing loading and therefore higher landing speed due to the aircraft's increased weight brought on by its 1,000 kg potential bomb load (V_{land} for the Bf 110 G = 156, for the Me 210A = 188 km/h). To improve aerodynamics, the bomb load was incorporated into a bomb bay within a fattened forward fuselage, which was also shortened by about a meter—perhaps to give the aircraft a higher angle of attack on landing. Defensive armament consisted of two 13 mm MG 131 machine guns, mounted in remotely controlled tear-

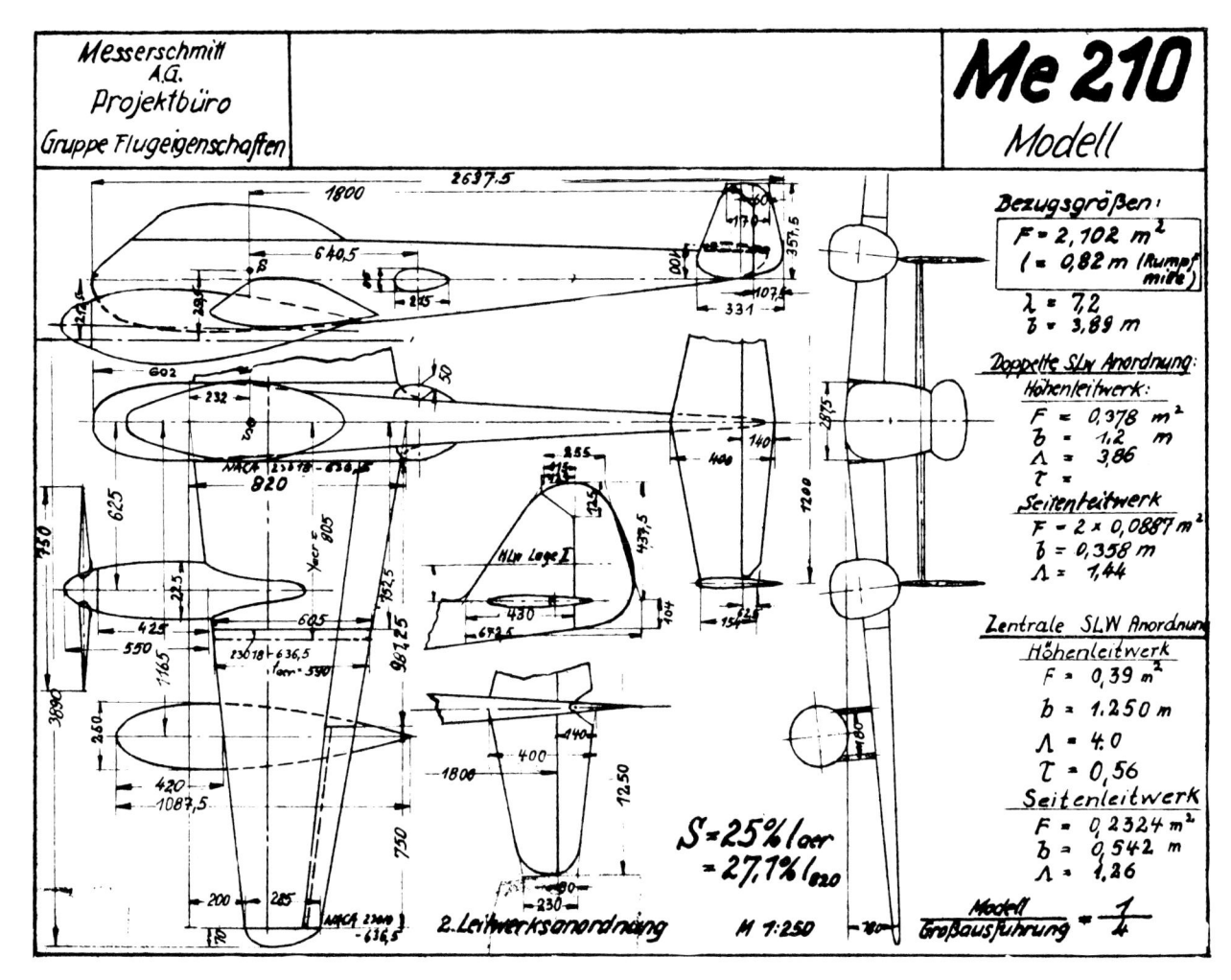

Overview of the wind tunnel model with alternative single or twin-tailed layout and wing drop tanks. The extremely tapered fuselage tip is clearly visible here.

drop-shaped barbettes on either side of the fuselage. A new development by the companies of Rheinmetall-Borsig and AEG, the system was also made availabe for other types like Arado's Ar 240 development (a direct competitor to the Me 210). The empennage at the end of the Me 210's remarkably short and narrow fuselage was initially of a twin rudder design as found on the Bf 110. Two DB 601F engines, each rated at 990 kW/1,350 hp, provided power for the aircraft. The first two prototypes were designed (by Walter Rethel's *Kobü*), built (by Hubert Baur's prototype construction department), and flown (by Dr. Hermann Wurster) within the space of a year. Me 210V-1, D-AABF took to the air for the first time on 5 September, and V-2, WL-ABEO, flew on 10 October 1939.

When flight testing the Me 210, Dr. Wurster's primary concern was to ensure the airplane was stable along all its axes. Even at this early stage, he recommended lengthening the fuselage in order to improve the handling along the aircraft's longitudinal axis and thereby give better landing and takeoff handling. It wasn't until nearly two years later that Willy Messerschmitt reluctantly had the fuselage lengthened. At the time, it was thought that there were simpler means of improving the aircraft's flight handling. As early as 17 October 1939 the *E-Stelle* had a pilot test fly the Me 210. Part of Gustl Neidhart's report reads:[1]

"The Me 210V-1 aircraft type was flown for the first time: the airplane's stall handling is quite good. Control forces are altogether too heavy. Handling in the vertical plane is unsatisfactory. Control harmony in all planes is difficult to achieve. The criticisms raised about the flight handling characteristics can be expected to disappear with simple remedies."

[1] E-Stelle Rechlin, Me 210 Nachfliegen V-1 am 12.10.39 in Augsburg. Report 2, 1669 from 17 Oct 39

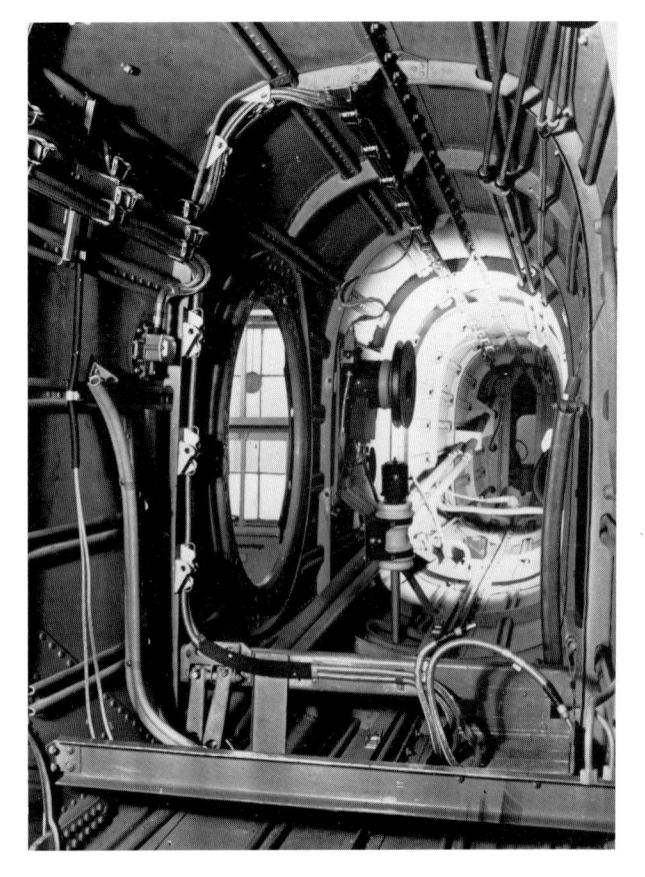

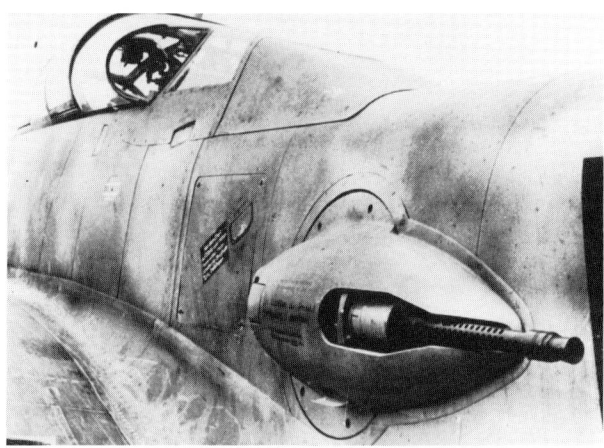

The left barbette showing the MG 131; the barbette could traverse upward 90 degrees and downward 45 degrees, while the gun could extend outward by 45 degrees—but was prevented from firing when crossing the area of the horizontal stabilizer.

In comparison with other Messerschmitt or German types at the time, the Me 210 was a much more complicated design, as shown by this view of the monocoque fuselage. The remotely-controlled barbettes for the MG 131 guns have not yet been fitted.

Me 210V-1, D-AABF, flown on its maiden flight by Dr. Hermann Wurster on 2 September 1939. The unarmed machine served exclusively as a testbed for flight handling characteristics and flew until 1941.

Me 210V-2 after being fitted with a single rudder, nicknamed "barn door" by the Messerschmitt team because of its size. The side gun blisters are only mockups.

170

When these "simple remedies" didn't fix the problems and the Me 210's flight handling continued to puzzle those involved in the program, Willy Messerschmitt got the DVL involved. Prof. Karl H. Doetsch recalls:[2]

"In July 1940 Prof. Messerschmitt, a member of the DVL's senate, asked its director, Prof. G. Bock, to advise him on the somewhat mysterious difficulties plaguing the Me 210's flight handling. E-Stelle Rechlin had criticized the aircraft for 'responding uncharacteristically to aileron input' after takeoff. Prof. Bock, Prof. A. W. Quick, and I set out for Augsburg to get to the bottom of the matter. I personally flew my Me 110, fully kitted out with flight data recording equipment, directly to the Mtt flight testing department, where I also found Rechlin's official-in-charge eagerly awaiting the results of our investigation.

My aircraft had a control stick which measured force application and was directly linked to a dial-type gauge. Hempel, my data recording monitor, removed the joystick and its gauge from our Me 110 and installed it in a Me 210, then we took off to check things out. It was then, just after lifting off, that I experienced something I'd never seen before; up to that point, it was generally accepted that in matters of control application to control effect, it was always the effect that suffered, yet suddenly I was finding the opposite to be true. With remarkably little application of force—confirmed by the gauge—the intended roll acceleration during the initial climbout (i.e. high lift coefficient) was amazingly high. This statement in my report was confirmed as the source of his expression 'uncharacteristic quality' used by my Rechlin colleague. My report was immediately passed on to Professors Bock and Quick, who were at the moment in discussion with Prof. Messerschmitt. The statement led Herr Schomerus, the aerodynamics director, to suspect that the error might lie in the wing profile blending.* Subsequent studies confirmed these suspicions. A fabric covered segment of balsa wood was attached to the wing near the aileron on a trial basis to correct the profile, and this measure did indeed fix the 'uncharacteristic quality' of roll control. Once the wing blueprints had been changed accordingly, this no longer posed a problem for the type's certification for production.

There was also another happy consequence rising from the simple, yet so successful test flight; at my suggestion, Prof. Messerschmitt immediately gave the go-ahead for a fundamental overhaul of the company's flight data recording equipment and facilities—the carbon recorders for measuring pressure and altitude went out the window, replaced by ultra-modern measuring devices and a laboratory furnished in oak (I was told that expenses totaled 2 million RM!). This trust I'd earned from him developed into what was for me a long, close, and quite friendly cooperative relationship."

Willy Messerschmitt later filed a "Report on the General State of Preparatory Work for Me 210 Production, 15 July 1940" wherein he stated that there were no problems with the aircraft's flight handling characteristics. The control effects and forces were mostly equal to or slightly less than those on the Bf 110C. Stall handling was good both with and without leading edge slats. It was up to the *E-Stelle* to decide whether to keep the slats on the design or not. There was only a little design work left in preparation for production, to include the dive brakes and the automatic pull-out system. Production versions were to have a central single rudder—probably for improving the aft field of fire—as had provisionally been fitted to the V-2 and incorporated into production from the V-5 onward.

The report went on to say that the situation of the proposed license manufacturers (Focke-Wulf, Weser, and MIAG) had not yet been resolved with regard to acquisition of equipment, semi-finished materials, finished components, and standard parts, which put startup of production in jeopardy. Of the 16 prototypes ordered, only four were flight testing. Still outstanding were the final flight performance evaluation and airborne weapons firing.

This report was far too optimistic in its outlook, as evidenced just a few weeks later when Fritz Wendel was forced to bale out of the Me 210V-2 on 5 September 1940. Pulling

Investigation into mistakes on the Me 210 led to an initial understanding that the wing profile had been "tweaked" at some point during its design and that the leading edge slats, dropped by the RLM for cost reasons, were indeed necessary for improving stall handling characteristics. Finished components or entire aircraft had to be overhauled accordingly, as on this Me 210A in March of 1942. The leading edge spars and wing underside had to be exposed in order to thicken the profile and fit the slats.

---
[2] Letter to the author from Prof. Karl H. Doetsch, Braunschweig, dated 16 May 1991.
*The aerodynamics engineers in Ludwig Bölkow's and Jochen Puffert's project department had already discovered an error in the blending of the wing profile beforehand.

On 5 September 1940 Fritz Wendel was forced to take to the silk due to an empennage fracture on the Me 210V-2. Suffering only minor injuries, he is seen here smoking a cigarette near the wreckage of his plane.

up at a speed of 605 km/h, the tail had begun oscillating, which in turn led to the empennage vibrating and eventually breaking. This revealed another flaw in the Me 210: its short, narrow aft fuselage was namely too weak. This accident was another strong argument for modifying the fuselage, which in any event would not only need to be lengthened but also beefed up as well. These were concessions which, because of the increase in weight, Willy Messerschmitt only reluctantly agreed to.

Nor were RLM senior test pilots G. Neidhart and Malz overly pleased when they test flew the Me 210V-9 (G7+4T), considered to be the first production model, in December

Preparations for production of the Me 210 and even full-scale production itself occurred while the aircraft was still in its testing stages. Production was to have taken place at the parent factory in Augsburg, in Regensburg, and at the license-building companies of Luther-Werke in Braunschweig and Gothaer Waggonfabrik. Although late, the first production aircraft had been completed in Augsburg and Regensburg by the summer of 1941.

1940. They discovered a new phenomenon: the Me 210 would stall out when flown in a steep turn in an unclean configuration, plus it was difficult to harmonize the aircraft in the lateral plane. In a fly-off against the Bf 110, Neidhart and Malz reported:[3]

"A comparison flight with Bf 110E/N G6+3T showed the Bf 110 to be clearly superior in a dogfight. The Me 210 was originally in a good firing position, but after two to three circles found itself the hunted instead of the hunter. It was discovered that the Bf 110 recovers more quickly from a near stall than the Me 210. The Me 210 stalls at 280 km/h, the Bf 110 at 220 km/h. Cranking the Me 210 around is not as strenuous as the Bf 110 due to the lesser control forces needed for the horizontal stabilizers. However, the Bf 110 can take more punishment and feels like it has more acceleration than the Me 210. It is recommended that another test flight take place as soon as possible. When cranking, the Me 210 doesn't seem to want stall as much as on the previous flight, but we intentionally avoided any opportunity of slipping."

On 19 February 1941 Carl Francke, the chief of the *E-Stelle,* wrote to Willy Messerschmitt:[4]

"From what Neidhart says, I get the impression that the 210 no longer has problems with the tailplane's forces and effect, except on landing. Specifically, with the tailplane trim linked to the landing flaps as they extend, at approach speeds and even with the trim setting in the rearmost position, the stick cannot be pushed forward enough to compensate and the plane therefore stalls itself. Because of this characteristic I have proposed a stopgap solution: do not link the trim adjustment with the landing flaps, but have the trim tabs controlled manually, thereby allowing the pilot to trim the plane for landing as needed and depending on the center of gravity. This solution has the disadvantage in that the aircraft must be trimmed nose-heavy on takeoff, probably at the forward-most setting in most instances, as the elevators need to be set quite tail-heavy in your case and for three-point landings. In my opinion, the whole matter can be traced to the fact that elevator effect is not entirely adequate for landing. I have already mentioned to you that we had similar difficulties with the power-driven trim adjustment on the 177 and that all problems were cured after the tailplane was enlarged and an internal trim adjustment fitted. I would therefore recommend that you continue with the intended development of a larger elevator with internal trim with all due haste."

[3] Excerpt from Me 210V-9 flight reports filed by pilots Malz and Neidhart on 18 and 19 Dec 1940
[4] Letter to W. Messerschmitt, Augsburg from *Fl.Oberstabs-Ing.* Carl Francke, Rechlin, dated 19 Feb 1941.

In addition to the difficulties encountered when evaluating the Me 210's flight handling characteristics, the type also had more than its fair share of "routine" problems, as shown in these two photos, taken in November 1941 in Rechlin, when the landing gear collapsed during taxiing. Ernst Udet had warned Willy Messerschmitt beforehand that such accidents—particularly during wartime—would not be acceptable.

In early 1941 the first production Me 210 was to be handed over to the contractor for a newly established combat evaluation detachment within the *Luftwaffe*. Altogether, 1,000 aircraft were expected over the course of the year. Yet aside from the prototypes, which went to Rechlin and Tarnewitz for testing and suffered numerous crashes and accidents, a production airplane had yet to reach its intended recipient. *Generalluftzeugmeister* Ernst Udet, also under pressure from senior *Luftwaffe* officials, complained to Messerschmitt both personally and in letters about keeping to his delivery timetables. He made known his disappointment in the fact that, with a war going on, airplanes were crashing due to landing gear failure simply because a few kilos of weight had been saved in the design. He was referring both to the problems during the Me 210's test phase, as well as to the numerous landing gear accidents with the Bf 109 in the field.

When no production examples had been delivered by May of 1941, on 5 June 1941 Udet called in a so-called "industrial advisory board." All those involved in the Me 210 program could present their problems to the board, and neutral observers would offer suggestions on alleviating the crisis. It was one of Udet's final visits to Augsburg. Prof. Messerschmitt was assigned the task of rectifying the following problems on the Me 210 *post haste*:

173

Stall and pitch handling studies were carried out on the Me 210 following a series of spin-related accidents in the field. Airflow pattern, revealed through the use of cotton strips attached to the wing, was monitored both visually and from a camera mounted on the vertical stabilizer. The tests were conducted at maximum military power for stalls and sideslips, at minimum cruise for level flight and sideslips, stalled vertical banks and landing profiles. The Me 210 shown here is on final approach to the Augsburg airfield. Plant 2 is located ahead of the left engine, with Plant 3 beyond it and Plant 4 at the far left in the photo. Plant 1 is obstructed by the airplane's canopy.

1. The Me 210 porpoises in the lateral plane, making it difficult to accurately sight when attacking using guns.
2. The power-driven tailplane is not an acceptable solution; an internally trimmed elevator must be installed.
3. The tendency to ground loop on takeoff and landing must be eliminated by lengthening the fuselage (first incorporated on W.Nr. 101 on 3/24/42)
4. The reason for the DB 601's recent engine fires on the Bf 110 and 210 must be found and eliminated in cooperation with the Daimler-Benz company.
5. DB's proposed turbocharger intake lip must be modified for aerodynamic reasons.
6. Elevator buffeting when the dive brakes are extended must be eliminated.
7. The automatic pull-out system does not meet specifications.

First Me 210s Delivered

Finally, in July of 1941 the first two production Me 210A-1s were handed over to the Messerschmitt company's quality control. However, they did not yet meet the standards recommended by the advisory board. This also ap-

The first two Me 210 production aircraft prior to delivery, seen here at Augsburg in July of 1941. At the request of the RLM, these machines were initially delivered without leading edge slats.

plied to nearly 50 other production aircraft sent to the test centers at Rechlin and Tarnewitz and to *Erprobungskommando 210*, the detachment which was to form the basis for the *Schnellkampfgeschwader 210* and which received its first 16 machines in November. Further Me 210A-1s built after January 1942 went straight from the assembly line at Augsburg to the airbase at Landsberg am Lech, where the necessary modifications were carried out, and then on to the field units. In the field, where the 210 was expected to replace the Bf 110, the strengths and weaknesses of the Me 210 now became public knowledge. At first, there were few opinions of the Me 210's flight handling characteristics as there was only a limited number of *Luftwaffe* pilots flying the type. One of the few pilots sent from the Eastern Front to convert from the Bf 110 to the Me 210 at Landsberg was Johannes Kaufmann, who in his book "Meine Flugberichte 1935-1945" wrote:[5]

"The Me 210 made a powerful impression on us. Unfortunately, we only had mimimal instruction due to the lack of training staff. One had to probe slowly to get a feel for the plane's innovations in comparison with the Me 110 and learn how to handle it properly.

The aircraft was significantly faster, and the bombs could be carried inside the fuselage. Climb rate was much better, and it dived according to specifications. The guns were more effective, and navigation and radio aids had been improved. In addition, a three-plane autopilot had been fitted, which we were unfortunately unable to try out. All in all, the Me 210 was a classy bird. We had much hope for the plane and eagerly looked forward to our first flight.

On 14 November it was time. Our first takeoff, made extremely cautiously due to the recognized danger of ground looping, was uneventful. The takeoff run was relatively short, and I flew a standard airfield circuit lasting seven minutes. The warning about ground looping was well-founded, for I needed to make several corrections with the rudder and the engines to keep straight. My second flight was better, and I stayed in the air for 26 minutes in order to try out the new type for a bit. It had good flight handling characteristics, better than those of the Me 110. Good maneuverability and rapid acceleration were of particular note. I ran into no special problems on approach and landing. Here it differed from the Me 110, of course, but this can in no way be construed as a shortcoming.

We practiced dropping bombs, air-to-air gunnery, diving, and formation flying and quickly became familiar with its peculiarities, and our expectations ran high.

Our trainee crews were the first to run into problems. We accordingly moved to Lechfeld to give us more takeoff and landing room.

[5] Johannes Kaufmann, *Oberstleutnant a. D.*, Meine Flugberichte 1935-1945, Schwäbisch Hall 1989, p. 135-136

Unfortunately, the snowy winter interfered with our plans. As a result, we were only able to fly out of Lechfeld from 13 January 1942 to 2 March, deploying that afternoon to Tours. Upon arriving at Tours from Regensburg, a most unusual sight met my eyes. There were several crashed planes scattered about the airfield, obviously new and recognizable as 210s. Even though I wasn't able to see it in this one glimpse, right after I'd landed I was told what had happened.

Without exception, these crashes were attributable to young, inexperienced pilots, and a new decision had to be made. This came relatively quickly.

The type was withdrawn from service. We weren't able to keep our mounts and were forced to revert back to our old Me 110s for the 1942 summer offensive against Russia. The entire base was deeply disappointed."

Me 210 Production Halted

Johannes Kaufmann's recollections bring up two points. The first is the *Luftwaffe*'s inadequate training, especially during the war. Combined with an airplane that had not yet reached maturity, this had a catastrophic effect. Even though Neidhart and Malz had identified the Me 210's negative characteristics, the young pilots were apparently not heeding the warnings and would go into a flat spin when turning in an unclean configuration (i.e. gear and flaps down). This was undoubtedly the case which had occurred with the trainee crews at Tours on their airfield circuits and landings. Company test pilots and experienced *Luftwaffe* pilots, on the other hand, were able to compensate for a plane's poor handling with better application of flying skills.

Although Kaufmann did not mention any numbers, there must have been at least 15 such accidents with the Me 210 in a short period of time, something which caused consternation among the *Luftwaffe*'s senior officials. The result of a *Luftwaffe* emergency meeting was a decision which had unfortunate consequences for all those involved in the program. On 10 March 1942, the GL/C sent the following telegram to Willy Messerschmitt in Augsburg and Theo Croneiß in Regensburg:[6]

"Based on the meeting of 9 March, the *Reichsmarschall* has determined that continued production of the Me 210 is to cease immediately. Any decision to resume production of a new type with lengthened fuselage, enlarged tailplane with internal trim adjustment, slats and correction of all faults discovered during field testing is to be made following evaluation by the GL and field units. It is planned to

[6] Telegram GL/C no. 617/42 (Secret) from 10 March 1942 to Prof. Messerschmitt, Augsburg, with duplicate sent to *Major* Croneiß, Regensburg

The production stop ordered by the RLM in March 1942 for the Me 210 had a particularly devastating effect upon the parent facility in Augsburg. In one fell swoop the production series, by this time running at full steam, was shut down. As a result, parts and components in various stages of completion—from cut raw metals for individual pieces to completed airframes—began to stockpile. All this material was stored in warehouse hangars at the Augsburg facility, in two hangars at the Gablingen airbase, and at the Regensburg site. By July 1942 the material brought to Gablingen alone, such as the jigs shown here, was enough to fill 600 fully loaded freight cars.

The approximately 50 Me 210s produced, many of which are shown here parked under artificial trees on the factory field, waited up to a year before their conversion to Me 410 standards.

increase production of the Bf 110 and Bf 109 in lieu of the Me 210. The *Reichsmarschall* expects that the Mtt company will use all conceivable resources in producing the Bf 110 and Bf 109 to fill the gap caused by cancellation of the Me 210. I expect Prof. Messerschmitt, Dir. Hentzen and *Major* Croneiß to report to *Oberst* Vorwaldt on 12 March at 1100 hrs for a meeting to discuss future plans. I request that you bring along detailed operational plans for personnel and materials so that decisions can be made about the continued use of the work force. Additionally, please bring a set of blueprints for the elevator trim adjustment."

Up to this point 77 Me 210s, including the V-series prototypes, had been delivered to the test centers and field units, and another 205 were in various stages of completion. Göring's decision hit the Augsburg and Regensburg plants hard. Production of the Me 210 was in full swing at Augsburg, a plant used to successes vice setbacks. The Lutherwerke and Gothaer Waggonfabrik license production companies were still in the process of converting over and had not yet supplied any airplanes. At Augsburg an air of uncertainty and unrest settled over the several thousands of workers there. Messerschmitt took the opportunity to address his employees, which he did via loudspeaker on 25 March.[7]

It took weeks to break down the production tooling and assembly jigs and store the materials for the Me 210. 600 railcars of parts were stored at Gablingen alone. However, the prototype construction and flight testing departments at Lechfeld continued working at a feverish pace.

On the 25th of March Willy Messerschmitt also wrote to Francke, the director of the test centers:[8]

"Dear *Herr* Franke,

You are undoubtedly aware of the difficulties the field units are having with the Me 210 and that these problems can be traced back to stalling. As a number of aircraft have already been built and are still under construction, it is most urgent that we fix the problem as quickly as possible. Those deficiencies already identified (ground looping on landing, unpleasant elevator control) have been corrected by lengthening the fuselage and installing trim control for the tailplane (both recommendations from the test center), and after just a few flights both the field and the test center have felt these areas are now satisfactory. The other criticisms about the landing gear, hydraulics, weapons, etc. basically just require engineering fixes and will be corrected shortly. The most critical point turns out to be one which we were least concerned with. The problem is basically this, that there has been no maximum authorized speed set for the aircraft in a dive. Obviously, we cannot define the requirement today and then hurriedly build it in, ensuring an aircraft in a spin can simply pull out at any rpm setting desired once control is regained."

Additionally, Messerschmitt asked Francke to send him the pilot Heinrich Beauvais for investigating the Me 210's spin characteristics, since his company pilots did not have any experience in this realm. Francke approved this request on 30 March, whereupon Messerschmitt thanked him and in a letter on 8 April added:

"You will be interested to know that, with an improved slat, everyone who has flown it says the airplane is now

---

[7] Address by W. Messerschmitt titled "Durch Lautsprecher an Gefolgschaft am 25.3.42. Mtt" (IWM, FD.4355/45, Vol. 1 [box p. 205]

[8] Letter from W. Messerschmitt to *Flieger-Oberstabsingenieur* Francke, E-Stelle Rechlin, dated 25 Mar 1942

In the early months of 1942 Erprobungskommando 210, based in Lechfeld, received a pre-production batch of Me 210A planes which had been upgraded to the initial conversion stage, exemplified by V-31 and V-36 shown here.

Much data was obtained during the summer of 1942 from flight tests conducted using innumerable Bf 109, 110, 163, 210, 309, 262, and 323 testbeds, all of which were being flown during the same timeframe. Willy Messerschmitt is shown here with his pilots Karl Baur and Fritz Wedel (left) standing next to one of the first Bf 109G variants.

more harmless than the 110. I still don't understand how it was possible that the unpleasant stall handling wasn't noticed by any of the pilots, not even our company test pilots."

[9] Letter from W. Messerschmitt to *Flieger-Oberstabsingenieur* Francke, at the time in Hohenlychen, dated 8 Apr 1942

Consequences for Willy Messerschmitt

In the RLM meeting on 12 March *Generalfeldmarschall* Milch yet again demanded that Willy Messerschmitt correct all the Me 210's mistakes without delay, and set a deadline by which time a number of converted machines were to be ready. When in April it became apparent that these planes would not be available by 1 May, Milch sent his senior engineer Roluf Lucht on an inspection of Augsburg. As he subsequently reported, Lucht found the mood in the factory to be tense and Willy Messerschmitt in poor health. This was undoubtedly due to events all adding up at once: Hermann Göring's action terminating production, Milch's establishment of a deadline and his often voiced threat to remove Messerschmitt from the company's management and replace him with a general director. Then there were the financial problems resulting from this misery, running into the millions of marks and putting the company in a serious financial predicament.

Freidrich W. Seiler and Theo Croneiß, members of Mtt-AG's board of directors, acted as go-betweens for Milch and Messerschmitt. At a 30 April meeting of the board of directors, it was unanimously decided that "Prof. Willy Messerschmitt will carry out his duties as senior design engineer of the company for the duration of the war." Theo Croneiß assumed the posts of chief executive officer at Mtt-AG and operations director for the Augsburg Works, while banker F. W. Seiler became chairman of the board. This solution was accepted by Milch.

Longer Fuselage, More Powerful Engine Change the Me 210 into the Me 410

In the meantime, special work shifts were organized to push through the changes to the Me 210 stipulated by the RLM. In mid-July the first modified aircraft began appearing, having a fuselage lengthened by 80 cm and powered by either the DB 601F engines or the DB 605. This version was also built in Hungary, which had obtained the Me 210 license in 1941 with the stipulation that the production run would be evenly split between the Hungarian Air Force and the German *Luftwaffe*. A total of 352 Me 210s were built in both Germany and Hungary.

On 26 August 1942 the design was first flown with the much more powerful DB 603 engines which had been planned for the Me 310. When the plane reached 525 km/h at low level during trials at Rechlin, Milch ordered that this aircraft was now to be designated the Me 410.

Production got underway at once, initially by converting the Me 210A airframes on hand. The majority of those built after the spring of 1943, however, were new. The Me 410 was manufactured at Augsburg and, from late 1943 onward, also at Dornier in Oberpfaffenhofen. A total of 1,160 machines were built before production ran out in the fall of 1944, and the type was employed as high-speed bomber, *Zerstörer* heavy fighter in the defense of the *Reich*, and as a reconnaissance aircraft.

With the exception of problems with lateral stability, the poor flight handling characteristics had been virtually eliminated on the Me 410. In this latter form, it can be considered the *Luftwaffe*'s best multi-role aircraft in its weight class.

Up until the spring of 1943 nearly all Me 410 airframes were converted from Me 210s (such as this former Me 210, originally built in the summer of 1941), or were built mainly from warehoused Me 210 components. Later, from about the 700th machine onward, Me 410s were built from new.

The cleaned-up Me 210 with lengthened fuselage and more powerful DB 603 engines was, on Milch's orders, designated the Me 410. This is Me 410V-1, based on a Me 210A-1 (Werknr. 027) from December 1941. The airplanes were brought up to Me 410 standards in two conversion stages: 1) modification of the wing and leading edge, extension of the fuselage, 2) converting the powerplant from the DB 601F to the DB 603. First flight of this design took place on 26 August 1942.

The Me 410 was employed as a nightfighter only on rare occasions Several reconnaissance aircraft were fitted with the FuG 200 Hohentwiel system for sniffing out shipping convoys.

Trials with the heavy-caliber BK 5 (5 cm gun) began in mid-1943 for the purpose of bringing down American B-17 bombers. From the spring of 1944 this weapon was fitted to production examples, which were delivered to front-line units as the Me 410B-2/U4.

Me 310 and Follow-On Developments

With the Me 210 entering production in early 1941 (even though flight trials had not yet been completed), the project and design sections began working on the type's successor, the Me 310. This was to basically be a larger and more powerfully engined Me 210. Thus, a design was projected and drawn up which incorporated outer wing sections extending by 1.50 m (17.9 m wingspan), a lengthened fuselage, more powerful DB 603 or Jumo 213 engines, a pressurized cockpit and improved systems. Its roles were the same as for the Me 210: dive bomber, heavy fighter, strike fighter, long-range reconnaissance. A contract was issued for two prototypes, but in early 1942 all work was halted in order to resolve the problems with the Me 210.

In September 1943 a Me 210 was fitted with the Me 310's wing and lengthened slats to evaluate the former's stall handling characteristics. The tests showed that the hybrid had less dangerous stall tendencies than the Me 210, but revealed an increase in aileron stick force.

The cleaned-up Me 210, going into renewed production as the Me 410, incorporated a lengthened fuselage and the DB 603 engines of the Me 310. When the *Probü* offered the RLM its Me 410 as a standard weapons system to

The enlarged wings which were to have been used on the Me 310 were fitted to this Me 210A-1, Werknr. 179, in August 1943 for testing the Me 410's stall handling characteristics.

include the new common DB 603 and BMW 801 fighter/ *Zerstörer* engines, not even the larger wings of the Me 310 could be used as the engines had outgrown them. These projects required another expansion of the outer wing panels, increasing the span to 20.5 m, and to counter the change in the center of gravity the inner wing sections were given a slight negative sweep. Only a single prototype, with two DB 603 standardized engines, ever entered flight testing before the entire Me 410 program was halted in the autumn of 1944.

The bombing raid of 22 February 1944 on the Augsburg Works seriously affected Me 410 production, which was in full swing at this time. Production of this aircraft only continued up until the autumn of that year.

The RLM's search for a high-speed bomber resulted in this proposal from the Messerschmitt project section in the autumn of 1943. It shows a Me 410 with either one or two performance-boosting turbojet engines mounted under the fuselage.

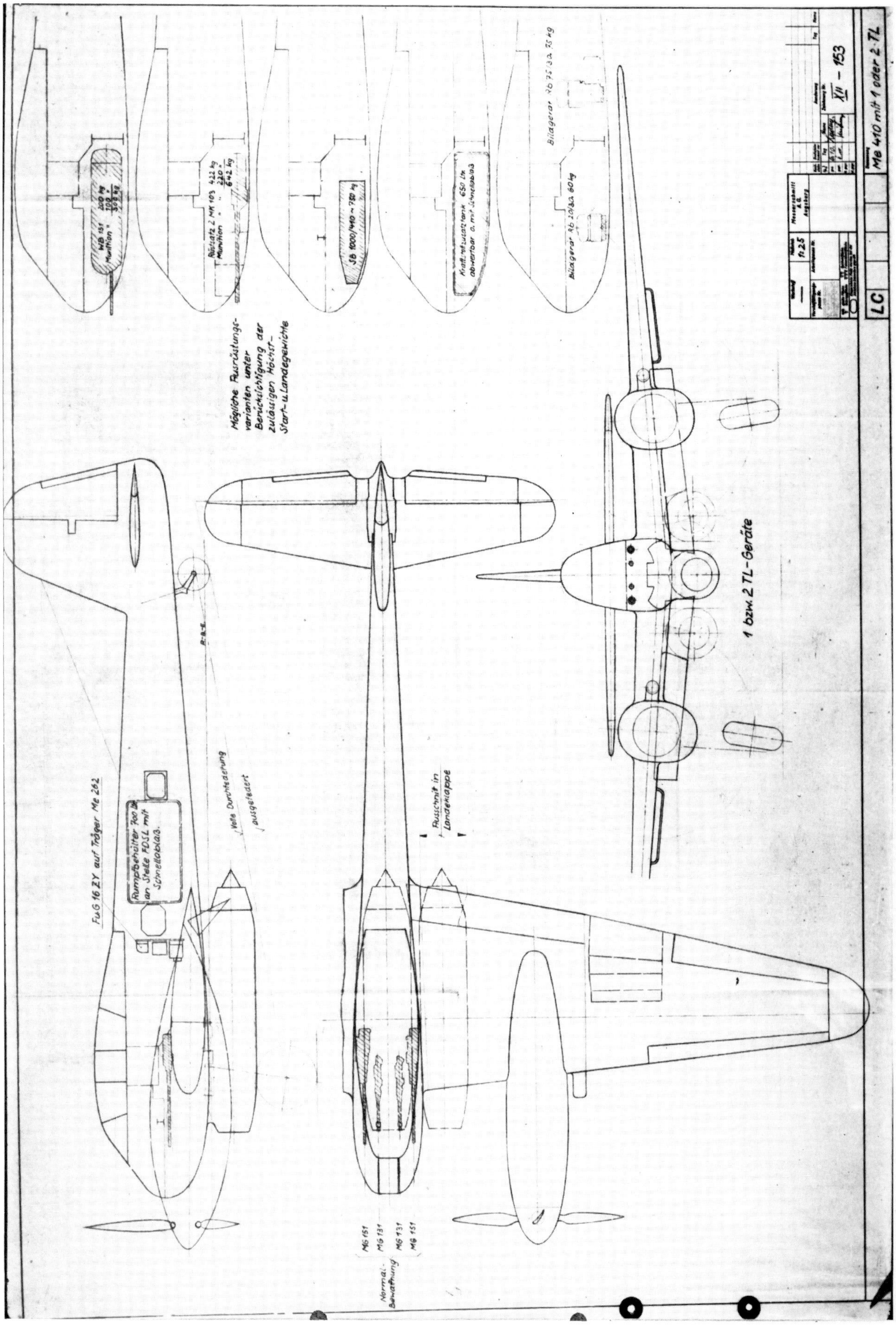
Mögliche Rüstungs-
varianten unter
Berücksichtigung der
zulässigen Höchst-
Start-u. Landegewichte
WB 151
Rüstsatz 2 MK 103
SB 1000/410 ~ 780 kg
Kraftstoffzusatztank 650 Ltr.
abwerfbar o. mit Schnellablaß
Bildgerät Rb 20/30 60 kg
Bildgerät Rb 75/30 75 kg
1 bzw. 2 TL-Geräte
FuG 16 ZY auf Träger Me 262
Rumpfbehälter 700 l
an Stelle FDSL mit
Schnellablaß
Volle Durchfederung
ausgefedert
Ausschnitt in
Landeklappe
Normal-
Bewaffnung
MG 151
MG 131
MG 131
MG 151
LC
Me 410 mit 1 oder 2 TL
XII - 153

Me 210/410 Variants

In order to obtain test data in the shortest time possible, contracts were issued for a large number of prototypes, both for the original Me 210 as well as for the conversion series and the Me 410. Each prototype was generally tailored to the evaluation of a specialized area. Prototypes for the Me 210 series included the V-1 through V-16, with the V-17 through V-38 for the conversion series, and the V-1 through V-22 for the Me 410 program. The most important variants were:

Me 210A-1	Dive bomber/heavy fighter with 2x DB 601F, armament 2x MG 17, 2x MB 151/20 fixed forward-firing guns and 2x MG 131 defensive guns in fuselage barbettes. Bomb load up to 1,000 kg.
Me 210B	Reconnaissance aircraft with reduced armament.
Me 210Ca-1	*Zerstörer* heavy fighter with 2x DB 605, armament and bomb load as for A-1. License built in Hungary.
Me 210C-1	Reconnaissance aircraft as Ca-1, with reduced armament. License built in Hungary.
Me 210S	Initial production run aircraft modified as strike fighters.
Me 410A-1	High-speed bomber, modified from Me 210A-1s with 2x DB 603A, armament 2x MG 17, 2x MG 151/20, 2x MG 131, bomb load up to 1000 kg.
Me 410A-1/U1	Converted reconnaissance plane
Me 410A-1/U2	*Zerstörer* with WB 151 pylon unit in bomb bay.
Me 410A-1/U4	*Zerstörer* with BK 5 and 20 rounds in bomb bay.
Me 410A-3	Reconnaissance aircraft
Me 410B-1	High-speed bomber. New-built Me 410 with armament same as A-1
Me 410B-2	*Zerstörer* with 2x MK 103
Me 410B-3/U1	Reconnaissance aircraft with GM-1 boost system
Me 410B-6	Reconnaissance aircraft with FuG 200 Hohentwiel search system
Me 410B-7, B-8	All weather reconnaissance aircraft
Me 410C and D	Planned variants with new forward fuselage, larger wings, and fighter/*Zerstörer* DB 603G/Jumo 213 standardized engines.

Search for Replacement Heavy Fighters and High-Speed Bombers

Failure in the form of the Me 210 disaster and success in the form of the Me 262's first jet-powered flight followed closely on the heels of each other during the years 1941 and 1942.

This was an incentive for the developmental teams headed by Willy Messerschmitt, Woldemar Voigt, Hermann Wurster and Alexander Lippisch to come up with many other concepts and ideas in addition to the ongoing work with the Me 163, 209, 309, 264, 323, and Me 328. The period from mid-1942 to mid-1943 can probably be considered the most technologically interesting timeframe of the entire war.

Projects by A. Lippisch and H. Wurster

Alexander Lippisch, who since early 1939 had been working on the Me 163 with his team (see section), became involved in the events surrounding the Me 210 and in early 1942 submitted a project for a high-speed bomber flying wing concept to the RLM, i.e. Erhard Milch. It was designated the Li P 10 and called for a tailless airplane with slightly swept wings and a DB 606 double engine driving a pusher propeller via an extended shaft. However, Alexander Lippisch was forced to withdraw his first design because of

Prof. Messerschmitt with Woldemar Voigt, head of the project section, at the drawing board set up in the department (1942).

Right: Original design of the Lippisch P 010 high-speed bomber project with dual DB 606 engines, from late 1941. The designers gambled with the meter-long extension shaft driving the large-diameter propeller. The design utilized mildly swept wings with a span of 16 meters; the fuselage was 8.85 meters long. Using the same number, a similar project with two turbojet engines was also drawn up.

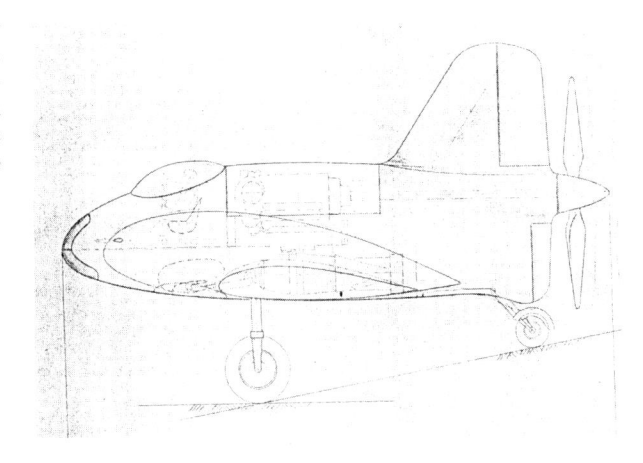

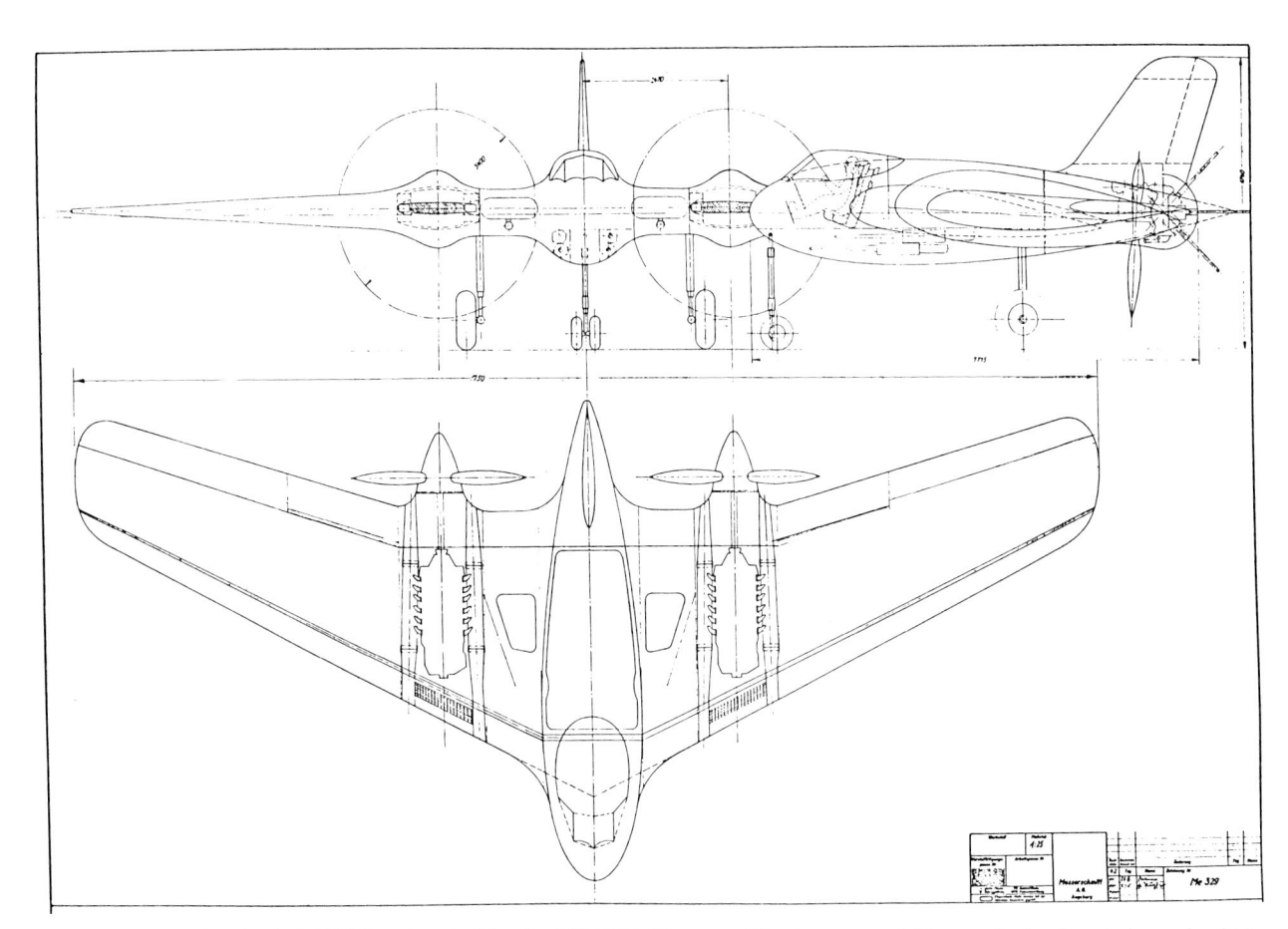

As a counter to the Lippisch P 010 project, in March of 1942 Dr. Hermann Wurster proposed his own design for a twin-engined Me 329. It was to have been employed as a heavy fighter, escort, nightfighter, dive bomber, fighter-bomber, and reconnaissance aircraft. Plans called for two DB 603 or Jumo 213 engines. Armament was to have included four MG 151/20 and two MK 103 guns, plus a maximum payload of 2.4 metric tons carried in a bomb bay and on underwing racks. Empty weight was to have been 6,950 kg, with a takeoff weight of 12,150 kg. Speed at maximum pressure altitude was estimated at 792 km/h, with a range of 4,450 km. Wingspan was 17.50 m, length 7.71 m.

major technical difficulties, especially with the extended propeller shaft whose development had cost much valuable time. He worked out a new variant which made use of two single DB 603 engines driving two pusher propellers and mounted on the wing trailing edge.

Lippisch resubmitted the project without the knowledge of Mtt's *Probü* or of Prof. Messerschmitt himself, and shortly later the RLM issued him a contract for a "flying model." The *Amt* must have recognized it was making a serious breach of protocol by giving such a contract to Lippisch directly, but without his friend Udet Messerschmitt had become something of a *persona non grata* within the RLM.

Dr. Hermann Wurster, after giving up his flying activities in 1942, had been acting as a sort of liaison between the Mtt *Probü* and "Department L," and independent of the Li P 10 project had developed a similar bomber *Zerstörer* project under the designation Me 329. Willy Messerschmitt already had this project in hand when he heard of Lippisch's Li P 10 design while at a meeting with the RLM concerning the Me 210.

There followed several internal meetings with those responsible parties within the project department, during which Theo Croneiß once again acted as a mediator for Messerschmitt with Lippisch. On 28 August 1942 Willy Messerschmitt called for a comparison of the two projects—the Me 329 and the Li P 10—with the Me 210, and suggested that they might possibly be combined into one design. Continuing on, he stated that it seemed the Me 329 was a more mature design, but he did not see how airplanes with wing areas of 55 and 65 m² (Me 329, Li P 10) would be faster—as claimed in the project documentation—simply because they had no tailplane, than one with a normal tail having a wing area of 36 m² (Me 210). Since this was a fundamental question, he asked that the two project departments undertake comparison studies with their aerodynamics engineers and performance calculators. At the same time models would be tested in the wind tunnel and mockups of the two designs built.

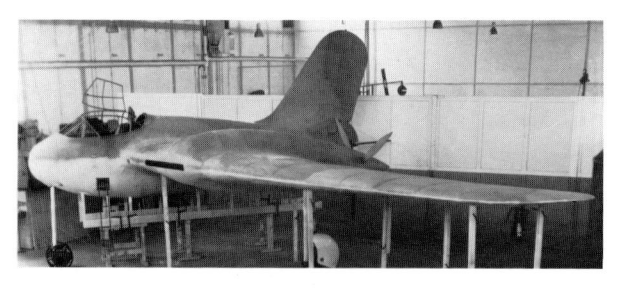

Prof. Messerschmitt preferred the Me 329, and work on the design—which included a 1:1 partial mockup—progressed throughout 1942 and into early 1943. Photos of this mockup were included in Messerschmitt's description in his bid proposal to the RLM.

Tailless or Conventional?

By early September an initial research paper had been submitted by the Lippisch team's Walter Stender under the somewhat provocational sounding title "Why does the tailless aircraft fly faster?"[2] The comparison between the Me 210 and the Li P 10 was done exclusively on paper, and Stender's introduction to the report made the following claim:

"In addition to the recognized benefits the tailless, swept-wing design offers when using jet propulsion at near-supersonic speeds, it also has its advantages in the speed ranges produced by the standard engines in use today. This fact is often treated with skepticism in expert circles, but can be proven using a specific example. The comparison was made between a Li P 10 high-speed bomber and a normal aircraft designed for the same role, whose dimensions correspond to those of the Me 210."

And the conclusion listed the numerous advantages of the "tailless" design:

"Altogether, the tailless aircraft is superior in all other circumstances as well, to include: surface friction, pressure drag, fuselage interference, nacelle interference, tailplane interference, induced drag, propeller wash drag, propeller efficiency, and propulsion system efficiency.

This superiority leads to higher speeds and subsequently greater efficiency from the propulsion system, the cooling system, and the surface friction. Additional advantages of the tailless concept include an unbroken view to the sides, ideal cable tow platform, elimination of all tailplane-associated problems, including damage from gunfire, suitable for wooden construction, and lower fuel consumption."

Hans Hornung and Woldemar Voigt submitted the *Probü's* report to the RLM in December. It was subsequently made available to the entire aviation industry as part of the series "Deutsche Luftfahrtforschung—Untersuchungen und Mitteilungen."[3] In the report, Hornung and Voigt compared the Me 410, the Li P 10, and the Me 329, with all three designs being calculated for the same role and landing speed

---
[1] Schnellbomber in schwanzloser Ausführung, TDM memorandum from 2 Sep '42 to Messrs Lippisch, Wurster, Voigt, Croneiß
[2] (W. Stender) Warum fliegt das schwanzlose Flugzeug schneller?, Messerschmitt AG, Augsburg, 7 Sep 1942 (Secret)
[3] Hornung/Voigt, Vergleich zwischen Normal- und schwanzlosen Flugzeugen, Messerschmitt AG, project dept., technical report TB No. 80/42 from 1 Dec 42. Later published by Oberbayerische Forschungsanstalt Oberammergau as "Deutsche Luftfahrtforschung, Untersuchungen und Mitteilungen No. 7838" (Secret)

(Me 410 = 170 km/h). For example, the figures for horizontal flight at maximum pressure altitude were as follows:

Me 410	672 km/h
P 10	682 km/h
Me 329	685 km/h
For service ceiling:	
Me 410	10900 m
P 10	12100 m
Me 329	12500 m
Range at best glide ratio:	
Me 410	2020 km
P 10	2480 km
Me 329	2520 km

Hornung and Voigt also came to the same positive conclusions regarding tailless aircraft and stated that this type of design had been considered for earlier Messerschmitt projects, but was never used because of the increased development risks.

"Using the most favorable aspect ratios for each type, the results are as follows:
- speeds are the same.
- service ceilings are 6% higher for the tailless design.
- ranges are 15% greater for the tailless design than for standard designs.

In conclusion, it seems that the tailless design is basically equal to the standard design at low angles of attack, but at high angles it is markedly superior. All in all, the tailless concept can therefore be considered a superior design in terms of performance.

Parity of speed performance and the superiority of the tailless configuration for long-range aircraft is in accordance with the data obtained from project studies undertaken with the Me 210, Me 261, Me 262, and Me 264 (the Me 261 and Me 264 were not developed as tailless designs at the time only because of the associated developmental risk). Whereas formerly a long-range aircraft's suitability would be attributed to making the most use of its extremely large dimensions, this quality is now attributed to a general superiority in the area of high C_a-values."

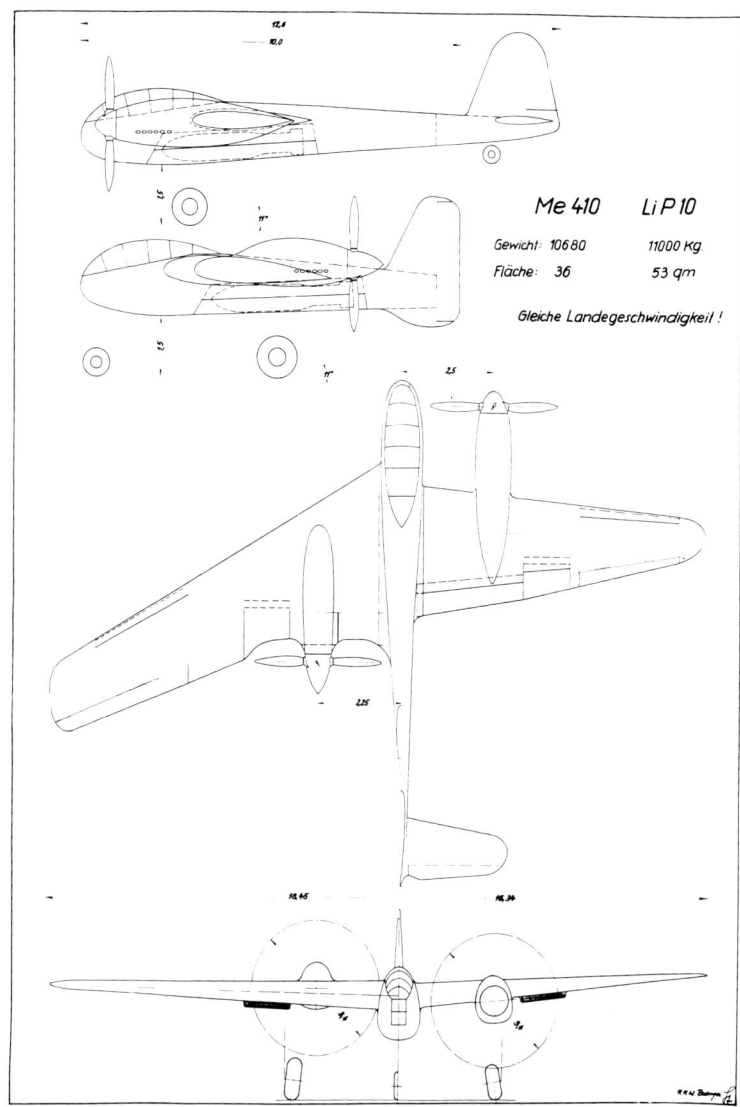

Kurt Schnittke built a flying model of the Li P 10 in Augsburg based on plans by Walter Stender. (Post-war publications have often identified this Lippisch project as the Me 265, a designation which cannot be substantiated from documentation.)

Comparison of a twin-engined high-speed bomber showing standard and tailless design, from November 1942, contrasts the Lippisch Li P 10 (as it is now designated) with the Me 410. The Li P 10 made use of a large percentage of the Me 410's fuselage, including internal equipment. As on the Me 410, the design was to have been powered by two DB 603 engines. Wingspan was 18.45 m, length 10 m, and takeoff weight 11 metric tons.

Notwithstanding the good results achieved by the Lippisch projects on paper, by late 1942/early 1943 the time and risks needed for development forestalled any hope of making these projects a reality. Also, production had resumed on the Me 210/410 in the interim. Although work on his projects at first continued within the company—using jet propulsion systems—Lippisch found the situation to be so untenable that it was one of the reasons behind his subsequent move to Vienna.

Swept Wings and Butterfly Tails

For Willy Messerschmitt, his technical differences of opinion with Alexander Lippisch had nothing to do with feelings of animosity, as it often seemed. Nevertheless, each side drew his line in the sand; Messerschmitt demanded an aircraft with fuselage, and for Lippisch it was a tailless flying wing with swept wings. Messerschmitt was not opposed to the use of swept wings. On 8 January 1943 he brought the matter up for discussion in a message to Messrs. Lippisch and Voigt, and at the same time mentioning the applicability of a butterfly tail:[4]

"The swept wing is an absolute must for high-speed aircraft. Without a doubt, the swept wing must of necessity be constructed inherently stable. Centrally located engines or engines located near the centerline would enable us to make use of a short fuselage as *Herr* Lippisch uses in his designs.

[4] Introduction of swept wings. Message to Messrs. Lippisch and Voigt from 8 Jan 43

Two brilliant competitors: Alexander Lippisch (left) and Willy Messerschmitt, here in a discussion in front of the test flight facilities in Augsburg with Fritz Wendel looking on. During their discussions, each defended his position admirably: Lippisch for the tailless design and Messerschmitt for a conventional aircraft layout. However, their views did not differ that greatly when it came to the future use of jet engines and swept wing designs.

The other thing is finding out whether splitting the vertical stabilizer into two control surfaces in the form of a V would give us the following advantages over a tailless configuration:

1. With the current designs, the aileron moves upward when banking or landing and thereby causes a reduction in lift for the entire wing at the exact moment when I want to have the most lift possible. Once the airplane has been designed to be inherently stable in the pitch axis, as we assume it will be, with a V-tail it might be possible to achieve aileron control using the two control surfaces on the V-tail in the same sense.
2. Such a butterfly shaped rudder would have the advantage of giving us a larger center of gravity area to work with than before, i.e. some of the trimming can be accomplished with these control surfaces, the rest with the landing flaps.
3. The use of this V-tail would be advantageous in that we can incorporate a more effective flap system over part of the span, or even the entire span, and thereby increase the lift coefficient. We may even be able to use a fixed or extending slat on the V-tail to counter the landing flaps' nose-heavy pitch moment using the minimal size of the empennage.
4. We should see whether it would be possible to use control surfaces without fins, similar to what was tried on the M 29, even though this may lead to problems with the varying aerodynamic pressure center at extremely high speeds."

Referring to the Li P 10 design and the Me 329 (which he favored), as well as the modified Me 210 with its large glazed cockpit canopy, Willy Messerschmitt also countered on January 1943 with the following argument:

"With the fighters using new tactics of generally making their initial attack run against a bomber from head-on and with the development of improved gunsights, it is questionable whether the all-round canopy idea will survive in the long term or whether for safety reasons it wouldn't be better to put the crew behind the engine. In this way the airplane and its centerline propulsion system would take on renewed significance in the bomber role. We must also see if we can put two engines in tandem, one with a pusher propeller and the other driving a tractor propeller."

A short time later the *Probü* submitted a high-speed bomber design incorporating an engine in the nose and one behind the pilot. In view of the firepower coming from the Allied bombers, however, a better solution was needed for protecting the pilot.

Aside from the previously mentioned projects, Willy Messerschmitt submitted even more proposals addressing the pressing matter of high-speed bombers and *Zerstörers*. There was the P 1090 project from early 1943, a twin-engined modular design drawn up in conjunction with a proposal for reducing the number of different frontline aircraft types (see section). An internal discussion on the P 1090 led to another new idea which would have been easy to develop and require minimal expenditure: Regensburg Works director Karl Lindner proposed the "Me 109 *Zwilling*," a design which joined together two Bf 109Gs and incorporated heavier armament. This was what Messerschmitt was looking for, and he had the project department research it in detail. Despite the fact that the design looked good on paper, particularly with regard to how quickly it could go into production, the RLM was not overly interested in this plane either, and not even a prototype was built.

Once the Me 210 had been cleaned up and had re-entered production as the Me 410, Messerschmitt offered this airplane to the RLM as a "weapons system," with potential roles ranging from *Zerstörer* to reconnaissance plane and—in a performance boosting measure—the fitting of an additional turbojet engine (see section).

When several companies submitted proposals to the RLM for a heavy fighter and high-speed bomber design in early 1943, Heinkel and Junkers recognized that the Me 109Z was the best solution. The Dornier company, however, stood by its Do 335 design. As a result, Prof. Messerschmitt instructed his project department to draw up a design for a high-speed bomber using both a tractor and a pusher propeller as a "counter project" for a comparison study with the Do 335. This produced a design which with two DB 605 engines would have a maximum speed of 757 km/h and with two DB 603G motors would attain 820 km/h at an altitude of 8,800 meters. Messerschmitt's study therefore offered improved speed performance over the figures submitted by Dornier at the time. Additional data for the project drawn up by Hans Hornig were: two MK 108 guns, empty weight was 6,620 kg, takeoff weight 9,040 kg, wingspan 15.75 m with an area of 36 m2, length 13.53 m.

The first proposal to come out of Regensburg for a high-speed bomber and heavy fighter was a quickly drawn up design for the "Me 109Z"—a mating of two standard production BF 109G airframes side by side. The principle behind this was based on RLM recommendations, some of which came from Ernst Udet himself, ironed out in preliminary trials and tested in practice with the He 111Z tug for the "Gigant" heavy-lift gliders.

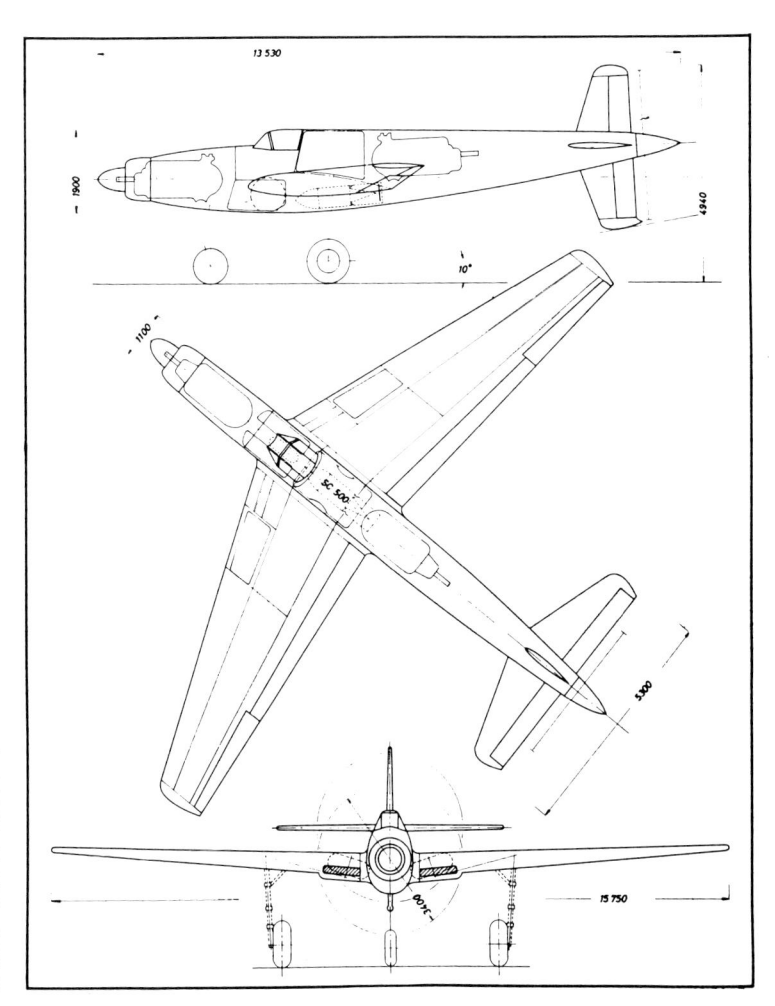

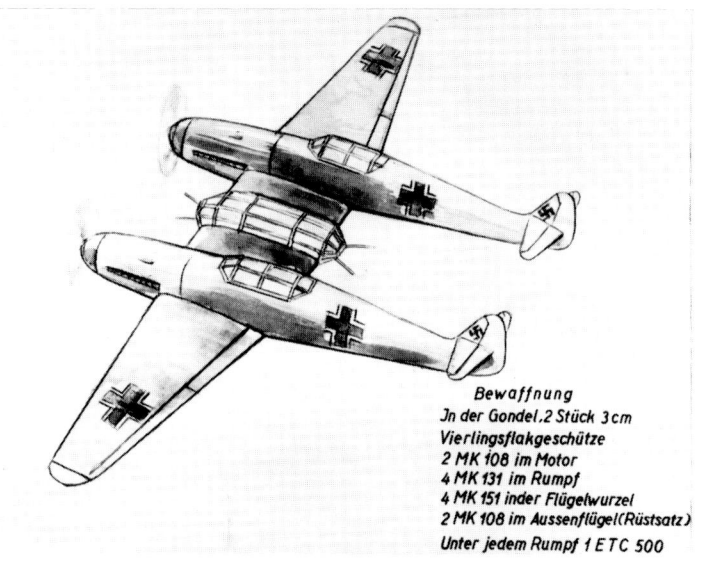

Bewaffnung
Jn der Gondel,2 Stück 3cm
Vierlingsflakgeschütze
2 MK 108 im Motor
4 MK 131 im Rumpf
4 MK 151 inder Flügelwurzel
2 MK 108 im Aussenflügel(Rüstsatz)
Unter jedem Rumpf 1 ETC 500

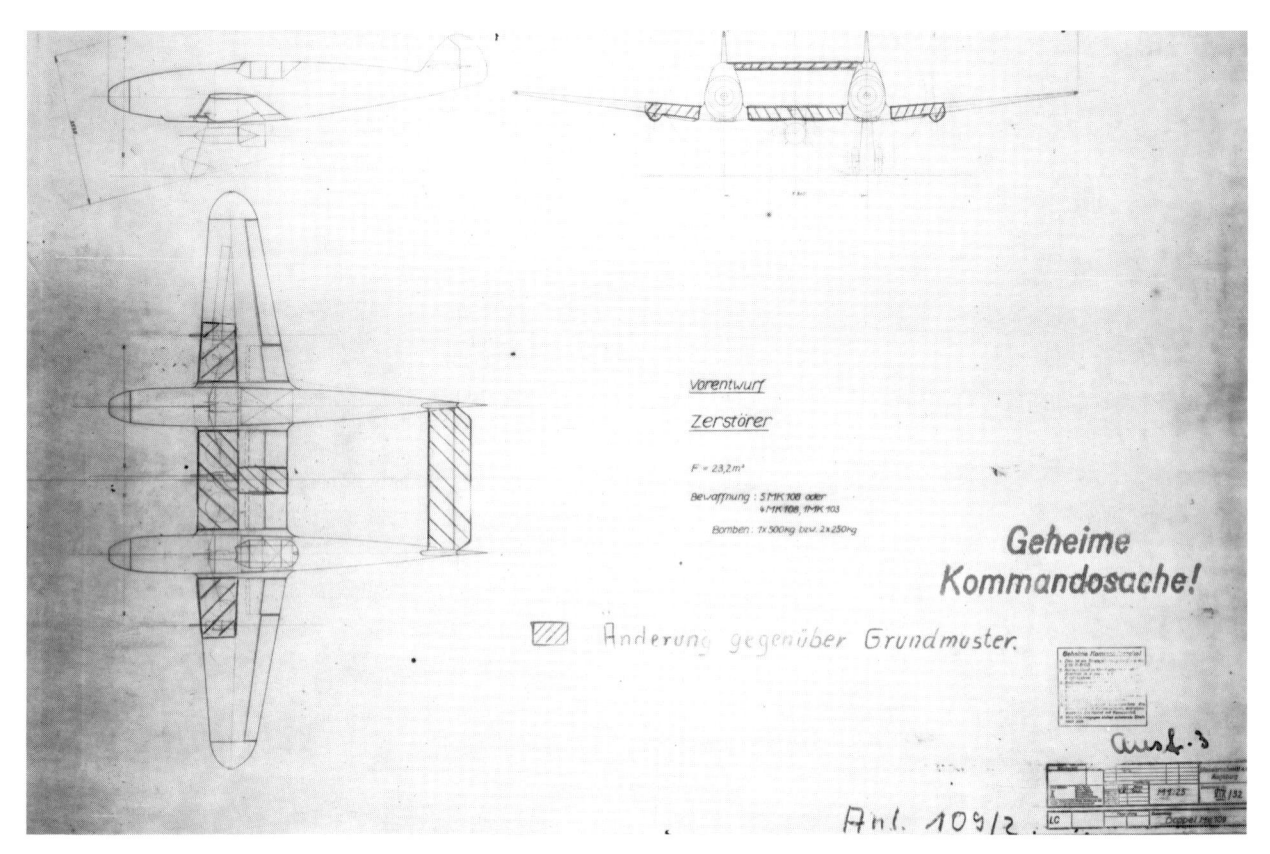

In December 1942 the project section drew up the Me 109Z as it had been proposed to the RLM for use as a high-speed bomber and heavy fighter. The "twin Me 109" was to have been flown as a single-seater powered either by the DB 605 standard to the Bf 109G or by two Jumo 213 engines. For the heavy fighter configuration, armament was to have consisted of five MK 108 or four MK 108 and one MK 103 cannons and have an empty weight of 4,900 kg with DB engines or 5,300 kg with Jumo engines. Takeoff weights were 6,200 and 6,600 kg respectively, while maximum speed at maximum pressure altitude was 690 and 750 km/h respectively. For the high-speed bomber, armament was reduced to two MK 108s, with a payload of 2,000 kg carried on external racks. This resulted in a speed loss of about 75 km/h. Wingspan was 13.27 m, length 8.92 m. Also included in the proposal was a "twin Me 309" with two DB 603G engines, based on the Me 309 which was undergoing flight testing at the time. In this case, armament was to have been two MK 108s and two MK 103s plus a 500 kg bomb load for the heavy fighter or two MK 108s and 2,000 kg bomb load for the bomber version. Empty weight was 6,380 or 6,100 kg; takeoff weight was 8,325 or 10,100 kg. Maximum speed at maximum pressure altitude was to have been around 760 km/h, reducing by about 75 km/h with a 1,000 kg bomb. Wingspan was 16 meters.

Aircraft Using the Modular Assembly Principle - The P 1090 Project

The Eastern Front's ravenous appetite for men and materials soon made it obvious to the highest authorities that, in addition to more and more inductions, a dramatic increase in weapons production was the only way to alleviate the situation. The armament industry was faced with the dilemma of seeing ever more of its qualified workers—engineers and technicians—being called up for military service, with their ranks either being filled by unqualified employees or going vacant altogether. More and more, the army of workers was being staffed by German women, prisoners of war, and foreign workers from both East and West; in the latter half of the war, an increasing number of concentration camp prisoners also began swelling the ranks of the work force. As a result, the industry would have to find methods of simplifying production while at the same time increasing output.

Streamlining and Commonality

For fighter production, this difficult task fell to the technical director at Messerschmitt GmbH Regensburg, Karl Lindner. Within a year, from 1942 to 1943, he introduced assembly line production for the Bf 109 and by streamlining was able to reduce assembly time by some 40 percent. The Regensburg plant increased its output from about 50 Bf 109s per month in 1942 to over 250 in 1943. Production man hours per aircraft dropped from 9,000 for the Bf 109F to 5,000 for the G model.

The matter was of particular concern to Willy Messerschmitt after production of the Me 210 was halted in March 1942, suspending deliveries to the *Luftwaffe* totaling hundreds of aircraft. Messerschmitt went to the RLM in early 1943 with a proposal for streamlining airplane production and "thinning out the variety of front-line aircraft being produced."[5] In light of the roughly 20 key types (not including trainers and sailplanes) then in service with the *Luftwaffe*, i.e. combat aircraft and transports with hundreds of variants, and factoring in another 20 or so types under development, Messerschmitt recommended that the number be reduced to just five basic types. These could then be used in a variety of different roles by fitting them out with field conversion kits (which would also be reduced in number). He broke the basic types down as follows and added several specific recommendations for those Messerschmitt aircraft already in production or development which would be suitable:

[5] Messerschmitt AG, Memorandum on streamlining the number of front-line types (P 1090), Augsburg, 6 Feb 43

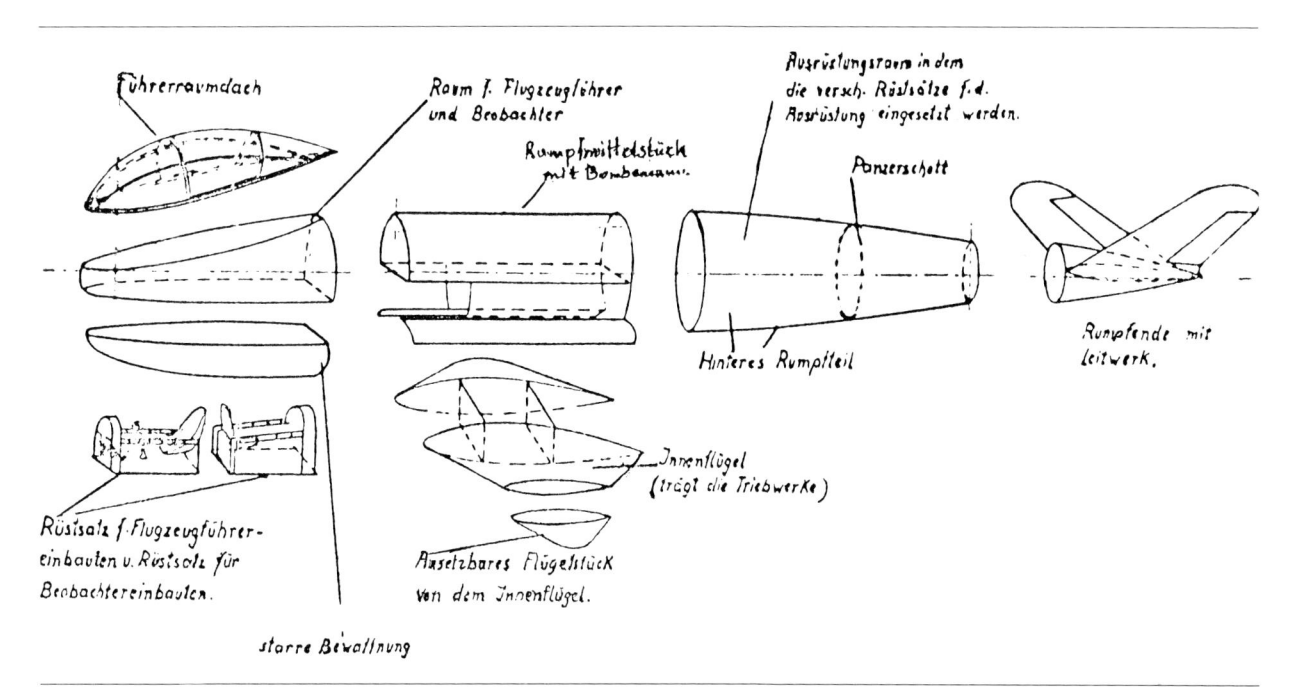

The only surviving drawing showing the P 1090 "modular plane" is this conceptual drawing from early 1943. The aircraft, roughly in the same size class as the Me 410, was to have been powered by two DB 603G engines or two turbojets. It was to have been a single or twin-seat heavy fighter, strike fighter, high-altitude fighter, nightfighter, high-speed bomber, dive bomber, torpedo bomber, or fighter-bomber. The aircraft's role was altered through the use of field conversion kits; the cockpit area could be swapped out for different armament configurations, the fuselage center section for photographic and/or ordnance systems, and the rear fuselage for radio, navigation, night-fighting, and rescue equipment such as rubber rafts. The fuel was to have been carried in the wings. By swapping out the outer wing sections, the wing area could be increased from 28 m² to 31 or 36 m². Empty weight varied between 5.5 and 8 metric tons, with a takeoff weight of up to 11.5 metric tons. Estimated speeds for the piston-engined versions fluctuated between 500 and 775 km/h, while the jet-powered versions could attain 1,010 km/h. The twin-wheeled landing gear retracted into the engine nacelles. Tailwheel was retractable; jet-powered aircraft had a tricycle gear.

Type I:
Light combat aircraft, single-engine, single-seat, e.g. light fighter (such as Bf 109, Fw 190) with potential as
- fighter, high-altitude fighter
- fighter-bomber
- strike fighter
- tactical reconnaissance aircraft
- carrier fighter/fighter-bomber/reconnaissance
- torpedo bomber

A candidate for this category was Messerschmitt's Me 209, just entering development, which could also utilize components from the Bf 109 production.

Type II
Medium combat aircraft, two-seat, twin-engined, e.g. *Zerstörer* heavy fighter (such as Bf 110, Ju 88) with potential as
- high-speed bomber
- dive bomber
- strike fighter
- night fighter
- reconnaissance aircraft

The existing Me 410 from Messerschmitt's stable would have made an ideal choice; at the same time Messerschmitt offered his P 1090 as a new development. Then there was the "Me 109 *Zwilling*" concept from the Regensburg factory.

Type III
Heavy combat aircraft, four-engined high-performance plane (such as the Ju 290) with potential as
- strategic reconnaissance aircraft
- strategic bomber
- ultra-long range bomber (over 10,000 km)
- high-speed transport (up to 20 metric tons cargo)

Messerschmitt here suggested his Me 264, just entering its testing phase (q.v.).

Type IV:
Multi-engined medium transport (such as the Ju 52) with application as troop and material transport
- flying hospital
- flying workshop
- tow tug

The Ju 252, Ju 352, and the Ar 232, all either in development or under construction, would fall into this category (as opposed to their predecessors with increased payload)

Type V
Heavy-lift transport (such as the Me 323, Ju 290) with the same roles as Type IV plus specialized tasks

Willy Messerschmitt expanded the proposal to include powerplants, limiting them to three or four basic types to be developed as quick-change standardized engines. Thus, in his opinion, types III through V could utilize the same engine model. He also drew attention to the idea that special role aircraft and trainers could be derived from their respective categories. For example, even before the war he had offered a lower-powered version of his Bf 109 as a trainer. Messerschmitt made similar recommendations for reducing the number of propeller, wheel, aircraft systems, and armament field kit types, as well.

Commonality Even for Jet Aircraft

Development of jet aircraft could proceed along similar lines, even though this initially would be limited to Types I and II and later possibly Type III. Although the P 1090 envisaged using turbojet engines in addition to conventional powerplants, the *Probü* simultaneously developed a whole line of single- and twin-jet airplanes incorporating a wide variety of weapons combinations. Even the official description for the Me 262 listed a plethora of potential roles, including those of fighter-bomber and high-speed bomber with a bomb load of up to 500 kg. In early 1943 Messerschmitt had his workshop build a type of wooden modular mockup specially for the P 1090 project as a tangible reinforcement when presenting his ideas to the RLM. The RLM, however, was unwilling to listen to such a wide-sweeping idea. The *Technisches Amt* nevertheless did indeed institute a general purge of aircraft types in 1943, so that the numerous roles were either distributed among the surviving types or dropped altogether.

Rolf von Chlingensperg, who had been involved in drafting the commonality proposal, initially shared these thoughts with the Messerschmitt Corporation's upper level management in the spring of 1943. At the Regensburg factory, Messerschmitt and von Chlingensperg found open doors at every turn. Director Linder saw his large-scale production plans confirmed in the proposal and even went so far as to call for a basic standardization of construction components, as well.[6] At the same time he recommended for Type II the "Me 109 *Zwilling*," a mating of two Bf 109Gs with the heaviest firepower the hybrid design would be capable of carrying. The idea piqued Willy Messerschmitt's interest to such an extent that even his P 1090 project was pushed into the background.

[6] Messerschmitt GmbH, Opinion on proposal for standardization of types, Regensburg, 30 Apr 43

From BFW to the Messerschmitt Company

On 24 August 1936 an open house was held on to celebrate the tenth anniversary of the establishment of the BFW—simultaneously the twentieth anniversary of aviation construction in Augsburg. Not only was the public treated to a collection of every Messerschmitt airplane from the M 17 to the Bf 110, but through this photo montage they were also introduced to BFW's board of directors and administrative staff.

Frau Lilli Stromeyer was a primary stockholder and, as such, the company's sponsor. Willy Messerschmitt married her in 1952.

Theo Croneiß, a long-time friend and supporter of Messerschmitt, was chairman of the board of directors and, until his death on 7 November 1942, the senior chief of Messerschmitt's Regensburg facility.

Banker Friedrich Wilhelm Seiler came to BFW in 1933, initially acting as deputy chairman of the board but, from 1942, became chairman of the board of the directors. Up until the last, he was an outstanding financial administrator for the company.

Willy Messerschmitt was the technical director and member of the executive board of directors and manager of the Augsburg facility until 1942.

Konrad Merkel, for many years Messerschmitt's lawyer, was a member of the board of directors and, following the death of Croneiß, temporarily operations manager of the Regensburg Works.

Mayor Josef Mayr represented the city of Augsburg on the board of directors.

Director A. S. Schwarzkopff represented the RLM from 1933 to 1937. He wrote of his experiences during the BFW's exciting years of expansion in the patriotic novel "Vertrauen ist alles" (Trust is Everything).

Rakan Kokothaki was the commercial director of the BFW and Mtt AG and, after the death of Croneiß, became senior executive manager at Augsburg.

As a member of the executive board of directors, director Fritz H. Hentzen was responsible for production, and from 1941 was manager of the F 2 production ring for all Messerschmitt aircraft.

191

Shortly after the compulsory composition had concluded in April of 1933, the Bayerische Flugzeugwerke AG was able to resume its activities. The RLM began passing on a steady flow of license production jobs for "foreign" designed trainers and tactical reconaissance airplanes from Heinkel, Gotha, Arado, and Dornier. In addition, the first development contract for the Bf 108 as a competitor for the 1934 Europa Circuit was awarded. This latter program had a volume of about 400 airplanes, which were to be delivered by 1937 in quantities gradually increasing by year.

As the BFW had manufactured barely 300 planes in total over the last eight years (1926-1933), it had neither the work force nor the space to deal with such an order. Thus, from the airfield at Haunstetter Straße soon could be heard the constant buzz of construction activity, which kept up incessantly until 1937/1938. This was hardly able to keep pace with the rapidly increasing number of workers occurring simultaneously. While the BFW had begun with 82 employees on 1 May 1933, by the end of the year this number had increased to 524, nearly tripling to 1,414 by the end of 1934, almost doubling again by the end of 1935 to 2,403, and reaching 5,182 by the end of 1936.

This explosion was in part necessitated by the RLM's decision in 1938 to produce the Bf 109 as the *Luftwaffe*'s standard fighter, an action which—despite the company's ongoing expansion—required other license-production companies to become involved in the program.

This expansion was also needed in order to meet development and production contracts pending or already awarded in quick succession for other Messerschmitt aircraft: the Bf 108B, the Bf 110 and its planned successor the Me 210, as well as for the Me 209 and 309 successors to the Bf 109 and the Me 262. And then there was the development for the larger aircraft, the Me 261 and 264. The "*Giganten*" were the exceptions to the rule; neither the factory nor the airfield could accommodate their production, and these giants were assembled at the Leipheim airbase starting in late 1940, thus making Leipheim the first of Messerschmitt's dispersed production sites.

Main entrance of the new central administrative building seen from the Haunstetter Straße.

The Augsburg Works underwent its main expansion phase during the years from 1934 to 1938. Four production facilities sprang up on the site, three of them entirely new:

- Werk 1: The BFW's main facility, with the buildings inherited from the old Bayerische Rumpler-Werke at the north edge of the airfield now expanded to include a large new main administration building for the Messerschmitt AG and the later corporation. The old testing hangar became prototype construction with newer, larger testing hangars being added.
- Werk 2: New hangars were set up for jig construction, pre-assembly and production to the west along

View from the factory airfield looking northward toward the newly completed central administrative building with the fabrication section wing (left). In the foreground are new Bf 109B-1 fighters, ready for their checkout flights (July 1937).

The Allied air raid of 22 February 1944 hit the Augsburg Works hard. This is the central administrative building during the attack. The building withstood the pounding and survives today.

Aftermath of the 22 February raid: rubble of W*erk* 1 with the ruins of the old Rumpler and BFW administrative building.

the Haunstetter Straße. These were joined by a sports complex with open-air swimming pool.
- *Werk 3*: Along the southern portion of the airfield came the hangars for production final assembly and production flight testing.
- *Werk 4*: Built along the eastern side of Haunstetter Straße were the machine shops, wing construction, and the warehouses.

In all, the site embraced 165 000 m² of space and included the main administration building, 14 production hangars, four warehouses, six flight testing hangars, buildings for stress analysis and, from 1944, a wind tunnel. There were also seven cafeterias, one doctor's office and two first aid stations.

At the same time, two large housing areas were built to accommodate the employees: the Hochfeld area with multi-storey apartments and the Haunstetten area with single-family houses.

An excerpt from the Messerschmitt totals compiled for the years 1933 to 1939 reads:[1]

"As these totals reflect, development has shot up so dramatically after being freed from the bankruptcy matter that it is virtually impossible to provide an accurate numbers comparison. Completion of the production facilities has ushered in an upswing which can plainly be seen in both output as well as the work force figure..."

The first statistics from these totals reveal this upswing in black and white:

Year	Profits in 1 000 RM	Employees (as of December)
1933	166.0	524
1934	2616.1	1414
1935	8224.6	2403
1936	12099.5	5182
1937	27448.7	5439
1938	38565.7	6491
1939	78220.5	879

Factory housing for Messerschmitt employees at Augsburg Hochfeld.

[1] (Messerschmitt AG), Messerschmitt-Statistik. Ergebnisse in Wort, Zahl und Bild. Augsburg 1941, Secret

Willy Messerschmitt's office in the new central administrative building.

Willy Messerschmitt and Hubert Bauer conversing in the visitor's corner of Messerschmitt's office.

By 1940 the work force at Messerschmitt AG had risen to 10,000, eventually reaching its peak of 22,000 employees (to include those working at dispersed plants) during the war.

On 11 July 1938 the board of directors at the Bayerische Flugzeugwerke AG voted to rename the company as the Messerschmitt AG with a starting capital of 4 million RM in 4,000 stocks. Over the last few months Messerschmitt had received many accolades and awards from outside agencies, and this was the company's way of recognizing Willy Messerschmitt's efforts. Messerschmitt had turned 40 years old on the 26th of June, and Augsburg's mayor Josef Mayr took the occasion to congratulate him, writing:[2]

"...I am certain that I reflect the mood of the entire population of Augsburg when I convey to you their sincerest

[2] Augsburg Municipal Archives, file no. 407/5001

The 1,000 Reichsmark stock of BFW showing its double modification through overprinting: company name change on 11 July 1938 to Messerschmitt AG and increase of the nominal value from 1,000 to 1,400 RM on 12 June 1941.

194

best wishes. Your great contributions to German aviation and to the Bayerische Flugzeugwerke and the city of Augsburg itself are so ingrained into the heart of each citizen that your name is spoken of everywhere with the utmost regard."

With the RLM's approval for the name change, Messerschmitt aircraft could then be designated with the abbreviation Me. This was first applied to the Me 209, which was just about to complete its maiden flight. (Despite this, the company had always informally called any aircraft "Me.") However, in the RLM the older types, such as the Bf 108, 109, and 110 continued to be listed with the abbreviation "Bf" (for Bayerische Flugzeugwerke).

Establishment of the BFW-GmbH Regensburg

With the mid-1936 decision to produce the Bf 109 as the *Luftwaffe*'s standard fighter there increased the need for production facilities for the aircraft. The RLM assumed that the aircraft would be produced in such numbers that not even the expanded Augsburg Works and other license production facilities would be able to handle the load.

Nor did the RLM want to burden any one city with such a large armament company, and it therefore recommended that some of the production be moved to another location. Theo Croneiß, responsible for expanding the aviation industry in Bavaria and a member of BFW's board of directors, proposed setting up a production facility in the relatively non-industrialized city of Regensburg. At the 21 August 1936 meeting of the board of directors it was decided to establish the legally independent Bayerische Flugzeugwerke GmbH (from 1941 the Messerschmitt GmbH). There were to be personnel ties between the management and the BFW AG, Augsburg. Initial investment of 14 million RM was to be put up by the RLM. Planning and preparation for construction immediately began at Regensburg-Prüfening at the hardly used airfield there, so that just nine months later, on 8 May 1937, it was possible to hold the opening ceremonies for the entire facility. In the autumn of that year the first production assembly jigs and materials for the Bf 108 were moved from Augsburg to Regensburg. At the same time, due to a lack of skilled labor

Company-owned wind tunnel was still used for testing in 1944.

Opening ceremonies for the all-new Regensburg factory of the BFW GmbH on 8 May 1937. This is a view looking over a new hangar toward the city, with the cathedral spires visible on the horizon.

Photo of a flight test hangar with new Bf 109E fighters in May of 1940. This new facility launched the industrialization of Regensburg. At the same time, this was an opportunity to construct a factory along the lines of the National Socialist concept of "Beauty and Labor." From an architectural standpoint the factory was one of the most modern in Europe.

Regensburg chief test pilot Wendelin Trenkle prior to checking out a Bf 109 export model in the autumn of 1939.

in the immediate vicinity, 800 technicians were sent from Augsburg to Regensburg to begin production. By December 1937 the new chief test pilot for the Regensburg Works, Wendelin Trenkle, was able to take the first Bf 108s built for the *Luftwaffe* on their initial flights.

Augsburg architects Wilhelm Wichtendahl and Bernhard Hermkes carried out planning and construction of the Regensburg Works. These two men had been responsible for the expansion at the Augsburg site and the building of the adminstration center there. At Regensburg they built a production center with large, airy hangars located on grounds that somewhat resembled a large park. In an environment utilizing the so-called principles of humanization and "beauty of labor," (although without an assembly line at first), by the end of 1937 there were already 1,000 personnel busily working as part of the first expansion phase. Further expansion work of the facilities was undertaken during the war by architect Max Dömges.

By 1938 the plant was forced to take over production of the Bf 109 from Augsburg, as well as the technical administration for all fighter production, plus take on the responsibility for providing instruction to the license production firms. In addition, the Me 210 was built at Regensburg until production stopped in March of 1942. At the end of 1941 the work force included over 6,500 employees, of which about 30% were women.

When in 1942 director Karl Linder expanded the company to accommodate large scale production, set up an assembly line and reduced production time for Bf 109 components by up to 40%, Regensburg's nearly 10,000 employees were working at what was probably the most modern aircraft factory in all of Europe.

The first heavy bombing raid on Regensburg took place on 17 August 1943, killing 400 workers. It was the second raid by the U.S. 8th Air Force on a Bf 109 production center and followed the 13 August attack on the Wiener Neustädter Flugzeugwerke. The Regensburg Works suffered

A meeting of the board of directors offered an opportunity for a visit to the airfield at Regensburg-Prüfening in the summer of 1940: F. W. Seiler, R. Kokothaki, T. Croneiß, W. Messerschmitt, Lilli Stromeyer, F. H. Hentzen, and the director of Leichtbau Regensburg GmbH (LBR), Friedrich Mayer.

New Bf 109G-6 fighters with gondola weapons shortly after final assembly during initial engine run-up, June 1943.

60% damage, and Bf 109 production dropped from a monthly average of 200 to 77 aircraft in September. The attack resulted in a wide scale decentralization program for production facilities, with some of the production going to the airfield at Obertraubling. Obertraubling had come under the scope of Regensburg since 1940, and it was there that Regensburg's part of the *Gigant* production took place (as mentioned earlier, Mtt-AG carried out its *Gigant* work at Leipheim). In addition, the 70 Me 163B V-series aircraft were built at Obertraubling, which also undertook special modifications and a large part of flight testing operations for the Mtt GmbH.

Decentralized production was responsible for the largest output of Bf 109s throughout the course of the war in 1944. October of that year alone saw 755 Bf 109s roll off the assembly lines. In September of 1944 Mtt-GmbH also began series production of the Me 262. Along with the Bf 109, production was dispersed among roughly 100 outstations to include parts assembly and warehouses, as well as 14 factories scattered throughout the whole of Bavaria, the Sudetenland, and Austria. Some production and assembly operations were sited in thick forests or underground caverns. The largest forest works was located in the Gautinger Forest and employed 1,200 technicians and workers for manufacturing the wings for the Me 262. At Kematen in the Tyrol, a large cave was configured as a production center, as was a tunnel on the highway between Murnau and Garmisch. As often as possible, a 2 km long stretch of straight highway was set up as a runway near the forest works. These forest centers were so well concealed that none of them was ever attacked from the air.

Generalingenieur Roluf Lucht was put in place by the RLM from the end of 1943 as the director of the Regensburg facility following the death of Theo Croneiß. Kurt Schnittke (right) was assigned to him as a pilot for the company's Bf 108B.

During the war, up to 30 percent of the workforce at Regensburg and in other factories was comprised of women.

Beginning on 22 April 1945, seven years of aircraft production in the Regensburg area came to an end with the arrival of American tanks at the factory gates of Regensburg, Obertraubling, and the other manufacturing facilities, some of which were located at the Mauthausen and Flossenbürg concentration camps. The final tally of aircraft built was 12,012, of which 10,893 were Bf 109s, 516 Bf 108s, 94 Me 210s, 70 Me 163Bs, 345 Me 262s and 103 Me 323s.

Messerschmitt AG planned and executed a similar dispersal program, particularly for its Me 262 production, and by October of 1944 there were over 50 dispersed facilities throughout southern Germany. The largest of these were Oberammergau with 2,200, Leipheim with 1,250, and Kottern with 2,000 employees. At Kottern, near Kempten, Max Stehle supervised the central jig and calibration equipment assembly facilities for Messerschmitt aircraft.

The use of women in the labor pool most certainly was not viewed so lightly as portrayed in this cartoon from the "Messerschmitt-Nachrichten" from 1943. The caption literally reads "My mother works at Messerschmitt but never brings me home any airplane."

In late 1943/early 1944 the entire project, design, and stress analysis departments were moved from Augsburg to the *Gebirgsjäger* barracks in Oberammergau. These were followed in 1944 by the prototype construction department, for which a cavern was dug into the mountain. In October of 1944 there were 2,231 workers employed at Oberammergau.

In 1944/45 several so-called "forest works" were set up, in which the Me 262 could be produced relatively free from bombing raids. This is one of those production centers for final assembly near Obertraubling, photographed in the summer of 1945.

Bf 109 Large-Scale Production at WNF

As part of the Hermann Göring network of plants, by late 1938 the Wiener Neustädter Flugzeugwerke GmbH (WNF) had evolved into one of the largest aircraft repair and production facilities in the entire Reich. Three factories were built sequentially at Wiener Neustadt from 1938 to 1940: Repair *Werk I* to the north, beginning work in 1938; *Werk II* to the northeast along the Pottendorfer Straße, with production starting in March and delivery of the first Bf 109E in June of 1939; and *Werk III*, a repair and components production center south of Fischamend with production starting in February 1940. Expansion and build-up of the Bf 109 production works (*Werk II*) occurred with the assistance and support of the Erla-Maschinenwerk Leipzig, which despatched its technical director, H. Steininger, to the WNF where he continued to serve in that capacity for a number of years.

Even though just 115 Bf 109s were shipped in 1939, by 1940 this total had jumped to 467, including the first Bf 109F on 30 September. There followed 856 machines in 1941, and 1942 saw the delivery of 1,400 planes, with the first Bf 109G rolling off the lines in the second quarter of that year. In 1943 the WNF reached the pinnacle of its production, its 15,000 workers (the highest number of employees in its history) cranking out 1,300 Bf 109Gs in the first half of that year alone. Subsequently, Ludwig Bölkow moved from Augsburg to manage the WNF's transition from the G model to the improved K model.

The first U.S. bombing raid against the WNF occurred on 13 August, destroying a Bf 109 production center. Additional attacks followed on 1 October and 2 November and led to cessation of production and a dispersement of the manufacturing facilities to 24 sites throughout the "Ostmark." This action served to boost monthly output to the highest ever, averaging 300 to 400 aircraft per month—about 50% of the total Bf 109 production. Altogether, the

A meeting at the Wiener Neustädter Flugzeugwerke in the autumn of 1943 brought these men together (left to right): Director Steininger from the WNF, W. Messerschmitt, F. W. Seiler, F. H. Hentzen, and R. Kokothaki.

WNF delivered 8,545 machines from 1939 to 1944, approximately one-fourth of all Bf 109s built during the war.

F 2 Production Ring for Messerschmitt Series Production

1941 saw the establishment of the "*Sonderausschuß F 2 im Hauptausschuß Zellen beim Reichsminister für Bewaffnung und Munition*" (literally, F 2 Special Sub-department to Fuselage Division, Reich Minister for Armament and Ammunition—abbreviated to F 2 Production Ring). This department was responsible for coordinating production of the Bf 108, Bf 109, Bf 110, Me 163, Me 210/410 and Me 323, as well as Messerschmitt's management of the Caudron C-445 and Potez P-63 in France. Chief of this organization was Messerschmitt's long-time companion Fritz H. Hentzen. The department was broken down into two groups, production and repair, and from a planning standpoint was responsible for 15 production centers and 25 repair operations throughout Europe, including sites in Paris, Brussels, Antwerp, Naples, Bucharest, Athens and Kharkov. A look at the planning from November 1942 provides an overview of the Bf 109 program with targeted monthly production quotas: prior to the production ring taking over management of the program, seven production centers had collectively built 180 Bf 109s per month (1939). After takeover, anticipated production rates were as follows:
1941:   3 production works with a total of 180 aircraft per month.

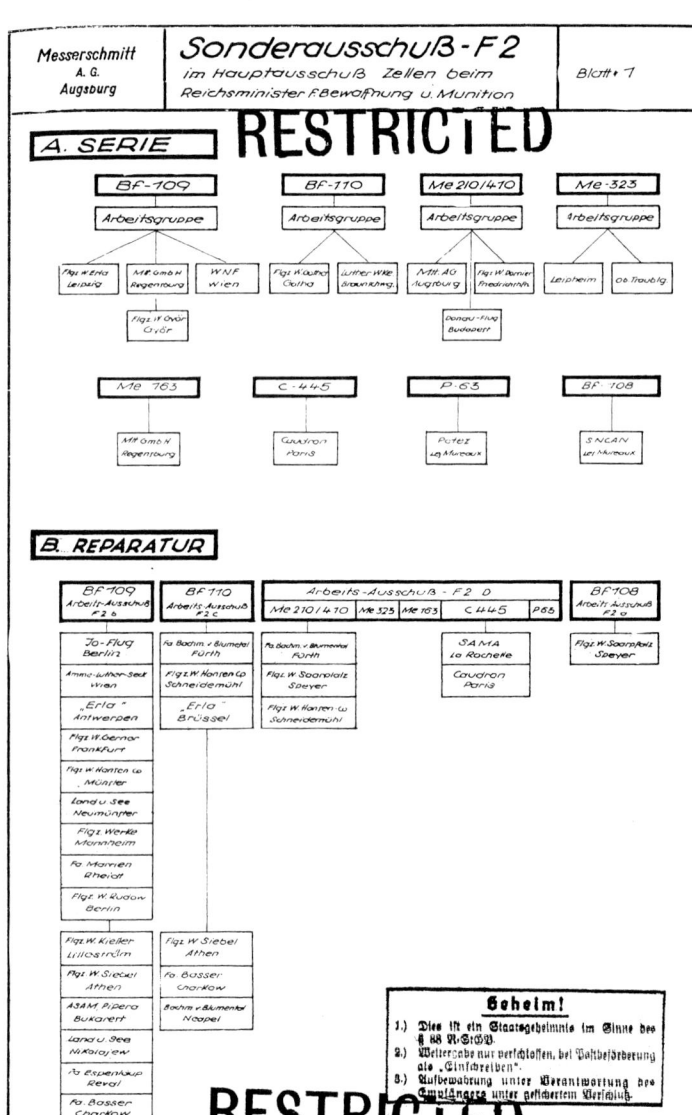

In 1941 the Reichsmarschall's industrial council for aeronautical equipment established so-called "production and maintenance rings" for aircraft, parts and accessories, engines, and equipment. One of the largest of these was for the manufacture of Messerschmitt aircraft and was administered by Messerschmitt director Fritz H. Hentzen.

1942: 3 production works with a total of 285 aircraft per month.
1943: 3 production works with a total of 720 aircraft per month.
1944: 3 production works with a total of 1,000 aircraft per month.

The three large-scale production facilities were the Messerschmitt-Regensburg complex, the Wiener Neustädter Flugzeugwerke, and the Erla-Maschinenwerk at Leipzig.

The production ring was also tasked with reducing manufacturing times, and in the fall of 1942 it projected an initial savings of 53% for the Bf 109 airframe, with further reductions of 10-15% "expected" in the future. In actual fact, production time for a Bf 109G was brought down to about 4,500 man-hours per aircraft—without subassemblies from state-owned storage centers, whereas for the Bf 110 for example it was 7,900 hrs.

The Fighter Staff Swings Into Action

Fighter production was sharply curtailed, with nearly 90% of the associated facilities suffering damage as a result of the increasing number of Allied bombing raids, which had begun targeting aviation industrial centers beginning in February 1944. Instead of the planned 2,000 fighter aircraft per month—mostly Bf 109s, Me 262s, and Fw 190s—February 1944 saw an output of only 1,100, with the numbers predicted to drop even further in the future. As the field units, the aviation and armaments ministries, and the aviation industry did not always see eye to eye when it came to finding solutions to this problem, a so-called "*Jägerstab*," or fighter staff, was established on 1 March 1944. This organization was in turn subordinate to the armaments ministry under Albert Speer. The chief of the fighter staff, who wielded considerable powers, was senior manager Karl Otto Sau.

He succeeded in stepping up output again despite the bombs and the war. Greater decentralization was planned (and partially carried out), encompassing about 300 operations and storage centers for the Messerschmitt company. The Regensburg Works group serves as an example of the effectiveness of the fighter staff's measures coupled with the efforts on the part of the industry itself. For example, it succeeded in producing over 700 Bf 109Gs in just the three summer months of 1944 alone.

Comparison Data from Messerschmitt Aircraft Production

After the end of the Second World War, the victorious powers made many interesting discoveries among the rubble of Germany's aircraft production factories. Company senior managers were questioned for weeks on end and had to file detailed reports on the activities of their respective operations. These surviving comparison tables, prices, experience data, etc. provide an insight into the high state of production achieved at Messerschmitt. The following is derived from Augsburg operations director Rakan Kokothaki's report of 15 September 1945:

The price, calculated by the company for the RLM, was based on the airframe, including assembly and flight evaluation, although it did not include components coming from state-owned subcontractors. These included engines and/or turbojets, propellers, armament, radio equipment, compasses, as well as other instrumentation.

Calculated costs for those airframes in production were:
Bf 109E	79,500 RM	Me 210	130,300 RM
Bf 109G	45,000 RM	Me 262	87,400 RM
Bf 110C	144,600 RM	Me 323	497,300 RM
Bf 110E	171,600 RM	Me 410	150,200 RM

If the price of the airframe is compared to the empty weight of the aircraft, the following figures result:
Me 210 (empty weight 7.5 metric tons):	17.40 RM/kg
Me 262 (empty weight 5.6 metric tons):	15.60 RM/kg
Me 323 (empty weight 34.0 metric tons):	14.60 RM/kg
Me 410 (empty weight 8.8 metric tons):	18.20 RM/kg

Production time for individual types was approximately:
Bf 109G	4,500 man-hours
Bf 110E	7,900 man-hours
Me 210	9,600 man-hours
Me 262	6,400 man-hours

The average hourly wage at Messerschmitt AG in Augsburg applied to both Germans and foreigners:*
Technicians:	1.21 RM
Skilled workers:	1.02 RM
Helper laborers:	.74 RM
Female skilled workers (heavy work):	.79 RM
Female skilled workers (light work):	.62 RM
Unskilled female workers	.50 RM

* The late 'thirties Reichsmark exchange rate with the Deutschmark is roughly 1 to 10. A few examples are: breadroll 0.03RM, cinema ticket 0.90 to 1.50RM, simple restaurant meal 1.20RM, furnished studio apartment 30RM/mo., and starting salary for engineer 325RM/mo.

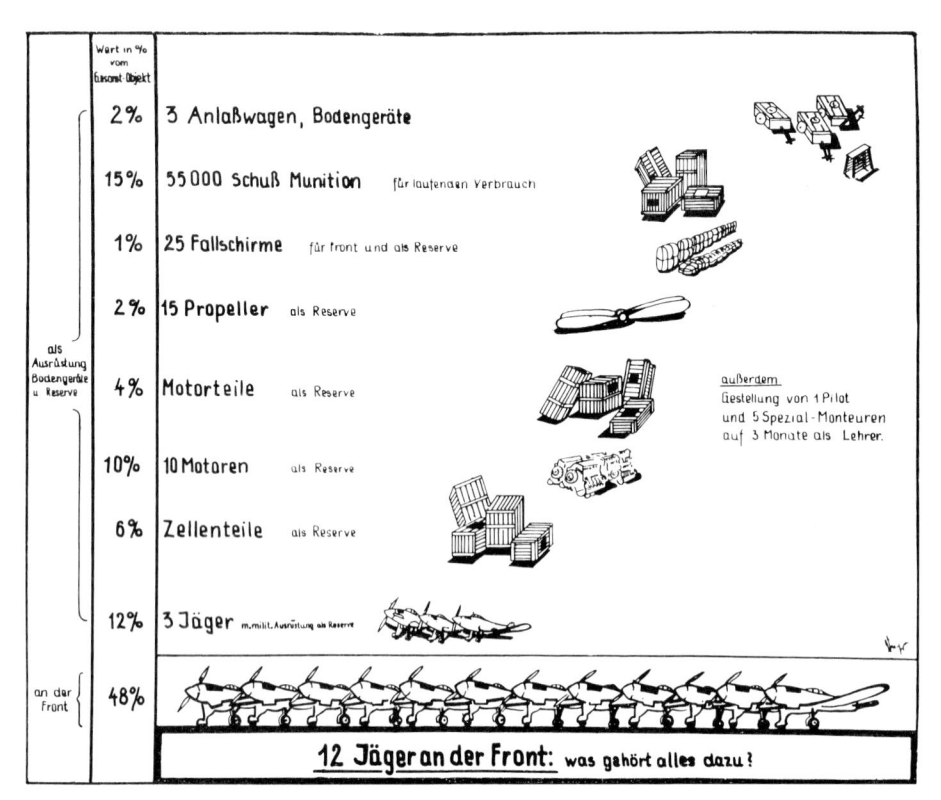

From 1939 onward the Mtt AG was at the top of the export sales charts. This shows what was involved in equipping a Bf 109 squadron. Although this RDLI graph illustrates an export contract, it is based on Luftwaffe experience—including operational conditions in Spain: the twelve aircraft themselves comprise 48% of the contract, while the requisite logistical support encompasses 52%.

A scene in front of the prototype construction center at Augsburg during the summer of 1942: while the Me 309V-1 is undergoing last-minute tweaking prior to its first flight (which took place on 18 July 1942), in the foreground are the wings for the "new" Me 209 under construction. In the hangar itself can be seen the Me 262V-1, unsuccessfully flown in March with a central Jumo 210 and two BMW turbojets. The hangar beyond houses additional Me 262 prototypes.

Messerschmitt Aircraft Production from 1934 to 1945

All told, there were about 45,000 aircraft built of the 17 different Messerschmitt models in the twelve years from 1933 to the end of the Second World War. Of this number, the Bf 109 easily outranked the other types with approximately 34,000, followed by the Bf 110 at over 6,000, the Me 210/410 at about 1,500, the Me 262 also at about 1,500, the Me 321/323 *Giganten* at approximately 400, and the Me 163A and B at about 400, as well. The variants and subvariants of these types, together with the factory and field conversion kits, numbered in the hundreds. The remaining nine types were only produced as prototypes, of which another four types would have been produced in small or larger numbers had wartime capacity been able to accommodate them.

Several air forces (Switzerland, Finland, Romania, Bulgaria, Yugoslavia) continued to fly Messerschmitt aircraft long after the war. Spain, Czechoslovakia, and France continued to produce the Me 109, 208, and 262 for a period of time and operate them in their air forces. Messerschmitt

A view of the airfield following the 22 February 1944 attack, looking toward the destroyed flight test hangar and the undamaged main administrative center. The aircraft in the background is a newly completed Me 410. The object in the foreground is the wing of a smashed Bf 109G, jutting skyward like some macabre memorial.

aircraft can be found today in museums and collections throughout the world. A few Me 108s and 109s are still flying as memorial airplanes, hearkening back to the age when Messerschmitt aircraft dominated the skies.

Special Role Developments - From High Altitude Strategic Reconnaissance Platforms to the "Giants"

In 1935 the BFW found itself caught up in a rapid growth period with development and production contracts for the Bf 108B (readying for production), Bf 109 (prototype stage), and the Bf 110 (under design). That same year saw the RLM's *Technisches Amt* award another series of contracts to the aviation industry for specialized airplanes. For the BFW, this resulted in the following roles and type designations:

High altitude strategic reconnaissance	Bf 161
High-speed bomber	Bf 162 (competing with Ju 88, He 119, Hs 127, Do 17M)
Liasion aircraft	Bf 163 (competing with Fi 156, Fh 201, later Si 201)
Round-the-world record setter	Bf 164 (competing with special versions of Ju 88, He 119)
Long-range bomber	Bf 165 (probably similar to Ju 89, Do 19)

Three of these aircraft types were built, although two were later canceled.

With input from the RLM, this type listing continued to grow at a steady rate from 1937 to 1940. At first it was by adding the Bf 109/110 intended heirs, the Me 209/309 and the Me 210/310, then by including those designs which could be considered special developments:

Me 261 long-range liaison/strategic reconnaissance
Me 262 jet fighter with turbojet engines
Me 263 heavy lift transport (later redesignated Me 321/323)
Me 264 strategic bomber/reconnaissance
Me 209 commuter plane
Me 308 commercial airliner

Nearly all the above-named types were built in one form or another, with three even going into production.

Bf 161 - Tactical and Strategic Reconnaissance Plane

Development of the P 1035 as a long-range strategic reconnaissance platform began in September 1934 under the direction of Robert Lusser in *Probü* at almost the same time as the Bf 110, as the project had been offered to the

Bf 161V-1, Werknr. 811, was flown for the first time by Dr. Wurster on 9 March 1938. The machine was unarmed, and in late 1939 was handed over to the Rechlin test center.

Powered by DB 600 engines, Bf 161V-2, Werknr. 812, D-AOFI, flew for the first time on 30 August 1938, and from 1939 was also used as a tow tug for the Me 163A program at Peenemünde West.

Below: Cutaway of the three-seat Bf 161 tactical reconnaissance aircraft with cameras located in the center fuselage.

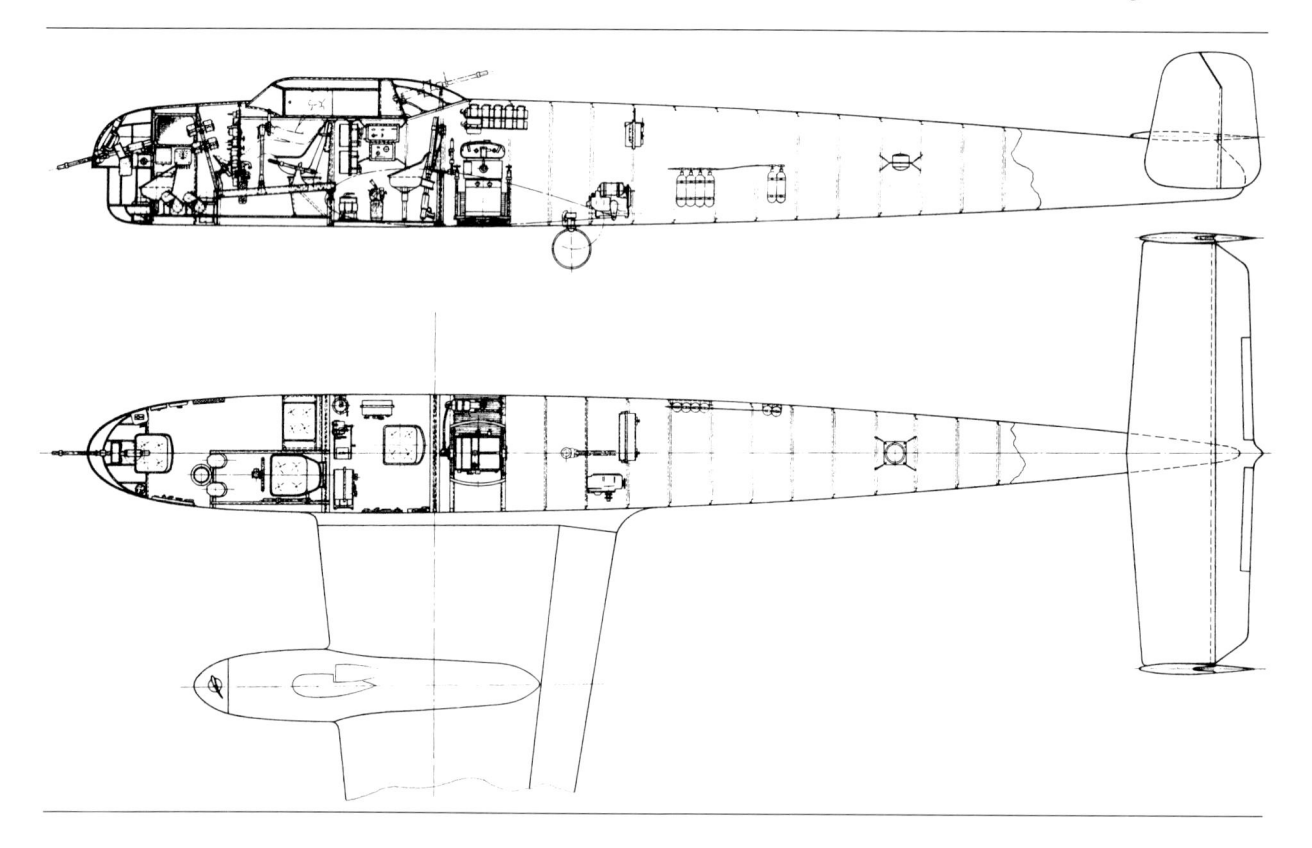

RLM as a system aircraft (see Bf 110 section). It wasn't until August 1935 that the RLM contracted for five prototypes (with 59 additional pre-production and production aircraft planned for the 1937/1938 acquisition program). The number of prototypes was tailored down to two, however. Of these two, the V-1 would be powered by two Jumo 210 (20-1) engines, while the V-2 would have two DB 600 (30-1) motors as powerplants. In terms of the airframe, the design differed primarily from the Bf 110 in the many changes to its fuselage, which was designed to accommodate a glazed canopy, cameras, and a three-man crew.

The long, drawn-out development and fabrication process, coupled with the drop in prototype numbers, showed that the RLM had little interest in the type's continued development. In the interim, the contract for a long-range reconnaissance aircraft had been awarded to Dornier (most probably due to the success of the Do 17 in Zurich).

Nor was there any further interest in a tactical, short-range version of the Bf 161, which the project department had drawn up in 1936. This design would have utilized yet-to-be-determined lower performance engines in the 400 hp/ 294 kW class and would have had a reduced takeoff weight from 4,600 kg to 4,400 kg, plus different recce equipment.

Of the two prototypes, neither of which was ever fully equipped, the V-1 went to the *E-Stelle* Rechlin in November 1939, while the V-2 remained at the Augsburg Works as a tow tug for the Me 163A. It was eventually transferred to Peenemünde West where it continued to soldier on in the same capacity.

Bf 162 - 500 km/h High Speed Bomber

The Bf 162 was the last of the three types stemming from the basic Bf 110 configuration (q.v.) and was offered under the P 1035 project to be tackled by the *Probü* and Robert Lusser. Despite this, its V-1 prototype was the second to fly, sandwiched between the Bf 110 and Bf 161 in early 1937.

In November 1935 the RLM passed its technical requirements for a high-speed bomber to the BFW, while at the same time officials from the RLM paid a visit to Augsburg and inspected a Bf 110 mockup configured in the form of the Bf 161 long-range reconnaissance plane. These were their impressions:[1]

"The Bf 110 mockup appears to be suited to the bomber role in the dimensions shown. Crew distribution is to be designed so that the pilot's seat is somewhat ahead of and to the left of the seat for the aircraft commander/bombardier, who lies prone or kneels when dropping the bombs.

[1] from Bf 110 record no. 12 from 14 Nov 1935

Three-view of the Bf 161 (H.F.) used as a tactical reconnaissance aircraft for close-support reconnaissance and combat photography, artillery spotting, and bombing. Armament was to have consisted of a nose gun and a dorsal gun station, each carrying a MG 15. Cameras or ordnance was carried in the fuselage bay. Flying weight was to have been 4,890 kg. Powered by two Jumo 210 engines, it was to have had a speed of 440 km/h, a range of 750 km, and a ceiling of 8,000 m. Wingspan was 17.12 m, length 12.85 m.

The instrument panel, throttle levers, and control systems must be laid out so that they interfere as little as possible with the view. The fuselage forward section may be made of plexiglas to provide the best possible view outside.

The two Maga 5 C/50s are to be located on the left sidewall behind the pilot's seat (Elvemags can be fitted in their place, if required), allowing easy access and room for the commander and radio operator in addition to the magazine. Commander and radioman are to be provided with sliding seats so that they can slide close to each other for purposes of communication. The normal combat station for the radioman, who is also the rear gunner, is behind the

205

The Bf 162V-1 high-speed bomber, Werknr. 817, later registered as D-AIXA, was flown on its maiden flight by Dr. Wurster on 26 February 1937. With two DB 600 engines, the plane had a takeoff weight of 6,500 kg, a wingspan of 16.69 m and a length of 12.45 m.

The Bf 162V-2, first flight on 31 October 1937 with Dr. Wurster at the controls, later served as a tow plane for the Me 163A program. Some of these pictures, taken on a photo flight in 1939, were later doctored to appear as operational flights and published by the OKL as a propaganda ploy, purporting to be a new type of bomber.

aftmost magazine and is to be of a similar configuration as that of the Bf 110.

The fixed machine gun is to be installed so that it can fire through the propeller shaft of either of the two engines. In this manner two machine guns may even be fitted.

Based on this initial discussion, BFW is making a project which shows the center of gravity and spacing to be such that the aircraft can actually be designed to meet the requirements and look as it does today. In light of this project, another meeting is to take place in either Berlin or Augsburg which will focus on establishing a fully equipped finalized mockup.

Current work on the Bf 110 as a heavy fighter and high-altitude strategic reconnaissance airplane remains unaffected by this."

Design work began in January 1936. There followed a RLM contract for five prototypes in 1936. A further 65 pre-production and production aircraft were planned for the 1937/1938 acquisition program. Powered by two DB 600D engines, Dr. Hermann Wurster flew Bf 162V-1, Werk-Nr. 817, D-AXIA, for the first time on 26 February 1937. The V-2 (Werk-Nr. 818, D-AOBE) followed on 31 October, and the V-3 (Werk-Nr. 819, D-AOVI with two DB 601A engines) on 7 July 1938. The other V-types and production aircraft were only partially completed or in component form when the program was canceled in favor of the Ju 88 on 9 March 1938, this in spite of the fact that the Bf 162 had demonstrated speeds of nearly 500 km/h during the course of testing. This made it faster than the Bf 109B and D variants then being produced (even though these latter types were powered by the lower performance Jumo 210). Termination of the program was the result of a shift in the aviation industry's production focus initiated by the RLM.

Despite the loss of the contract, at the 11 July 1939 meeting of the board of directors (with a report on fiscal year 1936, the year which saw the main work carried out on the Bf 162) Willy Messerschmitt was able to state with satisfaction:

"It is important to remember that the RLM's *Technisches Amt* had explained that the Me 162[2] was probably better performing than the Ju 88. However, because the BFW was already overtaxed with follow-on variants of the Me 109, Me 110 and newer fighters, the RLM decided to produce the Ju 88[3] in series in place of the Me 162."

Messerschmitt's system idea proved its worth in wrapping up the program: the material for the remaining V-types and the Bf 162 production aircraft which had been freed up by the program's cancellation was shifted over to the Bf 110C program. The test-flown V-1 was scrapped, while the V-2 (like the Bf 161V-2) went on to become a tow tug for the Me 163 program and remained at the plant. Following a photo flight by Dr. Wurster, retouched and photomontage pictures of this plane were freely disseminated with the caption claiming that this was a new Messerschmitt *Zerstörer*, the "Jaguar," which was supposed to go into action over England in the spring of 1940. This was purely propaganda, sanctioned by the *Luftwaffe*'s high command.

Bf 162V-3 was turned over to the *E-Stelle* Rechlin in 1939.

Bf 163 - Competitor for the Storch

In September 1935 the RLM awarded the BFW a developmental contract for three prototypes of a three-seat liaison and observation aircraft with the official designation Bf 163. The extremely short takeoff and landing characteristics specified for this aircraft meant that the Messerschmitt team would have to devote considerable effort to resolving aerodynamic problems, hitherto only encountered in the development of the Bf 108. For the Bf 163, therefore, the greatest emphasis was on the design of its wings. In addition to the requisite slats and flaps, the Messerschmitt team set themselves the task of designing the wing to pivot on its lateral axis, varying the angle of attack by 12% in flight. In addition, the wing would have to be folded for transport.

While the project got underway in Robert Lusser's project department under the designation of P 1051, profile studies were arranged at the AVA in Göttingen. Robert

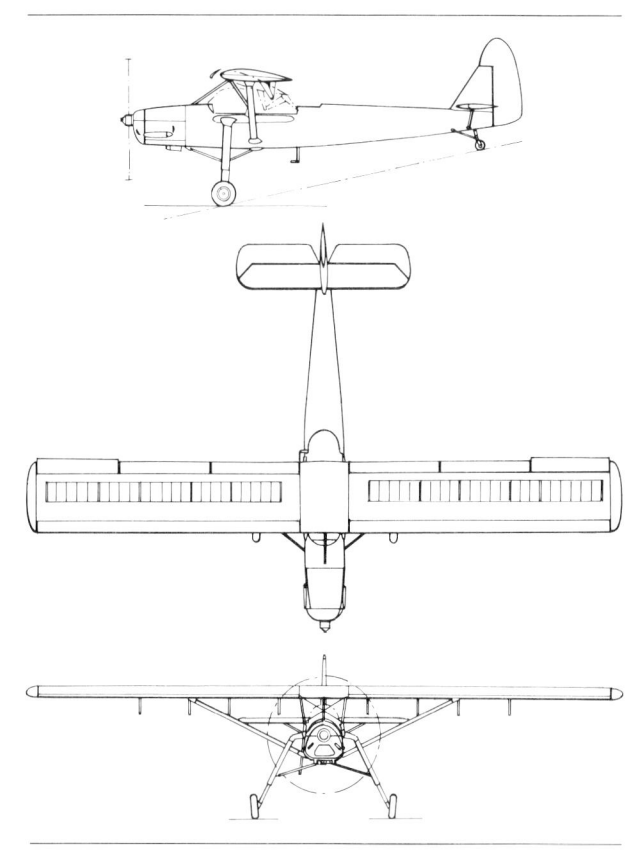

[2] Messerschmitt and all the company employees had always referred to those aircraft officially know by the abbreviation "Bf" (derived from Bayerische Flugzeugwerke) as "Me," despite the fact that this latter abbreviation didn't officially go into effect until mid-1938.
[3] This refers to the first version of the Ju 88, which was not yet dive-capable and had flown for the first time on 23 December 1936. The dive-bomber prototype didn't follow until 18 June 1938.

Three-view reproduction of the Bf 163. The aircraft was designed as an all-metal braced high-wing with single-spar rectangular wings and reinforced nose. The wing angle-of-attack could be adjusted 12 degrees through the use of a corkscrew gear. The three-man cockpit was partially enclosed. An Argus As 10C engine was fitted as the powerplant. The aircraft was also designed for the optional installation of the Hirth MH 508 engine.

Lusser addressed the issue at a conference for improving lift and performance at Göttingen in 1938:[1]

"The impetus for our comprehensive studies was a contract for building a liaison airplane—the Bf 163—with extremely high requirements for landing and takeoff runs. Had this requirement not been so great, we would not have devoted so much effort into the program. A $c_{a\,max}$ of 4.05 was quite good...We wanted to run the landing flaps across the entire trailing edge, i.e. designing the ailerons to be employed as flaps, as well."

As with the Bf 108, the Bf 163 wing made use of Willy Messerschmitt's patent-pending "Messerschmitt flap." The

[1] Lilienthal-Gesellschaft für Luftfahrtforschung, report 099/006 re: conference on "Improving Lift and Performance" in Göttingen on 14 January 1938, p. 27-28.

Although no photos exist showing the Bf 163 in its entirety, this view looking forward from the rear seat of the mockup has somehow managed to survive. As on the Bf 108, the side canopy sections opened outward and forward.

advantage of this flap over the Fowler flap was that, when retracted, it was only half covered by the wing, forming the wing's trailing edge. As such, it was double the overall height of a standard flap, had a more pronounced curvature, and was stiffer. Compared to the Fowler flap, its travel movement was halved, plus the nose heaviness was not as pronounced when deploying such a flap even at maximum lift setting.

Although the project department had almost completed its work—lacking only the dimensioning and calculation of the vertical stabilizer—Richard Bauer's design department wasn't able to handle the additional load, for it was working on the Bf 108, 109, and 110 almost simultaneously. Bauer therefore suggested that the work be passed on to an outside company. The contract went to the Rohrbach Metallflugzeugbau GmbH Berlin, in the process of liquidation and shortly to be taken over by Weserflugzeugbau GmbH, Lemwerder. Both Fritz Feilcke, director of Weserflug, and Dr. Adolf Rohrbach came to Augsburg in the spring of 1936 to inspect the mockup and gather up the relevant documentation. Detail work on the Bf 163 began at Rohrbach there in Berlin and was completed following the move to Lemwerder in the fall of 1937.

As it turned out, the conservative Rohrbach designers exceeded the airframe weight of this all-metal construction by a considerable amount. Proof of Willy Messerschmitt's highly-refined light metal construction methods, still viewed by most other companies with a certain amount of scepticism. Final assembly of the sole prototype built, registered as D-IUCY, was completed in late 1937 and flown on its maiden flight by Weser chief test pilot Gerhard Hubrich on 19 February 1938 at Lemwerder. Remembering this and other flights, much later Hubrich wrote:[2]

"During the machine's first flight I noticed its extreme nose-heaviness. The first corrective action was to change the tailplane trim setting. By the fourth flight, it could be flown normally, and with full use of landing aids (slats and Fowlers and adjusted wing angle) it had a minimum speed of 48 km/h. I don't remember the maximum speed. I can still recall, though, that the tailplane's angle of attack had to be increased several times. As I mentioned, the plane was altogether too heavy, and we understood that if we had been able to eliminate the mistakes the second plane could have been a winner."

After having been flown by an *E-Stelle* pilot who was also critical of its poor flight handling, the plane was turned over to Augsburg in mid-1938. There, BFW chief test pilot Dr. Hermann Wurster also flew the Bf 163V-1 and had his fair share of problems with the trim as well.

[2] Gerhard Hubrich, letter to the author on 21 June 1967

International visitors often called at Messerschmitt in the prewar years to witness the renowned flying demonstrations by Messerschmitt pilots Dr. Hermann Wurster, Willi Stör, and Fritz Wendel. This photograph, taken in the autumn of 1938, shows Italian officers General Porro and Capt. Gasperi with hosts Otto Brindlinger (from marketing) and Willy Messerschmitt (left to right). It was on this occasion that company photographer Margarete Thiel took several photographs which, by chance, partially caught Bf 163V-1, D-IUCY, on film in the background. This is the only photographic evidence produced to date.

A reconstruction model from 1970 showing the Bf 163V-1 as it was flown on its maiden flight at the Weserflugzeugbau Company by Gerhard Hubrich in 1938.

Because of its long, braced undercarriage and cabane structure over a well-glazed cockpit, it resembled none of the other sleek Messerschmitt aircraft, and the design soon had Messerschmitt workers nicknaming the type such things as "Lame Jack" and "Garden House."

By this time the Fieseler Storch, built to the same RLM requirements as the Bf 163, was already in production. The Storch had a year's advantage over both the Messerschmitt design and the third competitor, the Si 201 at Halle (which also didn't fly until 1938), and made both of these designs redundant. Richard Bauer, who also taught aeronautical engineering at Munich-Feldmoching's *Fliegertechnische Schule*, suggested that because of its innovative design the aircraft be given to the school as a teaching aid. This bestowal occurred in September of 1939. The machine stood outside the school's academic building for several years after the war when the Americans occupied the site, a fact relayed to the author by Theo Lässig—at the time a member of the anti-aircraft auxiliary forces.

The project was so thoroughly expunged, the purging of related documentation at Messerschmitt so complete, that not even a single photo of the Bf 163 survives today.[3] The type number was later assigned to disguise a completely new and secret "Project X," the later Me 163A, developed by Dr. Alexander Lippisch and his DFS team which had transferred to Messerschmitt in January 1939. Since three prototypes of the Bf 163 had officially been ordered (Weserflug had only built parts for the V-2 and the V-3), numbering of the Lippisch project consequently began with V-4 (see chapter on Me 163)

Bf 164 (P 1053) - Aircraft for a Round-the-World Flight

At one point a round-the-world aircraft project had apparently piqued the interest of the RLM, for it awarded a contract for such a design sometime around 1935. The project was canceled early on, however, and there is such little documentation remaining that evidence of its existence is limited to a contemporary RLM planning document and a few listings. It appears likely that the undertaking may have been some type of prestige program, proudly demonstrating Germany's aviation technology through a successful flight around the world. Those involved possibly had in mind the spectacular examples such as the round-the-world flight by four U.S. Navy Douglas PT-2 flying boats back in May through September of 1923, or Italy's wing of SM-55 flying boats which flew from Rome to Rio in 1930 and 1933.

In any case, the RLM awarded a developmental contract for a round-the-world competition airplane to Junkers (based on its Ju 88), Heinkel (based on the He 119), Dornier (probably based on the Do 17), and BFW (the Bf 164). Each company was expected to produce three aircraft powered by DB 601 engines, with designs ready for flight by the spring of 1937.

It is conceivable that the RLM expected Willy Messerschmitt to lean heavily upon the Bf 162 for this project, although it appears that—like with the Bf 109—he

---

[3] A reconstruction of the history of this plane can be found in the VFW Fokker aircraft monograph no. 3: Verbindungsflugzeug Weswerflug Bf 163, D-IUCY (1970), by Hans Justus Meier (with the assistance of the author). All publications since then stem from this initial source.

[1] RLM: Übersichtsplan der Bearbeitungsgebiete bei der Flugzeugentwicklungsgruppe LC II/1 (Lucht) from 1 Dec 35
RLM: Flugzeugbaumustertabelle der LC II/1, as of 1 Oct 1936

was given free rein to develop what he wanted. For, at a preliminary meeting with the *Technisches Amt*, Messerschmitt reputedly pointed out that he had planned something similar back in 1932 with the M 34. Thus, it seems that the thinking which went into the M 34 also influenced this new project.

After cancellation of the P 1053, elements of this project also seem to have played a part in influencing the long-range Me 261 and Me 264 aircraft (q.v.), designs which were realized.

Termination of development on the round-the-world project, which came around 1937, appears to have had several causes. One of these was the fact that the RLM saw that the aviation industry was overburdened with developmental contracts as it was. Then, it became obvious that the DB 601 engines would become available without having been fully tested. Plus, the upcoming International Flying Meet at Zurich-Dübendorf in July of 1937 would provide ample opportunity to showcase Germany's advances in aviation, something which did indeed occur with remarkable results.

Type number 164, no longer assigned, was later allocated to another Messerschmitt development, the Me C 164 commercial airliner begun by the French subsidiary Caudron in 1941.

Bf 165 - Strategic Bomber Project

The final project in the series is also a little-known design. With just two sources[1,2] to rely on, we nevertheless know that it was a large bomber having a thick fuselage and several gun positions, with a tapering tail boom. This configuration can also be derived from wind tunnel tests conducted at the AVA in Göttingen in 1937. Data was taken from fuselage model configurations with an extending dorsal turret and a ventral gun position which opened outward.

In addition, during fiscal year 1937 funds amounting to over 100,000 RM were allocated for construction of "a mockup for the Bf 165 long-range bomber."

This data and the relatively high sum for a mockup confirm that the project department had at least worked on the project in some detail. The mockup was most likely the one shown to Adolf Hitler during his visit to BFW in Augsburg in November 1937. Hitler's adjutant, Nicolaus von Below, describes this event (also attended by a high ranking RLM delegation which included Erhard Milch and Ernst Udet) in his memoirs:[3]

"In one of the works hangars Messerschmitt approached Hitler and offered to show him one of his new designs. Two massive doors leading to another hangar were then pushed aside, and we beheld a mockup of a four-engined bomber. At the same time, this caused quite a stir among the gentlemen from the RLM. The shock could clearly be seen on Milch's face. Messerschmitt gave further details of this airplane. Hitler seemed to show considerable interest. It was a project for a four-engined strategic bomber, predecessor to the Me 264 bomber designed during the war but which never went into production. Messerschmitt claimed a range of

[1] Zusammenstellung von Kabinenwiderständen, Mtt-AG, Augsburg (project dept.) 26 Feb 41
[2] BFW. Vorgenommene Prüfung des Jahresabschlusses 31 Dec 37, RLM sales
[3] Nicolaus von Below. Als Hitlers Adjutant 1937-45, Mainz 1980, p 52-53

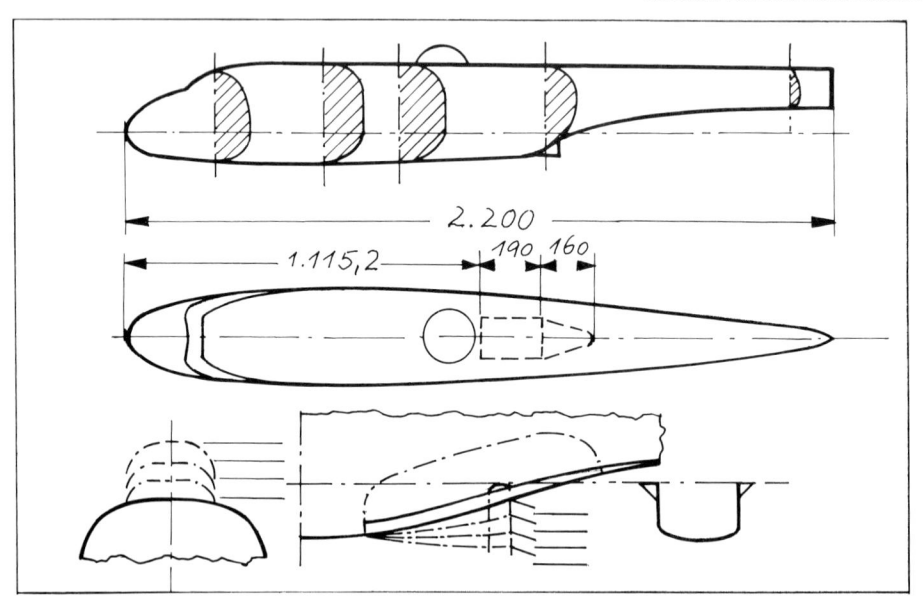

Fuselage model of a heavy bomber with two extendable gunner's positions, as evaluated in 1937 at the AVA in Göttingen. By converting the model scale it can be determined that the original would have had a length of 22 meters, ruling out any type other than a (four-engined) heavy aircraft.

During his sole visit to the BFW in November 1937, *Führer* Adolf Hitler was shown the mockup for the Bf 165, as noted by his adjutant Nicolaus von Below in his memoirs. Prof. Messerschmitt led Hitler and his large entourage on a tour of the factory. Next to Messerschmitt (from right to left) are Ernst Udet, Adolf Hitler, F. H. Hentzen, and Theo Croneiß.

6,000 km and a speed of 600 km/h with a 1,000 kg bomb load. The RLM officials expressed their doubts about these figures. Hitler was more reserved. When it came to aircraft construction, he was not as knowledgeable as he was with shipbuilding or the building of armored vehicles and guns. However, he thought that it must be possible to design a multi-engined bomber which could fly faster than fighters. But then if fighters would then be able to fly at speeds of 600 km/h, the bomber would need to be able to fly at no less than 650 km/h. Armor and defensive weaponry would of necessity be dispensed with on such a 'high-speed' bomber, for its speed would virtually preclude any aerial combat. Milch confirmed that the *Luftwaffe* had contracted for a bomber with just such capabilities a long time ago. Messerschmitt argued that twin- or four-engined bombers with such high speeds could not be built because Germany lacked the requisite powerplants. Hitler felt that the correct course of action was to give the four-engined bomber priority. Milch objected to this approach, claiming that the raw material situation limited the *Luftwaffe*'s options and thus tends to favor the twin-engined high-speed bomber. No real decisions resulted from this discussion."

The Bf 165 project thus seems to have been roughly similar in concept to the Ju 89/Do 19 heavy bomber designs. While these two latter types actually flew in prototype form (Ju 89: first flight 11 April 1937, Do 19: first flight 26 October 1936), the Bf 165 was probably terminated shortly after this visit if Milch's remarks about the high-speed bomber are anything to go by. The four-engined bomber had lost its only advocate with the premature death of its initiator, *Generalleutnant* Walter Wever (who had undoubtedly issued the contract for the Bf 165). Wever was killed when his plane crashed on 3 April 1936.

Me 308 - Me C 164, Commercial Airliner

In 1939 the *Technisches Amt* issued a requirement on behalf of Deutsche Lufthansa for a commuter plane and short-haul airliner. In order to provide the six to eight passengers with a good view, the specifications called for the design to be a high-wing, with comfortable interior furnishings and a sound dampened interior. The range was expected to be in the neighborhood of 1,300 km. It was planned to utilize two 360 hp Argus As 410 engines, with the option of fitting 500 hp engines in their stead.

Willy Messerschmitt participated in the requirement, in spite of the fact that his project and design departments were gainfully employed at the time. Once he had personally clarified details of the project, work began at Augsburg in late 1940 under the supervision of Hans Regelin, continuing under Walter Rethel starting in March of 1941. Internally, the project was initially carried on the books as the Me 308. But when the *Kobü* found itself running into bottlenecks, the project was passed on to the French company of Caudron at Issy-les-Moulinaux near Paris under the designation of Me C 164. Caudron was a subsidiary of the Messerschmitt AG where 50 designers worked for Augsburg under German supervision on various smaller projects or parts orders. Since military projects took precedence even here, work on the civilian Me C 164 proceeded at a somewhat sluggish pace. The program was eventually terminated at a point when approximately 30% of the design blueprints had been completed.

In any event, this was also the successful debut of another competitor and the winner of the competition, the Siebel Si 204A, which had flown for the first time back on 25 May 1940. It should be noted that the Siebel company had enjoyed a bit of a jump ahead in the development of the project, since Siebel was the one who had made the proposal for a short-haul airliner in the first place.

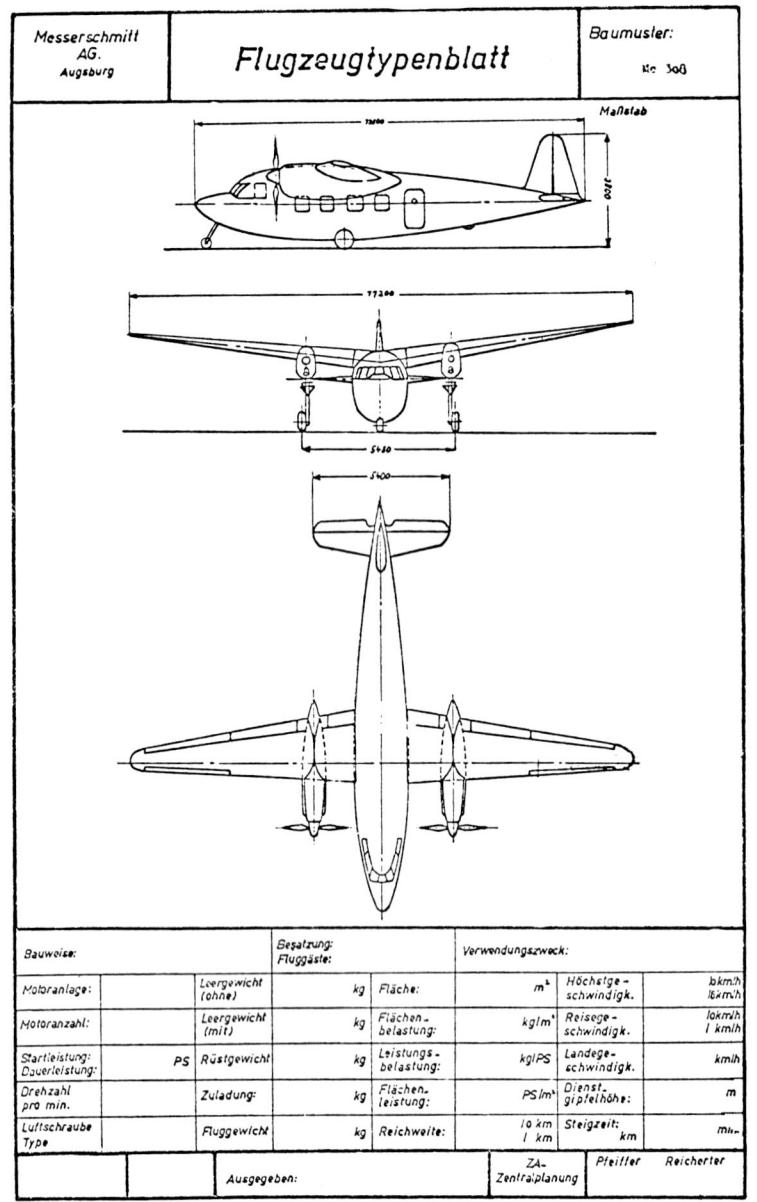

The Me 308 project for an eight-seat commercial plane was begun in Augsburg in 1940 and continued at Caudron in France until it was dropped in about 1942. The airplane was to have been powered by two Argus As 410 engines or other powerplants in the same class. Wingspan was 17.20 m, length 12.60 m, height 3.80 m.

Me 261 - The 11,000 km Airplane

"Shortly after the 1936 Summer Olympics in Berlin there were some rumors going around Messerschmitt that a courier plane was to be drawn up which would then be made available to carry the press corps to the next Olympics, the 1940 games in Tokyo. This was similar to what Otto Brindlinger had done between Sweden and Berlin for the Berlin Olympics," noted Hans Kaiser, at the time a project engineer in the project department under Robert Lusser. The idea of a long-range plane was not a new one for Messerschmitt; previously there had been his own project for an "antipodal aircraft" from 1932, followed by the RLM contract for a round-the-world plane from 1936, although the latter was halted in 1937. These formed a solid foundation for a preliminary proposal for the nonstop Tokyo plane which Hans Kaiser began working on during the winter of 1937/1938. One individual who apparently took a personal interest in the proposal submitted to the RLM was Adolf Hitler, who wanted to use the airplane for a long-distance flight to the 1940 Olympics in Tokyo. Somewhat disrespectfully, then, the Messerschmitt workers nicknamed the project the "Adolfine."

In addition to the Bf 165 (q.v.), this project also seemed to be a topic of discussion during Hitler's only visit to the Messerschmitt Works at Augsburg in November of 1937. From this encounter Hitler and Messerschmitt appeared to have learned to value each other, a claim supported by the numerous meetings between the two which lasted almost until the end of the war. Most likely, both were able to profit by these discussions. Messerschmitt was able to present his numerous ideas, and Hitler could broaden his technical knowledge base, something which never ceased to amaze those he was conversing with.

The RLM's decision to develop three prototypes was reached in early 1938. Willy Messerschmitt himself worked intensely on the project, but gradually passed on the bulk of the work to Paul Konrad and Konrad's stepbrother, Karl Seifert, both of whom Messerschmitt had hired in 1937/1938. The brothers brought their experience in sailplane construction to the company, and had also already drawn up a long-range plane of their own. However, the two were forced to liquidate their own company at Rosenheim in 1937.

The project department tackled two designs at the same time, working out optimal data for project number P 1061 with twin coupled engines and project number P 1062 with four single engines. Of the two, P 1061 was given priority. It was expected to have made its first flight in prototype form in early 1939, giving it a developmental time of over two years—a relatively long period in comparison with other Messerschmitt types. The reasons behind this were twofold; on the one hand the developmental and prototype construction teams were burdened with military projects, particularly the Me 210 program in development at the time. On the other hand, with the outbreak of war on 1 Septem-

The Me 261 as a long-range reconnaissance plane with a five-man crew.

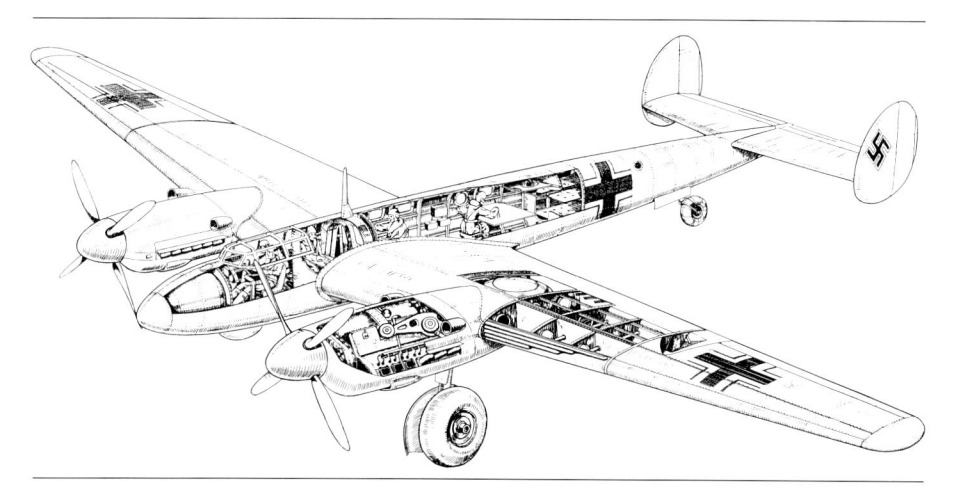

A typical initial sketch by Willy Messerschmitt from May of 1938. The actual airplane, completed in 1940, was quite similar to this draft design.

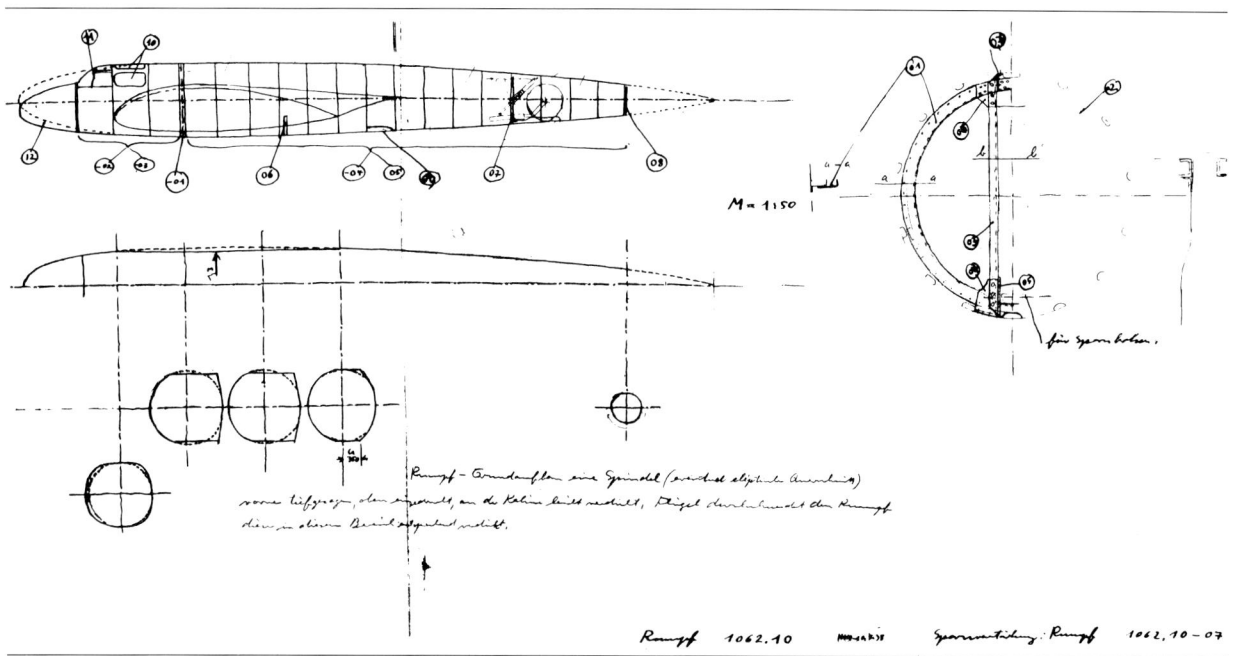

ber 1939 it was doubtful the 1940 Olympics would even take place.

In August 1940 Messerschmitt therefore offered the project to the RLM as a long-range aircraft with a range of 11,000 km and an average cruising speed of 400 km/h, adding the following roles to its reportoire:

1. courier airplane for mail or with eight removable passenger seats (takeoff weight 28,680 kg).
2. strategic reconnaissance plane with two automatic cameras (takeoff weight 28,030 kg).

[1] Messerschmitt AG Augsburg, Angebots-Baubeschreibung Me 261, August 1940

3. freighter with reinforced cabin flooring
    700 kg payload up to 13,200 km range
    900 kg payload up to 11,000 km range
    2,900 kg payload up to 8,000 km range

These payloads were to be carried with an all-up weight of up to 28,630 kg. Other combinations with external loads were also possible. The crew would consist of four to six men.

The aircraft was an all-metal monocoque design, with the unique feature of having the majority of its wing sealed to accommodate roughly 16 metric tons of fuel so that the slender fuselage could be left free for the payload. Two DB

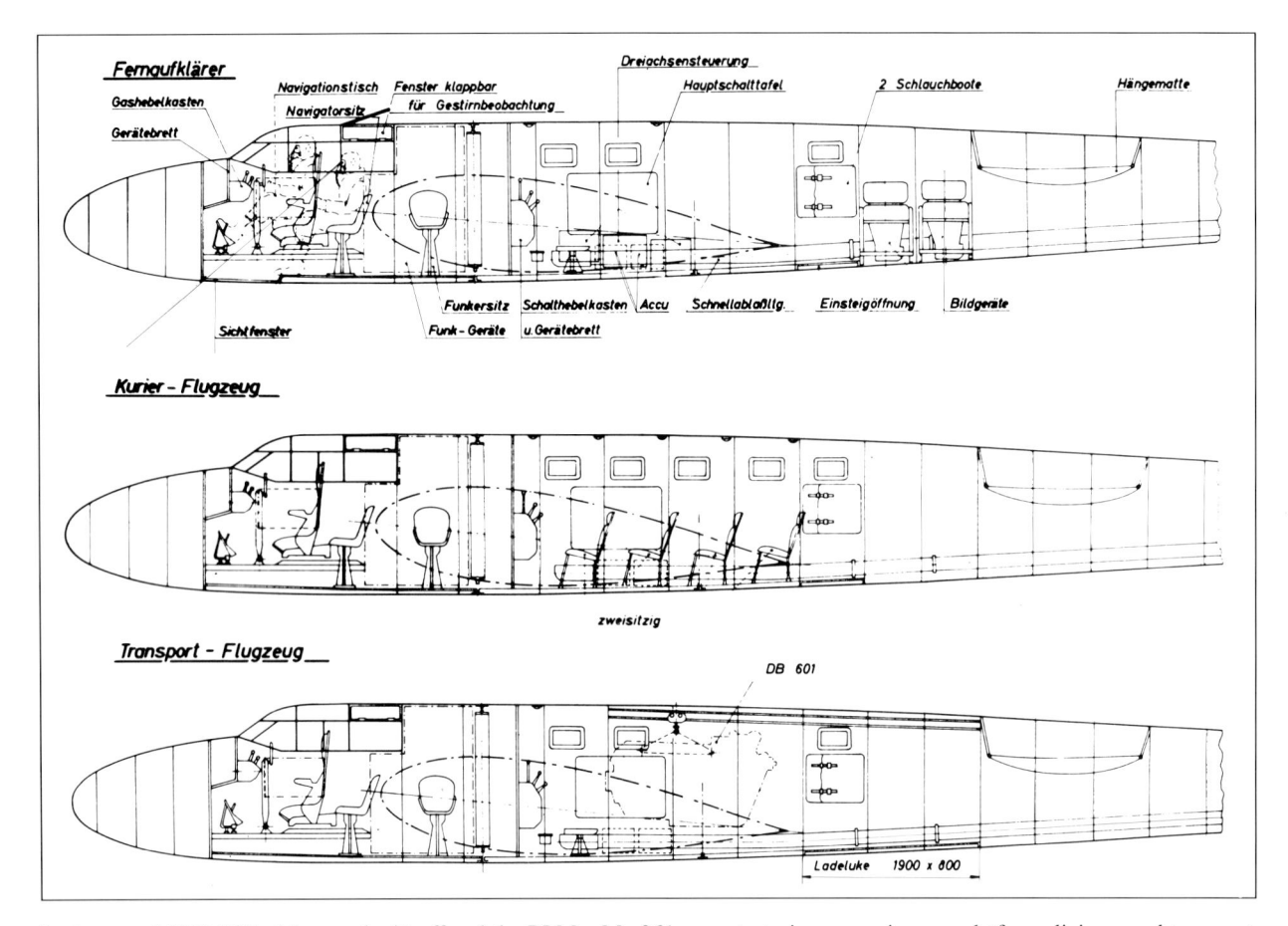

In August of 1940 Willy Messerschmitt offered the RLM a Me 261 as a strategic reconnaissance platform, liaison, and transport airplane. These drawings show the utilization of the aircraft's narrow fuselage to good effect. Fuel was carried in the sealed wings.

606 coupled engines, each outputting 2,000 kW/2,700 hp, drove two four-bladed propellers having a diameter of 4.5 meters.

Now labeled as the Me 261V-1, the prototype construction department was on the verge of completing the airplane when Willy Messerschmitt announced in December 1940 that the machine was to be prepared for a record-breaking flight, as well. Karl Baur flew the plane on its maiden flight just one day before Christmas and further trials continued into January 1941. It was soon discovered that the plane's flight handling characteristics were well under control. By its seventh flight on 24 January Fritz Wendel was able to state that the control surfaces in all flight attitudes were acceptable and that control forces were either good or within tolerable limits. What was also noted, however, was a problem area which was both typical and not inconsequential for the Me 261: the undercarriage. The landing gear suffered abnormal problems with its hydraulics, strut breakage, bowing of the gear doors, and buckling of its legs. The loss of a contract for 20-30 reconnaissance machines, personally relayed to Messerschmitt by the RLM in January 1941 (probably orally) is most likely attributable to these problems, as is the lack of continued interest in a record-setting flight.

As early as the fourth flight, on 14 January a broken strut caused the wheel of the right landing gear to rotate at a right angle to the direction of flight. Pilot Karl Baur did not want to risk further damage to the aircraft and so made a belly landing in the snow at the nearby Gablingen airfield. The damage was so slight that the V-1 was able to complete its fifth flight just eight days later.

In March of 1941 the V-1 was pulled from flying operations after 26 flights and employed in vibration testing before being refitted with a wing increased by five square meters for the long-range reconnaissance role. In early October it was delivered to the *E-Stelle* Rechlin. On 14 May 1942 the V-2's landing gear buckled when braking its engines during taxi trials prior to its maiden flight. Damage was so severe that it wasn't until 25 November before the second prototype took to the air for the first time.

The Me 261V-1 after belly landing in the snow at the Gablingen airfield on 14 January 1941. The aircraft was repaired and flying again in just eight days.

The Me 261V-1 prior to its delivery to Rechlin in 1942. This frontal view shows off the twin DB 606 engines and the thick, 27 m long wings, capable of holding 15 metric tons of fuel.

The V-3, with larger wings and the potential of taking off in an overloaded configuration of 29.5 metric tons and fitted out for a range of 11,500 km, successfully completed its maiden flight on 9 October 1942.

In March 1943 this plane was transported to Lechfeld, where Messerschmitt flight testing was now taking place, not in the least due to the fact that the Augsburg field did not have a concrete runway and was too small for larger aircraft.

On 16 April this particular machine accomplished what Willy Messerschmitt had planned over two years before—an endurance flight. At 1045 hrs the V-3 took off with an all-up weight of 25 metric tons (of which 15 tons was fuel). Karl Baur was the pilot, supported by radioman *Kapitän* Voß, flight test engineer Gerhard Caroli, advisor for flight handling matters Wegner, and crewmen Pöhlmann and *Oberfeldwebel* Förnzler. Covering a distance of 4,500 km over southern Germany and Austria at altitudes varying between 500 and 4,000 m, with speeds sometimes exceeding 500 km/h, the machine landed back at Lechfeld at 2035hrs, nearly ten hours after it had first taken off.

In his flight report Karl Baur noted:[2] "Landing was made on the left undercarriage leg with 5,000 liters of fuel remaining in the tanks. After a short rollout the airplane leaned to the right, damaging the right outer wing section and the propeller. Total damage was approx. 5%." (Hydraulic failure prevented the right undercarriage leg from extending.)

With this flight came the proof needed to show the type's suitability for long-range reconnaissance work. In

[2] Messerschmitt AG Augsburg, flight report no. 917/3 from 20 Apr 43

The Me 261V-3, BJ+CR, at Lechfeld prior to its ten-hour flight on 16 April 1943. Between 1,000 and 2,030 hrs the plane covered a distance of nearly 4,500 km over southern Germany and Austria.

July the aircraft was turned over to the *Versuchsstelle für Höhenflüge* (VfH, High Altitude Flight Test Center) under the supervision of *Oberst* Rowehl.

Me 264 - The 15,000 km Airplane

Although the project department was not urgently pursuing the P 1061, the parallel design to the P 1062 (Me 261), those involved had wanted to avoid relying on the yet undependable coupled DB 606 engine by using four single powerplants. Through extreme aerodynamic refinement, it was hoped that the long-range aircraft, long favored by Messerschmitt, would be able to fly for distances up to 20,000 km. The short term goal was to design a "high-speed transport for economically hauling high-value perishables from the tropics." The Messerschmitt project engineers thus dubbed the concept "the banana plane." However, since it appeared that no contract for such a design would be forthcoming in the near future, in 1937 the program was put on the back burner out of sight.

In late 1940 Messerschmitt, convinced of the success of his Me 261 design (which was close to its maiden flight at this point), dusted off the P 1061 project again and this time tailored it for military roles. In a 12/20/1940 memorandum[3] to his project engineers Woldemar Voigt, Paul Konrad, and Wolfgang Degel, Messerschmitt outlined his ideas for taking the Me 261 and in stages developing an "optimal plane" powered by four recently announced DB 603 engines with a range of 20,000 km. It would be marketed in both the military and civilian roles, with a potential payload of two to three metric tons carried in the fuselage, as well as under the wings. Weapons or other observation features could be accommodated by means of folding panels or retractable turrets, thereby reducing fuselage

[3] W. Messerschmitt, Optimalflugzeug, memo to Mr. Voigt, project dept. from 20 Dec 40

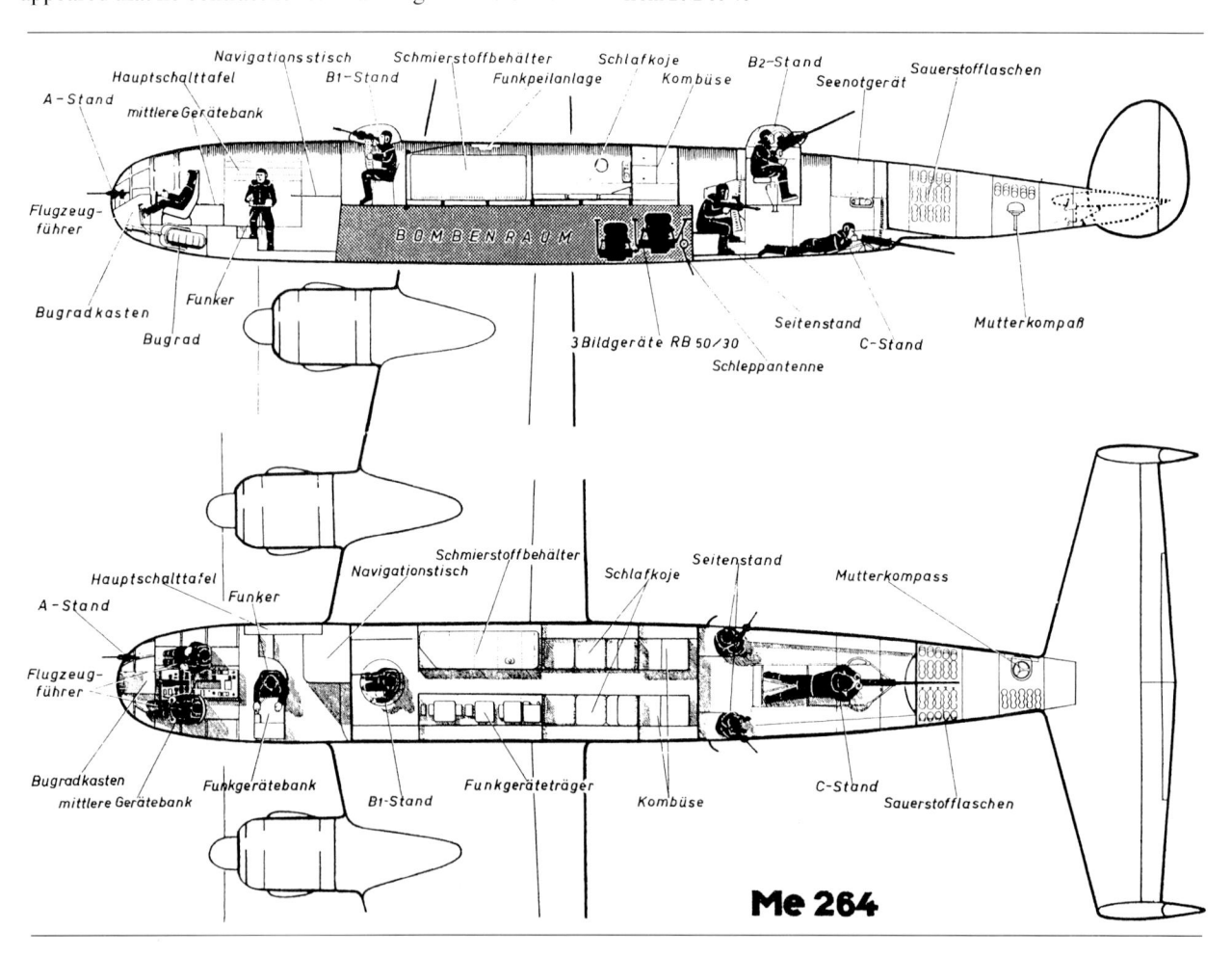

The Me 264 strategic bomber with a crew of seven.

216

drag. Range and maximum speed were to be the aircraft's distinguishing characteristics.

Messerschmitt was able to call upon this design when he was given an official contract in early 1941 by the RLM for a "four-engined aircraft with two-ton payload for harrassment flights against the U.S."

The RLM contract, issued while Ernst Udet was still in office, called for 30 Me 264 (V-1 through V-30) planes, with materials authorized for an initial six aircraft. The V-1 through V-3 went directly into design and assembly in early 1941, apparently with a view toward using the V-1 exclusively as a flight handling testbed, i.e. with no military equipment on board, as the type's role definition was still in a state of flux. This approach was in direct opposition to Messerschmitt's views about building pure test planes from which operational types would then be developed. The Me 264V-1's first flight was scheduled for May of 1942.

The *Probü* conducted an exhaustive study for optimizing the design from 1941 to early 1942. The resulting statistics showed that if special measures were taken (e.g. using a second Me 264 to tow the plane on the first leg of its mission, or fitting two additional engines) the range with a 5,000 kg bomb load could be increased from 13,000 km to 18,000 km (ultra-long-range strategic bomber). Medium range operations up to 8,000 km could be carried out with a load of 14,000 kg. High altitude flights at 14,500 m in the pressurized airplane would give the type a range of over 5,000 km.

Armament was to have consisted of four remote-controlled turrets firing twin MG 151 or MG 131 *Zwilling* machine guns.

By utilizing Jumo 213 engines, only recently advertised, it was expected that the range would increase by an additional 3,000 km over the above named figures.

The RLM Vacillates

This was the situation as 1941 turned into 1942, when Milch canceled the Me 264 (a much more complicated machine than other Messerschmitt aircraft) program in January of 1942. Its development had increasingly come under the shadow of the ongoing Me 210 program and its imminent disaster of 13 March 1942. The project's termination was also a good opportunity for Erhard Milch, who had taken over Ernst Udet's functions upon his death, to tangibly demonstrate his aversion to Messerschmitt stemming from his BFW days. Hermann Göring also criticized Messerschmitt and his airplanes, wanting to know where the promised *Amerika Bomber* was that Messerschmitt had talked to Hitler about. By stopping the Me 264, Milch had likely hoped that he could give Messerschmitt the boot and bring in his competitors: Kurt Tank with his Ta 400 project and Ernst Heinkel with an enlarged version of the He 177. But Messerschmitt lodged a protest with the *Amt*, arguing that a lot of work had already gone into the Me 264, and on 15 April 1942 succeeded in getting Milch to carry out "a review of the Me 264" through his department chief *Freiherr* von Gablenz.

The results of the investigation, conducted under the direction of the commander of the *Erprobungsstellen*, *Oberstleutnant* Petersen, were released on 7 May. Petersen and a commission had visited Messerschmitt at Augsburg, Lechfeld, and Leipheim on 24 and 25 April. He was presented with blueprints and documentation for the design—now nearly 90% complete—and was also shown the V-1 through V-3 prototypes. Petersen's conclusions are as follows:[4]

"In its current configuration, the aircraft is designed as a four-engined long-range airplane without pressurized cockpit, with bombs and armor as well as manually controlled gun stations.

Engines are either 4 x 801D or 4 x Jumo 211J.

Using the 211J powerplant, estimated range is 13,000 km; with the 801D it is 14,000 km. Takeoff weight is 45 metric tons in both cases. Thus, the east coast of America is just within range.

A 2,000 kg bomb load is possible if the armor and window guns are dispensed with.

With regard to performance, using the 801D the aircraft just meets the specified requirements. The airframe's

[4] Kommando der Erprobungsstellen der Luftwaffe, *Obstlt.* and *Kommandeur* Petersen, re: Überprüfung der Arbeiten am Flugzeugmuster Me 264 an Verteiler, dated 7 May 1942, No. 15200/42, Secret.

The Me 264V-1's cockpit glazing and provisional equipment layout.

suitability for overloaded takeoffs can only be determined after evaluation.

The first prototype will be ready to fly in the autumn of this year, with V-2 and V-3 following sometime during the winter. Full scale production would get underway in the winter of 1942/1943, with operations using single aircraft possibly beginning in the fall of 1943.

We know of no other aircraft which has the same range with a similar speed and payload.

In conclusion, the consensus is that it would be worthwhile to build the project, submitted by Mtt.A.G. under Record No. IV/20/42 Me 264 from 4/24/42, and utilize the type for Atlantic operations.

Regardless of whether the type would be able to reach the American east coast in all circumstances, there is still an urgent need for a long-range aircraft with the ability to scout the Atlantic as an armed reconnaissance platform and act as a reconnaissance aircraft and convoy shadower for He 177 formations.

Other important roles for such an airplane include:
1) joint operations with submarines.
2) shadowing convoys.
3) dropping radio buoys behind convoys to plot their locations for bombers and submarines.
4) long-range maritime scouting operations.

In view of the current shortage of high-speed bombers it is recommended that a limited number of aircraft be built, with the caveat that the aircraft should only be employed operationally if it is fitted with the new all-round canopy glazing and instrument suite."

The findings were clearly in Messerschmitt's favor, and although it had never fully stopped, he was now "legally" able to resume work on the project and prototypes.

Four Engines or Six?

Immediately upon recognizing the desire for an airplane with even greater range, Messerschmitt had instructed Woldemar Voigt to draw up a variant with six engines (something the team had been anticipating for a long time), which Messerschmitt then presented to the *Amt*. Voigt had wanted to call the design the Me 364, since it was basically a Me 264 with added plugs in the fuselage and wings, but the *Amt* never authorized this designation since a contract was never issued for the type. Consequently, the project was given the number P 1085. Petersen had not mentioned this project in his report because he felt that it was too labor intensive for a short-term solution. Nevertheless, he felt that this variant stood a chance of flying to America and back without having to refuel.

Because the Augsburg developmental team was overburdened, back in 1941 Messerschmitt had turned the design for the 264's empennage (based on that of the 261) over to the Fokker company in Amsterdam.

Now he did the same for the wing of the six-engined variant, which would also be developed using the 549-460 long-range NACA profile as a basis.

Although he may have allowed work on the Me 264 prototypes to resume, Milch remained adamantly opposed to full-scale production, favoring the Tank Ta 400 project instead. As Messerschmitt demonstrated in a comparitive study, however, in many respects the Ta 400 was inferior performance-wise to the more advanced Me 264.

The Messerschmitt team hoped to stay one step ahead of the competition by producing newer and newer variants. Based upon the six-engined P 1075 and P 1085 projects they created ever more fantastic designs for a wide variety of roles, including:
- simultaneous towing
- aerial refueling
- carrying parasite fighters
- overloaded takeoffs with weights ranging from 60 to 105 metric tons using jettisonable undercarriage
- takeoff using RATO packs and catapults

The nose gear installation in the Me 264V-1. This type was the fourth German aircraft model equipped with a nose gear.

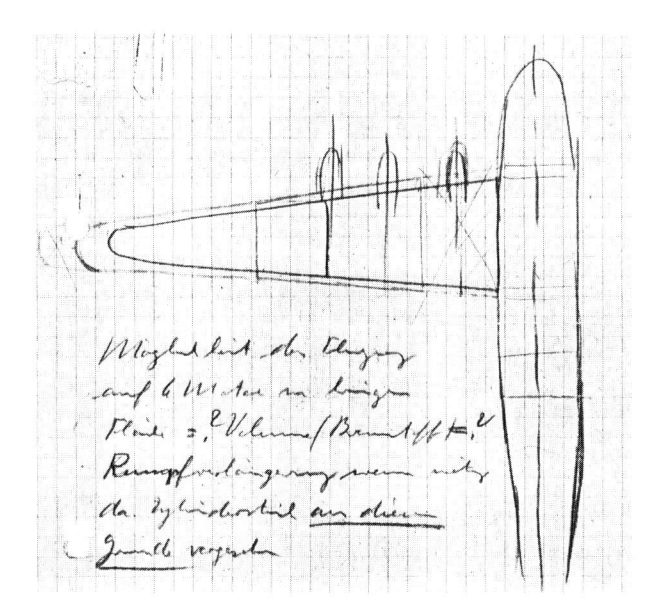

Willy Messerschmitt's initial hand-drawn sketch, dated 16 February 1942, with the comment that it was to be brought up to six engines by increasing the wingspan and fuselage length. The project section converted this drawing into project numbers P 1075 and P 1085.

Model of the six-engined Me 264. The designation "Me 364" was not approved by the RLM due to the fact that no developmental contract ensued.

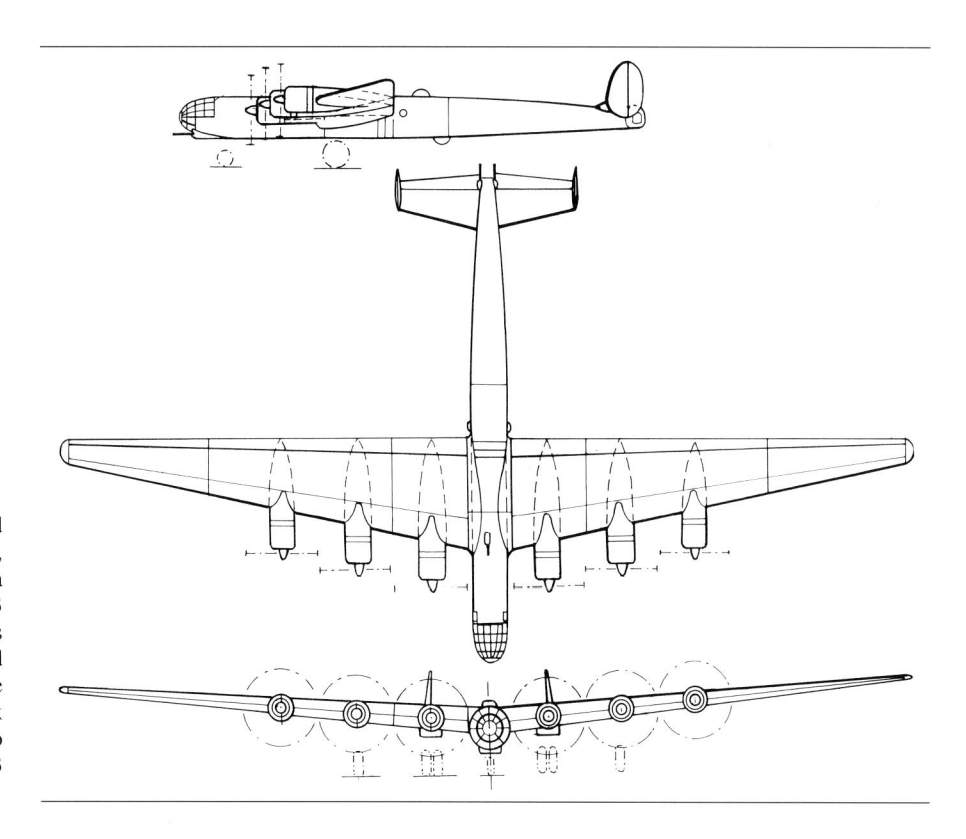

Three-view of the six-engined Me 264. With a crew of eight, the aircraft was to have had a maximum takeoff weight of 83 metric tons. Its wingspan was 47.50 m, wing area 170 m², and length 25.50 m. It was to have been powered by either six BMW 801s, DB 603s, or Jumo 213s (based on a 12 May 1943 drawing).

219

(The "Graf Zeppelin" Research Insititute at the University of Stuttgart was carrying out proof of concept studies in this area at the time.)

All these efforts were taking their toll on the project department to the point where consideration was given to transferring these projects to partner companies. Negotiations with Claude Dornier revealed that Dornier was only willing to take over the work if he would be able to build his own design, something which Messerschmitt naturally declined. The closest was Dr. Hugo Eckener from the Luftschiffbau Zeppelin, who had already taken over follow-on development of the Me 323; he was willing to assign the work to his French subsidiary of SNCASO at Cannes. Negotiations were also carried out with the Weserflug company in Bremen.

During this time, assembly of the three prototypes continued under increasingly difficult work conditions, a trend which would become more pronounced as time went on and dictate production operations for the remainder of the war (without even factoring in the large-scale Allied bombing raids). The following two examples from the monthly reports of the "ring production director" Fritz Hentzen confirm that Messerschmitt was unjustifiably accused of missing deadlines in aircraft production. In the July 1942 report, Hentzen wrote of the Me 264:[5]

"Three prototypes are under construction. Estimated date available first flight for the V-1: 10 October, V-2 and V-3 unknown. Manufacture of V-4 and subsequent aircraft stopped. All prototype and pre-production versions are suffering from a lack of externally supplied parts and finished components. In particular, the Elektrometall company, Cannstadt, is in serious arrears. Much of the Elektrometall company's employee base was drafted, and the company was forced to curtail its prototype construction and pass its contract on to subcontractors. The subcontractors, in turn, had difficulties in meeting demands because some of their work force had also been called up."

[5] Messerschmitt AG Augsburg, report of *Sonderausschuss F-2* for July 1942, Secret.

And the August report announced the latest setback:

"Construction of the V-1, currently in its final stages, cannot be completed in October of this year as previously reported. The undercarriage, which the VDM company had promised to deliver by 9/1/42 at the latest, has not yet been supplied and no new delivery date has been set. In general, it should be noted that acquiring externally supplied parts and, more recently, even engines, for the prototype airplanes is proving most difficult and is resulting in serious delays in the program."

The Me 264 Flies

Finally, on 23 December 1942, Karl Baur was able to begin flight testing the Me 264V-1 at Augsburg. Baur had become the specialist for Messerschmitt's larger designs, and after the Me 321/323 the Me 264 was the second largest Messerschmitt airplane ever, with a wingspan of 39 meters. Testing moved to Lechfeld in early 1943, since Lechfeld had an asphalt runway which provided better traction when braking. After only a few flights, the Me 264 was positively appraised by test director Gerhard Caroli: "The nose gear is a distinct advantage on such large aircraft. Landings are much easier." (After the Me 309, the Me 264 was the second Messerschmitt airplane with a nose-wheel and the fourth German design).

Flight testing revealed two problem areas: reducing the rudder forces and harmonizing the two rudders, plus the numerous mechanical problems in the cockpit and with the landing gear normally encountered on a new type. Such hydraulic problems led to a wheels-up landing on the 23rd of March, damaging the plane.

The Jumo 211 engines fitted were actually a stopgap measure since the originally planned Db 603 engines were not yet available. Thus, with the Jumo engines the design did not reach its estimated performance potential.

The previously mentioned problems and the decision to carry out the difficult job of refitting the Me 264V-1 with the more powerful BMW 801TJ engines (plus the *Amt's* decision to forego production) meant a temporary interrup-

To the observer, the Me 264V-1 had a fascinating shape and size (43 m wingspan). RLM advisor *Oberstleutnant* Siegfried Knemeyer called it "the Ship...."

This photo almost conveys the 500 km/h speed of the Me 264V-1, here with BMW 801 engines.

Powered by four Jumo 211 engines, the Me 264V-1 takes off on its first flight on 23 December 1942 with Karl Baur at the controls.

tion in testing, lasting from August 1943 to March 1944. Fitting the BMW engines subsequently led to a better assessment of the Me 264, primarily in terms of improved takeoff performance. When Lechfeld became the focus of an Allied bombing strike on 18 March 1944 (following Augsburg in February), portions of the test program were dispersed further afield. The Me 264 was ferried to Memmingen on 16 April, and it was at this location where it turned in the "best performance of its career," received the most favorable assessment from the *Amt*, and also met its demise. One year after Erhard Milch's investigation of the Me 264 project, *Flugbaumeister* Scheibe and *Oberstleutnant* Knemeyer of the RLM were able to fly the plane and then travel to Oberammergau to discuss the design with Willy Messerschmitt and his project team. This assessment from 16 April 1944,[6] retained in the Messerschmitt archives, was positive, as well. But it also highlighted the fact that, even in the fifth year of the Me 264's development, the RLM remained indecisive regarding the role of large aircraft designs, vacillating on the award of contracts and overestimating the capacity of the aviation industry.

"1) **Test Flying**

Oberstleutnant Knemeyer and *Flugbaumeister* Scheibe are in fundamental agreement in their assessment. *Oberstleutnant* Knemeyer expressed the sentiment that the airplane 'could be bought right off the rack.' Criticism was made of the rudder forces, which were somewhat too powerful and unharmonized.

These deficiencies were already well known and have yet to be corrected only because of the lack of capacity. *Oberstleutnant* Knemeyer is of the opinion that given the current state of aircraft construction, there should be no problems in correcting the previously named deficiencies once the matter of capacity has been resolved.

---

*Knemeyer and Scheibe were aware of all German heavy planes at this time: the six-engined BV 222 flying boat (first flight Sep 1940), the six-engined Me 323 *Gigant* transport (first flight Apr 1941), the four-engined Ju 290 transport (first flight summer 1942), the six-engined Ju 390 transport/strategic reconnaissance aircraft (first flight Oct 1943), and the largest six-engined flying boat, the BV 238 (first flight Mar 1944).

[6] Messerschmitt AG Augsburg, memorandum for record dated 16 Apr 44, Me 264 No. 101

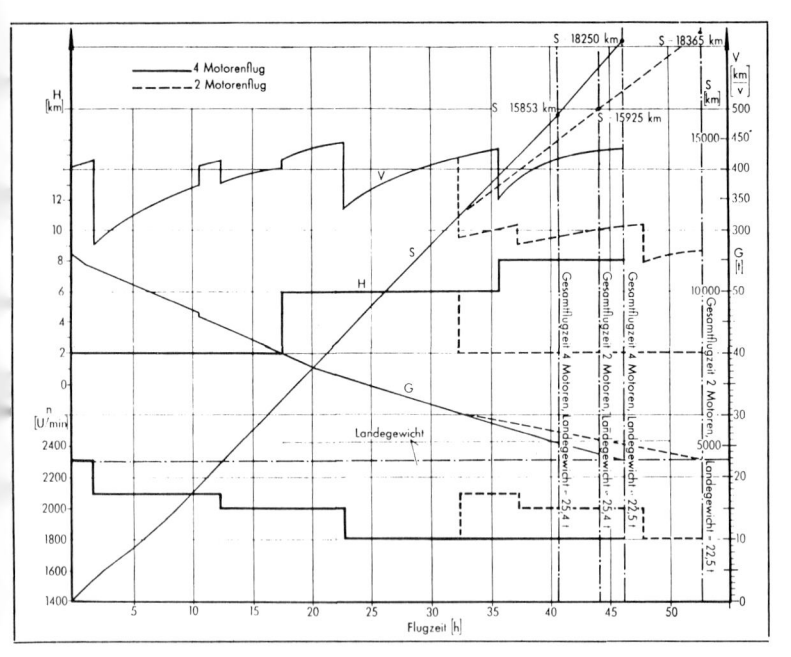

BMW engineers calculated that a Me 264 powered by BMW 801 engines would have a range of 18,000 km at variable altitudes, enabling it to reach America and return to base.

Oberstleutnant Knemeyer rated the airplane's layout as follows:

'Weapons fitting is completely adequate based on current needs and should pose no problems. The aircraft's interior dimensions are perfectly suited for carrying ordnance. The only complaint regarding its design is the long-known matter of visibility.'

If the aircraft goes into production, both men called for the pilot to be seated higher up with a design which allowed a vertical view through the canopy.

2) **Meeting**

Flugbaumeister Scheibe recommended fitting more powerful engines (Jumo 222) and was in favor of the six-engined version. At the meeting in Oberammergau, *Oberstleutnant* Knemeyer stressed that an airplane with the size and performance of the Me 264, were it to go into production today, would be *THE* plane. However, no information on the capacity to handle such a project can be given, since the RLM has not fully determined which of the designs should be continued or which types should be replaced by the Me 264 follow-on developments (4-engined with more powerful motors and 6-engined).

In any event, the 264 must serve as the basis for any future large aircraft development. *Oberstleutnant* Knemeyer expressed the opinion that the company which the *Amt* intends to make responsible for continued development of the large aircraft concept should be the one to test the Me 264V-1 and build any further prototypes. The RLM has not yet reached a decision in this matter, but is expected to in the immediate future.

For the Mtt.AG. this means that it must continue with the Me 264's flight testing and will remain responsible for the design until such time as the RLM makes its decision."

During the Me 264V-1's flight testing in Memmingen it was fitted with a Patin autopilot, the installation of which briefly interrupted the trials program. On 27 April, flying at a height of 450 m, the aircraft reached a speed of 490 km/h at maximum military setting and 470 km/h at maximum continuous power. With only about 30 hours clocked in the flight test program, the Me 264V-1 became a casualty of an Allied bombing raid on Memmingen on 18 July 1944.

The Me 264V-2 and V-3, which were to have been fitted with a bomb bay and defensive armament, were only partially completed when both had been moved to the dispersed facility of "Metallbau Offingen" near Neu Ulm. They were never finished, although in late 1944 there was another attempt to rescue the program and build the planes as long-range courier machines with a range of 12,000 km and a payload of 3-4 metric tons. By this time all efforts were devoted to the production of the Me 262.

Under such circumstances, the idea of using the Me 264 (like the Me 323) as a testbed for steam turbine aircraft engines stood virtually no chance at all of becoming a reality.

The Last of the Futuristic Projects

In spite of these setbacks, and in view of the favorable feedback from the *Amt*'s technical experts and results from the V-1's flight testing, Willy Messerschmitt tenaciously clung to his "pet project."

In the spring of 1944 the Me 264 reached the pinnacle of its design development when Messerschmitt brought the latest technological advances on offer into play: turbojet engines and swept wings.

The turbojets would be utilized as supplemental powerplants, much like the two additional propeller engines of the P 1085 project, and several studies were made exploring a wide variety of attachment locations and installation configurations (see drawings on p. 220).

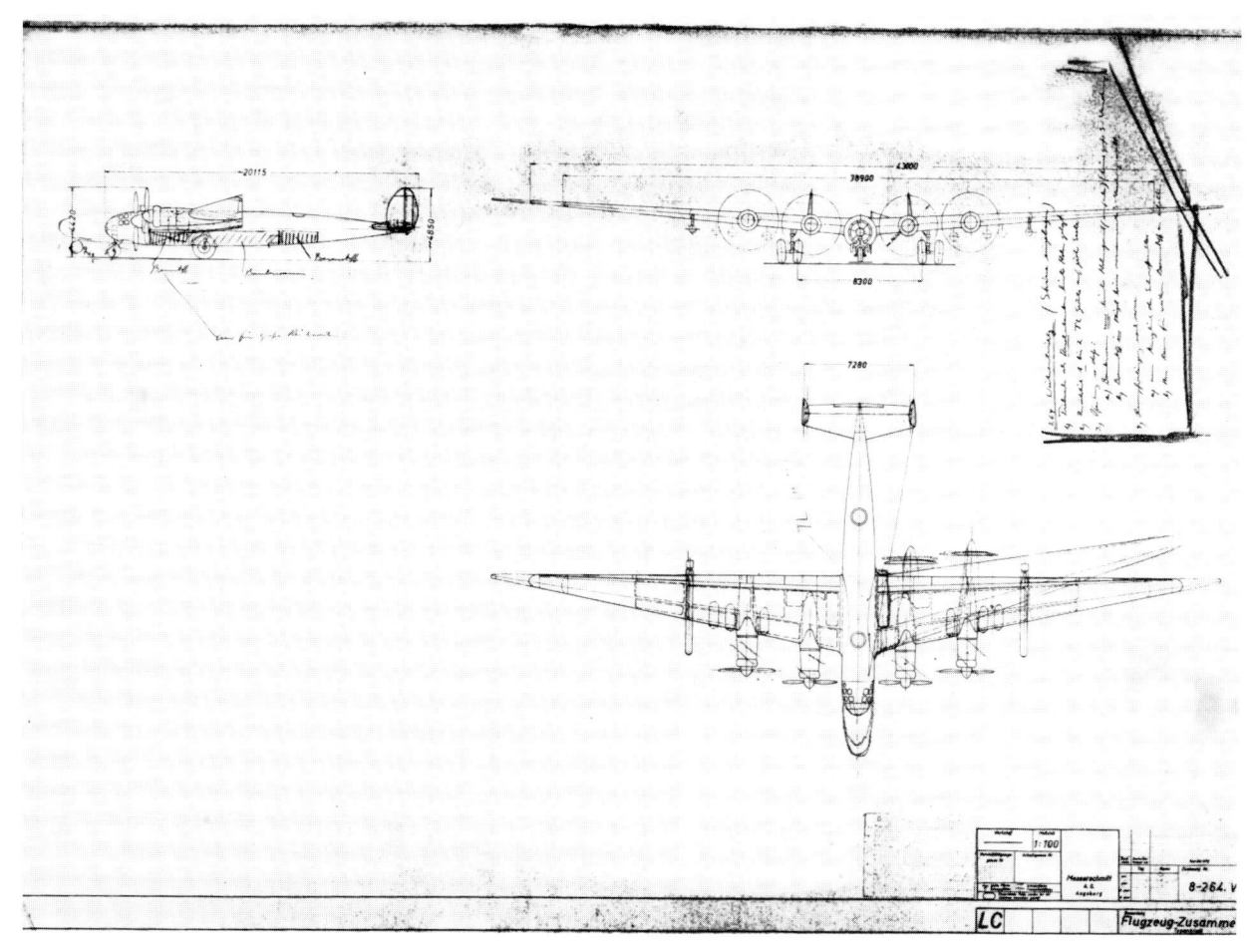

Initial sketch by Willy Messerschmitt showing the conversion of the Me 264 with swept wings, four pusher-type propellers, and one or two supplemental turbojet engines in the wing roots, plus a lengthened fuselage forward section (spring 1944).

These projects were so fascinating that even Hitler, in one of the final meetings with the "ultra-long-range bomber" on its agenda, expressed the thought that: "with its turbojet/piston engine combination, this Me 264 gives us a potential of utterly revolutionary importance. Due to its payload capacity and penetration range it appears to be capable of carrying out any mission, and in small numbers, as well."[7]

It wasn't until after the war that projects like this Messerschmitt design would become a reality, both in the U.S. (Convair B-36, first flight 1946) and in the USSR (Tupolev Tu-20, first flight 1954).

In Germany, however, design and construction of large aircraft unquestionably fluctuated between hot and cold, depending on the requirements and priorities of the contractor (i.e. the RLM). It was never able to come to grips with the urgency and significance of long-range aircraft. To be sure, just eight years before it had decided against such deveolpments when it canceled the long-range bomber project (see Bf 165), the only major difference in the case of the Me 264 being the time it took to make the same decision.

[7]quote by the director of the *Jägerstab*, Saur, in W. Baumbach, Zu spät, p. 159

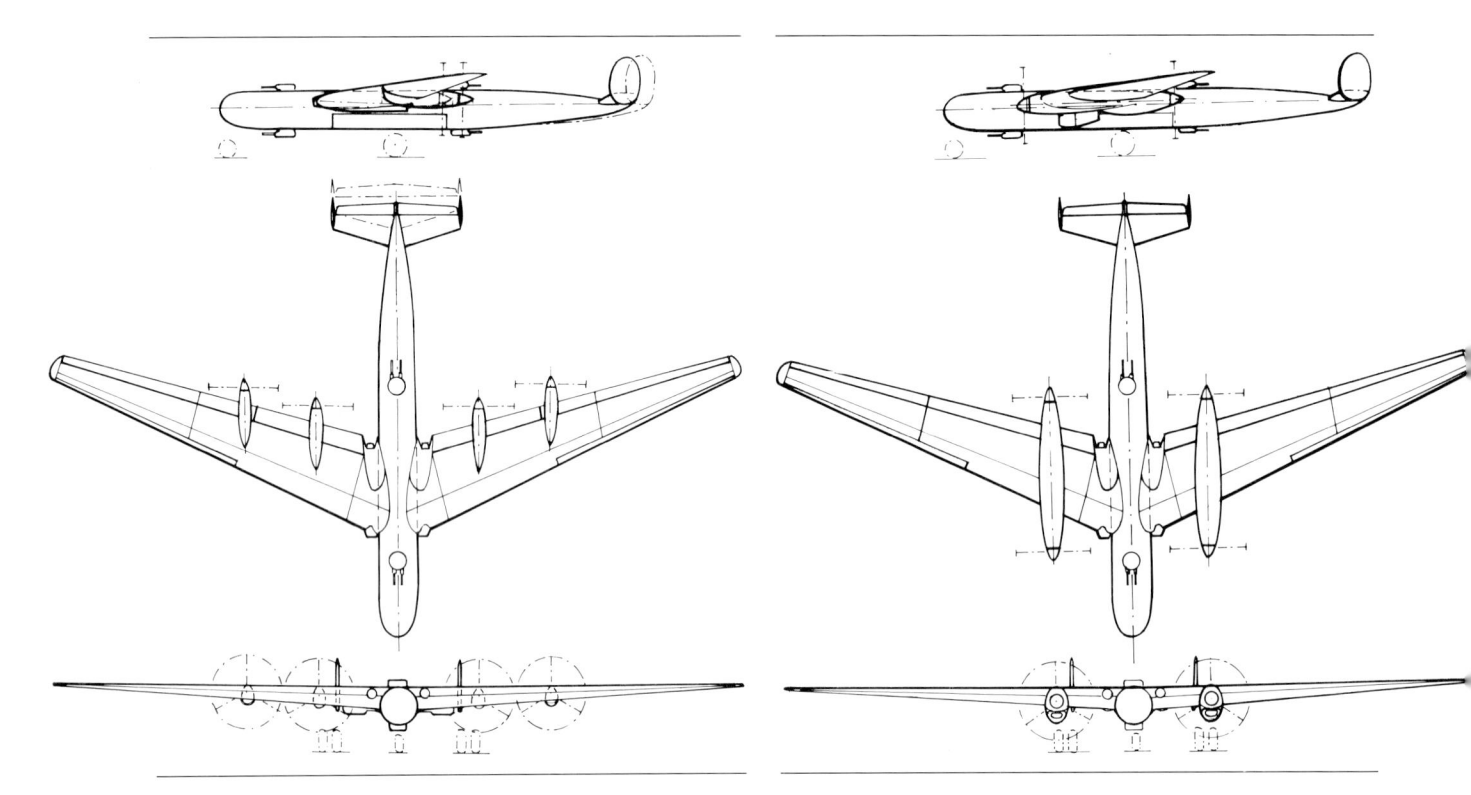

The project section took Messerschmitt's ideas and came up with two variants: a Me 264 with four pusher propellers and two jet engines, and a layout with two engine gondolas, each with a pusher and tractor propeller and two turbojets. Wingspan was to have been 39 m, length 25 m, and height 5 m (based on drawing XVIII from 27 July 1944).

The Me 321 and Me 323 Giants

In the early 'thirties there were many proposals for aerial tow convoys in commercial aviation, i.e. transporting cargo in large sailplanes towed by powered aircraft on long cables. Glider towing was a common occurence at the time and was also used for ferry flights, for example. In 1933 Hans Jacobs at the Deutsche Versuchsanstalt für Segelflug (DFS) developed the DFS 230 transport, a glider with a 1,280 kg payload capacity pulled behind a Ju 52/3m. In 1937 the *Luftwaffe* showed an interest in the type, as well as in the concept of towed flight in general. A contract was issued for the development of a military version of the DFS 230 for one pilot and nine airborne soldiers. By the spring of 1938 the first training detachment for glider pilots had been formed and the DFS 230 ordered into production, with the first operational unit following a short time later. 41 of these transport gliders and their paratroopers were used operationally at the start of the campaign in the West, landing at Fort Eban Emael in Belgium. Transport gliders thus proved their effectiveness as a military tool and thereby became a functional weapon in air warfare.

Heavy Lift Gliders for Operation "Seelöwe"

From early summer until the autumn of 1940 the Wehrmacht was busily making preparations for the invasion of England, an operation which was primarily to be carried out using water craft (transport ships, ferries, etc.) under the protection of the *Luftwaffe*. However, in early October 1940 Hitler decided to postpone the invasion in the face of an increasing awareness that the *Luftwaffe* would not be able to gain the necessary air supremacy in the Battle of Britain.

Willy Messerschmitt, who was confronted with the problematic nature of the invasion during a visit with Hitler, began thinking of a way to airlift heavy equipment (tanks and artillery pieces). He talked the matter of aerial towing over with Prof. Walter Georgii, director of DFS Darmstadt, and on 4 October 1940 went to Ernst Udet (in his capacity as Chief of the RLM's *Technisches Amt*) with the proposal for "guidelines for towing heavy cargo, in particular heavy tanks":[1]

"An unusually effective means of positioning troops behind enemy lines is through the use of paratroopers, airmobile infantry, and by employing gliders towed behind metal aircraft. Generally speaking, this involves lightly armed forces. A much more effective approach, however, would be to successfully deploy heavy weaponry, especially heavy tanks, behind the enemy's lines.

The following is a new proposal:

1. The heavy tank is fitted with bolted on attachment points to which the wings and aft fuselage with empennage could be fitted, i.e. at a fixed weight the object to be towed would form the kernel of the aircraft.
2. Wings, empennage, and aft fuselage sections can be diagraphically enlarged from current transport gliders to prevent any aerodynamic risks.
3. Mating of individual components to the tank would take place at the point of departure.
4. For landing, the tank would be fitted with heavy wooden skids on its tracks, which would break away once the tracks began revolving.
5. A jettisonable axle with two correspondingly large wheels is planned for takeoffs.
6. The Ju 52 is available for towing. Since a single aircraft is not up to the task, multiples of four aircraft would be coupled together as shown in the accompanying sketch. Differences in speed can be countered through the use of pulleys. In case of difficulty, any of the aircraft can be uncoupled.
7. Four Ju 52s can easily pull a towed aircraft weighing 40 to 50 metric tons into the air and to cruising altitude.
8. Discussions with the Glider Research Institute at Ainring have confirmed that there are no reservations regarding the method of towing. Practice flights with smaller towed airplanes can start at once.
9. Primary material for the glider is to be pine and thick plywood in addition to welded steel fittings. Only simple wooden jigs are required for assembly.
10. Since the onboard equipment is at a minimum and the aerodynamic shape of common gliders is to be used, assembly should not take longer than a few weeks.
11. Since the construction of such a glider is much simpler than other aircraft, the spin-up time and run duration can be kept to a minimum, so that with proper organization a large number can be manufactured in just a few months.
12. It goes without saying that other heavy, restrictive objects can be airlifted in such a manner. In certain instances the transport would require a proper fuselage for carrying the objects, e.g. heavy artillery pieces. (The tank would not require such a concession.)"

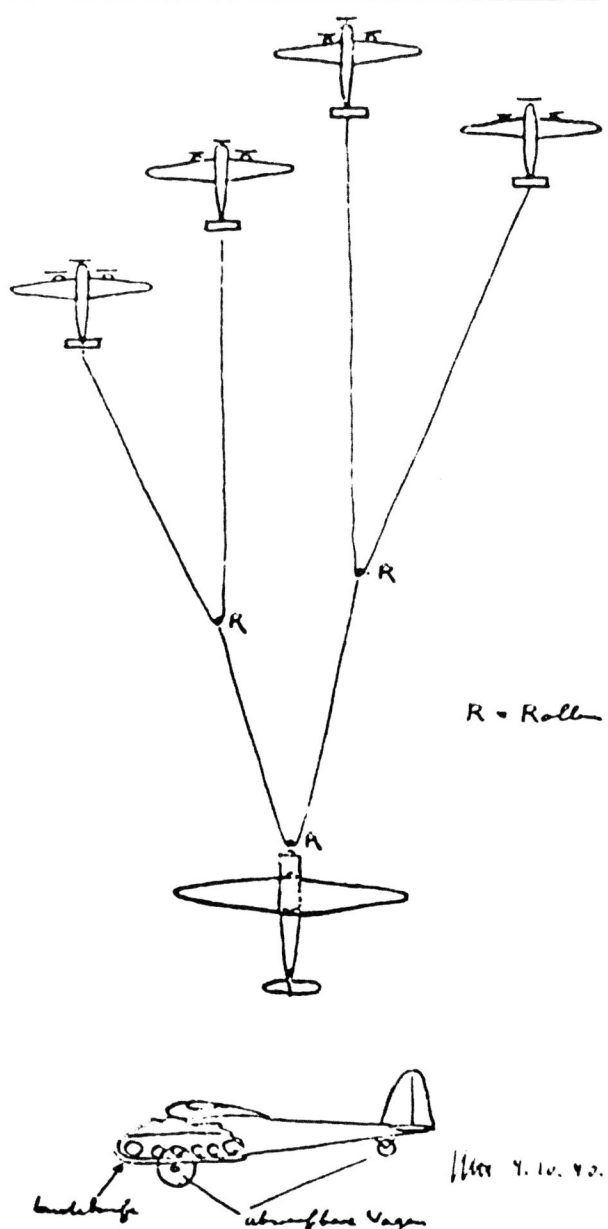

Messerschmitt's original idea of a tank-carrying glider transport and a tow team of four Ju 52 planes, drawn on 4 October 1940 and included in a letter to Ernst Udet on the same day.

[1] letter from Willy Messerschmitt to the *Generalluftzeugmeister Generaloberst* Ernst Udet, *Reichsluftfahrtministerium*, Berlin, dated 4 Oct 1940, Secret

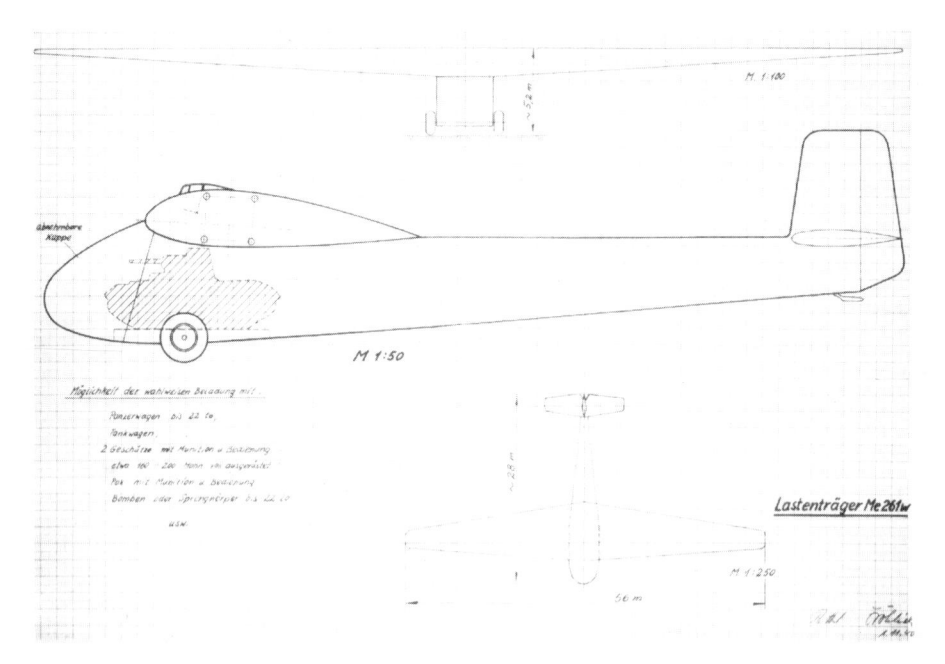

Earliest known drawing of a Me 261 glider transport, designed for carrying 22 metric tons of cargo and having a wingspan of 56 m and a length of 28 m, from the design department of Josef Fröhlich and Walter Rethel, dated 1 November 1940.

Convinced of the soundness of his proposal, Willy Messerschmitt pulled a handful of people from his developmental group to form a small team under the supervision of Josef Fröhlich; on 8 October they began working the rough drafts, and on 15 October started on the design proper. One of Josef Fröhlich's designs with a fuselage, dated 11/1/41 and designated Me 261w, served as the basis after being given a more aerodynamic shape. The RLM awarded developmental contracts to Messerschmitt for the Me 321 (under the code name "Warsaw South") and Junkers for the Ju 322 ("Warsaw East"), and on 6 November Messerschmitt (in Berlin at the time) cabled Josef Fröhlich: "start the program immediately, but with double the numbers. Execution most imperative." Within a few days a wind tunnel model of Fröhlich's modified design had been built, which Ludwig Bölkow carried to the DVL's wind tunnel in Berlin-Adlershof. He returned to Augsburg a week later with the most critical data results. While the team built the glider, Woldemar Voigt was already working on the figures for a Me 263 powered version with four DB 601 engines, which was offered to the RLM on 26 November. In January 1941 final assembly work got underway on the glider at Leipheim, an airfield which the RLM had authorized both for security purposes as well as because Augsburg no longer had enough room to accommodate building such a large airplane.

At the same time Udet approved continued project studies of the powered version, although only if captured French engines were used—there were other plans for German engine production in the long term. On the 25th of February 1941—140 days after the start of design work—Karl Baur and observer Curt Zeller flew the Me 321 on its maiden flight as it was lifted into the air by Ju 90V-7. The aircraft had a takeoff weight of 15 metric tons. The flight ended after 22 minutes with a pinpoint accurate landing. Adjust-

Wind tunnel model of the Me 321 as evaluated by Ludwig Bölkow in the 3 meter tunnel of the DVL in Berlin-Adlershof in November and December of 1940.

The first Me 321 (V-1) prior to its maiden flight in tow behind the Ju 90V-7, on 25 February 1940 in Leipheim.

226

ments were made to the tailplane, controls, and instrumentation so that on 7 and 8 March pilots Carl Francke, Bernhard Flinsch, and Hanna Reitsch from *E-Stelle* Rechlin were able to conduct evaluation flights. Their verdict was that the airplane flew easily, but demanded considerable strength because its control forces were too high. True glider pilots with experience flying transport gliders would be able to master the aircraft after a short period of preliminary and conversion training.

By the 15th of April four more prototypes had flown, and on 7 May 1941 the first production glider took to the sky at Obertraubling. Of the 200 aircraft ordered by the RLM, an equal 100 were built at both Leipheim and Obertraubling. The contract was filled by late 1941, although the design went into battle on the Eastern Front rather than against England. Messerschmitt may have been able to meet the delivery timetables, but problems arose with storing those aircraft over the winter which the *Luftwaffe* had not yet taken on charge. No storage areas had been arranged due to bureaucratic delays. Thus, in the spring of 1942 several machines had to either be repaired again or chopped up; glued together and covered in fabric, the airplanes had originally been designed for a single flight and had not survived the weather. Nor had the *Luftwaffe* been able to keep up with training pilots and crews for the towed planes.

An all-wood design, the competing Junkers Ju 322 "Mammut" (Mammoth), was never put into production. The prototype was found to be unflyable after just two takeoff attempts in April 1941. The glider had to be prematurely uncoupled on both attempts, with the pilot being forced to set down outside the airfield perimeter at Merseburg. The program was abandoned.

Me 321 Assembly in Leipheim and Obertraubling

With its 55 meter wingspan and 28.15 meter length, the Me 321's dimensions placed both the designers as well as the assembly line workers in situations never before encountered. At the Leipheim airfield, construction supervisor Fröhlich was first assigned 15, then 20 designers, who were forced to work under considerable time constraints. Additionally, so as not to affect ongoing production programs no other aircraft manufacturing company could be brought into the program for assembling the components. Assembly and initial flight testing took place at the Leipheim (near Ulm) and Obertraubling (near Regensburg) airfields under the supervision of Messerschmitt engineers. Final assembly of the major components took place under open skies since no hangar had the capacity to accommodate the

Open air production in Leipheim, summer 1940. The tubular steel fuselage sections in the foreground are being fitted with flooring made of wooden planking.

large design. Many improvisations had to be made with the unskilled work force assigned to the program.

The aircraft's load bearing components were made of welded steel tubing framework, produced by Mannesmann-Röhrenwerke in Düsseldorf and shipped as prefinished tubular frames to the final assembly point by land under the covername of "lattice masts." Individual components were produced in carpentry shops, cabinetmaker's workshops, furniture factories, locksmiths, etc., all of which first had to become knowledgeable about aeronautical testing methods, gluing standards, coding systems, etc. Planning and execution of the entire production program took place under the supervision of officials from the Organization Todt, and were conducted under the code name of "Warsaw South."

The three piece wing had an area of 300 m^2 and utilized a main spar in the form of a rectangular tubular steel lattice frame with wooden ribbing. The wing's leading edge was covered in plywood, whereas the remainder was fabric covered. The fuselage was a fabric-covered latticework design. The single-seat cockpit later evolved into a three-seater, situated some 5.50 meters above the ground. The braced tailplane was made of fabric covered wood. A hydraulic power servo-motor enabled the tailplane to be adjusted. The Me 321's undercarriage consisted of four skids, supplemented by a dual undercarriage for takeoffs comprising a main gear of two Ju 90 balloon tires of 1.60 m diameter and two swiveling Bf 109 wheels on the forward skids. Jacking the aircraft up to rest on the wheels was no simple matter and was time consuming, accomplished by purpose-designed and built cranes and/or bridging vehicles.

The cargo hold, loaded from the front through the two nose doors, was 11 m long, 3.15 m wide, and 3.30 m high. Loads of up to 22 metric tons could be loaded, or up to 27 metric tons if operating from larger airfields, and might in-

Ju 90/Me 321 tow configuration, with the Me 321 taking off using four RATO packs. Additional "Gigant" gliders are visible in the background. For a time during the autumn of 1940 there were over 70 Me 321 gliders in Leipheim awaiting their acceptance flights.

clude an 88 mm Flak with tractor and crew, two trucks, a Panzer IV or V, fuel drums, ammunition, etc. Fitting a mid-level floor made possible the transport of 175 soldiers and their weapons. Having an empty weight of 12,400 kg and a maximum takeoff weight of 40,000 kg, the Messerschmitt Me 321 was the largest glider in the world and could lift more than double its own weight into the air.

Problems with Glider Towing

Although not always available, a four-engined Junkers Ju 90 fitted with American Pratt & Whitney Twin Wasp radials was used to pull the Me 321V-1 on its maiden flight. The standard Ju 90 with its less powerful BMW engines didn't have the muscle to lift an empty "Gigant" to altitude, much less one with a payload. Prof. Messerschmitt's initial proposal for four Ju 52 towplanes transformed itself into three He 111s or three Bf 110s, with the center aircraft having a tether 20 meters longer than the two outer planes. Although this was indeed a viable method, it posed considerable risks for the pilots involved. Ernst Peter logged 48 *Troika* flights in his flight book and wrote of his experiences:[2]

"The *Luftwaffe* was responsible for testing the heavy lift gliders and despatched a special unit of the XI Fliegerkorps to Leipheim for this purpose. The finished planes were first flown by test pilots behind the Ju 90 and then parked at Leipheim until such time as they were handed over to the *Luftwaffe*. It was during this time that the *Troika* tow was experimented with and provided acceptable results. This method involved three modified Bf 110C-1/U-1s pulling a Me 321....The first *Troika* tow using three Bf 110s with N engines occurred on 3/8/1941 with Flinsch and an observer by the name of Zeiler in the *Gigant*. The flight in the Me 321 lasted 18 minutes at a flying weight of 15 metric tons. Not everything went according to plan, however, and the subsequent seventh flight of the Me 321V-1 was made behind a Ju 90 again. During this flight, on 14 March, with two *Luftwaffe* men at the controls for the first time, one of the tow cable couplings on the Me 321 tore loose at a height of about 150 meters and caused the second cable to separate. The Me 321 overturned and crashed outside the perimeter of the airfield at Leipheim.

The first *Giganten* which rolled off the line at Leipheim (A-1 variants) were designed for just one pilot, but during

[2] Ernst Peter,...schleppte und flog Giganten, Suttgart 1976, pp. 17-19

"Troika-tow": three Bf 110C planes tow a fully laden Me 321, which has ignited eight RATO packs. This was a complicated and risky method of towing.

By late 1941 towing methods had considerably improved with the use of the five-engined He 111Z.

the course of the test program this was found to be quite risky and something which caused some serious misgivings. One pilot alone could not be expected to cope with the great physical effort needed (in spite of the installed servo-motors) to control the plane, particularly over longer flights or in gusty weather; this gigantic plane required the strength of an ox, something beyond the capabilities of a normal pilot.

Thus it was that on the second machine (Me 321V-2) Messerschmitt saw fit to install side-by-side controls in the single seat cockpit, with a second 'pilot' assisting in reducing the tailplane's control forces. This solution proved inadequate, and Prof. Messerschmitt widened the cockpit to accommodate two pilots sitting shoulder-to-shoulder. As a result of this modification the production machines (with dual control) were designated Me 321B-1s.

Flight testing of the production aircraft began on 5/7/1941 at Messerschmitt's airfield in Obertraubling. After ten test flights the evaluation team had improved the *Giganten* and gathered enough experience so that the enormous transport glider could be mastered even by those glider pilots familiar with the DFS 230 and Go 242. Even fully-loaded takeoffs using eight RATO packs and the *Troika* tow method were soon being flown and proved to be no more difficult than taking off when empty....Yet, for all the experience and knowledge, improvements and modifications, the *Troika* tow generally remained an extremely complicated and dangerous way to pull the machine into the air."

In spite of heavy losses during training, ferrying, and in the field, the *Troika* tow method using three Bf 110Cs was kept up for a long time. Eventually, however, the *Erprobungsstelle* Rechlin and *Generaloberst* Ernst Udet successfully arranged for the development of a five-engined He 111Z twin fuselage plane, which entered flight testing in late 1941. The He 111Z, of which twelve were built, comprised two He 111H-6s joined together by a 6.15 meter long center wing section and fitted with a fifth motor. This made towing the Me 321 much less dangerous, although a fully loaded *Gigant* still required six Walter RATO packs to nudge it into the air.

Transport Gliders in Operation

In June of 1941 a *Großraum-Lastenseglergruppe* (Heavy-lift transport glider group) was formed with three squadrons and another three motorized tow tug squadrons. Composition totaled 18 *Giganten* and 36 Bf 110s. This unit was assigned to three sectors of the Eastern Front shortly after the start of the Russian campaign and was responsible for supplying fuels and lubricants to armored units. On the ferry flights from Leipheim or Obertraubling to the Front the teams were forced to make five stopovers—landing, refueling, and taking off in a *Troika* tow again—due to the limited range of the tugs. Airspeed in a towed configuration was 180 km/h.

Operational conditions soon revealed weaknesses in these large heavy-lift gliders.

The turnaround time between takeoffs was considerable, and its huge transport capabilities could not be fully exploited. In the spring of 1942 four squadrons were then deployed to areas near Riga, where they supplied army units with fuel and ammunition. In some cases, the cargo amounted to 27 metric tons per *Gigant*. When taking off with a full payload, eight Walter RATO packs were used, each rated at 4.9 kN/500 kp thrust. They were situated beneath the wings, four to a side, and fired off in pairs.

From 1942 onward the Heinkel He 111Z served as the primary tow tug, increasing the towed speed to about 220 km/h at 4,000 m. Towed flight time sometimes ran to six hours. Other operational areas in Russia included supplying the Kuban bridgehead from the Crimea, and in the Mediterranean Theater operations were flown from Naples. In the summer of 1944 the last Me 321 squadrons were employed for supplying ground units cut off from supply lines.

A Me 321B-1 ready for takeoff with 8 Walter RATO packs (each with 500 kg of thrust).

The majority of these Me 321s subsequently were unable to be flown out of these pockets or the frontline airstrips where they'd landed.

Projects using Steam Turbines and Pulse-Jet Engines

At the same time development was underway on the Me 321, there were also plans for designing a powered version using piston engines. The aircraft also offered possibilities as a testbed for two variants of jet propulsion then in design or development. When word spread throughout the aviation community of Messerschmitt's giant plane (despite all the secrecy surrounding it), Prof. Lösch and *Dipl.-Ing.* Pauker from the *Technische Hochschule* in Vienna offered their steam turbine for installation in the Me 321. The proposal was disseminated in April 1941 within the RLM's *Technisches Amt* by Prof. Adolf Baeumker, director of aeronautical research, but the Me 321's intense evaluation program precluded any type of real follow-up, even for operations in the East.

The same fate befell a Messerschmitt proposal made by *Probü* chief Woldemar Voigt in June 1942. His idea called for retrofitting the Me 321 with either 12 of Argus' As 014 pulse-jet engines—each with 300 kp thrust—or 24 smaller engines with 150 kp thrust each, plus supplemental RATO packs for a total of 12,000 kp of thrust.

The Me 323 Heavy Lift Transport

Just a few short weeks after the glider's first flight, on 13 March 1941 the RLM established the type number for a motorized version of the Me 321, the Me 323, and awarded a preliminary contract for the development of this new plane. The first prototype, the Me 323V-1a, was born from a glider airframe fitted with four captured Gnôme et Rhône 14N engines whose performance and fuel consumption had been determined previously on the BMW's test bench in Spandau. Karl Baur was at the controls on the prototype's first flight at Leipheim on 21 April 1941—four months after the first glider. The four-wheeled undercarriage was the same as on the glider version, the only difference being that this time it was permanently attached. Taxi handling was good. Over the coming weeks Karl Baur flew the plane to its load limits, which was at about 42 metric tons. Based on these successful flights, in June of 1941 the RLM contracted for the construction of 14 prototype Me 323s, evenly split between four- and six-engined variants. Karl Baur also flew the first six-engined Me 323V-1b (later V-11) in early August 1941. This aircraft had a reinforced fuselage and wing and was fitted with a lower-sitting ten-wheeled landing gear, a feature which facilitated easier loading of the cargo hold. Taxiing along the ground was simple, and full circles could be made with diameters of 70 meters or less. All remaining V-types had flown by April 1942, and in July the V-12 completed the 100-hour milestone for the program. Various other engines were tested as well, including the Alfa Romeo and the Jumo 211.

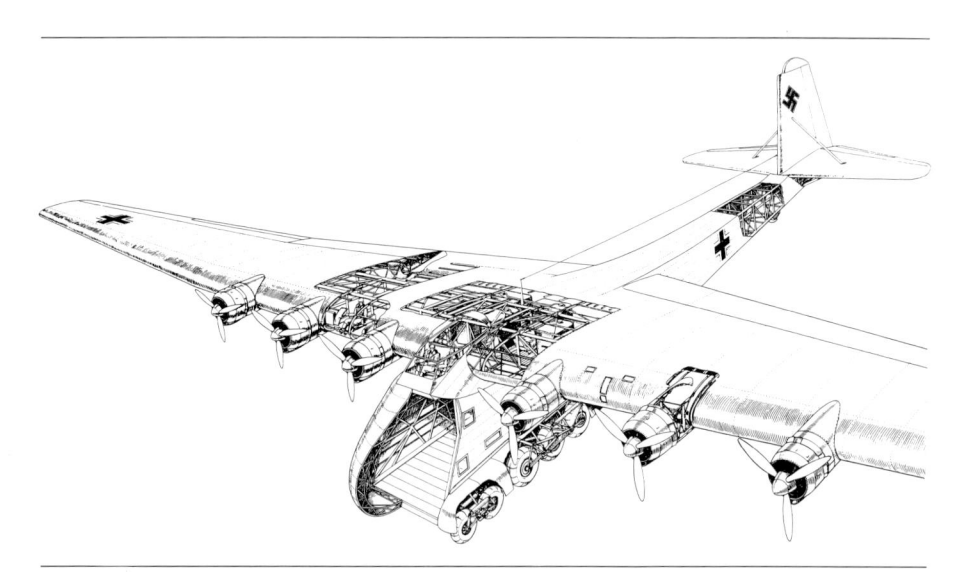

Structural cutaway of a Me 323. The crew was housed in the two-seat cockpit and in the wings.

Three-view of a production Me 323 example with four engines. The first RLM contract called for production to be divided between four- and six-engined versions.

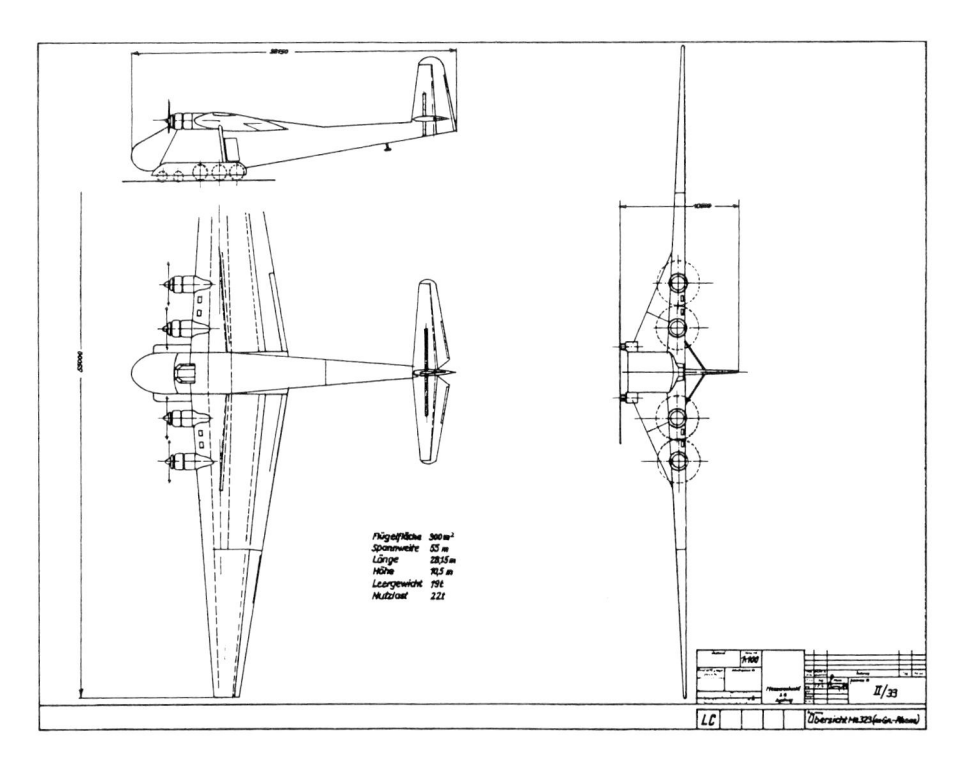

The first Me 323 was a glider modified with four engines, seen here during testing in April 1940 at Leipheim.

The entire "Gigant" program is captured in this photo of the flight test program from the summer of 1942. Taking off is one of the six four-engined models, while a six-engined version banks in on final. A glider variant waits patiently at the end of the runway.

As it was not practical to pull large numbers of 1,000 hp engines from Germany's ongoing production program, both the prototypes and production machines were fitted with "captured" French Gnôme et Rhône 14N engines, rated at 870 kW/1,180 hp. Complete engine units including nacelle, oil cooler, and engine mounts from the Bloch 175 and Lioré et Oliver Léo 45 bombers could be fitted to the Me 323. There were even engines available which rotated in different directions (14N-48 and 14N-49), which were used to offset the tendency to ground loop on takeoff. Those engines on the right wing rotated to the right, while those on the left rotated to the left. Gnôme et Rhône began producing 14N engines again before the stocks of captured engines were depleted.

Design responsibility for the *Giganten* was given to Josef Fröhlich and a small team of about 50 employees at Leipheim. But by the fall of 1942 virtually every qualified worker became urgently needed for the new Me 209, 309, and 262 programs. As a result, ways were sought to hand further development of the Me 323—which Willy Messerschmitt felt still held great potential—over to a partner company. Talks with Dr. Hugo Eckener and Dr. Dürr at the Luftschiffbau Zeppelin eventually led to the LZ initially taking over the developmental department in the spring of 1942, then (in August) the test construction of the Me 323, while production remained at Leipheim and Obertraubling. The LZ carried out project studies of G and H variants of the Me 323 with significantly improved payload capacities. In addition, the LZ researched a *Gigant* with its French partner SNCASO in Cannes which would have double the capacity of production Me 323 planes. This project, ZSO 523 from 1944, was to have had a takeoff weight of 95 metric tons and a wingspan of 70 meters. Further development of the *Giganten* was prevented by termination of production and Germany's withdrawal from France.

Me 323 Production

The first contract from August 1941 called for 300 Me 323s, of which 100 were to be the four-engined version. In February of 1942 the contract was increased to 453 machines, all of which were to be the six-engined type. Ultimately, however, only about 200 aircraft were built, of which approximately 180 were actually ever delivered to the *Luftwaffe*; the rest were mostly destroyed on the ground in bombing raids.

Production began in the spring of 1942 at Leipheim and Obertraubling, and the *Luftwaffe* took delivery of its first planes during the summer. Production of the Me 323D-1, D-2, and D-6 variants ran concurrently, depending on the supply of the French engines and other accessories. Like the Me 321, manufacture of the main components and smaller assemblies took place at decentralized locations. The fuselage framework and wing spars came from the Mannesmann Works in Komotau and Düsseldorf, the ribs from furniture factories in southern Germany, and the landing gear from Skoda in Pilsen. These components were sometimes assembled in the most primitive conditions. While the main components could be built in their own airplane hangars, final assembly had to take place in the open. A small Messerschmitt core team supervised an army of about 2,000 unskilled laborers, many of whom were foreign workers.

Me 323 - Technology and Operations

The Me 323 was able to transport unusually heavy and cumbersome military loads, i.e. equipment which would normally have been shipped by rail or water. Like that of the Me 321 and corresponding to the standard rail gauge, the cargo hold was 11 meters long, 3.15 meters wide, and 3.30 high. It had a solid floor so that armored vehicles could onload and offload under their own power. Typical loads included two medium trucks carrying a two-ton load, an 88 mm Flak with its crew and ammunition, more than 50 barrels of fuel, 9,000 loaves of bread, 60 wounded personnel, and with a middle deck fitted, 130 soldiers in full gear. Maximum additional load was 12 metric tons, or 13 tons for short-haul flights.

The Me 323 D-6 variant had a maximum speed of 280 km/h, a cruising speed of 250 km/h and took 16 minutes to climb to 2,000 meters. All D-variants were rigged to carry eight Walter RATO packs of 4.9 kN/500 kp thrust each, as were all Me 321 gliders. However, the Me 323 only used the RATO packs when taking off in overloaded conditions or from extremely short runways. Maximum takeoff weight was 45,000 kg.

The fuel tanks, holding 5,340 liters of fuel, were located in the wings. The 750 km range on internal fuel alone could be increased by supplemental rubber tanks or fuel drums carried inside the fuselage—with a corresponding loss of payload capacity—to 1,100 km. The centerwing section had to be reinforced to take the engines and its leading edge covered in sheet metal, since the engines often

Construction of the wing center section for a Me 323E with cutouts for the gun positions. The wing in the background has already been partially covered in plywood and fabric.

Taking off at full military throttle and an extra 2,000 kg of thrust, the fourth production Me 323D (RD+QD) roars off on its acceptance flight on 18 September 1942.

A Me 323D (RF+XM) is shown here flying on five engines. A comparison of this photo with a similar air-to-air shot of the M20 reveals the influence that Willy Messerschmitt had on this design as well.

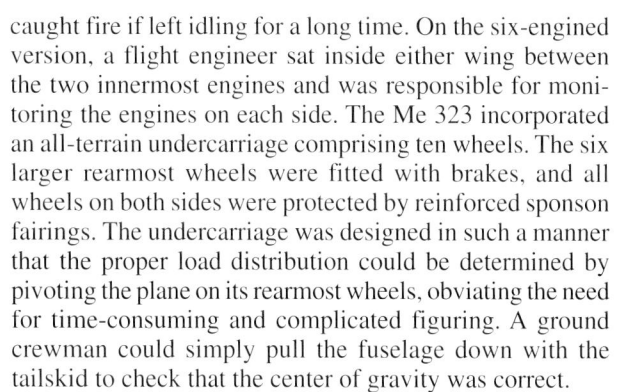

A delegation under Dr. Hugo Eckener (3rd from left) from Luftschiffbau Zeppelin, which visited Leipheim in the summer of 1942, is being given a tour by developmental manager Josef Fröhlich (2nd from left).

caught fire if left idling for a long time. On the six-engined version, a flight engineer sat inside either wing between the two innermost engines and was responsible for monitoring the engines on each side. The Me 323 incorporated an all-terrain undercarriage comprising ten wheels. The six larger rearmost wheels were fitted with brakes, and all wheels on both sides were protected by reinforced sponson fairings. The undercarriage was designed in such a manner that the proper load distribution could be determined by pivoting the plane on its rearmost wheels, obviating the need for time-consuming and complicated figuring. A ground crewman could simply pull the fuselage down with the tailskid to check that the center of gravity was correct.

Defensive armament consisted of five machine gun stations, each with one or two MG 15s. Eight to ten MG 34s could also be mounted in the side windows when transporting infantry. In some instances, field units supplemented the firepower even further by coming up with their own solutions.

The first Me 323 *Giganten* were turned over to the *Luftwaffe* at Leipheim in September 1942, initially for training and schooling the crews. Two transport groups which had previously flown Ju 52s went through conversion training on the Me 323 in October 1942, followed by others,

"Rhinoceros" was the name given to this escort version (weapons platform), which carried 1x HD 151/2, 2x MG 131 and 2x MG 151/2 in the fuselage and 4x HD 151/2 in the wings. At least three of these Me 323 E/WT were delivered to the Luftwaffe.

Loading trials using a Me 323D showing its ability to carry a heavy prime mover and a 15 cm field howitzer. Numerous types of army equipment made up a comprehensive evaluation program for testing the transport capability of the "Giganten." With a length of 11 meters, a width of 3.15 meters and a height of 3.30 meters, the cargo bay could accommodate just about anything.

and these were all sent to Naples in November of 1942. From there they flew to Tunis and Bizerte in support of the Afrika Korps. A round trip flight time from Naples was about seven hours. They carried with them 2,400 liters of fuel, and on the return leg the crews used this to refuel at an airfield in Sicily. To a great extent the *Giganten* were at the mercy of British fighters stationed in Malta, and accordingly suffered increasing losses at their hands. The supply flights to Tunis had to be abandoned in March of 1943; operations to Sardinia then began in May, followed by withdrawal operations from Corsica. The remaining Me 323 units were sent to the Warsaw area, and later to airfields in Hungary and Romania, as well. By the end of 1943 the fuel situation had become so critical that several squadrons were disbanded. In May of 1944 there were only two *Gruppen* operating the Me 323 *Gigant* left. By the autumn of 1944 even these units had to be dissolved.

The life of the airplane had originally been designed for 200 flight hours or 100 takeoffs, yet even after the first few operational missions this was revised to 250 hours following a partial overhaul, eventually jumping to 350 hours. The Messerschmitt Me 323 heavy-lift transport was the largest, most economical, and probably the most profitable transport plane of the Second World War.

This comparative diagram from 1944 shows the economic advantages of the stop-gap Me 323 over the Ju 52/3m, the most common transport aircraft of the day.

Field maintenance on the Me 323 could be carried out using specialized maintenance vehicles and cranes.

There are two sides to the brief history of these "giants," the largest glider transports in the world and the largest transport airplanes of World War II. For Willy Messerschmitt and his company the type's development and construction was a great feat of organization and improvisation, with the whole project being integrated in short order into the other production programs then underway.

For the *Luftwaffe*, however, the Me 323 was

"...nothing more than a stopgap solution. The huge plane was slow and quite cumbersone, and despite its heavy armament more often than not drew the short straw in combat situations, as evidenced by TG 5's experiences in supplying Tunisia in 1942 and 1943. In any event, the aircraft did not meet up to the *Luftwaffe*'s expectations. It was not equipped for IFR flights and could only be flown by pilots

This advertisement, drawn up in 1943, shows that Willy Messerschmitt also had dreams of marketing the "Giganten" in the civilian transport arena following the war.

with considerable experience. Takeoffs and landings required an exceptional amount of physical exertion."

This was the verdict in a postwar assessment by Fritz Morzik, *General der Transportflieger* and the person responsible for overseeing the operations of these aircraft.[3]

The *Giganten* were nevertheless remarkable forerunners of today's military heavy-lift aircraft, from the Transall (first flight 1963) to the Galaxy (first flight 1968).

Me 321 and Me 323 Variants and Projects

Me 321A-1	Heavy-lift glider with single-seat cockpit.
Me 321B-1	Heavy-lift glider with two- to three-seat armored flight deck, 8 Walter RATO packs, braking parachutes, 4x MG 15 machine guns.
Me 323A	Four-engined model (Me 323V-1a).
Me 323B	Six-engined unarmed model (Me 323V-1b).
Me 323C	Production version, four-engined (except for five prototypes not built)
Me 323D-1, D-2, D-6	Production version with 6x Gnôme Rhône 14N (from Bloch 175 and LeO 45), 3 right-rotating and 3 left-rotating, with Ratier variable-pitch or Heine fixed-pitch wooden propellers. Armament: 7 positions with MG 15 and MG 131 guns. Takeoff weight 43 metric tons, of which 11 tons was payload.
Me 323E-1	Production version with 6x 14N engines. Armament increased by two HDL 151 turrets in the wings.
Me 323E/WT	Heavily armed escort, only one built from modified Me 323E. Armament: 7x MG 151/20 and 2x MG 131, 17 man crew, no payload capacity.
Me 323F	Follow-on development of E with 6x Jumo 211F engines for 11.6 metric ton capacity and 54 ton takeoff weight, trials with V-14.
Z Me 323G	Designed by Luftschiffbau Zeppelin in 1943 with 6x Gnôme Rhône 14R engines and a capacity of 12.4 tons and 54 ton takeoff weight.
Z Me 323H	New LZ design from 1944 with 6x BMW 801 engines for a capacity of 16.2 tons at a takeoff weight of 58 tons.
Z SO 523	Heavy-lift transport designed by LZ and SNCASO in 1944 with 6x Gnôme Rhône 18R engines, 70 meter wingspan, 40 meter length, 456 m² wing area, 46.8 ton payload capacity, and 95 metric ton takeoff weight.

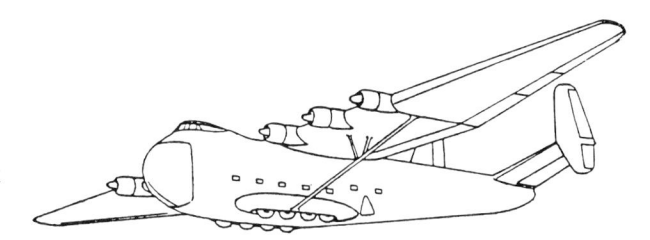

The basic lines of the Me 323 can be seen in this Z SO 523 project from 1944.

---

[3] Fritz Morzik, Gerhard Hümmelchen, Die deutschen Transportflieger im Zweiten Weltkrieg, Frankfurt/Main, 1966, p. 33/34

Messerschmitt Jets

Me 262, First Operational Jet Fighter

On his 80th birthday Willy Messerschmitt was asked which of his airplanes he considered to be the most important, and he replied: "the Me 262"—because the work on this design provided the aviation world with such a rich wellspring of knowledge. In the annals of aviation, the Me 262 is unequivocally considered to be a milestone pioneering development and a role model for numerous aircraft types subsequently appearing on both sides of the Iron Curtain. The Me 262 enjoyed a considerable speed advantage, flying over 200 km/h faster than the fastest piston-powered fighters of the day. The path from preliminary study in the fall of 1938 to its first combat in July 1944 was punctuated by many technical delays (particularly with engine development), errors in judgment, problems caused by the war and, not least, through Adolf Hitler's absurd decree that this thoroughbred fighter be converted into an ultra-high-speed *Blitzbomber*.

The P 1065 Pursuit Fighter Project

At a meeting with the RLM in Berlin in early 1938 Willy Messerschmitt first learned of Heinkel's ongoing developmental work with turbojet engines in Rostock. By this point in time, Dr. Hans von Ohain's HeS 2 design had already run during bench testing, and trials with the combustion chamber on the 5 kN/500kp HeS 3 were getting underway. During the same discussion, Willy Messerschmitt was also informed of similar work going on at BMW and Junkers.

It goes without saying that Messerschmitt and his project team, at the time still headed by Robert Lusser (Woldemar Voigt took over when Lusser went to Heinkel in 1939) immediately began formulating plans on how to make use of the new jet engines. By that fall the *Probü* had drawn up several preliminary drafts of single- and twin-engined fighter aircraft, and it quickly became apparent that the main question would be whether to install the powerplants into or onto the airplane. The foundation for these studies was the BMW P 3302 engine, still in its developmental stage, which was expected to have a thrust of 6 kN/600kp and a diameter of 600 mm. This development work took place at the Brandenburgische Motorenwerke (Bramo, a member of the BMW group from July 1939 onward) in Spandau under the supervision of Hermann Oestrich.

With the help of the RLM's "Preliminary Technical Guidelines for Jet-Powered High-Speed Fighters" publication from 1939, the *Probü* drew up a pursuit fighter project

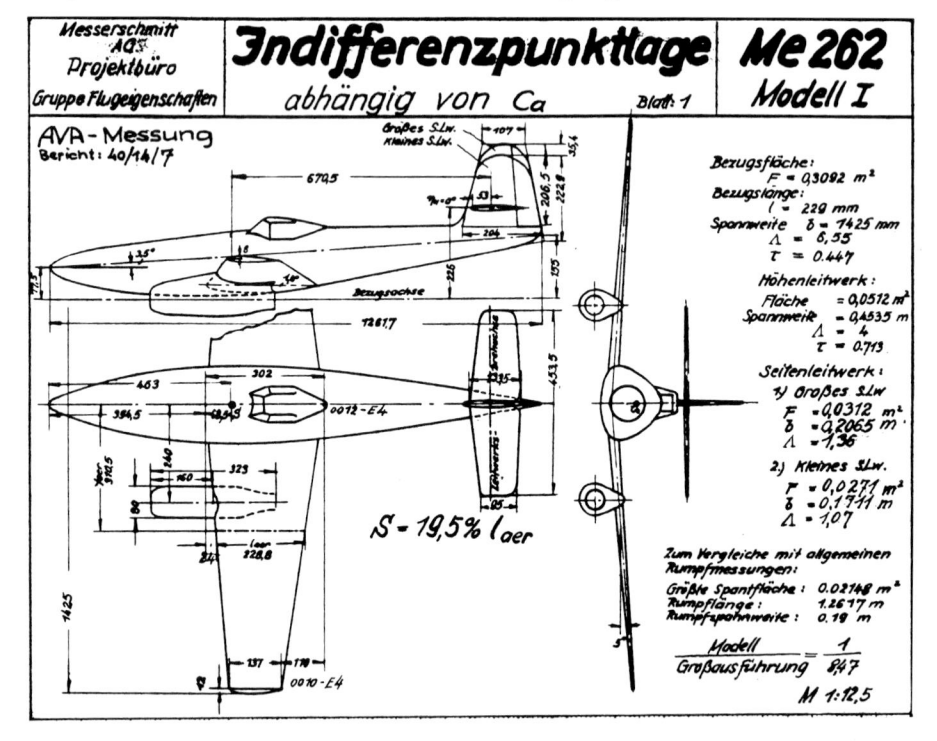

Initial design of the P 1065 as it was tested in the wind tunnel at AVA Göttingen in 1939/1940. The tapered wing design had a rectangular fuselage cross section, which the evaluations confirmed as being the best form. During these trials, the engine nacelles were fitted both beneath as well as above the model's wings.

Wind tunnel model 2 during tests in August 1941 at AVA Göttingen. The model here has the later wings with the outer sections swept 18%. The engine layout being evaluated here, with the nacelles buried in the wings, was not featured on the final design. Instead, the designers took the simpler approach (from a design standpoint) of suspending the engines beneath the wings.

with two BMW turbojet engines. At full thrust it was expected to have a speed of 900 km/h. The aircraft was to incorporate a tapered wing with an area of 18 m² and a span of 9.4 m, with a fuselage length of 9.3 m. This was the project submitted to the RLM on 7 June 1939 under the designation P 1065 (at first sometimes abbreviated to P 65).

Work on the project intensified even further once updated data on the BMW engine became available in the latter half of 1939. Two options were studied in detail: a midwing design with oval fuselage cross-section (roughly equivalent to Willy Messerschmitt's hand-drawn sketch of 17 October 1939 [see previous section]). The other design, strongly reflecting the influences of Messerschmitt *Probü* members Wolfgang Degel, Karl Althoff, and Rudolf Seitz, which envisioned a low-wing plane with triangular cross-section. It was soon obvious that this approach offered so many more advantages that it was used for the project and was what later gave the Me 262 its typical "face." Compared with the oval fuselage, this project had a surface area smaller by some 2.3 m², it was structurally more robust, and provided better options for installing equipment and undercarriage, especially for the nose gear being considered even at this early stage.

In conjunction with the flurry of activity going on in the *Probü*, intensive work on the P 1065 was occurring at other locales, both within the company and outside of Augsburg. The mockup department built a visual mockup of the airframe and a cockpit mockup, which representatives from the RLM examined for the first time on 19 December 1939 and found to their liking. With BMW there was discussion of converting a Bf 110 into a flying testbed for the new jet engine. As the dimensions and weight of the P 3302 had yet to be finalized, there were problems with incorporating it into the P 1065 design. Accordingly, BMW drew up two variants: the P 3302 for an underwing configuration and the P 3304 for a nacelle imbedded in the wing.

Starting in November 1939 the *Probü* began feeding the *Kobü* with individual subassembly blueprints for the planned V-1 prototype. These initially called for the P 65 (with its initial straight tapered wing and 10.40 configuration) to be fitted with a tailwheel undercarriage. Wolfgang Degel recalls that the team didn't want to have to deal with the risks involved in two new technologies simultaneously, namely the nose gear and the jet engine, for which only limited data was available. In the "project transfer memo," the engines and nose gear are referenced as follows:[2]

"The V-1 is fitted with two engines (TL 600, P 3302), each weighing 533 kg including starter units....Which prototype is to receive the nosewheel and what loads the nosewheel is to be stressed for has not yet been decided."

The P 1065 Officially Becomes the Me 262

On March 1, 1940, under the RLM type designation Me 262, the Messerschmitt AG was officially awarded the contract for development and assembly of three sample planes plus one airframe for stress testing. The cockpit mockup was the site of discussions examining the cockpit's armor, the layout of the pressurized cockpit, and the installation of the ejection seat. Fitting of dive brakes and the location of braking parachutes were also addressed. In May 1940 Messerschmitt provided the RLM with a modified design for the P 1065 project which showed the turbojets in nacelles situated beneath the wings. The arrangement of the spar bridge had proved problematic with the engines imbedded in the wings. Furthermore, it had been discovered that the BMW engine diameter would be 690 instead of the previously claimed 600 mm due to changes made to the compressor. In addition, BMW intimated that the engine mass had increased, meaning that the outer wing sections would have to be swept to maintain the center of gravity.

[1] (Messerschmitt AG, project dept.) Vorteile der dreieckigen Rumpfform. Augsburg 27 Oct 39
[2] Messerschmitt AG Augsburg, Projektübergabe P 65, Projektrichtlinien V-1 dated 27 Nov 39

Me 262 Straight Wing Aerodynamic Trials

A series of aerodynamic studies and wind tunnel trials took place for the P 1065 project at the AVA in Göttingen from 1939 to 1941. Department director for aerodynamics at Messerschmitt, which also encompassed flight handling and performance, was Riclef Schomerus. He was joined by Ludwig Bölkow in March of 1939, who then became responsible for aerodynamics in the transsonic region. One of his first tasks was to select the wing profile for the Me 262. He first recommended and studied an extremely thin profile, running from 9% at the root to 6% at the tip. Willy Messerschmitt doubted whether such a thin wing could accommodate the retractable undercarriage and the mechanisms for operating the leading edge slats. Ludwig Bölkow solved the problem by inventing the so-called *Kippnase* leading edge. Results of the trials in the Göttingen wind tunnel using the thin profile—called 0009-E-4—may have been positive, but this did not prevent the Me 262 from being built with a conventional profile of 12% thickness and with standard leading edge slats. The reason behind this was that the fighter's development was under enough time pressure as it was without compounding the risks further.

The same thinking was behind the swept wing. Although wind tunnel testing had convincingly shown that wings with sweeps between 30 and 45 degrees had much better drag characteristics when approaching the speed of sound, it was decided to take the more conventional route. Consequently, only the outer section of the Me 262's wing was swept, and this was done only to readjust the center of gravity. *Probü*'s alternative triangular fuselage proposal also derived its final shape from the wind tunnel test results. The wing/fuselage join area of the model was filled with plasticene to counter the local airflow's overspeed.

Assembly and First Flights of the Me 262V-1

With regard to staying on schedule, Willy Messerschmitt had told the RLM in early 1940 that there would be no problems with the V-1 airframe flying by the end of July 1940 as long as BMW could meet its promised delivery date of 1 July 1940. It soon became apparent, however, that BMW would not be able to meet its deadline. The engine had run into a string of problems with the compressor and combustion chamber while on the test bench, making design changes a necessity.

This in turn delayed construction work on the Me 262V-1, which didn't begin until July 1940. At this time, the Schmidt-Argus pulse-jet engine seemed suitable for powering aircraft, and Rudolf Seitz in the *Probü* suggested installing six to eight of these (each with 150 kg of thrust) in the Me 262V-2, also under construction. Initially, three jets would be hung beneath each wing parallel to the wing chord. Later three to four pulse-jets would be bundled together into nacelles. (Takeoff weight with eight jets would have been 3,870 kg).

Another option for testing the Me 262V-1 under jet propulsion was discussed between the RLM, the *E-Stelle* Rechlin, and the Messerschmitt and Walter Companies in April 1941: installing two Walter rocket engines, each with 750 kg of thrust. However, this engine was still in development and would not be available as soon as originally hoped. It wasn't until 1944/45 that the project was realized in another form as the Me 262 *Heimatschützer* designs. In the event that the BMW engine would suffer further setbacks, Prof. Messerschmitt had decided to fit the Me 262V-1 with a conventional engine so that at least the team could test the flight handling qualities of the design. This took place in early 1941; it involved taking the complete engine assembly of a Jumo 210G from a Bf 109D, including its vari-

[3] W. Messerschmitt, Termine P 1065. Letter to the RLM, dept. LC2 of Hauptstabsing. Reidenbach, Berlin, dated 19 Feb 40

Me 262V-1 (PC+UA), with a Jumo 210G powerplant taken from a Bf 109D, as flown by Fritz Wendel on 18 April 1941 in Augsburg. Takeoff weight was 3,155 kg, wingspan 12.35 m, and length 10.46 m. The first flight with BMW 3302 jet engines fitted in addition to the piston engine, on 25 March 1942, was unsuccessful. The first jet-powered flight using Jumo 004 engines took place on 19 July 1943. Fate caught up with the V-1 when it crash landed at Lechfeld on 7 June 1944 after its 95th test flight.

Me 262V-3 during the program's first all-jet powered flight on 18 July 1942 in Leipheim. Again, Fritz Wendel was at the controls. The two Jumo T-1 engines were also prototypes, being the V-8 and the V-9.

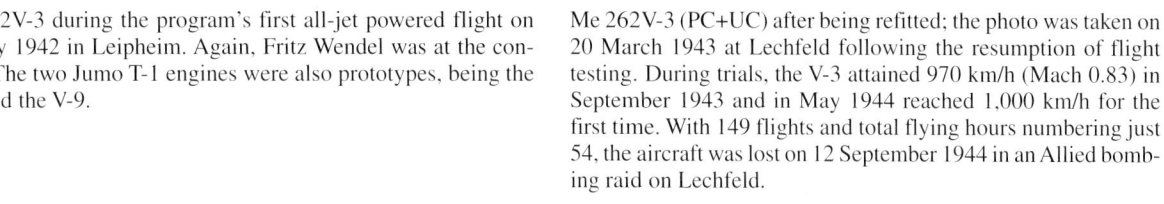

Me 262V-3 (PC+UC) after being refitted; the photo was taken on 20 March 1943 at Lechfeld following the resumption of flight testing. During trials, the V-3 attained 970 km/h (Mach 0.83) in September 1943 and in May 1944 reached 1,000 km/h for the first time. With 149 flights and total flying hours numbering just 54, the aircraft was lost on 12 September 1944 in an Allied bombing raid on Lechfeld.

able-pitch propeller, and fitting it into the nose of the Me 262V-1 under a new nacelle. The Jumo was rated at 550 kW/750 hp, and in this configuration the plane's takeoff weight was 3,155 kg.

On 18 April 1941 Fritz Wendel took off in the Me 262V-1 for the first time from Augsburg. He subsequently reached a level speed of 450 km/h and 800 km/h in a shallow glide. Following this flight, test pilot Karl Baur carried out several further runs to the end of July in order to assess the type's flight handling, particularly its rudder effectiveness. In mid-November the BMW P 3302 finally arrived. It was a prototype, and had an output of just 4.5 kN/450 kp instead of the promised 6 kN/600 kp. Initially, there was little faith in the BMW engines, and the Jumo 210 motor stayed put. Flying with both jets and piston engine, Fritz Wendel took off on 25 March 1942 at the controls of the V-1 on its second "maiden flight" from Augsburg-Haunstetten. Just shortly after liftoff, however, at an altitude of about 50 meters, both BMW engines suddenly stalled out. With only the Jumo 210 running, Wendel was forced to make a hard landing on the airfield and the aircraft was damaged. BMW's engines, it seemed, were still in need of major improvements, which basically necessitated a complete overhaul of the design and ruled out any talk of setting delivery dates.

Evaluation with Junkers Jet Engines

With the BMW engine's disheartening results, the RLM reduced its July 1941 order from five test planes and 20 pre-production models to just the five V-types. A decision on the pre-production aircraft would only be reached after flight testing had been completed. The design team then focused on Junkers Jumo T 1 turbojets (later designated the Jumo 004), then undergoing development and testing under the supervision of Dr. Anselm Franz in Dessau.

The first successful test bench runup of a Jumo T 1 had already taken place on 11 October 1940. Its first flight occurred on 15 March 1942 in a Bf 110 testbed modified with the assistance of Messerschmitt. The first engines, themselves prototypes, ready for installation in the Me 262 were shipped to Augsburg in the following months. These engines, with a larger diameter of 765 mm and heavier at 720 kg had an output of 8.2 kN/840 kp static thrust. They were fitted in the Me 262V-3. The aircraft was brought to Leipheim for secret testing (where the Me 321 had been produced and where there also was a concrete runway available for use).

On 17 July 1942 Fritz Wendel began taxi trials and soon found that the elevator controls were ineffective because the delay of the wake behind the wings prevented sufficient air from flowing over the tailplane (not encountered with the Me 262V-1 because of the propeller wake from its piston engine). However, Wendel discovered a trick to overcoming this problem: after a run of about 500 m he would briefly punch the brakes hard, causing the tail to rise up and bite into the airflow. Using this rather unconventional method, on the following day (July 18th) he flew the first two flights in the Me 262 under jet power alone. During the second flight he attained a speed of 720 km/h. Other than the problem with the tailsitter undercarriage, a vibrating stick caused by compression and rudder position, and minor problems with braking on rollout due to the engine's powerful idle thrust, Fritz Wendel could find nothing wrong in the design. Wendel once stated after the war that no first flight filled him with such enthusiasm as did that of the Me 262.

239

On 11 August 1942 the *E-Stelle* Rechlin was scheduled to test fly the Me 262V-3 on its seventh flight. At around 12 o'clock on what was a hot and muggy day, Heinrich Beauvais took off according to Wendel's instructions. Although the trick with the brakes worked, Beauvais was unable to build up enough speed. He shot off the end of the runway and the plane, its undercarriage and engines shorn off, eventually came to rest in a cornfield. The pilot was uninjured, but the plane was a write-off with 70 percent damage. The cause of this abortive takeoff attempt was most likely due to the fact that turbojet engines produce less thrust in hotter weather, something which was not generally known at the time. Whether a nosewheel (employed for the first time on the competing He 280, first flight 30 March 1941) would have helped in this instance is a matter for conjecture. In any case, Willy Messerschmitt had planned to fit a nose gear beginning with the V-5 (the *Amt* had stipulated a nosewheel from the V-11 onward, and refitting the V-1 through V-4 would have been a costly undertaking at this stage).

Because of this accident and a lack of turbojet engines—even Junkers could only supply them on a sporadic basis—no further jet flights with the Me 262 took place until 30 September 1942. Willy Messerschmitt summed up the Me 262's first year in a 1944 investigation into the Me 262 construction delays as follows:[4]

"The first phase of development is chiefly highlighted by the fact that engine manufacturers as well as those offices responsible were overly optimistic about the development of turbojet engines and that setbacks were not taken into account with the necessary farsightedness. Notwithstanding, the RLM was unable to come to an early decision for making production preparations."

Fritz Wendel was able to continue the Me 262's evaluation program with the first flight of the V-2 on 1 October 1942 at Lechfeld. Lechfeld had a longer concrete runway than Leipheim and thus offered more safety when taking off and landing. For nearly six months the V-2 was the only test plane available, and in early 1943 it was temporarily pulled from the program in order to fit it with a leading edge insert between fuselage and engine, thus facilitating better lift distribution along the inner wing. This fillet gave the entire leading edge the same sweep angle the length of the wing. This made for tighter turns as well, and dispensed with the need for tapping the brakes on takeoff for those aircraft equipped with a tailwheel. This improvement was retrofitted to all V-types and incorporated into the design of production models.

The V-2 crashed on 18 April 1943 on its 48th flight. Back from repairs, the Me 262V-3 had rejoined the test program on 20 March, and the V-4 expanded the program even further on 15 May (it flew until 25 July). The V-1, refitted with Jumo 004 engines, also joined the team on 19 July. It was the third maiden flight for this, the first Me 262 airplane, and a fourth followed on 29 July when the V-1 became the first Me 262 to fire its MG 151/20 guns in the air, proving the type's suitability as a gun platform.

[4] W. Messerschmitt, letter to *Oberkriegsgerichtsrat der Luftwaffe* Dr. Schleich at the RLM on 31 Mar 44

On 18 July 1942, while the V-3 was making its first flight with turbojets, V-2 (PC+UB) waits here for its engines outside the test construction hangar in Augsburg. Initially to have been fitted with six Schmidt-Argus pulse jets, these were dropped in favor of Jumo turbojet powerplants. It wasn't until September, however, that two Jumo T-1 engines became available. Once these were installed, Fritz Wendel flew the V-2 on its maiden flight at Lechfeld on 1 October. On 17 April 1943 Hauptma*nn* Wolfgang Späte became the first Luftwaffe officer to fly the type in the V-2; on 18 April company pilot Wilhelm Ostertag crashed in the V-2 on its 48th flight at Lechfeld.

Closely involved with the history of the Me 262 were the company as represented by Willy Messerschmitt (from right), the Luftwaffe with Adolf Galland (General of the Fighters, left), and the contractor, represented by pilot engineer Heinrich Beauvais from the Rechlin Test Center (center), seen here at a meeting in April of 1942.

On 6 June the V-5 flew, the first to be equipped with a permanently fitted nosewheel (taken from the Me 309). The first Me 262 with retractable nose gear was the V-6, which followed the V-5 four months later on 17 October (thereby beating the *Amt*'s deadline, which had called for it to be fitted by the V-11). Adding in the Me 262V-7's first flight of 20 December 1943, seven V-types had flown up to this point, and Junkers had supplied not quite 20 jet engines.

Luftwaffe Officers Enjoy Flying the Me 262

Four years after development had begun, Me 262 jet operations were beginning to take on a sort of routine at Lechfeld by the spring of 1943. Willy Messerschmitt was now able to offer *E-Stelle* and *Luftwaffe* pilots the opportunity to fly the plane. In April and May of 1943 alone, the V-2, V-3, and V-4 were flown by six non-company pilots in Lechfeld, all of whom expressed their enthusiasm for the new flight experience which jet propulsion provided compared to propeller driven aircraft. First up had been *Hauptmann* Wolfgang Späte from the *Erprobungskommando 16* (Me 163) on 17 April. He recalls:[5]

"Hardly were you in the air before the Me 262 gave the impression that it could effortlessly clear all previously existing speed barriers, that you'd discovered a new, most graceful way to move through the third dimension.

...It was like a revelation to me: this could be our salvation in defending the Reich! Within a short period of time, I'd made my decision. I made a top priority long-distance call to the office of the *General der Jagdflieger*. By good fortune, *General* Galland was able to be reached immediately. With a few brief sentences I gave him my impressions and indicated that decisions had to be made at the highest levels. 'Therefore,' I continued, 'I suggest that you come to Lechfeld as quickly as possible and fly the plane for yourself. Then you can put things in perspective for the *Reichsmarschall* based on your own experience, so that he can then exercise all the authority with the Führer which is now required.'"

The next day revealed how close enthusiasm for the new technology and its imponderability were to each other when Wilhelm Ostertag crashed due to unknown circumstances in the V-2 just a day after Wolfgang Späte had flown it. A new company test pilot, Ostertag had only been instructed on the V-2 on 16 February 1943. He thus became the first victim of the Me 262 program.

On 22 May *General* Adolf Galland had the opportunity to fly Me 262V-4. Like *Hauptmann* Späte before him,

[5] Wolfgang Späte, Der streng geheime Vogel. Erprobung an der Schallgrenze. Munich 1983, p. 149

Galland was almost euphoric in his enthusiasm. It was he who made the historic comment "It is as though an angel were pushing me along." Immediately after his flight, Galland sent a telegram to *Feldmarschall* Milch with the following contents:[6]

"The Me 262 is a major success which guarantees us an operational advantage of unimaginable proportions, assuming the enemy continues flying with piston-engined aircraft. From a flying standpoint the airframe makes quite a good impression. Its engine is quite satisfying, except on takeoff and landing. This aircraft opens up completely new tactical possibilities for us."

He discussed the matter further in his book "Die Ersten und die Letzten:"[7]

"The Me 262 jet fighter, the fastest fighter plane in the world, was a fact. I'd just flown it. And I knew that with it we could beat any other fighter. Naturally it had its 'teething troubles'....As a result, a joint proposal was drawn up to immediately start production of an initial series of 100 aircraft which would then be used for concurrent technical and tactical evaluation. Such an undertaking contradicted the methodical approach common to German aircraft construction up to that time. We wanted to immediately start paving the way for eventual series production, in the meantime utilizing the time with the 100 planes to gain the experience needed until production actually got underway....

This proposal, along with my flight report, was drawn up and signed on the spot. A copy was sent to Milch, and that very day I took the original with me to Göring at Castle Veltenstein. The report concluded with the statement: This almost unbelievable technological superiority is the means by which we can tip the odds in our favor in the battle for air superiority over the Reich and later over the front lines, despite our fewer numbers. No effort or thought should be spared in making immediate arrangements for series production, and that production should begin as soon as possible. Instead of an Me 262, we can do without two or three Me 109s for air defence if it becomes necessary from a production standpoint."

A somewhat more tempered approach—particularly with regard to series production—was taken by the *E-Stelle* Rechlin's Heinrich Beauvais in his reports on flights with the V-2 and V-3:[8]

"The tremendous speeds attainable makes the immediate introduction of the Me 262 seem urgent. However,

[6] Adolf Galland, Die Ersten und die Letzten. Die Jagdflieger im zweiten Weltkrieg. Darmstadt 1953, p. 348-349
[7] Galland, p. 351

241

the state of the engine's and airframe's development is such that, although the aircraft can be flown quite well (e.g. the easy running of the engine is most welcome!), its full suitability for frontline operations cannot be confirmed if we apply the same criteria as we did for the Me 210 in the fall of 1942. We cannot rule out the possibility that serious problems will arise during production if a solid information base isn't gained from testing and/or introduction."

Galland's elated portrayal of his Me 262 flight to the commander-in-chief of the *Luftwaffe* was contagious, and Hermann Göring allowed himself to be infected by it.

Göring, after all, had been a fighter pilot during the First World War. (He had not, however, seen the Me 262 fly for himself, a situation which he remedied during a visit to Rechlin-Lärz on 24 July when the Me 262V-4 flew as part of a demonstration of new aircraft.) Göring telephoned Milch. Both were relatively certain that Hitler, whom it was necessary to consult in such situations, would agree. Milch—whom Galland had also informed beforehand—acted at once: at a high-level conference of the RLM's office chiefs on 25 May, it was specified that the Me 262 would go into production immediately and the Me 209 would be set aside in its place. *Oberst* Petersen was named commissioner for the Me 262.

Just a short while later, in June, Milch was forced to take back much of his earlier decision. For the time being, the Me 209 was to remain in the program, thus tying up considerable developmental and construction resources which would otherwise have been available for the Me 262. Hitler had been strongly opposed to the idea; his disappointment with the many promises Hermann Göring and the *Luftwaffe* leadership had made—particularly with regard to the He 177 heavy bomber—prompted him to decree that there were to be no preparations made for Me 262 production and that technical evaluation was to be accomplished exclusively with those prototypes already built or currently under construction. (This order was soon revoked, however.)

The Long Road to Series Production

RLM planning had been far from consistent ever since the Me 262V-3 first successfully took to the air on 18 July 1942. It ranged from five to ten V-types and a pre-production run of 20 to 50 planes per month starting in 1944, with development and construction taking place outside Messerschmitt (at Siebel). At the same time, similar plans also affected the He 280 competing design. Fitted with Heinkel engines, this jet-powered aircraft had been in testing since the 30th of March 1941, but was crossed off the development program by Erhard Milch on 27 March 1943 in favor of the Me 262 (not least because the Me 262 and its Jumo engines had proven to have better performance and a greater range).

On the other hand, since late 1942 Messerschmitt had repeatedly told the RLM that he would not even be able to keep to the timetable for the ten prototypes because the Jumo turbojets were not being delivered as scheduled. The undercarriage, made by the Elektron-Co (EC), was also not being shipped on time. Additionally, he was short 600 workers for the planned pre-production series, expected to start up in January 1943.

Messerschmitt also drew the *Amt*'s attention to the fact that the Me 262 design, now over three years old, would have to be reworked for production and be brought up to the latest standards in terms of its systems and armament, something which would necessitate additional technical engineers. Accordingly, he submitted new specifications incorporating these changes on 25 March 1943. These included plans for four to six MK 108 cannons to be carried in the nose gear equipped Me 262.

This, too, was the first time Messerschmitt offered the Me 262 as a fighter-bomber as well, carrying a single 500 kg or two 250 kg bombs, or a single BT 700 torpedo on streamlined pylons to be situated beneath the forward fuselage. This version would require the removal of two of the plane's guns. The paper was reworked yet again for the specifications of 10 August, followed by a supplement on 10 September 1943 outlining the entire spectrum of Me 262 variants, many of which became a reality before the war's end with only minor changes to the basic model. This concept was to a large extent in harmony with Messerschmitt's proposal at the beginning of the year which called for a single basic model spawning all variants needed for operations (see P 1090).[10]

The Me 262 program immediately found itself running into resource difficulties in light of the fact that Erhard Milch had stopped the Me 309 (the Me 109's intended successor) and ordered production of the alternative Me 209 (q.v.) in early 1943. As the Me 262's significance slowly began to dawn, two lines of reasoning began to emerge from those involved—both inside the company as well as outside: the one group did not want to dispense with the current stock of fighters (Bf 109, Fw 190, Me 209). Willy Messerschmitt in particular took the viewpoint—even expressing it to Hitler personally on 27 June—that the Me 209 must go into production, especially in its high-altitude variant, because jet engine development had not yet matured. Not only that,

---

[8] E-Stelle Rechlin, Me 262 - Stand der Flugerprobung und Nachfliegen. Report no. 3 from 29 Jun 43

[9] Mtt-AG, Augsburg, (W. Degel/W. Voigt) Me 262 project description, dated 25 Mar 1943

[10] Mtt-AG, *Probü* (K. Althoff), Me 262 Verwendungsmöglichkeiten, Projektbaubeschreibung, supplement to description from 10 Aug 43. Augsburg, dated 11 Sep 1943

but high-altitude flight was an entirely open field. In addition, jet engines consumed fuel at an enormous rate. The other line of thinking championed a concentration of efforts on the Me 262. Adolf Galland fell squarely in this camp following his Me 262 flight on 22 May and only wanted to have those piston-engined fighters built which were absolutely necessary. Milch fell in behind Göring's and Galland's wishes and stopped the Me 209. Consequently, Messerschmitt began the Me 262 "program planning" at once, which was completed by mid-June 1943 and, to avoid any further delays, made tough demands on the RLM. By way of introduction it stated:[11]

"On 5/25/43 *Generalfeldmarschall* Milch decided that the Me 209 is to be set aside in favor of the Me 262 and that special measures are to be taken with Me 262 construction with the goal of producing 100 machines by the end of 1943. The Mtt. AG investigative report in the enclosed portfolio shows that this target cannot be attained, but that 100 aircraft may be produced by mid-May 1944 provided certain conditions are met. The basic prerequisites for this planning are:
1. Prioritizing the Me 262 above all previously running programs (including plant operations and indirect suppliers)
2. Complete protection for the work force (against military induction), including management.
3. The RLM to accept all risks inherent in such a rapid production startup."

Other stipulations included assigning 265 designers, toolmakers, and jig builders for pre-production, expanding the production facilities to include making the Gablingen airfield available for manufacturing the wings, expanding the flight test program at Lechfeld, getting Messerschmitt GmbH Regensburg involved, planning material requirements and transportation requirements, and ensuring availability of turbojet engines. For the subsequent large-scale production at Augsburg and Regensburg and at the WNF and Erla plants, Augsburg/Gablingen would require 4,700 laborers, and Regensburg would need a further 6,300. Of these, 20 percent would have to be skilled workers.

Only under these conditions would full-scale production of the Me 262 be possible without seriously impacting the Me 410's ongoing production at Augsburg and the Bf 109's production at Regensburg.

The RLM approved this plan in July of 1943, but was unable to fulfill it in detail. One critical demand which could not be met was that of skilled and unskilled workers for jig construction, which had been moved to Kottern. Production got underway using temporary assembly jigs, some of which were made of wood, causing precision and interchangeability to suffer accordingly. This in turn impacted on the flight handling and performance of the Me 262, and field units complained incessantly about these inadequacies, which required much corrective action on their part.

The final layout for the first 100 aircraft (pre-production series)—all fighters—was fixed at a meeting with the RLM on 28 May 1943. Features of the all-metal monocoque aircraft were tailored back in comparison with the original design. These involved:[12]

- *engines:* Jumo 004A, later 004B
- *fuel system:* two 600 liter armor protected tanks
- *armament:* 3 MG 151/20 with 320 rounds each. Parallel with this would be construction of a removable section with 6 MK 108 (30 mm) cannons, or just 4 Mk 108s with installation of the Jumo 004B
- *armor:* protection at the back, plus head and upper torso
- *airbrakes:* dropped
- *ejection seat:* dropped
- *undercarriage:* 770x270 mm for the main gear and 650 x 150 mm for the nose gear initially. Thus, 5.5 metric ton authorized load at rest. Improved landing gear with 830x300 mm tires.
- *assisted takeoff:* Rheinmetall-Borsig solid fuel rockets.
- *pressurized cockpit:* airtight components and configuration, but without pressure generators.
- *structural integrity:* H5 with 5,000 kg weight after reaching an altitude of 6,000 m. Maximum permissable speed at low level limited to 900 km/h.
- *flight handling:* in comparison with Me 262V-3, 40% reduction in elevator forces, 50% for ailerons.

Momentous Involvement

The first operations using a fighter in the fighter-bomber role (*Jagdbomber*, or *Jabo* for short) took place during the Battle of Britain with Bf 109 Es due to the fact that German bombers had suffered major losses at the hands of the British air defense network. Now, in early 1943, Adolf Hitler again mandated that all fighters be configured for the fighter-bomber role (probably in anticipation of an Allied invasion of the Continent). Willy Messerschmitt had envisioned this role for the Me 262 in any case, and in March 1943 first drew up a *Jabo* variant with a single 500 kg or two 250 kg bombs. He wasn't given an RLM contract, however, and production planning for the Me 262 fighter was causing him enough grief as it was.

In light of the increasing intensity of Allied bombing raids on German cities and industrial complexes, it was only

---

[11] Messerschmitt AG Augsburg, central planning (J. Reicherter), Programm Me 262 from 15 Jun 43

[12] Messerschmitt AG Augsburg, Me 262 Bauzustand, Me 262 report no. 11 from 28 May 1943

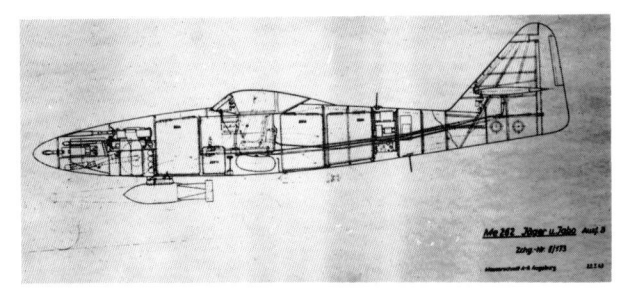

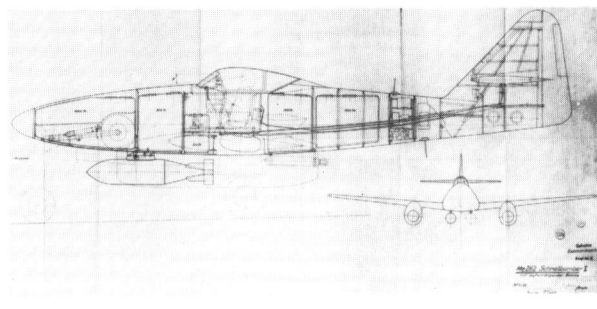

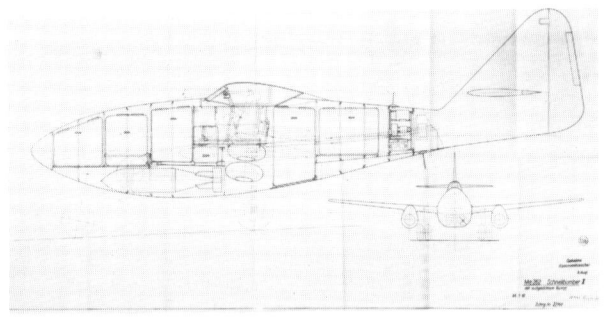

eryone present, he asked if this aircraft would be suitable as a bomber. Messerschmitt answered in the affirmative, and added that *Feldmarschall* Milch was hindering him and an insufficient number of workers was being made available to him. This was the result of a dispute between Messerschmitt and Milch which had been smoldering for many years. I was able to explain this to Hitler and tell him that Messerschmitt was constantly demanding too much without actually reaching the general standards which would have justified these demands. He liked giving out single performance figures as though they were reflective of production. I asked Hitler to take the matter up with Milch again."

[13] Nicolaus von Below, als Hitlers Adjutant 1937-45. Mainz 1980, p. 350

The Me 262V-6, first flown on 17 October 1943, was the first to be fitted with a retractable nose gear and generally had the features of production aircraft (then in the planning stages). This aircraft was the first flown in a demonstration for *Reichsmarschall* Hermann Göring on 2 November 1943 at Lechfeld, and on 26 November performed for Adolf Hitler at Insterburg. On both occasions the pilot was Gerd Lindner. For unknown reasons, this plane was lost on its 28th flight when the famous glider and factory test pilot Kurt Schmidt crashed at Lechfeld on 9 March 1944.

In addition to a fighter type, Willy Messerschmitt offered the Me 262 as a fighter-bomber as early as the project description from 25 March 1943. The fighter-bomber would be able to carry a single 500 kg bomb, two 250 kg bombs, or a single BT 700 torpedo on shackles under the fuselage (a). In an expanded design description dating from the summer of 1943, Messerschmitt proposed two unarmed high-speed bomber variants (b and c): No. I with an external 1,000 kg or two 500 kg bombs and a range of up to 1,700 km, and No. II with a fatter fuselage and internal bomb bay capable of holding one 1,000 kg or two 500 kg bombs and a range of up to 1,900 km.

the fighter pilots who really understood what potential the Me 262 had as a fighter. Which makes all the more incomprehensible three episodes surrounding the Me 262 which occurred in the fall of 1943:

1. Hitler's *Luftwaffe* adjutant, Nicolaus von Below, recalls:[13]

"More and more frequently Germany was becoming the helpless victim of British air raids. On 7 September Hitler called Prof. Messerschmitt to him and inquired into the state of affairs with the jet plane's development. Surprising ev-

Reichsmarschall Göring (center), Willy Messerschmitt and pilot Gerd Lindner after the demonstration of the Me 262V-6 on 2 November 1943 in Lechfeld. Previously, Göring had paid a visit to the Regensburg factory (being rebuilt) where company directors had solicited him for greater support for the Me 262 program.

244

2. On November 2nd *Reichsmarschall* Göring inspected the Regensburg plant following its rebuild after the destruction of 17 August. Impressed with what he saw, Göring was also treated to a demonstration of the Me 262V-6 at Lechfeld. During the visit, he asked Messerschmitt if the Me 262 could also carry bombs. Messerschmitt answered yes and estimated that in the event of a contract he could have the wiring and bomb racks fitted within two weeks.

3. Three weeks later, on 26 November 1943, this same Me 262V-6 was shown to Hitler at Insterburg. Nicolaus von Below continues:[14]

"All those who were responsible for aircraft manufacturing were gathered there, Göring, Milch, Speer, Saur, Messerschmitt, Galland, Vorwald, etc. In my opinion, the *Luftwaffe* had again made the mistake of virtually only showing those weapons and equipment which had not yet reached full operational maturity. Hitler walked down the long row of parked aircraft at a leisurely pace, taking in the latest Me 109 and Me 410, the Ar 234, the Do 335, and the Me 262, among others. Milch accompanied him, providing Hitler with details and answers to his questions. Here Hitler saw the Me 262 for the first time and was quite impressed with its appearance. He called Messerschmitt over to him and asked him pointblank if this plane could also be built as a bomber. Messerschmitt replied yes and said the machine could carry two 250 kg bombs. Hitler's only response was: "This is our *Schnellbomber*," and commanded that the Me 262 be designed exclusively for this role. Milch tried to temper Hitler's decision by only authorizing some of those aircraft to be manufactured as bombers, but was defeated in the face of Hitler's intransigence. Göring, too, after a few days broached the subject with him again, only to be abruptly dismissed. On the return journey to the Wolf's Lair I had another opportunity to talk with him about the matter and tried to save the Me 262 as a fighter. Although he agreed with me in principle, and even wanted more fighters for the Reich himself, he justified his insistence by the brewing political problems. He saw the greatest threat in the immediate future to be an Allied landing in France. We would have to do everything in our power to prevent this."

None of these events had any immediate effect on the Me 262 being planned for production as a fighter. It wasn't until May of 1944, when Hitler asked about the status of this ultra-high-speed bomber and learned that no such aircraft was available, that there was a blizzard of decisions. On May 28th Göring passed along Hitler's order that the Me 262 be built exclusively as an ultra-high-speed bomber to all parties involved, and on 29 May personally gave the order to Willy Messerschmitt, Adolf Galland, and others at

[14] von Below, p. 355

Me 262A-2 *Blitzbomber*, loaded with two 250 kg bombs. The *Blitzbomber*, which Hitler had called for in November of 1943, began trials in May 1944 and entered production in the summer of that year. The version designated as the Me 262A-2 was the same as the fighter-bomber version from March 1943. A 250 kg bomb could be carried on each of the "Wik*ingerschiff*" (Viking Ship) pylons developed by the Messerschmitt company. Takeoff weight was 7.5 metric tons. With one or two bombs, the plane's top speed fell by 100-135 km/h.

the Obersalzburg. Consequently, the Me 262 program was put under the care of *Oberst* Walter Marienfeld, the commander of bomber pilots. Galland's only comfort was that fighter trials would be allowed to continue with a limited number of planes in the hope of being able to eventually use the Me 262 in its intended role. This was a bitter disappointment for the fighter corps, which at this stage was still highly motivated. They attempted to sway Hitler using every trick in the book, such as bringing the matter up directly in conversation during award ceremonies (which he usually presented himself).

In late June 1944 Hitler had ordered the Fighter Staff to concentrate production on fighters, *Zerstörer*s, and nightfighters. Since this included fighter-bombers, Fighter Staff director Karl Otto Saur now found it possible to divert every other Me 262 as a fighter. Nonetheless, Hitler's order to build the Me 262 as a *Schnellbomber* remained in force, and eventually he forbade discussion of the Me 262 in any other sense.

Trials with the fighter-bomber version got underway in May of 1944 with the formation of a group of *Luftwaffe* pilots under *Major* Schenk at Lechfeld. Flying a test program himself, Schenk demonstrated that precision bombing was only possible with the *Jabo* from a dive, with pullout at about 2,000 m. However, any decisive blow the *Erprobungskommando* Schenk may have been able to deliver on the Western Front (where it began operations in August) was frustrated by restrictions to their operating profiles. Higher authorities had ordered that the Me 262 was not to be dived and ordnance was to be released at altitudes no lower than 4,000 m. Two other weaknesses hampered the *Jabo*: To avoid shifting the center of gravity after bomb release, the fuel tanks had to be emptied according to a complicated plan. The range was only about

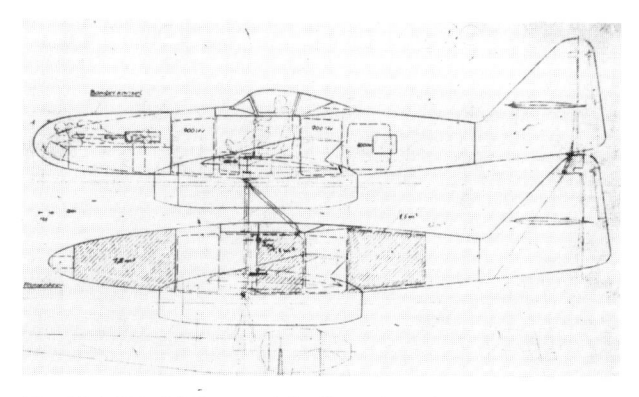

The Me 262A-2/U2 was originally projected as a control aircraft for a Mi*stel* configuration; the bombardier was to have guided the second Me 262 (filled with explosives) into the target after release from the mother craft.

200 km. It goes without saying that initial operations achieved little success. The Allied invasion in June happened without the Me 262 being able to thwart it. A special bomber version of the Me 262, the A-2/U2, with a Lotfe 7H bombsight operated by a bombardier lying prone in the extreme nose section, was successfully flown in late 1944 but was abandoned shortly thereafter. Ultimately, approximately one-third of the 1,433 aircraft built were completed as Me 262A-2 fighter-bombers.

As it was virtually impossible to accurately hit a target with the Me 262A-2's Revi 16D gunsight, other possibilities were explored for improving accuracy: a bombardier, lying prone, was to have been accommodated in the nose for operating the Lotfe 7H bombsight. The wooden nose section was to have been delivered as a conversion kit. Two of these high-speed bombers, designated Me 262A-2/U2, were actually constructed: Werknr. 110484 was turned over to Rechlin on 7 January 1945 following successful bombing sorties, while Werknr. 110555 (in photo) was refitted beginning in December 1944 and flown at Lechfeld in March 1945. The antenna-like protrusions are sensors for acoustic measuring equipment.

Flight Testing Continues

Flight testing of the Me 262 at Lechfeld continued unabated while these problems were being resolved, mostly by those in high-ranking positions. All ten V-types originally ordered had flown by 15 April. However, since half of these prototypes had been lost by this point, the trials program was constantly being fed new aircraft from the recently started pre-production run. By the war's end there were still about 15 planes remaining in the program. A further 15 aircraft went to the *Erprobungskommando* in Rechlin. The remaining pre-production airplanes were assigned to the *Luftwaffe*'s *Erprobungskommando 262* unit established at Lechfeld in early 1944 under the command of *Hauptmann* Werner Thierfelder.

With no prior experience to fall back on, it was this unit which the *Luftwaffe* used to begin pilot training as well as learning air combat maneuvers, and from July *Ekdo 262* started flying their first missions against Allied reconnaissance aircraft. At Lechfeld, company and field unit testing worked closely together to study and rectify the numerous problems, difficulties, and setbacks inherent to such a fast airplane and its newfangled propulsion system. Some worth mentioning were:

- Pilots were not used to the relatively long takeoff runs using turbojet engines. This was alleviated somewhat by employing RATO packs to assist in takeoff (first tested with the Me 262V-5 on 13 July 1943).
- The Me 262's landing speed was at first quite high, at about 190 km/h. By attaching cotton strips, it was discovered that the boundary layer was separating at high angles of attack in the area of the inner wing. As a result, the inner wing's leading edge was extended and fitted with an automatic slat.
- The Jumo 004 engines were proving to be most problematic. When an engine gave out in flight, a relatively frequent occurrence initially, it often could not be restarted and the pilot had to attempt a landing on one engine or sometimes even in an unpowered glide. An even more dangerous situation was when the exhaust needle came loose. Controlled by the engine regulator, the cone (also known as the "onion") would then fly backward and plug the exhaust orifice, and the engine would suffer a sudden flameout. If the pilot failed to recognize the asymmetrical loss of thrust and immediately compensate by applying opposite rudder, the aircraft would wing over into an uncontrollable dive. It wasn't until after the war that this was found to have been the case in the crash of Osterflug's V-2 in April of 1943 and those of other *Luftwaffe* pilots. When rapidly throttling back at altitudes over 8,000 m the turbojet engines would often simply quit. During engine startup the turbojet frequently caught fire, particularly if several attempts had been made beforehand. The engines also occasionally caught fire during flight, as well.
- Damage to the undercarriage, bowed struts, blown tires, and worn treads all pointed to the fact that the stress and high landing speeds necessitated an improvement of the landing gear system. As it evolved through development the Me 262 became heavier through more armament and additional equipment, and the undercarriage was srengthened and fitted with bigger tires.

Up until the end of the war, the second most common cause of accidents (behind engine failures) for the Me 262 was undercarriage strut failure, as shown here. The nose gear in particular suffered from a combination of hydraulic, mechanical, and material problems. Many a Me 262 had to be recovered when the gear failed to extend or collapsed on landing, such as shown in this scene from the summer of 1944 at Lechfeld. Those taking part in the recovery (from left to right) are Moritz Asam from the prototype construction department, Hauptman*n* Werner Thierfelder from *Erprobungskommando 262*, Gerhard Caroli, director of flight testing, and test pilot Gerd Lindner.

- Drop tests from an Me 323 using Me 262 airframes were conducted in February and October of 1943 at altitudes of 6,000 m and 7,000 m over the Chiemsee in order to study the tailplane porpoising encountered at high speeds. Flight paths were tracked by a squad of AAA plotters, and their data showed that the plane achieved a maximum speed of 870 km/h shortly before hitting the water.
- An adjustable control stick was experimented with to offset the high control forces encountered, particularly when pulling out of dives at 800 to 900 km/h.
- The control surfaces of the V-types and the first pre-production aircraft were initially fabric-covered. At about 750 km/h the fabric began to flutter and generate air vortices, which could lead the aircraft to begin tumbling out of control. At higher speeds the problem could not fully be solved even when the control surfaces were reskinned with metal sheeting.

Impetus From the Fighter Staff

Early in 1944 the aircraft armament program was subordinated to Albert Speer's armament ministry and the specially formed *Jägerstab* (Fighter Staff), established on 1 March (run by Karl-Otto Saur and simultaneously reducing the influence of Erhard Milch. It wasn't until this time that Willy Messerschmitt and his chairman of the board F. W. Seiler were able to bump the Me 262 up to a higher priority class. Seiler was convinced that the Me 262 must have sole priority. Despite the reduction of older aircraft types in May (something which Messerschmitt had called for back in early 1943), there were still plenty of airplanes in the construction program, and newer ones were being added all the time. These included the Ar 234, the Ju 287, the Do 335, and the Ta 152. In the fall these were followed by the He 162 single-engine jet fighter which Saur had added to the program with Hitler's blessings and, to the horror of Messerschmitt and Seiler, would drain even more resources

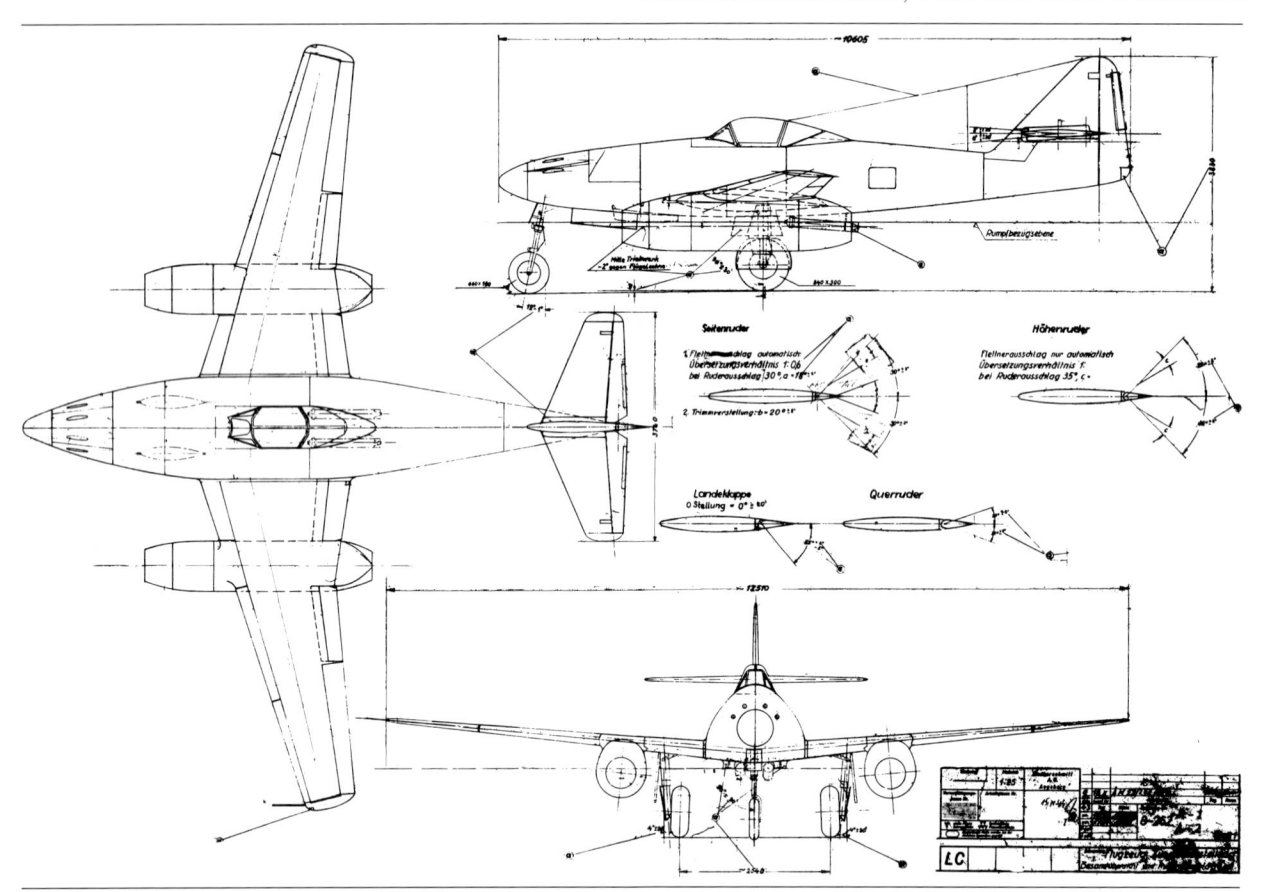

Production three-view of the Me 262A-1 fighter and A-2 fighter-bomber (the A-2 configuration was achieved by fitting pylons under the forward fuselage and removing two of the four MK 108 cannons). Wingspan was 12.51 m, length 10.61 m and height 3.83 m (drawing from 15 November 1943).

Unusual perspective of a production Me 262A-1 in the autumn of 1944. Clearly visible are the extended leading edge slats, which ran the entire length of the wing up to the engine nacelle.

away from the Me 262. Furthermore, Messerschmitt was convinced that, given the state of jet aircraft development at the time, it was foolish to consider building a single-jet airplane.

It was under the *Jägerstab*, which first set the Me 262 program on its feet, that the massive dispersal program for moving the entire aviation industry out from under the bombs of the Allies got underway. Production sites, both aboveground and underground, were either occupied outright or were tackled. For the Me 262 two manufacturing networks were set up, with one falling under the control of the Messerschmitt AG in Augsburg and the other being managed by Regensburg. For a time, the capable Regensburg director Karl Linder had overall responsibility for the program. Production sites covered the entire area of southern Germany.

One example of the ability that Messerschmitt engineers had when it came to improvisation was the forest works, which enabled production up to and including final assembly to take place under the natural camouflage effect of the forest. The Augsburg district had four such centers:

- near Burgau for final assembly of the Me 262, 4,700 m² area.
- near the Leipheim airfield for final assembly of the Me 262, 4,200 m² area.
- near the Schwäbisch Hall airfield for final assembly of the Me 262, 3,800 m²
- near Horgau for building Me 262 wings, 2,500 m².

Underground production centers in the region of Augsburg included Schwaz in Tyrol and the highway tunnel near Leonberg, with an area of 14,750 m². Beginning on 1 April 1945 20,000 m² of the planned 95,000 m² *Weingut II* bunker began operations for producing component parts.

[15] Messerschmitt AG Augsburg (R. Kokothaki), production facilities as of 30 Apr 45

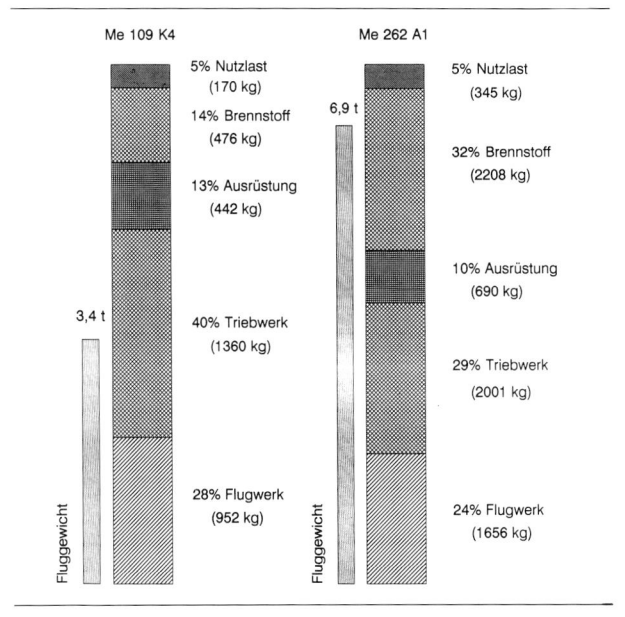

Weight distribution and comparison chart showing the differences between the Bf 109K and the Me 262A. The advantage of the jet engine over the piston engine is quite evident when comparing powerplant weights. In addition, note the better overall performance of the turbojet.

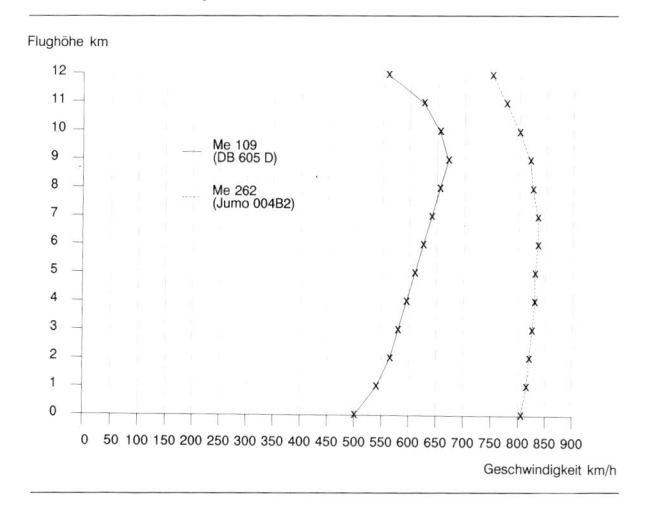

The speed difference between the Bf 109K and the Me 262A was about 200 km/h in 1944/45. As the performance of the production model Jumo 004 turbojet engines varied considerably at this time, the performance of individual Me 262s also differed from one to the next. Messerschmitt guaranteed a maximum speed of 870 km/h for a well-constructed Me 262 powered by good Jumo 004 powerplants.

Located at Kaufering near Landsberg, this was the program's largest bunker complex, and when ultimately completed, the site was expected to have produced up to 1,200 Me 262s each month.[15]

The dispersal efforts at first sparked a rise in production, culminating in the winter of 1944/45, but collapsed in early 1945 when the transportation system these plants depended on was ground to a halt by Allied ground attack fighter-bombers. The absence of Jumo 004 engines was particularly felt. Unfinished aircraft could be seen everywhere. Other planned manufacturing networks, such as Avia in occupied Czechoslovakia and Rheimag in the Harz Mountains, also encountered similar difficulties and—other than sample aircraft built from pre-delivered components—were unable to produce a single Me 262.

In all, Messerschmitt/Augsburg (beginning in March) and Regensburg (from September 1944) produced the following numbers of Me 262s: March 1944 - 1, April - 15, May - 7, June - 28, July - 58, August - 15, September - 94, October - 118, November - 101, December - 131, January 1945 - 228, February - 296, March - 240, and April - 101, for a total of 1,433 Me 262 aircraft completed. Furthermore, approximately 500 more could be found in various stages of assembly, and an additional 2,000 or so were awaiting materials or in planning. Only about 600 to 800 ever reached the various *Luftwaffe* units. Altogether, roughly 100 testbed planes were flown from 1941 to 1945 at Lechfeld and Rechlin.

Me 262 In Combat

Erprobungskommando 262, formed under *Hauptmann* Werner Thierfelder at Lechfeld in early 1944, became the nucleus for future combat operations using the Me 262. The unit was given its first machines from the pre-production run in the spring of that year. Most of the pilots retraining on the Me 262 came from Bf 110 or Me 410 units. There followed air-to-air gunnery training and tactical maneuvering, for which there was no prior experience to draw from. In July the pilots made their first contact with the enemy when they encountered Mosquito reconnaissance aircraft on lone patrols and shot down three of the intruders. *Hauptmann* Thierfelder was shot down on 18 July when engaging an American bomber formation for the first time. On 15 August *Feldwebel* Helmut Lennartz became the first to bring down a bomber, a B-17. Willy Messerschmitt had been following these operations with interest and sent a personal letter of congratulations to the successful pilot. He wrote Lennartz:[16]

[16] Willy Messerschmitt, letter to *Feldwebel* Lenhardt (sic), *Erprobungskommando Lechfeld* dated 5 September 1944

The cockpit of the Me 262A-1 was spartan, yet was fitted with all the necessary instrumentation. The left side panel has the throttle levers for the Jumo 004 engines. Since piston engine throttle control was much more jerky, the sensitive turbojet controls caused the most problems for pilots who had formerly flown the Bf 109 or Fw 190.

"You have flown our latest German fighter on its first combat missions, and in your hands the Me 262 has emerged victorious from its baptism of fire. I congratulate you on this great achievement, which brings for me a two-fold sense of joy—both a confirmation of our latest design and the certainty that we can now wrest back air superiority over our homeland. It is in this spirit that, in addition to congratulating you on the Me 262's fourth victory, I add my best wishes for many other successful battles."

In conjunction with Thierfelder, the *Erprobungskommando Schenk* began bombing trials with the Me 262 starting in May. This unit, drawn from *Kampfgeschwader 51* was deployed to Juvincourt near Reims and operated on the Western Front. In October it was absorbed into KG 51 again, which flew combat operations from Rheine, Giebelstadt, Leipheim, and Schwäbisch Hall in the following months with two of its *Gruppen*.

Based on *Erprobungskommando Theirfelder's* successes, *Reichsmarschall* Göring ordered Walter Nowotny, a *Hauptmann* and successful fighter pilot, to establish an operational detachment and demonstrate that the Me 262 was more suitable as a fighter for intercepting large enemy formations than it was in its unnatural role as a bomber.

In March 1945 several of JG 7's Me 262A-1s at Parchim were armed with 24 unguided R4M rockets (5 cm caliber), making the Me 262 the single most effective and groundbreaking fighter interceptor of the World War II. Operations with the rockets against American bomber formations began on 18 March 1945, but were too late to effectively change the outcome of the war.

Following a brief orientation on the Me 262 at Lechfeld, the unit began operating in October 1944 from Achmer, Hesepe, and Osnabrück with 30 planes. It scored 22 kills for seven losses of its own, including *Major* Nowotny on 8 November. By the time the unit was withdrawn in November the number of operational aircraft had been whittled down to just four. Results from this operation showed a need for thorough training and finding new tactics. The airplane's weakness lay in its powerplants and the undercarriage; affecting one-third to one-fourth of the planes, these deficiencies were of paramount concern.

After receiving reinforcements, the unit was reestablished under *Oberst* Johannes Steinhoff as III *Gruppe* of JG 7. Supply problems prevented the unit from going into operation until the middle of February 1944, now under the command of *Major* Theodor Weissenberger. In late February one of III/JG 7's *Staffeln* became the first to be fitted with supplemental armament in the form of 24 R-4M air-to-air rockets per aircraft. This new weapon virtually guaranteed that each Me 262 sortie would be successful. The Me 262's superiority with a well-trained pilot at the controls, something which Steinhoff and Weissenberger strongly advocated, was made apparent following February's operations, when III/JG 7 recorded 60 kills for a loss of just 6 of their own number.

The loss of aircraft as a result of non-combat related problems was much higher, however. Of the 14 Me 262s lost by JG 7 to such causes in February in March, 40% were due to engine malfunctions, 20% to landing gear problems, 20% to pilot and navigational error, and 20% as a

[17] Messerschmitt AG Augsburg, Technischer Außendienst, TA report/H. Böhler, re: Besuch bei JG 7 in Brandenburg-Briest, no. V-011-45 from 20 Mar 45

The bomb-tow configuration shown here was developed in cooperation with the DFS; attached to the wing from a V1 missile was a 500 kg or 1,000 kg bomb, or a 900 liter fuel tank, towed behind the Me 262. Once ground trials had been completed, Gerd Lindner took to the air in Me 262V-10 on 30 October 1944. This initial test was conducted without a load, but on 18 November he towed a 500 kg bomb (photo) and on 21 November a 1,000 kg bomb. During this flight, the trailer began oscillating so violently that it broke away. No tests were carried out with the fuel tank. There was also a project on the drawing boards for towing a P1103 micro fighter (q.v.)

result of FOD on landing and takeoff, as well as tire damage.[17]

As the war drew to a close, the *Geschwader's* few remaining operational Me 262s were assembled at Prague-Ruzyne, but a shortage of fuel and replacement parts all but grounded the aircraft for the remainder of the conflict. In January 1945 Göring authorized *General* Galland to establish a jet-fighter unit (JV 44) with about 50 hand-picked pilots and 25 Me 262 fighters. Flying in tight formation, in March of 1945 the unit traveled from Brandenburg-Briest to Munich-Riem, from whence it attempted to engage American bombers until the cessation of hostilities. The unit also occasionally operated from blocked off sections of highway.

Around this time Hitler ordered a reduction of bomber units in favor of fighter units, *Kampfgeschwader 54* was accordingly retrained as KG(J) 54 and rearmed with the Me 262 beginning in October of 1944. Those pilots who were trained in IFR operations were first tasked in the role of all-weather interception. To this end, some of the Me 262s were fitted with the FuG 125 and the new EZ 42 lead computing gyroscopic gunsight. First operations agains U.S. bomber formations began in late 1944 from Giebelstadt, Zerbst, and Neuburg on the Danube. By the war's end the *Geschwader* had taken approximately 175 Me 262s on charge. 40 pilots were killed in action or in accidents while flying the Me 262.

A reconnaissance unit, equipped with the Me 262A-1a/U3 sub-variant, was formed in December of 1944 at

Lechfeld and carried out probing missions. The first Me 262 night-fighters were used by *Kommando Welter* in January 1945. But modified two-seat trainers with night-fighting systems (FuG 218), as well as an Me 262 with oblique-firing guns did not reach the unit until April of that year, in the closing weeks of the war.

Fritz Wendel, who visited several of these units up until the war's end, felt the biggest obstacle to successful Me 262 operations was pilot training. With *Kommando Nowotny*, Wendel discovered that several of the pilots only had two takeoffs under their belt. He considered, however, ten flying hours to be the minimum time required for getting to grips with the plane. And for ground schooling, something which the units had fully dispensed with, Wendel thought that 100 hours was reasonable (particularly in view of the turbojet engines). Qualitiy training would have helped prevent many of the accidents. As proof, he pointed to the low accident rate among company pilots compared to their field unit counterparts.[18]

Several of those few *Luftwaffe* pilots who were lucky enough to fly the Me 262 and were able to master the new technology were so successful that within the war's final weeks they were able to shoot down large numbers of American bombers, now blanketing Germany's skies *en masse*.

Several former opponents also paid respect to the Me 262; British fighter pilot and test pilot Eric Brown wrote:

"If asked to nominate the most formidable combat aircraft to evolve in World War II, I would unhesitatingly propose Messerschmitt's Me 262."

[18] Messerschmitt AG, Tech. Außendienst, report from *Flugkapitän* Wendel, re: Me 262, from 5 Jun '45
[19] Eric Brown, Wings of the Luftwaffe, Shrewsbury 1987

French pilot Pierre Clostermann, who flew Tempest fighters with the Royal Air Force up to the war's end and occasionally encountered Me 262s, called the jet "the most sensational combat plane I've ever seen," and christened it the "Queen of the Fighters."[20]

Me 262 Follow-On Developments

In light of resource difficulties the aviation industry was facing in the war's fifth year, Willy Messerschmitt proposed such ideas for the Me 262 as he had recommended in early 1943 regarding commonality of aircraft types. Supplemental to the original fighter version with four to six machine guns and cannons, he proposed the following variants—all of which were simple modifications to the basic design:

Interzeptor I with two Jumo 004 turbojets and a Walter R-II/211 rocket engine with 1,700 kp takeoff thrust. An altitude of 12,000 meters could be reached in 4.5 minutes. (On one flight, the prototype was able to reach 8,000 m in 3 minutes). *Interzeptor II* was to have utilized the BMW 003R rocket engine in combination with the turbojets (giving it 1,800 kp total thrust). With the two powerplants, 3.9 minutes was the time expected to reach 12,000 m. During a flight in early 1945 the prototype attained 6,000 m with a climb rate of 74 meters per second. *Interzeptor III* was to have been fitted with two Walter R-II/211 rocket engines in place of the two Jumo 004s and thus would have had a simi-

[20] Pierre Clostermann, Die große Arena. Das Erinnerungsbuch des berühmten Jagdfliegers. Munich 1960, p. 116

A second interceptor variant was planned, making use of the BMW 003R rocket turbojet engine then in development. Powered by two of these engines and with a takeoff weight of 7,000 kg, the aircraft would have reached an altitude of 12,000 meters in 3.9 minutes. Flight endurance would have been 50 minutes. One prototype, designated Me 262C-2 "He*imatschützer II*," was built. These two photos show Werknr. 170074 during engine run-up; it was during such testing that the combustion chamber exploded on 25 January 1945, damaging the wing. At least one flight occurred using this engine combination, taking place at Lechfeld on 26 March 1945.

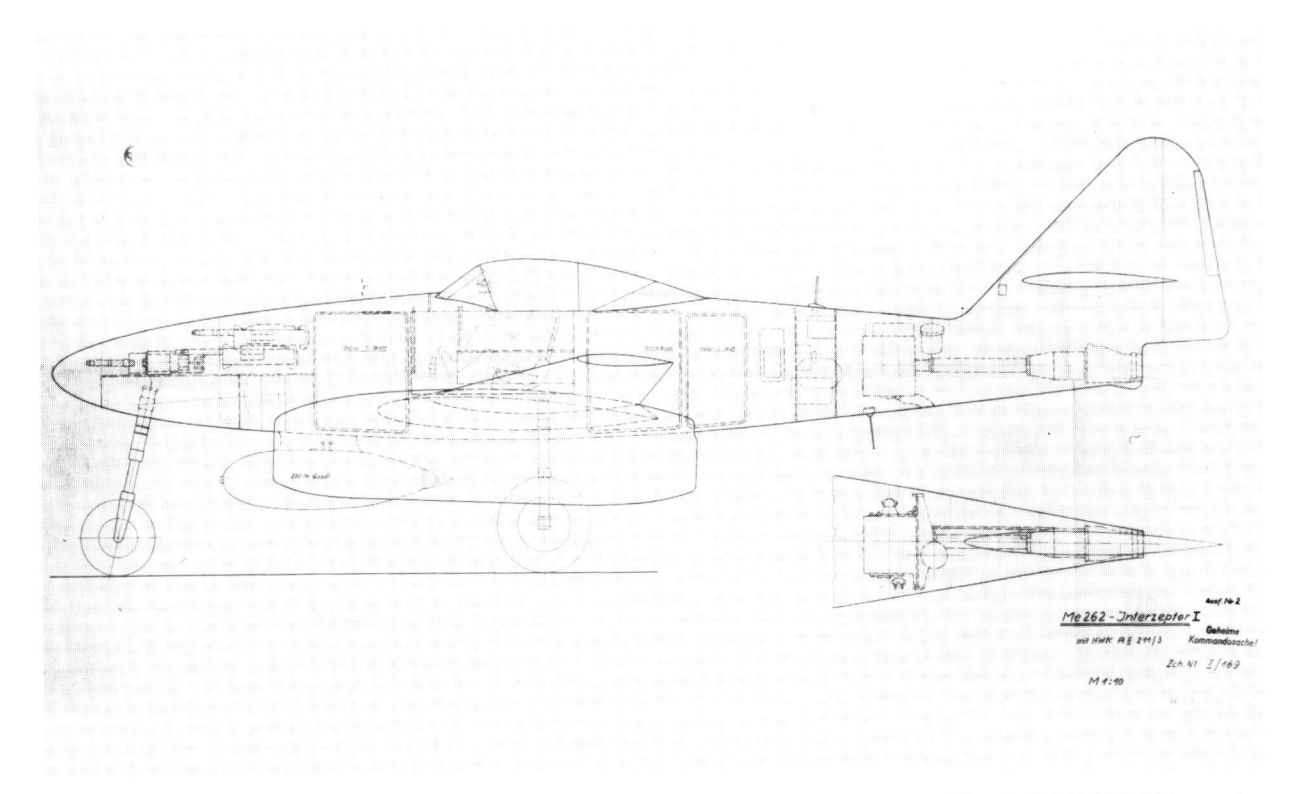

As early as the summer of 1944 the Me 262 was offered as an "*Interzeptor I*" variant with an auxiliary R-II 211(509) Walter rocket engine. Together with the thrust from the two Jumo 004 engines, the interceptor would have been able to climb to 12,000 meters in 4.5 minutes, giving it virtually the same climb rate as the Me 163B. However, it had an endurance of 42 minutes, much more preferable than that of the Me 163. Takeoff weight was to have been 7,885 kg. One example of this variant was actually built, the Me 262C-1a "*Heimatschützer I*"; Werknr. 130186 was modified in 1944/45 and flown by Gerd Lindner for the first time on 27 February 1945.

lar climb rate as the Me 163 (12,000 m in 2.43 min), but never advanced beyond the project stage.

From early on the *Amt* had shown a keen interest in a reconnaissance version of the Me 262, and Willy Messerschmitt offered three alternatives: *Aufklärer I* was the simplest approach, and envisioned swapping out the armament plug on the Me 262A-1 with a nose housing two fitted RB-75/30 cameras. The vertical installation of the cameras gave this variant two distinct bulges. Several examples of this variant were produced as the Me 262A-1a/U3. For *Aufklärer Ia* the cockpit was moved further forward to improve pilot visibility and the cameras situated in the aft fuselage section. *Aufklärer II* called for a fatter fuselage able to hold the cameras and extra fuel tanks, giving it a range of 1,000 to 2,300 km.

Along with the reconnaissance versions, Messerschmitt submitted three high-speed bomber versions: *Schnellbomber I* was the most straightforward, carrying a single 1,000 kg bomb, two 500 kg bombs, or a BT 700 torpedo externally. This version was produced as the Me 262A-2 with 2x 250 kg bombs and utilized in the *Blitzbomber* role Hitler had mandated. The *Schnellbomber Ia* was similar to the *Aufklärer Ia* with the pilot moved forward and external bomb load as with the Me 262A-2. Takeoff weight was to have been 8.9 metric tons. *Schnellbomber II* corresponded to the *Aufklärer II*, with a fuselage of increased capacity for accommodating 2x 500 kg bombs or 1x 1,000 kg bomb and fuel to give it an increased range of between 790 to 1,790 km. Estimated takeoff weight was 9,070 kg.

Also on offer was a trainer version having a second cockpit, created by inserting an extra plug in the fuselage center section and reducing the fuel capacity down from 900 liters. Depending on altitude, endurance was expected to fall between 40 and 70 minutes. By converting A-1 airframes to the B-1, manufacture of the trainer version took place at Blohm & Voss beginning in the fall of 1944. This variant also served as the basis for the Me 262B-1a/U1 stopgap night-fighter, a variant which also saw limited production.

The Me 262 trainer version resulted from a lengthened fuselage and canopy to accommodate the extra seat. The aft fuel tank was reduced in capacity from 900 liters to 400 liters. Takeoff weight was 5,240 kg, with a flight time of 50 minutes. The first of approximately 15 Me 262B-1 trainers converted at Blohm & Voss took to the air in the autumn of 1944.

This Me 262B-1 also served as the basis for a stopgap night fighter design, the Me 262B-1a/U1, which was fitted with night fighting equipment at Lufthansa's Staaken plant.

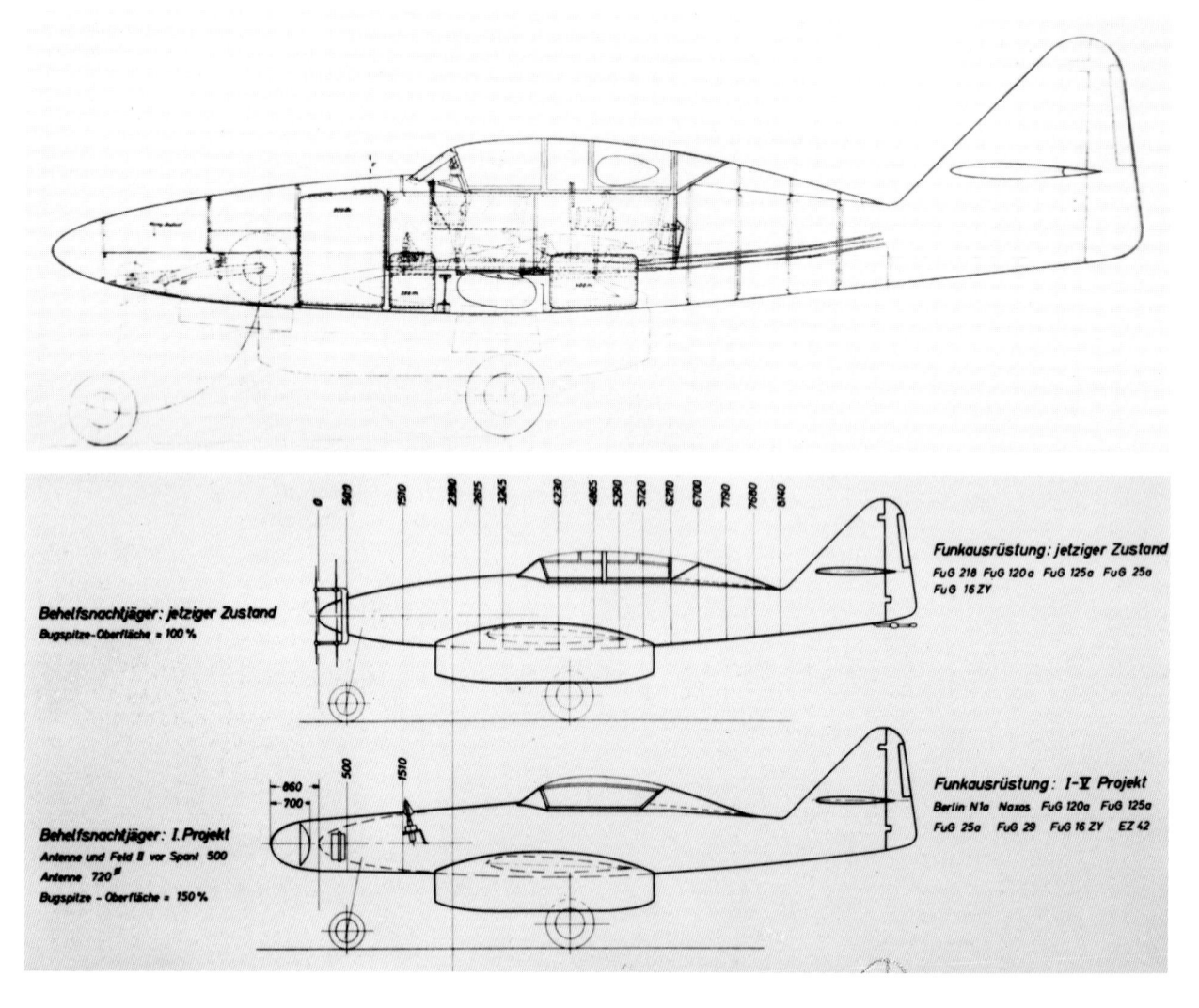

254

High Speed Development Projects

Aside from the Me 163 rocket fighter, the Me 262 was the first high performance aircraft which dared to probe the upper subsonic region between Mach 0.83 and 0.86, a realm for which there was extremely limited knowledge with regard to aerodynamic handling. In the autumn of 1943 Gerd Lindner attained a speed of 970 km/h (Mach 0.83) in the Me 262V-3. On 26 May 1944, while testing for structural soundness up to 6 g, pilot Herlitzius pulled 7.1 g at V_a = 750 km/h at an altitude of 1,050 m and a takeoff weight of 5,500 kg. The 1,000 km/h barrier was first broken with the V-3 in June, while at almost the same time other production aircraft within the Messerschmitt test program and at *Erprobungskommando 262* at Lechfeld were doing exactly the same thing.

To study the Mach effect in a better light, the *Probü* set up a special high speed test program (HG, from '"*Hochgeschwindigkeit*"- high speed) for the Me 262 in April 1944. The first stage, HG I, called for a testbed to be flown with the following modifications: changing the wing's contour by deepening the inner section, sweeping the elevators and extending the vertical stabilizer, and lowering the canopy profile. These changes were carried out on the Me 262V-9 in the fall of 1944, but results proved unsatisfactory, as the pilots found the low canopy intolerable and the elevators led to the aircraft's instability along the pitch axis.

HG II, projected for the summer of 1944, was to have employed swept wings. A production wing was to be cranked back by 20 degrees, resulting in a sweepback of 35

Some of the Messerschmitt flight testing team under Gerhard Caroli remained at Lechfeld well into the summer of 1945 for the purpose of training Allied pilots on the Me 262. When test pilot Gerd Lindner was killed in a car accident during this period, Ludwig Hofmann (photo) took over the training for British and American pilots and eventually ferried several Me 262s to France for shipment to the U.S. In all likelihood, this made him the last German pilot to fly the Me 262.

At least two Me 262s were armed with the heavy caliber MK 214 50 mm cannon. It was thought that this long-range weapon would enable the fighter to shoot bombers at greater distances, out of the range of the formation's withering defensive fire. In June 1945, while ferrying one of these planes from Melun to Cherbourg for shipment to America, the Me 262's engines failed and pilot Ludwig Hofmann was forced to bail out. (A-1a/U4, Werknr. 170083)

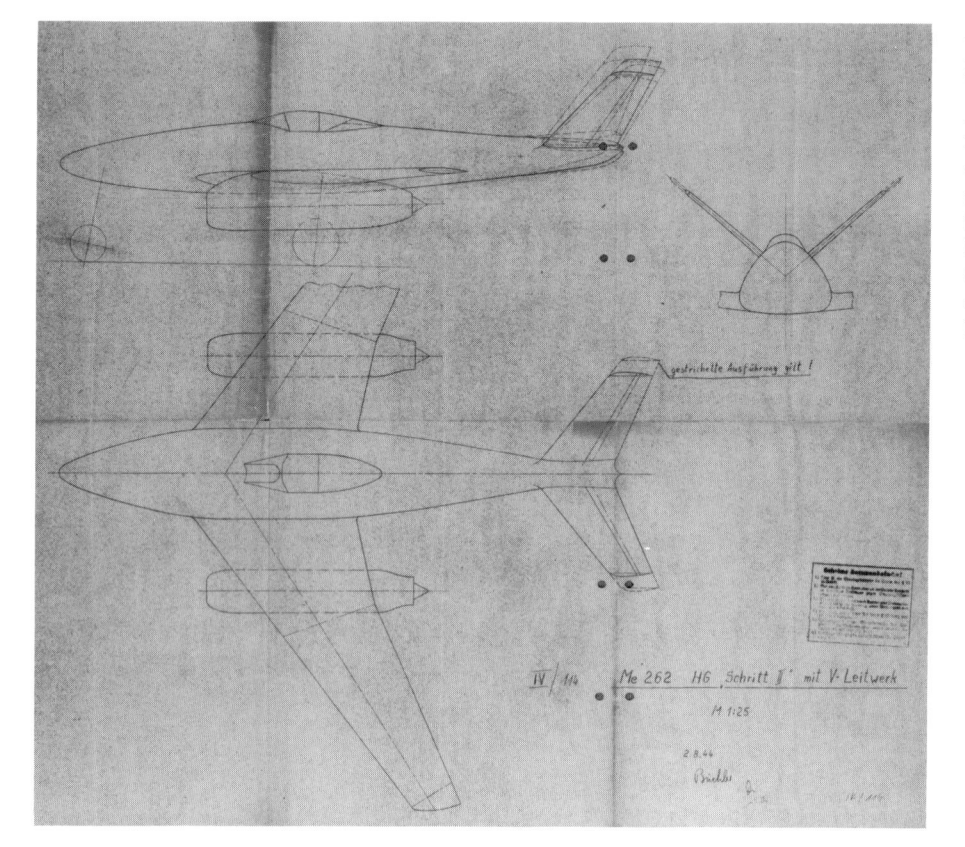

Another design was almost completed as well, this being the "H*ochgeschwindigkeit-Schritt II*"(High-Speed Step II). It was a modified Me 262 testbed having a wing sweep of 35 degrees and a standard empennage configuration. A study of the design with butterfly tail was also made at Oberammergau, as shown in this drawing from 2 August 1944 by Peter Büchler.

degrees (based on the 25% line). This reduced the wingspan of the aircraft to 12.16 m. This project was expected to take awhile, as it necessitated changes in design for installation of the engine, undercarriage, and landing flaps. By early 1945 two pairs of wings had indeed been built by the prototype group, but since no engines were forthcoming for the testbed aircraft by March of that year, no testing ever actually took place.

In December 1944 the high speed program was expanded to include a HG III stage, which would have resulted in virtually a new aircraft design. HG III's fundamental changes involved:[21]
- fitting a wing having a sweep of 35 degrees using the Me 262's production outer wing section.
- engines mounted centrally under the wing at the roots (optional installation of Jumo 004B-2, 004D, BMW 003, He S 011).
- main landing gear attached at the wing root; retracting forward; wheel wells located in center fuselage area under cockpit.
- 1,000 kg bomb centrally located under fuselage on a recessed mount.

The plane was expected to attain speeds of 1,050 km/h at sea level and 1,100 km/h at an altitude of 6,000 m.

A three-seat night fighter requested by the RLM in early 1945 was designed on the basis of the Me 262 HG III using two He S 011 engines.

Opposite Page: One of the new homes for captured aircraft in the U.S. was Freeman Field, where the airplanes were evaluated. Here Col. Harold E. Watson buzzes the field in a Me 262A-1 (Foreign Equipment number FE-110) at a speed of 520 mph. Col. Watson was in charge of Operation Lusty, a program to retrieve and collect aircraft in Germany and ship them to the United States. Numerous American pilots praised the Me 262's performance handling, which was markedly better than many contemporary American jets.

[21] Messerschmitt AG (*Probü*, Oberammergau), Me 262 Projektübergabe V (Hochgeschwindigkeitsentwicklung) from 18 April 1944, 28 Jul '44 (stage II) and 22 Dec '44 (stage III)

In July 1945 the British aircraft carrier HMS Reaper set out from Cherbourg with a collection of the final products of Germany's aviation industry as war prizes for the United States. Among the most significant was the Me 262, of which one of the few Me 262A-1a/U3 reconnaissance planes is shown here (recognizable by its camera bump on the nose). Many of the aircraft in this photo survive today in the National Air and Space Museum in Washington, including the Do 335 and He 219, as well as the Me 262.

Transferring Me 163/Me 262 Technology To Japan

The interlocking relationship of the Berlin-Rome-Tokyo Tripartite League had led to closer economic ties between Germany and Japan. Germany's aviation industry in particular had shipped many aircraft and powerplants to Japan during the 'thirties. The Messerschmitt company had been involved, too, shipping several Bf 108s and, between 1939 and 1943, four Bf 109E and Gs and four Me 210As.[22] In 1941 Willy Messerschmitt had sent Willi Stör (chief test pilot for production aircraft) and engineer Herbet Kaden to Japan to assist in test flying these planes. Throughout the war, Japan expressed considerable interest through its military attaché, General Otani, in Messerschmitt's newer designs, such as the Me 309, Me 321/323, Me 264 and—once Japan had learned of them—the Me 163 and Me 262 jet-powered planes. Japan's interest in the jets was so great, in fact, that Adolf Hitler ordered the types to be authorized for export and approved sending German engineers to assist Japan in license building these two types. In December of 1943 Willy Messerschmitt discussed this matter with General Otani and stated that he was ready to send workers to Japan when needed.

It was decided at a meeting in the RLM in April 1944 to give the blueprints for the Me 262A-1, the Me 163B, the Junkers, Walter, and BMW engines, as well as the documentation for the chemical fuels to Japan as part of a license agreement.[23]

The first information which the Japanese received at this time was the technical descriptions for the aircraft and engines (which were probably the only documents to have reached Japan via submarine in 1944). Although the documentation for the Me 163B (blueprints) was available and had been turned over to the RLM in June, production documentation for the Me 262A-1 was not yet complete (production had started using documentation for the V-types) and weren't ready for shipment until August.

The difficulty in transporting materials from Germany to Japan, only possible using submarines as couriers, was the primary culprit in delaying the project until well into December of 1944. Just as late was the signing of the license contract for production of the Me 163 and Me 262. It had a value of 20 million Reichsmarks. In December, two submarines converted to transports were loaded with production blueprints for the Me 262 and Me 163, as well as some of the more complicated components and aircraft systems. Weeks before, Willy Messerschmitt had said his goodbyes to four of his colleagues, who would be expected to work in Japan for about two years. On board one of the *U-boots* was Rolf von Chlingensperg, deputy director of the *Probü*, and Riclef Schomerus, director of aerodynamics in the *Probü*. They were to be responsible for project matters and aerodynamics for license production in Japan. The submarine, which set out in early 1945, was lost at sea. August Bringewald, former Me 262 program manager, and Viktor Caspar had been sent to Japan for supporting series production. They left in the second submarine (U 234), which was forced to put into a harbor in Norway for repairs. After weeks of delays, it eventually set sail again in March of 1945 and found itself in the middle of the Atlantic when the war ended. The crew capitulated in May and were interned in Canada.

Using the few handbooks which were to be had in the country in 1944/45, the Japanese succeeded in copying the Me 163 and 262 (albeit in modified form) and flying both types in August, shortly before the end of the war in the Pacific. This effectively made Japan the fifth country to make use of the new technology of jet propulsion.

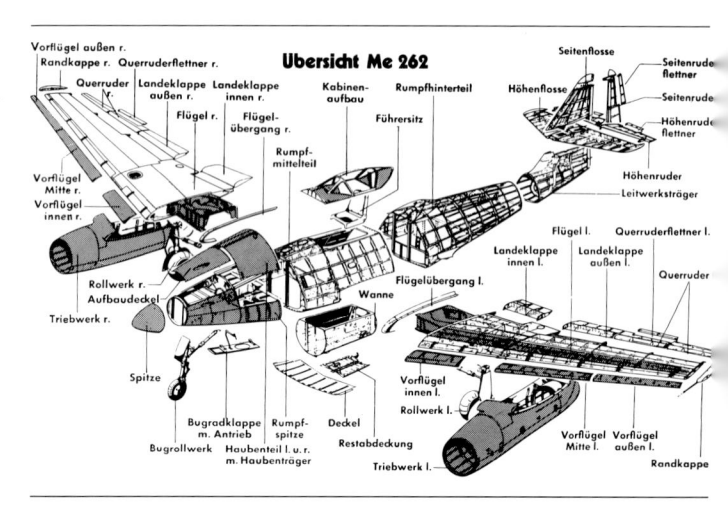

Me 262 components. The ever increasing shortage of raw materials led to steel being used as construction material on the Me 262 for the first time in Germany (shaded components). The main wing spars, nose section, engine nacelles, and the leading edge slats were all made of steel. The steel plating, developed in cooperation with Krupp, had a tensile strength of 110 kg/mm. As a result, the aluminum content of the Me 262 was only 0.55 t per ton of takeoff weight (airframe) compared with 0.95 t for the Bf 109K.

---

[22] (Messerschmitt AG) 1939-1944 Company Report. Jettingen, 31 Dec 1944
[23] (Messerschmitt AG, operations), Lizenz Japan, memorandum for record. Augsburg, 27 Apr 44

Overview of Me 262 Variants

("a" suffix denotes Jumo 004, "b" suffix denotes BMW 003 engine)

Me 262A-1a	fighter with 4x MK 108
Me 262A-1a/U1	fighter with 2x MK 103 and 2x MG 151 or 2x MK 108
Me 262A-1a/U2	all-weather fighter with 4x MK 108, Siemens K-22 autopilot, FuG 120, FuG 125, EZ 42
Me 262A-1/U3	stopgap reconnaissance plane with two RB 50/30 cameras. No guns.
Me 262A-1a/U4	fighter with 50 mm MK 214 cannon
Me 262A-1a/U5	fighter with 6x MK 108
Me 262A-2a	fighter-bomber/*Blitzbomber* with 2x MK 108, *Wikingerschiff* bomb pylons or ETC 503 racks, later 504 for 2x 250 kg or 2x 500 kg bomb
Me 262A-2a/U1	fighter-bomber/*Blitzbomber* with 2x MK 108 and TSA bomb sight in the nose, otherwise same as A-2.
Me 262A-2a/U2	ultra-high-speed bomber with second crewman lying prone for the Lotfe 7H bomb sight in the nose.
Me 262A-3	heavily armored Me 262, otherwise same as A-1.
Me 262A-5	production reconnaissance aircraft with 2x RB 50/30s and 2x MK 108 cannons.
Me 262B-1a	two-seat trainer with dual controls, 4x MK 108 cannons.
Me 262B-1a/U1	stopgap night-fighter based on trainer fuselage with 4x MK 108 and night-fighting systems.
Me 262B-2	night-fighter, production with extended fuselage.
Me 262C-1a	*Heimatschützer I*, (sometimes also designated as Me 262D-1), as A-1 with add'l Walter rocket engine in tailpiece.
Me 262C-2b	*Heimatschützer II*, as A-1 with 2 BMW 003R engines.
Me 262C-3	*Heimatschützer IV* with supplemental Walter rocket engine under fuselage.

Foreign production:

Avia S 92	Czech copy of Me 262A-1a, 12 built from 1946 to 1949.
Avia CS 92	Czech copy of Me 262B-1a, 1 built.
Nakajima J8N1	Copy built in Japan based on limited available Me 262 documentation, flown for the first time on 7 August 1945. Me 262 license production planned in Japan.

The Lippisch-Messerschmitt Me 163 Rocket Fighter

The Me 163 can only be indirectly considered a member of the Messerschmitt family, as the basic concept behind this rocket-powered interceptor came from Dr. Alexander Lippisch. Preliminary work on the project had begun back in 1937 with the *Deutsche Forschungsanstalt für Segelflug* (DFS) at Darmstadt as a result of a contract issued by the RLM's research division under the supervision of Dr. Adolf Baeumker. The contract was for a tailless aircraft using rocket propulsion, which was given the designation "Project X" and carried out under extreme secrecy. But the roots go back much further than this.

Prehistory Up to the DFS 194

The DFS had been awarded the research contract for "Project X" based on the many years of preliminary research carried out by Alexander Lippisch on tailless designed aircraft. Back in 1921 Lippisch, together with Gottlob Espenlaub, had built a tailless glider. This was followed by other flying models and test planes, such as the Storch I, II, and III (1927/28), the Ente (1928), and Storch IV through X (1929 to 1936), plus the Delta series (1930 to 1937)—the latter as tailless flying wings. Subsequently, Lippisch went on to design numerous projects and made proposals for high-speed aircraft, including one in 1934 for a single-seat delta winged fighter.

In aviation circles his designs were considered "difficult to fly," and so many of these experimental planes ended up on the trash heap. From 1936 onward it was left up to Heini Dittmar, the renowned glider pilot, to fully master the Delta-series Lippisch planes, and his skills did much to promote follow-on developments over the next several years. The two-seat DFS 39 sportplane (Lippisch's Delta IVc), powered by the 55 kW/75 hp Pobjoy radial engine and driving a tractor propeller, was certified airworthy in 1936, but the type never went into series production. Lippisch was therefore quite pleased when in late 1935 the RLM awarded him a research contract for two prototypes of a tailless airplane, officially designated the DFS 194. The diminutive, single-seat aircraft was originally to have been powered by an Argus As 10c engine driving a pusher-type propeller. However, when both aircraft entered flight testing it was minus their powerplants.

The DFS 194V-1 was initially designed as a tailless fighter to be powered by the Argus AS 10C engine. In 1940 it was modified to accept the R I-203 Walter rocket engine and was flown to speeds of 550 km/h in preliminary tests for the Me 163.

Project X Becomes the Me 163

A visit to DFS by Dr. Hermann Lorenz from the RLM's research department in 1937 resulted in a contract for a tailless design suitable for testing rocket engines at higher speeds. The engine production company of Hellmuth Walter in Kiel (HWK) was given the associated contract for a controllable liquid-fuel rocket using hydrogen peroxide as the basis for its fuel. By early 1939 management of the contracts passed from the RLM's research department to its developmental section, where *Flugbaumeister* Hans Antz became responsible for the airplane and Helmut Schelp for the powerplant. In many respects the developmental plans exceeded the capabilities of the DFS, and following talks with both Ernst Heinkel and Willy Messerschmitt, the developmental work on Lippisch's Project X high-speed experimental plane moved to the Messerschmitt company at Augsburg. On 2 January Alexander Lippisch arrived at the Augsburg plant accompanied by an initial core group of 12 of his closest employees. Other workers followed later and formed Department L of the Messerschmitt AG under the supervision of Lippisch himself.

Project X was a derivation of the DFS 39/Delta IVc, and was initially drawn up in two variants, both of which underwent wind tunnel testing as models in 1939. Version II with a wingspan of 8.85 m had the typically large rudder, but wings which still retained a pronounced dihedral. Comprehensive aerodynamic and structural analysis had been completed before Lippisch and his small team came to Augsburg, and the design was frozen with this general layout.

For security reasons Project X was now renamed the Me 163, taking over the designation of the terminated Bf 163V-1 through V-3 Fieseler Storch competitor (q.v.). Numbering of the next two (later ten) Me 163A planes under contract consequently began with V-4. The two DFS 194 prototypes were then integrated into Project X under the designation Me 194. V-1 was fitted with a central rudder and configured for accepting the proposed Walter engine. V-2 retained its original form with wingtip plates and was flown as a glider by Heini Dittmar for evaluating flight handling characteristics.

The beginning of the war on 2 September 1939 was a setback for the Me 163 and many other new developmental designs, dropping it to the lowest priority classification. Nevertheless, *Generalluftzeugmeister* Ernst Udet continued to remain a proponent of the Lippisch rocket interceptor. It wasn't until mid-1940 that the so-called "cold" Walter R I-203 rocket engine (thrust rating 3.9 kN/400 kp) was installed in the Me 194V-1, which had been shipped to Peenemünde-West in the interim. In August Heini Dittmar made the first "live" takeoffs. By late 1941 a total of 45 flights had been made under power. It was during these flights that Dittmar first achieved abnormally high rates of climb up to 3,000 meters' altitude and maximum speeds of 550 km/h. These were important preliminary trials for the Me 163s under construction at the time, of which the first, the Me 163A V-4, was completed at Augsburg in February of 1941 following several delays.

On several occasions, Lippisch complained to the plant management that the workshops were discriminating against the Me 163A in terms of meeting its deadline. There had been problems with gluing the plywood layers on the wing

The Me 163A V-4, its maiden flight flown unpowered by Heini Dittmar on 13 February 1941 at Augsburg. But on 2 October of that year, Dittmar became the first to fly over 1,000 km/h (Mach 0.84). This Me 163 remained the fastest aircraft in the world for six years, until 14 October 1947, when Charles Yeager became the first to break the sound barrier in level flight with the Bell X-1, reaching a speed of Mach 1.04.

leading edge and sealing the main tanks in the wings. Lippisch and the foremen he'd brought from Darmstadt were still used to the sailplane workshops using wood construction methods, while the Messerschmitt operations had gone completely over to metal construction. The composite Me 163A design (wooden wings, metal body) had a wingspan of 8.85 m, a length of 5.60 m, and a wing area of 17.5 m^2. The wingsweep was fixed at 23.4 degrees, and it had an empty weight of 1,400 kg. Takeoff was accomplished using a jettisonable wheeled dolly, and the Me 163 landed on an extendable skid.

While awaiting delivery of the Walter engine the type started its trial program as a pure glider. On 13 February 1941 Heini Ditmar took off from Augsburg's field behind a Bf 161V-2 specially modified for towing and was pulled to Lager Lechfeld, where the initial phase of the flight test program was completed. By spring the data for the flight characteristics had been compiled, and problems such as aileron and rudder flutter were eliminated, along with center of gravity difficulties. By gradually increasing the speeds, 950 km/h was eventually reached in a dive, the fastest speed any airplane had flown up to that point. Heini Dittmar considered the flight handling characteristics to be excellent, especially the plane's response to control inputs and its smooth ride in the air, a verdict which was subsequently borne out by other *Luftwaffe* pilots who flew the Me 163B. Ernst Udet, who was witness to a flight demonstration at Augsburg, became infected by Dittmar's enthusiasm and actively supported the program. The number of Me 163A prototypes increased to ten, and Lippisch drew up preliminary studies for an armed version.

After waiting for more than six months for the promised Walter rocket engines, the V-4 was towed from Augsburg to Peenemünde. Test pilots Paul Bader and Heinrich Beauvais had the opportunity to fly the aircraft in its glider configuration during a stopover in Rechlin, and confirmed Heini Dittmar's earlier assesment. At Peenemünde-West the 7.35 kN/750 kp Walter HWK R II 203 engine was finally installed, and on 13 August 1941 Dittmar undertook the first "live" takeoff with jettisonable undercarriage. Fully tanked, the Me 163A had a maximum takeoff weight of 2,400 kg, and wing loading on takeoff was 137 kg/m^2.

Takeoff and flight of the Me 163A V-4(KE+SW) at Peenemünde, powered by the Walter RII 203b engine.

The 1,000 km/h Speed Barrier Is Shattered

Cinematic theodolites were set up on Peenemünde's beach for measuring the speed during subsequent trials with the Me 163. Heini Dittmar succeeded in increasing the speed of the Me 163A V-4 by increments, and by the 20th flight he had already broken the 755 km/h mark set by Wendel in the Me 209 in 1939. The Me 163's climb rate was enormous, but the 4.5 minute burn time of the engine at full throttle was so short that taking off from a runway and climbing to 3,000 m left insufficient time for accelerating to maximum speed. Following three attempts at high-speed flight, it was therefore decided to fill the V-4s tanks only three-quarters full and tow it to 4,000 m behind the Bf 161. Heini Dittmar was then able to fly the plane to its limits at 3,600 m and observe the airspeed indicator as it just brushed the 1,000 km/h mark. Suddenly, separation phenomenon set in and the aircraft became uncontrollable. Dittmar pulled the stick back, cut the engine out, and was quickly able to get things under control again and make a smooth landing. The gauge subsequently showed 1003.67 km/h, i.e. Mach 0.84, thus making Dittmar the first person in the world to break the 1,000 km/h barrier. For reasons of secrecy, however, the record was never announced. Alexander Lippisch felt himself fully justified in his thoughts, ideas, and calculations. One of the things he mentioned at a speech given before the *Deutsche Akademie der Luftfahrtforschung* in 1943 was:[1]

"On 2 October 1941 Heinrich Dittmar achieved a horizontal speed of 1003 km/h in the tailless Me 163A V-4. In so doing, the tailless airplane which we designed has concluded the first developmental phase of a new class of aircraft, at the same time ushering in the first step in the realm of high speed flight."

By early 1942 the V-4 had completed 28 flights at full thrust. The best rate of climb measured was approximately 70 m/s. A second Me 163A, the V-5, was made available for testing from November 1941, as well. Rudolf Opitz was assigned as an additional test pilot for the program. Deliveries of the remaining eight Me 163s, mostly to Peenemünde, took place throughout the next year. Yet for the Lippisch-Messerschmitt Me 163B operational rocket interceptor a new chapter was just opening up, filled with many obstacles, setbacks, and delays.

The Long Road To The Operational Me 163B

The RLM's interest in the Me 163 grew considerably with Heini Dittmar's successful flights during the summer of 1941 at Peenemünde in the Me 163A V-4, culminating in the record-setting flight of October 2nd. Following an increase in the number of "Antons" (A-versions) to be acquired from two to ten (most of which were sent to Peenemünde throughout 1942, where they served as testbeds and trainers with *Erprobungskommando 16*), in the early autumn *Generalluftzeugmeister* Udet issued a contract for a further pre-production batch of 70 Me 163B types, as the military variant was officially designated. Delivery was to be completed by the summer of 1942, and the program was given top priority. In one of the last planning conferences with Udet on 22 October 1941, director Hentzen (whom

---

[1] Alexander Lippisch, über die Entwicklung der schwanzlosen Flugzeuge (on the development of tailless aircraft). Lecture from 20 May 43 (later published as a manuscript.)

Metal fuselage of the Me 163A on its frame (1941).

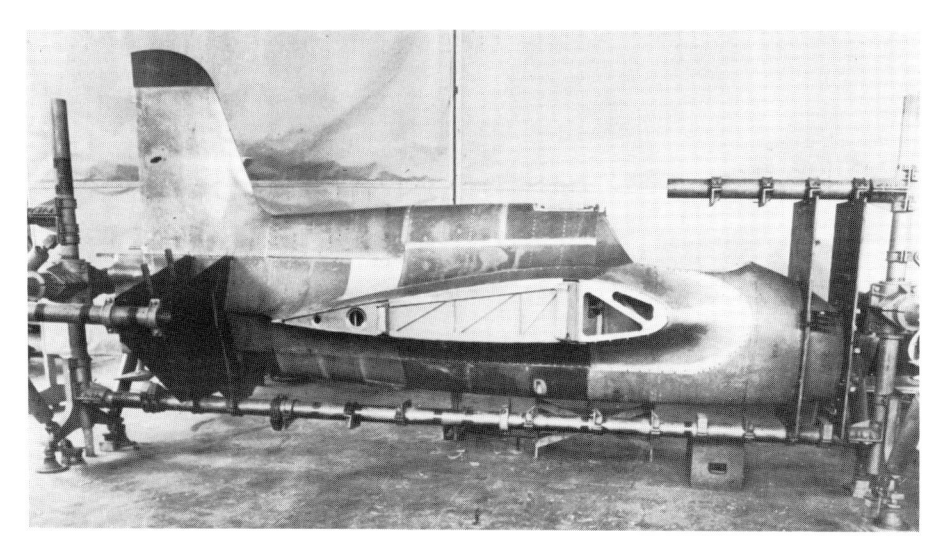

Lower fuselage section of the Me 163 during production at Junkers (1944).

Messerschmitt had sent to Berlin) was reluctant to guarantee an operational airplane in such a short time and obtained Udet's concession that the 70 machines would be given prototype status, allowing for necessary changes to be incorporated as needed. Regarding the RLM's stipulation of metal wings, Hentzen was able to get Udet to agree that these should be built in conjunction with the wooden wings, material availability permitting—something which Udet was skeptical of. (And indeed, these were never built).

In Augsburg, Department L had already begun construction work on the Me 163B by early October, for the designers were able to make use of the comprehensive drawings and blueprints which Alexander Lippisch and his colleagues had created in conjunction with the Me 163A. Taken from the "Anton" were the tested and proven wooden wings and their 23.4 degrees of sweep. The fuselage, however, was a new and much larger design to accommodate a more powerful engine, the military equipment, and the fuel, with the armament (initially 2x MG 151, later 2x MK 108) being installed in the wing roots. The all-metal fuselage was a monocoque design based on Messerschmitt construction principles. As designed, the plane had a takeoff weight of 3,300 kg. According to the consumption data provided by the engine manufacturer, it was estimated that the rocket motor would have a burn time at full throttle of 12 minutes, providing a mission profile of 3 minutes climb time to 12,000 m, then approximately 30 minutes' operating time in level flight at reduced throttle and 950 km/h, giving it a combat radius of 255 km.

Cutaway of the Me 163B showing the HWK 109-509 engine (9 July 1943)

1 Luftschraube für Generator	21 MG 151
2 Generator	23 Druckdichte Durchführung (Elt.-Ltg.) (bis V 46)
3 Entstörer	24 Umformer FuG 16
4 Regler	25 Vorratsgeber
5 Sammler	26 Boschhorn
6 Ausgleichgefäß	27 Gurtzuführung
7 FuG 16	28 Elt. Durchladeschallkasten
8 Fußsteuerung	29 Sicherung u. Verteilerkasten
9 Schleppkupplung	30 Schnellablaß
10 Panzerglas	31
11 Revi	32 großer Entstörer
12 Gerätebrett	33 kleiner Entstörer
13 Gerätebank	34 Regler
14 T-Behälter	35 Verteilerkasten
15 Knüppel	39 Erregungsschleife FuG 16
16 Handsteuerung	40 Antennenanpassungsgerät
17 Kufe	42 Sperrwelle
18 Arbeitszylinder	
19 Sitz	
20 Flettner	

This photo was taken at Bad Zwischenahn and shows a Me 163A trainer and a Me 163B operational machine of Erprobungskommando 16. Notice the straight wing leading edge and fatter, longer fuselage of the "Berta."

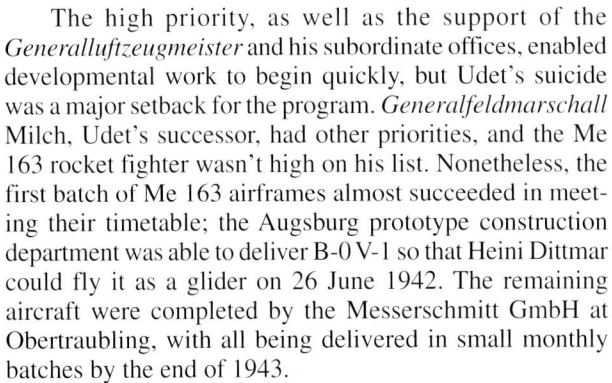

One of the unpowered Me 163B-0 prototypes on approach to the airfield at Lechfeld (1943).

The high priority, as well as the support of the *Generalluftzeugmeister* and his subordinate offices, enabled developmental work to begin quickly, but Udet's suicide was a major setback for the program. *Generalfeldmarschall* Milch, Udet's successor, had other priorities, and the Me 163 rocket fighter wasn't high on his list. Nonetheless, the first batch of Me 163 airframes almost succeeded in meeting their timetable; the Augsburg prototype construction department was able to deliver B-0 V-1 so that Heini Dittmar could fly it as a glider on 26 June 1942. The remaining aircraft were completed by the Messerschmitt GmbH at Obertraubling, with all being delivered in small monthly batches by the end of 1943.

Both the Walter company's engine (the HWK R II 211) and the BMW engine (P 3390A) failed to materialize as scheduled, and virtually the entire test program for the Me 163B not requiring a propulsion system was carried out in glider configuration. This included the undercarriage/skid tests, weapons testing, experimentation with an airbrake, assisted takeoff and braking parachute trials, and radio equipment testing. Of least concern was the flight handling characteristics of the Me 163, which—as Willy Messerschmitt easily recognized—was problem free from the beginning. Company testing took place in Lechfeld, while engine and operational testing were carried out at Peenemünde-West and Karlshagen. Additional prototypes were supplied to the engine companies at Kiel and Spandau.

The aircraft's few deficiencies which cropped up during testing were soon remedied. This applied to locating the skid in the most favorable position with regard to center of gravity and the proper cushioning for the undercarriage dolly, the skid, and the pilot seat. Until corrected, these problems were the cause of not a few accidents. In October of 1942, for example, Hanna Reitsch was forced to attempt a landing with the dolly still attached; the plane flipped over on landing, and she suffered serious head injuries as a result. Those changes resulting from the test program were retrofitted to all V-types at Klemm in Böblingen, which took over series production of the Me 163B following Messerschmitt's completion of the V-series, and supplied them to the field units.

The biggest headaches were caused by development of the new engines and getting them to production standard. Month by month, the manufacturers kept pushing back the delivery date. With no Walter engine yet available by early 1943, the team at Peenemünde took the less powerful R II 203 from the Me 163A and installed the Me 163B-0 V-8, which Rudolf Opitz flew for the first time on 21 February. It wasn't until June that the first R II 211 (the more powerful engine, now designated the HWK 109-509) was fitted into the Me 163B-0 V-2, and by August 1943 the Klemm company was able to begin delivering additional prototypes refitted with the new engine.

The high fuel consumption rate of the Walter engine proved to be a big disappointment. Fully fueled with 1,550 kg of T-Stoff and 468 kg of C-Stoff, operating time at full thrust was less than 6 minutes, i.e. only half of the original military requirement. In addition, the maximum takeoff weight with armament and ammunition was now 4,300 kg, meaning the airplane had become 1,000 kg heavier. This gave the "Berta" model a wing loading on takeoff of nearly 220 kg/m^2. Numerous problems cropped up during engine

Heini Dittmar in the cockpit of a Me 163B-0 prototype. Flying for much of the time without engines, this photo reveals that these machines were nevertheless fully kitted out with armored glass and Revi 16 gunsights for the two MG 151s.

Professor Willy Messerschmitt greets Hanna Reitsch, who flew the Me 163 at Lechfeld and was seriously injured when the aircraft flipped over. In the center is Alexander Lippisch, father of the Me 163 (October 1942).

testing. One of its undesirable qualities—never fully resolved even after the Me 163 entered operations—was the tendency to flame out when pulling relatively low negative g. Once the engine flamed out, the pilot had to wait one to two minutes before attempting a relight by means of a rather complicated process, and even then there was no guarantee that the engine would restart. In such circumstances the pilot would then be forced to land with a large fuel remainder or—if he had time—dump the fuel.

It took considerable effort to get the Me 163 into the air, and the process was totally unlike the standard operation for other aircraft in the *Luftwaffe*. It involved affixing beneath the skid a two-wheeled dolly undercarriage, which was jettisoned on takeoff and recovered by the ground crew. Landed Me 163s had to then be loaded back onto their dollies prior to being towed back to the operations area. Specialized forklifts were later used to recover the aircraft. Filling the tanks with the two dangerous chemical fuel components required utmost care, as there was considerable danger of explosion and acid burns for the ground crew. The T-Stoff (80% hydrogen peroxide) reacted to the slightest disturbance with a powerful explosion and could only be stored in aluminum containers. The C-Stoff (a mixture of 30% hydrazine hydrate, 57% methanol, 13% water and minor amounts of potassium-copper-cyanide) could only be poured into enamel-coated containers. Even the smallest amounts of this chemical spilled on the ground had to be rendered harmless by diluting it with large quantities of water. The noise generated by the plane in the air and on the ground was something to be heard. Even a static testing of the en-

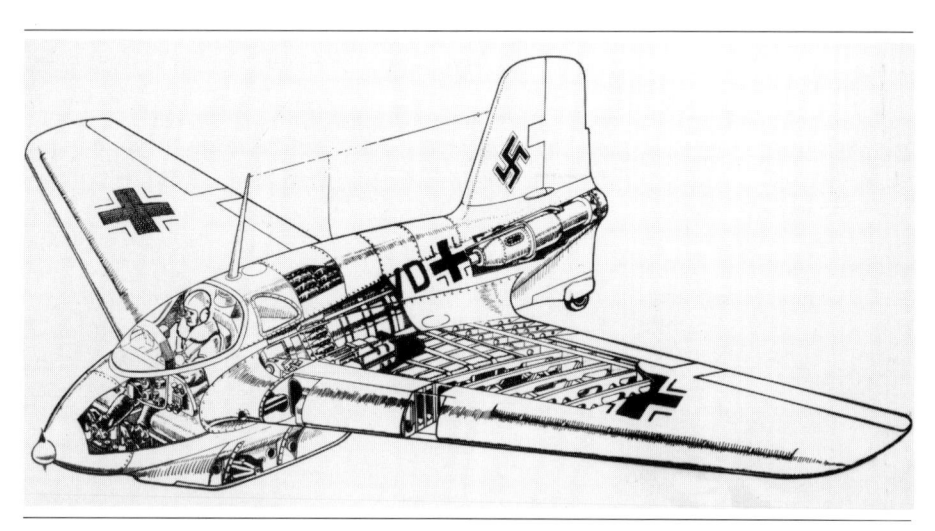

Structural cutaway of the Me 163 dating from the war years.

Installation of the Walter 109-509 engine in the Me 163B of the Deutsches Museum during restoration work in 1965.

Me 163B-0 V-9, Werknr. 10018, VD+ER, was initially flown in January 1943 as a glider, but in September was fitted with a Walter engine and was delivered to Erprobungskommando 16 in December of that year.

gine was an experience, as graphically recounted by Mano Ziegler, a pilot with *Erprobungskommando 16* and later author of the book "Raketenjäger Me 163:"[2]

"Two to three Me 163s were arriving almost daily. After Karl Voy and his team had checked them out in flight, they were given the once-over and then handed over to one of the three replacement *Staffeln*. At first Venlo had just four planes, Wittmundhafen six, while the Zwischenahn *Staffel*, not yet officially established, had none assigned to it and flew any that were available. One of these had been cordoned off for a test runup, and I found myself climbing into its cockpit. Otto Oertzen clambered up beside me on the boarding ladder. A push of the button, a slight forward nudge on the throttle, and the concert began. These static tests were a major event. Basically, a pilot jumped onto four or five thousand bridled horses, whose whinnying suddenly became a thousand demons all whistling through their teeth at once. An unparalleled concert from hell, a tortured and agonized bellowing followed by a long, drawn-out hissing, a primeval noise from the bowels of technology comparable to no other sound on earth. It seemed as though the air cringed in fear of the horrific pressures built up in the fire-spewing combustion chamber, pressures which shook the hangar's sturdy walls and quaked the stumpy concrete blocks, the tie-down cables straining to break free. An indescribable feeling of mastery overcame me as I held this mighty hurricane at my beck and call with the tiny throttle lever as Thor had once done with his thunder and lightning. All around me and just behind my back flowed the lifeblood of this demonic fire, the T- and C-Stoff, coursing through its thin tubing, a marriage made in hell, order from chaos only being established once the two partners embraced in the steel walls of the combustion chamber. The order of mathematical laws which created the forces only to harness their power. The watchmen of this power were the instruments on the panel, whose twitching needles revealed

[2] Mano Ziegler, Raketenjäger Me 163. Ein Tatsachenberiecht von einem, der überlebte. Stuttgart 1961, p. 116

267

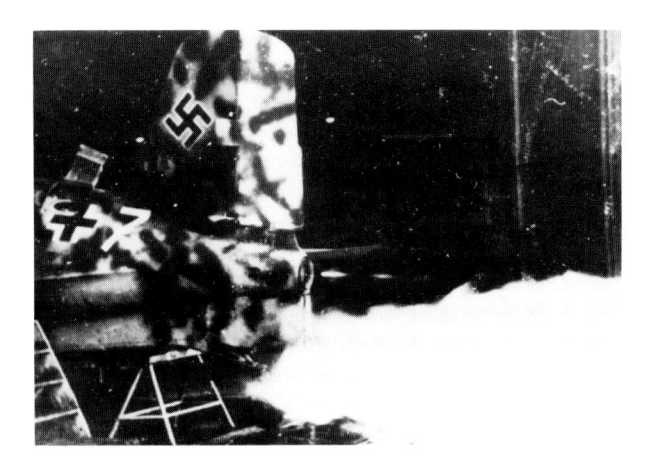

Static run-up of the HWK 109-509 engine in a tied down Me 163B, a scene vividly described by Mano Ziegler.

the numbers, their phosphorous glow brightly illuminating the cockpit in the half-lit atmosphere of the hangar....

After just four minutes the engine shut down with a loud thump"

In order to extend the Me 163's operating time, tests were carried out using rocket-powered takeoff trolleys on rails and with solid-fuel underwing booster rockets. A pressurized cockpit was not fitted; instead, the pilots were sent to the Zugspitze for a few weeks to acclimatize themselves to high altitudes. Normal operating ceiling was 12,000 m, but during one of the test flights an altitude of 15,500 m was attained.

The Me 163B's top speed was somewhere around 965 km/h, and climb could be maintained at a constant 700 km/h. The plane could go from takeoff to 12,000 m in just three and a half minutes. Operational machines were not able to reach the 1,000 km/h mark exceeded by Heini Dittmar in 1941, as compressability problems set in well before then. Test pilot Heinz Peters discussed this in a 1964 paper documenting his experiences:[3]

"As everyone knows, in its combat version the Me 163 was unable to fulfill the pilots' wishes of breaking the 1,000 km/h barrier. This was through no fault of the engine, but simply due to the fact that the Mach effects set in sooner. When flying too fast (over 900 km/h) the forward fuselage was noticed to lift slightly in spite of the control stick not moving. This was almost immediately followed by severe wing rocking and the rapid onset of a nose stall. The engine also suddenly cut out. These events generally happened so quickly that the separate movements of the airplane were imperceptible. The latter was most likely due to the unannounced conditions, causing the pilot to be thrown about and lose vision for a few moments."

Heinz Peters was also the one who, in December of 1944—with the plane already in operation with field units, provided the test facility's data for its 6 g structural soundness requirement. The results of the trials went well beyond the requirement, showing 7.1 g at 650 km/h and 7.5 g at 890 km/h.

[3] test pilot *Oberleutnant a. D.* Heinz Peters, Bericht über eine Sondererprobung, document, 1964

Operational Me 163B planes of JG 400 at Brandis in the autumn of 1944.

Erprobungskommando 16 for the Me 163B

In April 1942 *Hauptmann* Wolfgang Späte, a renowned sailplane pilot and highly decorated fighter pilot, was tasked with establishing the *Erprobungskommando 16* and at the same time became the type overseer for the Me 163B rocket fighter development program. In his latter capacity, he was responsible for ensuring that the military's requirements were given sufficient attention even before the plane was introduced into the field. After being briefed by Heini Dittmar, he made his first takeoff in a Me 163A on 11 May 1942. *Erprobungskommando 16* began introductory and test flights at Peenemünde-West. Many experienced fighter pilots, particularly former glider pilots and two company pilots, were ready for service within just a few weeks. Flights were made using the planes on hand or with the ten powered Me 163As delivered over the course of time. In late 1942 these were joined (once Lechfeld had completed the most critical portion of the flight test program) by the first Me 163B non-powered gliders. In any event, it wasn't until 1943 that aircraft were available in any appreciable numbers.

Readying for Combat

The *Erprobungskommando* shifted to the airfield at Bad Zwischenahn near Oldenburg following the massive bombing raid mounted against Peenemünde on 18 August 1943. In addition to flight testing, the bulk of pilot conversion training was also accomplished at this site. New trainees at a glider pilot's school progressed from the Kranich's dual controls to the Grunau Baby and Rhönsperber, then on to the aerobatic Habicht. The Habicht came with three different wingspans: 13.60 m, 8.00 m, and 6.00 m; the latter was called the Stummel Habicht. When it came time to advance to the Me 163A, the trainees were first taken aloft under tow, then moved on to gliding flights with low wing loading, and finally to flights with a supplemental water ballast. These were followed by "live" takeoffs with the Anton and advanced to the Berta. In late 1944 these types were supplemented by the unpowered Me 163S two-seat trainer, which could be used for glider training after being towed into the air.

The fisrt three operational Me 163B-0s arrived at Bad Zwischenahn in January of 1944. These were used to fly simulated attack profiles and formed the basis for the first *Staffel* (1/JG 400) of twelve airplanes, but the unit soon transferred to Wittmundhafen. It was from there that now-*Major* Späte flew the first operational sortie, on 13 May, against an American bomber formation escorted by several Republic P-47 Thunderbolt fighters. Approaching a P-47, the Me 163 built up too much speed, compressability set in and the engine flamed out. In August 1944 1/JG 400 was reinforced and transferred to Brandis to protect the Leuna plants. It was also in August that the unit scored its first successes. *Jagdgeschwader 400* was beefed up to two *Gruppen*, and the largest number of planes ever assigned to it peaked at 91 Me 163Bs during the last weeks of 1944. All operations ceased within a few short weeks of the sole remaining factory producing C-Stoff being destroyed in a bombing raid. The Me 163 rocket fighter logged 16 kills in all, with a loss of just six of its own number at the hands of enemy fighters.

The ten Me 163A models built served as trainers in 1944/45 for the operational Me 163Bs of Erprobungskommando 16. At least one "Anton" was also armed with R4M rockets.

Lippisch Leaves the Messerschmitt Company

Differences of opinion between Lippisch and Messerschmitt became more and more frequent. Director Theo Croneiß was forced to act as an intermediary on several occasions. A file memorandum from 11 March 1942 indicates that a planned move of Department L to Obertraubling was canceled to prevent undue loss of time. Lippisch, however, offered another slightly different explanation at the time:

"I want nothing more than my own shop within the Messerschmitt AG, or my own factory, if necessary, in which I can build the prototypes without any type of outside influence. Because only I and my people have the experience needed to bring the prototype aircraft to the point to where they are really combat capable in the air."

When his various proposals for tailless heavy fighters, high-speed bombers, and Me 163s with alternative propulsion systems were not followed up, Alexander Lippisch and five of his closest colleagues quit the Messerschmitt company and assumed the post of director at Vienna's Aeronautical Research Institute (*Luftfahrt Forschungsanstalt Wien*). The remaining 80 or so employees in Department L were reassigned to Messerschmitt's project, design, and stress analysis departments, to flight testing and prototype construction, depending on their skills.

Since work on the Me 163B continued, many employees had decided to stay on at Messerschmitt with "their" airplane, now under the supervision of Rudolf Rentel. Even Prof. Messerschmitt showed real interest in the work of Dr. Lippisch.

Nevertheless, he was forced to give priority to "normal planes" and their follow-on developments. Back in December of 1942 he had responded to Dr. Lippisch's requests by stating that in view of the fact that they were on the verge of a comprehensive program to get the Me 163B operationally ready, work on a variant with DB 605 engine would have to be canceled. Producing the long shaft for the pusher propeller and making a new fuselage was unjustifiable in light of an industry as overtaxed as it then was. Nonetheless, a comparison of this Me 334 with his normal planes would have been most interesting.

Nor was attention given to designs for a turbojet powered Me 163, which would have obviated the dependence on the still-immature Walter rocket engine. These included J. Hubert's P 12 project from September 1942 which envisioned a BMW turbojet engine and the P 20 design of Dr. Wurster from April 1943 with a Jumo 004. Unlike Hubert's concept, the Wurster project called for a retractable tricycle undercarriage. Alexander Lippisch remained contractually associated with the Messerschmitt AG as an advisor to the Me 163B program.

By the end of 1943 Augsburg and Obertraubling had divested themselves of Me 163B development and prototype construction, which were then transferred to the airbase at Laupheim. Following a heavy bombing raid on Laupheim on 19 July 1944, prototype construction under the management of Johann B. Kaiser moved again, this time to the fortress at Kronach, but work did not resume before the war's end.

Further Development Leads to the Me 263

The two greatest problems facing the Me 163 were its limited endurance and range, as well as immobility on the ground due to its skid. Improving the first problem was tackled in late 1942 when Lippisch had considered lengthening the B airframe by 110 cm. But as this would have led to changes to the wings and tail, it was decided to enlarge the entire design—takeoff weight rose from 4.2 metric tons for the Me 163B to 5.4 tons on the Me 163C, as the design was now annotated. Design work and component construction began sometime in mid-1943, but was delayed later that year when the Messerschmitt development program, including the Me 163 under the direction of Rudolf Rentel, was packed off to Oberammergau and the prototype construction relocated to Laupheim. Assembly of the 70 Me 163B prototypes wound down in Obertraubling, and series produciton and the ongoing refitting of the prototypes began at Klemm in Böblingen.

During the summer of 1944, as Messerschmitt was forced to concentrate more of its developmental and production resources on the Me 262 and Willy Messerschmitt yet again called for an increase in his labor force, the RLM

For initial tests the Me 163B V-18 was modified to the standards of the 248/263 and flown by Heinz Peters beginning in December 1944. In addition to takeoff and landing characteristics, Peters also praised the aircraft's good visibility from the cockpit.

The Me 263V-1 with fabric strips affixed to the wing, the purpose of which was to monitor the airflow over the wooden wing. The wing itself made use of much of the Me 163B's wing design.

The Me 263V-1 flew for the first time in February 1945 at Dessau, albeit without its MWK 109-509C engine and with fixed undercarriage. Pilot was Heinz Peters. Additional flights took place at Brandis.

arranged matters so that the Junkers company—where cancellations had freed up the necessary resources—would lighten Messerschmitt's load by completely taking over the Me 163's follow-on developments, its production, and responsibility for the program.

Prof. Messerschmitt and Junkers development chief Prof. Heinrich Hertel worked out the details between themselves personally. On 1 September 1944 the Junkers company assumed responsibility for and production of the Me 163B under the code name "Scholle" (flounder). Part of the development and testing teams were also transferred to Junkers and went to work at various dispersed sites around Dessau and Brandis. Even the Me 163B production resources of Klemm were subordinated to Junkers.

With regard to follow-on developments of the Me 163, Prof. Hertel decided—with the backing of the RLM—to altogether skip the Me 163C then in development and replace it with a new aircraft having a retractable undercarriage. It would make use of the wings and rudder of the Me 163B, just entering production, as well as a goodly portion of its avionics systems, a measure expected to reap a 40% savings in developmental and production resources. It was to have an all-new metal fuselage with pressurized cockpit and retractable tricycle gear, plus the more powerful Walter HWK 109 509C engine with larger fuel tanks holding 1,600 liters of T-Stoff and 840 liters of C-Stoff. It was anticipated that the plane would thus have an endurance of 15 minutes at 700 km/h and 11,000 meters' altitude.

In October Junkers was in possession of the blueprints for what was initially designated the Ju 248 rocket fighter, but was then renamed the Me 263. A "flying mockup" was created from Me 163B V-18, refitted with a longer fuselage and fixed undercarriage and test flown as a glider by Heinz Peters in tow behind a Bf 110 in late December. Peters praised the good flight handling characteristics inherent in even this provisional design, especially the soft landings on its tricycle undercarriage (compared to the hard landings on the Me 163A's and B's skid, which had already been the cause of numerous spinal injuries among pilots).

Heinz Peters also test flew another variant in December of 1944, the Me 163S trainer. This was a modified version Junkers had built in the Berlin workshops of Lufthansa and was exclusively for training the rookie pilots of JG 400, which had since moved to Brandis (at least one of several examples built was captured and flown by the Soviets).

With the cancellation of the Me 163C, Junkers was awarded a top priority production contract for 20 prototypes and 952 production variants of the Me 263. The first V-type was completed at Dessau in February 1945 and—as Junkers' smallest aircraft up to that point with major involvement of its work force—was also test flown by Heinz Peters as a towed glider with fixed landing gear. The planned 509C engine had not yet materialized.

The heavy bombardment of Central Germany in early 1945 brought work on the Me 263 airplanes to a virtual standstill. Only the dispersed factory sites were able to keep operating, assuming, of course, that materials were available or could be delivered. It took extreme effort just to produce the first three prototypes by March. The planned transfer of these to JG 400 for operational testing collapsed in the chaos of the war's final weeks. Despite a destruct order, many blueprints and aircraft survived the end of the war at Dessau and Brandis.

Even though the Americans were the first to arrive in Dessau in April of 1945, when they pulled out they left the Soviets much in the way of captured material, including

163 B
F = 17,3 m²
G = 4,2 to

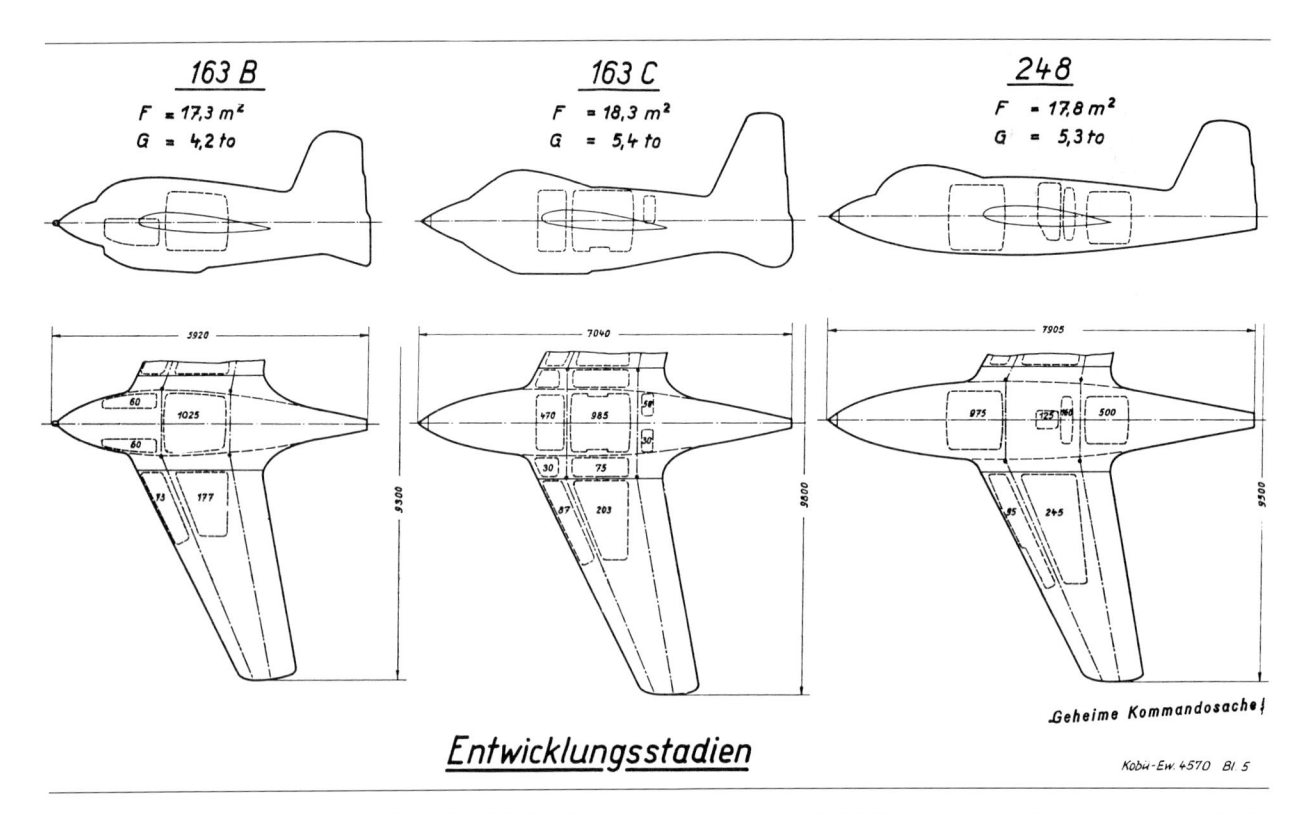

163 C
F = 18,3 m²
G = 5,4 to

248
F = 17,8 m²
G = 5,3 to

Geheime Kommandosache

Entwicklungsstadien

Kobü-Ew. 4570 Bl. 5

Metamorphosis of the Me 163B into the Ju 248/Me 263. Excepting the wings, the Me 163C was a new airplane design, as was the Ju 248/Me 263. In the case of the latter, however, wing and vertical stabilizer were borrowed from the Me 163B, saving developmental time by some 40%.

prototypes of the Me 263, for the type continued flying at Dessau under Soviet supervision. In June 1946 Junkers test pilot Mathis was killed in the crash of a Me 263.[4] Testing subsequently shifted to the Soviet Union. A few months later, on 22 October 1946, those Junkers engineers and testing teams remaining at Dessau were carted off together with their families to the Soviet Union as part of the *"Nacht-und-Nebel Aktion"* (Night and Fog Campaign), where they were forced to work on follow-on developments of the Ju 287 and Jumo 004 jet engine, among others, until well into the 1950s. The Me 263 concept was visibly reflected in Mikoyan and Gurevich's I-270 rocket plane flown in 1946/47. Unlike its role model, however, the Russian plane did not have swept wings; Soviet aircraft designers had little experience in this area and in the first postwar years were quite reluctant to implement this technology.

In the West, too, the British captured several Me 163Bs. Of these, the only rocket powered intercepter ever to have entered production and be used in combat, about a dozen have survived and can be seen in museums today throughout the world.

It is a little known fact that, in Japan, the Me 163B was copied by Mitsubishi in both a glider and powered version.

[4] statement from long-time Junkers engineer Reginald Schinzenber during an interview with the author (1970)

Me 163/263 Variants

There were 308 B-types manufactured in addition to the ten Me 163A and 70 B-0 series, plus a few Junkers-built B-0/R in 1944/45. A lack of available HWK 509B-1 and -C engines, however, meant that not all of the following variants had reached the stage listed below, established on 4/26/1944, and some of them may also have been delivered *sans* engine.

Me 163A	V-4 through V-13 (built by Messerschmitt AG and others). Engine HWK R II- 203, no armament.
Me 163B-0	V-1 through V-70 (V-1a, V-1b built at Messerschmitt AG, V-2 through V-68 at Mtt GmbH, Obertraubling). Engine HWK 109-509A (R II-211), except V-10 which had BMW P 3390. Armament 2x MG 151/20, from V-46 onward 2x MK 108.
Me 163B-0/R-1	20 aircraft (built at Klemm in 1944). Engine HWK 109-509A. Armament 2x MK 108.
Me 163B-0/R2	30 aircraft with B-1 wing (license transfer to Japan). Engine HWK 109- 509A. Armament 2x MK 108.
Me 163B-1/R1	70 aircraft with forward fuselage same as B-0. Engine HWK 109-509B- 1. Armament 2x MK 108.
Me 163B-1	390 aircraft planned (begun at Klemm and Junkers, incompleted). Engine HWK 109-509B-1 (with cruise thrust chamber). Armament 2x MK 108.
Me 163B-2	Version based on reworked drawings. Engine HWK 109-509B (no cruise thrust chamber). Armament 2x MK 108.
Me 163C-1	Enlarged Me 163B, new development except for wings. Development curtailed in 1944 in favor of Me 263 with four prototypes under construction. Engine HWK 109-509C with cruise thrust chamber. Armament 2x MK 103 in wing root, 2x MK 108 in fuselage.
Me 163S	Two-seat unpowered unarmed trainer. Of approximately five planes converted from Me 163B airframes, at least one was delivered to JG 400.
Me 263 Ju 248)	New design with undercarriage. Of the 972 (formerly prototypes and production versions ordered, only V-1 through V-3 were built at Junkers. Engine HWK 109-509C. Armament 2x MK 108 in the wing roots.

Japanese version: Mitsubishi J8M1 Shusui with Toku Ro 2 engine (copy of Walter HWK 109-509), first flight as glider - 8 January 1945, with engine power - 7 July.

Me 328 - Pulse-Jet Powered Multi-Role Aircraft

Starting in early 1941 the *Probü*'s Rudolf Seitz generated a variety of preliminary designs for a high-speed, single-seat, twin-engined light bomber with the capability of operating as a fighter, parasite fighter, long-range fighter, and reconnaissance aircraft, all under the selfsame project number P 1079. Because of its thrust performance, its engine was to be the Schmidt-Argus As 014 pulse-jet. Expected thrust was to have been somewhere in the neighborhood of 500 kg from a propulsion system weighing 80 kg initially, later rising to 140 kg. Based just on these impressive figures alone, building the P 1070 would have required a fraction of the man-hours and materials needed for the Bf 109F (cost of an As 014 was about 2,000 RM, that of a DB 601 was estimated at 43,000 RM). Rudolf Seitz saw the project's potential as a universal fighter which, like the *Giganten* production just getting underway, would be produced quickly and in large numbers.

The project was also in line with the thinking of Willy Messerschmitt who, in addition to discussions with the RLM, was able to use Adolf Hitler's statement that "If we had 200 bombers which could carry out daylight precision bombing in England, England would have been wiped out by now" as an argument for his P 1079.

An offer to the RLM resulted in the Messerschmitt company receiving a preliminary development contract for three prototypes, with work beginning on these starting in September 1941.

However, construction and workshop operations came to a standstill in early 1942 due to a lack of available resources, and the entire project was passed on to the DFS at Ainring. DFS acquired the welded sheet metal fuselage and skid and built the wooden wings and tail (the elevators were obtained from the Bf 109G) according to data provided by Messerschmitt's Ko*bü*. On 3 August 1942 Karl Baur flew the V-1 at Ainring as a glider after Rudolf Ziegler from the *E-Stelle* Rechlin had previously taken the airplane up on its maiden flight.

Flight Testing The Me 328 In Hörsching

Following these initial flights of the Me 328 at Ainring the remainder of the flight test program—including the second DFS-built machine—shifted to Hörsching airfield near Linz. There the As 014 pulse-jets were fitted in various configurations (under the wings, either side of the fuselage). In late November 1942 Haupt*mann* Wolfgang Späte, the per-

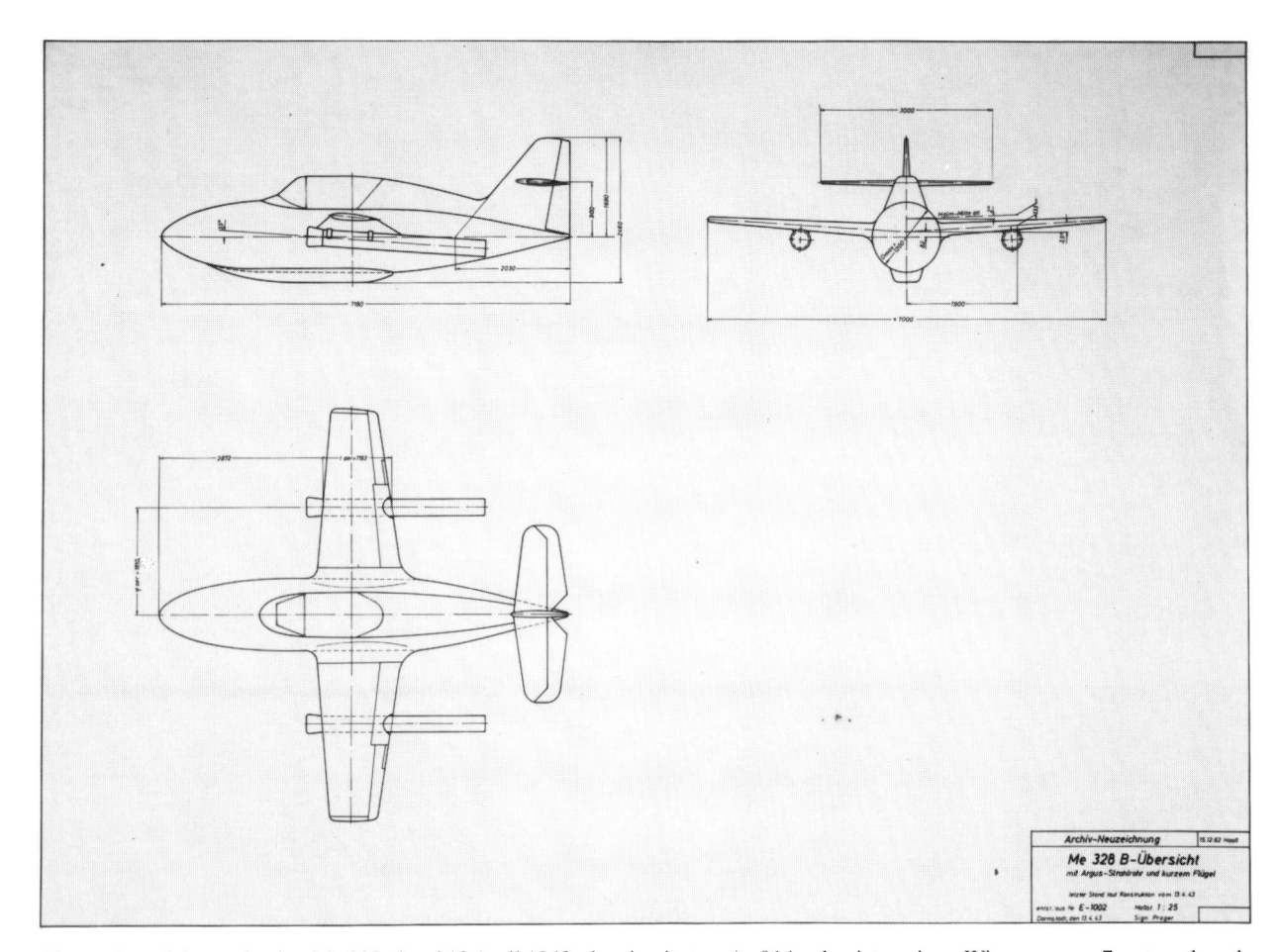

Three-view of the production Me 328, dated 13 April 1943, showing its two As 014 pulse-jet engines. Wingspan was 7 meters, length 7.18 meters.

Me 328A V-1 taking off at Hörsching under tow. Pilot Fritz Stamer—a pioneer from the Rhön sailplane days—always took care to fly with his hat on.

son responsible for jet aircraft at the time, filed the following report at Hörshing following an inspection visit:[1]

"The results of flight testing thus far can be summed up to the effect that an aircraft with a wing loading of 150 kg/m^2 can take off like a glider, under tow from a He 111, and can maintain 260 km/h in towed configuration. Takeoffs at higher wing loading can only be made when carried piggyback style or launched from a catapult. With the currently fitted landing gear, a towed takeoff is quite difficult since the He 111 tug can't be seen from the Me 328 for the first 100 meters of the takeoff run. During free flight (after cable release) the Me 329 glides at a rate of 7 m/s and 4 m/s when using the Argus jets, each of which has a thrust of 150 kg. A He 111 acting as a tug couldn't be expected to

[1] *Hauptmann* Späte, commander of *Kommandoführer Späte*, (Bericht über den Stand der Strahlflugzeug-Erprobung) to the *General der Jagdflieger* from 3 Dec 42

Initial arrangement of the As 014 pulse-jets on Me 328A V-2 during the summer of 1942.

Another Me 328 prototype with the engines mounted on the aft fuselage, here on a Do 217 as part of the Mistel piggyback configuration.

reach much higher than 3,000 meters' altitude....Maximum speed attained in a dive thus far has been 500 km/h. Gliding flights are normally done at 200-300 km/h....On landing, flare out must occur at 200 km/h. A skid landing on dry grass uses up about 500 m; on wet grass the slide run is 1,000 m....

The Argus jets used up until now have been measured at a maximum thrust of 150 kg in the blocks. Flight testing at the Argus company had initially led to the belief that performance would increase in conjunction with speed, not drop off as originally thought. Tests with the 150 kg pulse-jets on the Me 328 and a DFS 230 testbed at Hörsching, however, have lately shown the opposite to be the case. Based on the data gathered there, the jets seem to not only be dependent upon altitude, but on speed, as well, i.e. the performance of the jets drop off dramatically as speed increases. If this is actually the case, the whole matter of the Me 328 is now called into question if it turns out that the 300 kg jets planned for the final version are dependent upon speed."

Even though this open question regarding the As 014 engine's performance had been aired over two months before, in September 1942 the Probü's Woldemar Voigt sounded quite optimistic in a report he filed on the type:[2]

---

[2] Voigt, Messerschmitt AG project dept, Me 328. Sonderaktion zur Entwicklung und Fertigung. Technical Report TB no. 70/42, from 1 Sep '42

"Development and testing of the Me 328's airframe and engine are now so far advanced that, with minimal risk, there is strong evidence that a high-speed bomber can be made frontline ready in an exceptionally short period of time, one which would be a deciding factor in this war ..."

And the report also addressed the claim of its estimated maximum speed of 800 km/h without bombs at sea level and 660 km/h with bombs at a total thrust of 600 kg:

"Its most important tactical feature is the fact that at low levels, beneath radar coverage, a Me 328 with a heavy bomb load can reach cruising speeds well in excess of the maximum speeds of enemy fighters.

Another critical item is that it's always possible to boost the plane's speed at a later date by installing more powerful engines without having to undertake special modifications to the airframe, even increasing speeds to its limiting Mach number (greater speeds are possible with the introduction of the swept wing); range can be maintained by increasing the size of the fuel tank."

The report also hints that that Sta*atsekretär* Milch had already issued a top priority contract for the Me 328 by this time. In this case the RLM had undoubtedly expanded its high-speed bomber requirement, to which Lippisch, Messerschmitt, and other companies had all submitted conventional projects, to include a turbojet-powered aircraft. The Me 328 report continues:

"The type is a specialized development based on the Me 328A. Whereas the 328A is still undergoing development with a view towards a parasite fighter, the high-speed bomber with limited tactical operational capability is being advanced per current war-critical requirements."

These tactical capabilities were: "flying coastal artillery, counter-invasion, and advanced strike."

The RLM's (GL/C-E 2 office) technical guidelines from the fall of 1942 included the following stipulations for the Me 328B high-speed bomber:

"Crew: 1
Role: Coastal defense within a given penetration sector. Due to engine characteristics, operations at low level (below 1,000 m) against enemy maritime and ground targets (southern England). Level or toss bombing (from a dive angle of approx. 20 degrees).
Powerplant: 2x Argus pulse-jets, initially with 300 kg thrust each, later with 400 kg. Engine changes must be an extremely easy task due to the short life of the powerplants.
Performance: With 2x 300 Argus pulse-jets at sea level with 500 kg bomb: 600 km/h
            With 2x 400 kg Argus pulse-jets at sea level with 1,000 kg bomb: 700 km/h
            Max. landing speed on skid: 180 km/h. Special brakes for landing on wet surfaces and snow.
            One hour's endurance
            600 km/h range
Takeoff: The same dolly developed for the Me 163B is proposed for takeoff, i.e. takeoff using a dolly with a thrust propulsion system on a standard gauge track section. Average acceleration is 2 g. Track takeoff sequence at least one minute. Additional methods are piggyback, lift, and airfield takeoff (using Me 163 dolly).
Materials: The airframe is to be built primarily of wood.
Loads: Stress class H 5. Safe load factor is n = 5 for targeted weight with bomb."

Manufacturing Program for 300 Aircraft

Messerschmitt central planning drew up a program, dated 15 December 1942, for the manufacture of 300 Me 328B high-speed bombers designed as a "cantilever mid-wing plane of wood construction powered by 2 Argus 300 kp pulse-jets, enclosed cockpit, landing skids and external bomb load."[3]

According to the program, 20 V-types and 280 pre-production aircraft were to be built by the company of Jacobs-Schweyer Flugzeugbau GmbH, Darmstadt (JSF). Deliveries were projected to begin in June of 1943. A supplemental contract called for a follow-on monthly output of 75 to 100 aircraft. The JSF had considerable experience in the construction of sailplanes; the Rhönsperber and the Weihe were just two of the company's better known products.

Development and testing of the Me 328 was to remain with Messerschmitt and be carried out in conjunction with the DFS. DFL in Braunschweig served as the venue for wind tunnel testing. Several tests were undertaken to explore engine arrangement with a view toward protecting the fuselage from the heat and exhaust. Additionally, flight testing showed that acoustics from the quite loud, pulsating engines tended to damage the aft fuselage section.

It was these problems which led to the evaluation program dragging on well into 1943, so that production preparations and construction of a total of 10 V-types was just

[3] Messerschmitt AG, central planning, Programm Me 328B. Bau von 300 Flugzeugen, 15 Dec '42

One of the numerous models tested in wind tunnels, this one having the engines mounted above the wings.

getting underway when the program's termination order came down on 3 September 1943 (most likely as a result of the *Luftwaffe*'s standardization efforts). This, by the way, came at a time when Rudolf Seitz was involved in intense negotiations with the Junkers company regarding installation of the Jumo 004 engine, which would have been fitted in a nacelle under the fuselage.

To accommodate this new powerplant, the Me 328C variant would need to have a long-legged undercarriage with nose gear, something which wasn't exactly the best solution. It never went beyond the project stage, unlike the Me 328B which found renewed interest a few months later.

Me 328 SO For Kamikaze-Type Operations

By early 1944 the Pr*obü*'s Rudolf Seitz was able to generate interest again in "his" airplane for a role which had never before existed in Germany.

At a time when Germany's leaders were expecting the Allies to launch an invasion from England somewhere along the English channel, a small group of L*uftwaffe* pilots formed a unit dedicated to a "weapon of total commitment" (*Totaleinsatzwaffe*), i.e. using a manned glide bomb as a one-way weapon. The pilot would be expected to point the aircraft at a landing ship just before impact and either bale out at the last moment or perish in the impact. A group of 70 young volunteers was given flight training to this end. The Me 328B was selected as the weapon of choice by a narrow margin since it was felt that it would provide a suitable airframe in short order. In place of a bomb a 500 kg torpedo warhead was to have been carried in the nose.

In her book "Fliegen - mein Leben" Hanna Reitsch wrote:[4]

"Along with Heinz Kensche I was asked to take over flight testing of the planes. This initially took place with the Me 328 at Hörsching near Linz. The Me 328 had originally been planned as a fighter or *Zerstörer*....Now it was to be used without its engine as a sort of manned glide bomb in the suicide role. It was a single-seater with quite stubby wings of 4 to 5 meters span. Its glide ratio was about 1:12 at 250 km/h, and about 1:5 at 750 km/h. The Me 328 couldn't take off on its own, instead it was carried piggyback-style on the wings of a Do 217 to an altitude of 3,000 to 6,000 meters for its evaluation. The Me 328 could be uncoupled from the Do 217 from the pilot's seat, separating from the wings in flight without any difficulty whatsoever. The flight handling characteristics sufficed for its intended role. We required good visibility, comfort, good maneuverability, lateral stability, and directional stability. These conditions were met.

Testing concluded in April 1944. The Ministry then assigned a factory in Thuringia to assume series production. For reasons I've never clearly understood, series production never properly got off the ground."

Production startup of the Me 328 SO (*Selbstopferung* - self-sacrificing/suicide) spurred intense negotiations in June 1944 between the *Kobü* in Oberammergau, the Jacobs-Schweyer Flugzeugbau, and the Gothaer Waggonfabrik (GWF), which had the resources available. The JSF com-

[4] Hanna Reitsch, Fliegen - mein Leben, Stuttgart 1951, p. 280

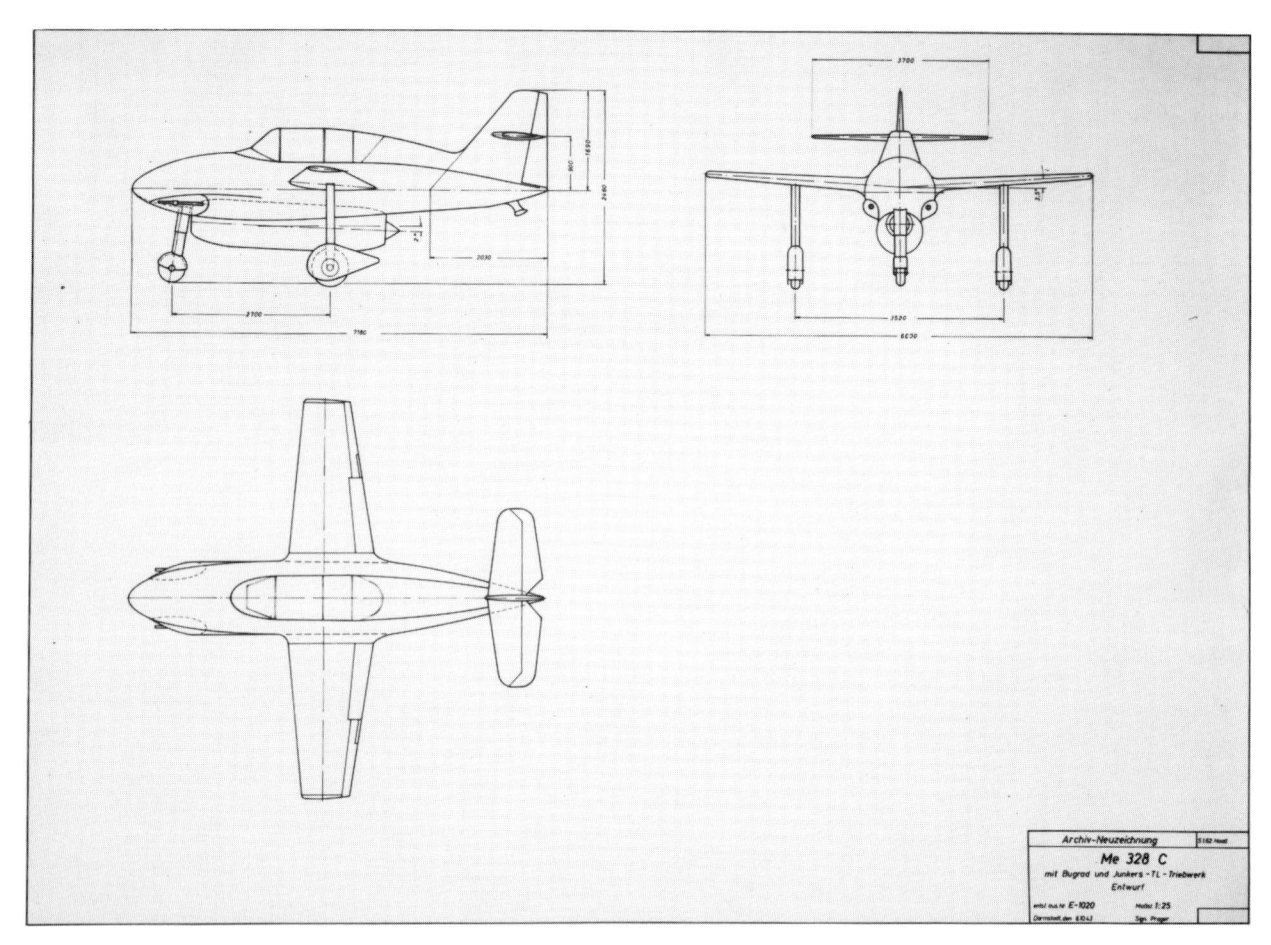

Three-view of the Me 328C with a Junkers Jumo 004 turbojet engine, dated 6 October 1943. The wingspan of this design was just 6.60 m, length 7.18 m. Work on this twin gun fighter variant had already begun by the time development was ordered halted on 3 September 1943. Remaining work was carried out as part of an offer to the RLM, including temperature measurements at Junkers using specially-treated wooden models which had the same lower fuselage as that of the Me 328C.

pany was to finish the V-3 through V-10 then under construction and incorporate the changes necessary to modify them to glider trainers, at the same time preparing GWF for series production (beginning with the V-11). The operational aircraft, gliders fitted with warheads, were to be towed to the vicinity of the target by tow tug planes.

It was during this period that the RLM made a new decision regarding a suicide plane in favor of a manned version of the Fi 103 flying bomb (also known as the V1, this missile had been undergoing flight testing since December 1942 and was also powered by the As 014 pulse-jet). Code-named Reichenberg, it would require relatively little expenditure to produce considerable numbers of these aircraft using available V1s. However, a "total commitment" was never required, as the suicide team formed as part of *Kampfgeschwader 200* was disbanded in February 1945 "due to a lack of fuel."

Thus ended the saga of the Me 328, an airplane which was never able to live up to its expectations due to the fact that its pulse-jets were incapable of delivering the performance claimed for them. Its weaknesses and disadvantages had been apparent almost from the moment it began testing in mid-1942. From a historical perspective, the Me 328 was the airplane with the greatest wing loading of its day, at 530 kg/m2, and the first manned aircraft using pulse-jet engines.

[5] The project P 1095, which Rudolf Seitz also worked on in 1943, also had several of the same guidelines as the Me 328C

The Messerschmitt P 1101 Swing-Wing Fighter

In July 1944 the *Oberkommando der Luftwaffe* (OKL, *Luftwaffe* High Command) called upon five aircraft manufacturers to submit proposals for a single-engined jet fighter capable of reaching a maximum speed of about 1,000 km/h at an altitude of 7,000 m. The types were to be designed around the Heinkel He S 011 engine, a powerplant with a static thrust of 12.7 kN/1,300 kp. The initial requirement of 30 minutes' endurance at full throttle was later raised to 60 minutes. Armament was planned to be four MK 108 cannons with sixty rounds each.

The fighter aircraft which this "emergency fighter program" called for was to be the successor to the Me 262—especially if the shortages of raw materials became even more acute—yet it would be faster than the Me 262 and be powered by only one engine. In order to meet the requirements all the project teams based their designs on the swept wing; by this time sufficient data pertaining to lift and drag effects resulting from sweepback had been accumulated from wind tunnel testing.

On 10 September 1944, at the first joint conference stemming from the OKL requirement, the Messerschmitt company submitted a design for an "Ein-*TL-Jäger*"(single-jet fighter), the P 1101, with a 40% sweepback.

History of the Swept Wing

The swept wing was first mentioned by Adolf Busemann at the Volta Congress in Rome in the fall of 1935 as part of his presentation on "aerodynamic lift at supersonic speeds," later published in the bound volume containing the congress's addresses. However, the idea subsequently slipped into the mists of obscurity, and it wasn't until several years later that Albert Betz stimulated systematic measurements at the Göttingen high-speed tunnel for wings with 45% sweepback and a tapered wing for comparison at speeds of Ma = 0.7 and Ma = 0.9. The first results were published in late 1939 in an AVA report by H. Straßl and Hubert Ludwieg. These findings, which showed a clear advantage of the swept wing concept at drag coefficients above Ma = 0.8, were disseminated throughout the German aviation industry. In September of that year Albert Betz and Adolf Busemann applied for a secret German patent for the swept wing.

The Messerschmitt company was quite interested in these results and arranged for a series of additional tests at the Göttingen wind tunnel (test cross-section 110 x 110 mm) using models of various sweepbacks (15% to 45%) and profiles, plus models of forward swept wings as well as those with nacelles and fuselage. The results of these tests were compiled by Mtt staff member Ludwig Bölkow in his technical report No. 22/40 from 10/21/40:

"The evaluation program shows that it is possible to fly at least up to Mach Ma = 0.9, i.e. 1,100 km/h at sea level using NACA profile 0012-64 and a sweepback of between 33% and 38% without any measurable effects of compressability on drag."

On 26 and 27 September 1940, Professor Ludwieg presented the results of high-speed sweepback tests to a large audience of experts at the *Lilienthalgesellschaft*'s Committee for General Airflow Research in Göttingen. By this point over 16 models of swept wings had been evaluated, some

Explanation of the Swept Wing Concept in Comparison with a Slipping Wing

The main reason for using a swept wing is to take advantage of the effect of the delayed critical Mach number at higher speeds. The airflow over a swept wing behaves in the same manner as it does over a wing in a slip, i.e. the velocity component parallel to the leading edge has little effect and the wing behaves as though the only flow affecting it comes from the velocity component $V \bullet \cos\varphi$ perpendicular to the leading edge. It therefore follows that the swept wing—except in the vicinity of the wing root—acts in relation to the effects of compression as though only this reduced Mach number affected it. In the high subsonic region, the use of sweepback delays the otherwise rapid increase in drag, and the plane is therefore able to fly faster with the same amount of thrust. Sweepback, however, has no bearing on frictional drag.

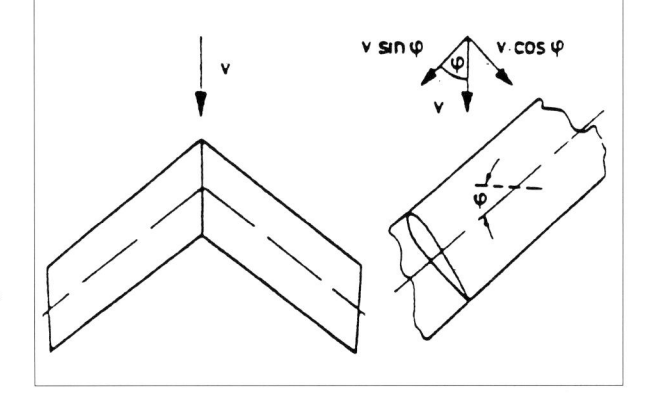

with fuselage and nacelles. In light of the test cross section of the wind tunnel, these models had to be kept quite small; their wingspan was about 7 cm.

At first, the aviation industry only reluctantly embraced the swept wing concept for its projects, as sweepback offered only limited advantages given the state of engine development and the maximum speeds possible at the time. The industry encouraged further studies using swept wings and models in various wind tunnels throughout the country, with a particular focus on understanding the effects of sweepback on flight behavior at slow speeds.

Sweepback studies and their results were kept so tightly under wraps that the swept wing concept and its high-speed advantages were unknown to the outside world until the end of the war and the arrival of American and British experts following the Allied troops into Göttingen, Braunschweig, and Oberammergau.

Planned Increase of Mach Number Leads to P 1101

Among the 335 takeoffs and landings and the 120 hours flown by the seven Me 262 test aircraft by late 1943, the high-speed flights of the V-3 in September and October are the most notable, with a top speed of 970 km/h (Mach 0.83) being attained. Although the goal of increasing the Me 262's critical Mach number took priority at an internal conference on 5 January 1944, Willy Messerschmitt also discussed offshoot proposals. A memorandum for record, dated 5 January, includes the comments:[1]

"As it is fundamentally recognized that housing the engines within the fuselage is better from a drag standpoint, Prof. Messerschmitt proposes constructing an ultra-high-speed airplane as a testbed for evaluating the current information with regard to increasing the critical Mach number. The fuselage of this aircraft is to be enlarged enough so that the engines can be installed side-by-side, one above the other, or in tandem, and be able to accommodate a wing with approximately 35% sweepback.

Studies are also in preparation to address the matter of engine air intake and exhausts. As the results of these will also be expected to take several months, Prof. Messerschmitt suggests locating the intake at the fuselage's aerodynamic stagnation point, with the exhaust opening as close to the tail as possible, as this would most likely be the configuration involving the least amount of risk. Intake flow loss and potential exhaust flow loss can be clarified on paper."

The meeting was a jumping off point for what became known as high-speed programs (Ho*chgeschwindigkeit*, or HG), with the swept wing concept up to and including variable geometry being the prerequisite. During the course of 1944 work began on six various HG programs, but by the end of the year cutbacks had to be made due to tightened resources:[2]

- Me 262 HG I - Aerodynamic refinements such as fitting a swept tailplane and reduced profile canopy (racing hood) were carried out on Me 262V-9 and tested from late 1944 to early 1945.
- Me 262 HG II - Testbed fitted with a wing having 35% sweepback. Destroyed in 1945 just before beginning trials.
- Me 262 HG III - Me 262 with engine nacelles blended in to fuselage and 45% sweepback.
- P 1101 - Experimental design with up to 45% sweep for the *Ein-TL-Jäger* concept.
- P 1106 - *Ein-TL-Jäger* with 40% sweep.
- Preliminary project work for supersonic developments.

Under the direction of Woldemar Voigt, Riclef Schomerus and a project group under the supervision of Hans Hornung were specially assigned to these projects in early 1944. Within the first half of 1944, this group produced a wealth of studies and projects.

Several twin- and four-jet *Zerstörer* and bomber projects initially dominated the project number of P 1101. During the middle part of that year, however, these were gradually eclipsed by various fighter designs under the same number, from which eventually crystalized the final Messerschmitt aircraft, the P 1101V-1, a design which almost became a reality.

P 1101 Configurations for the Requirement

Hans Hornung's P 1101 design, No. XVIII/113, from 30 August 1944, was selected for the requirement and prepared for submission. The brief description for this "smallest turbojet fighter with minimal costs and maximum performance" (using a He S 011 engine) included:[3]

---

[1] (Mtt-AG) *Probü*, memorandum for record, Me 262 Mach increase; discussion with Prof. Messerschmitt on 5 Jan 1944, Oberammergau, dated 22 Jan 44

[2] (Mtt-AG) project dept. management, memorandum for record, re: restrictions on high-speed development, Oberammergau, 20 Dec 1944
[3] Obb Forschungsanstalt [OFA, i.e. covername for the Messerschmitt development group moved from Augsburg to Oberammergau], P 1101 (*Ein-TL-Jäger*), brief description, Oberammergau, 13 Sep 44

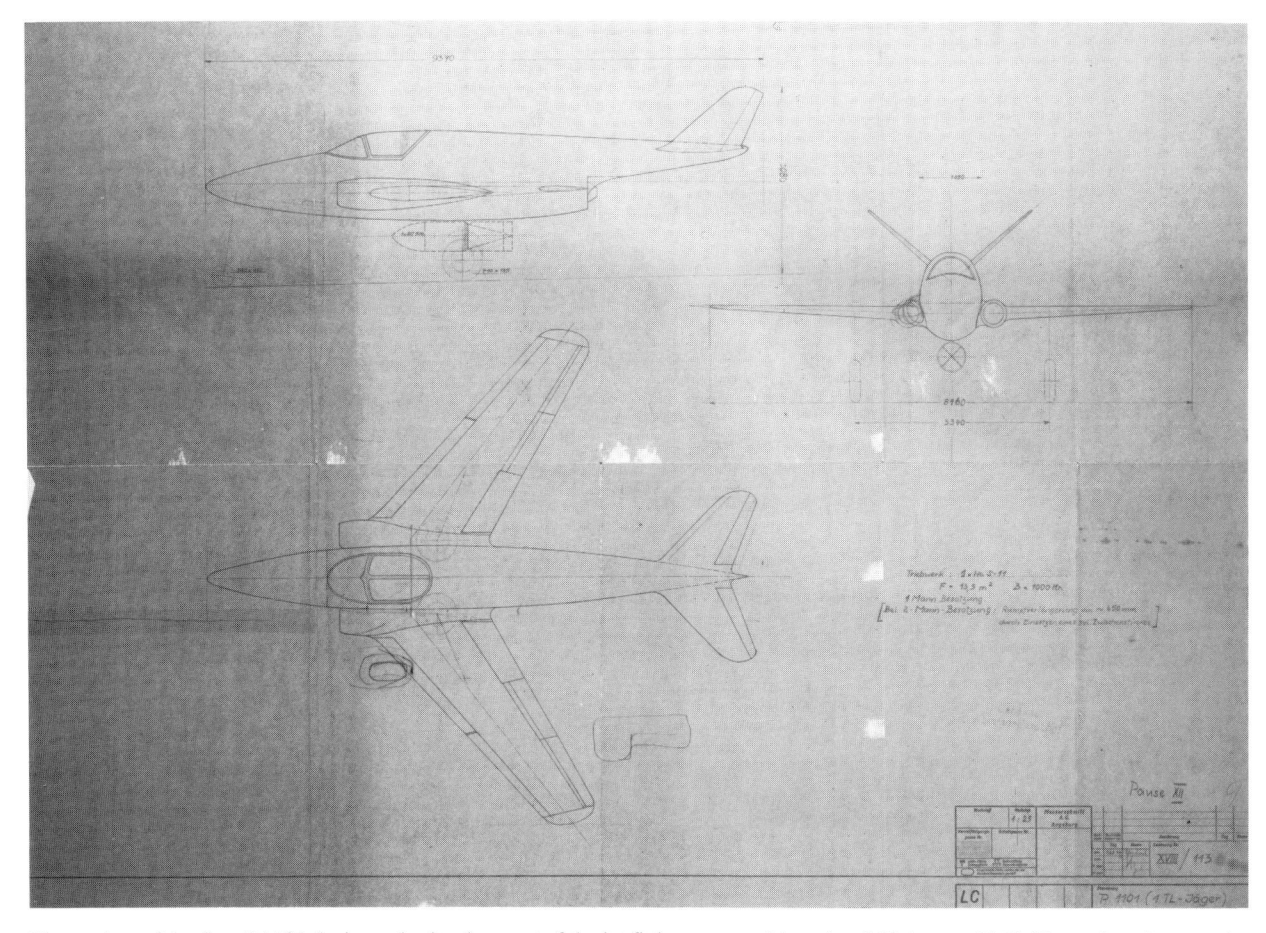

Three-view of the first P 1101 design submitted as part of the jet fighter competition, dated 30 August 1944. Planned engine was the He S 011. Armament was two MK 108 machine cannons. Wingspan 8.16 m, length 9.37 m.

- Sharply swept mid-wing design with retractable undercarriage, butterfly tail and with a turbojet engine in center fuselage section with "intake pants," whereby the air flowed through inlets on the wing leading edge and along the fuselage.
- Reasonable limitation of weapons and fuel weight with the focus on keeping overall weight down. General reduction of interference drag, with the engine installed in and blended into the fuselage.
- Use of a sharply swept wing and empennage with high-speed profiles with minimal wing thickness.
- Outer wings of Me 262 used and installed in swept configuration.
- Flight performance cannot yet be set, as there is only sketchy data available for the wings and none whatsoever for the fuselage regarding the effects of the Mach number on drag.

The controls, equipment, and armor was also expected to be inherited from the Me 262. And there were potential variants as well: a two-seater by installing a 650 mm long fuselage plug as well as fitting an auxiliary rocket engine for rapid climbing or combat boost.

At the instigation of the RLM project engineers from the companies of Blohm & Voss, Focke-Wulf, Heinkel, and Messerschmitt met together in Oberammergau on 10 September with the goal of setting comparable bases for calculating the performance of the turbojet fighter designs. Three E*in-TL-Jäger* designs with the He S 011 engine, in addition to a handful of variants with supplemental rocket engines, were presented and matched up with regard to fuel supply (830 kg) and armament (2x MK 108 with 60 rounds

[4] Mtt-AG, comparison of the *TL-Jäger* designs from Focke-Wulf, Heinkel, and Messerschmitt. Working group conference of the B.V., F.W., EHAG, and Mtt., report no. 1, Oberammergau 19 Sep '44

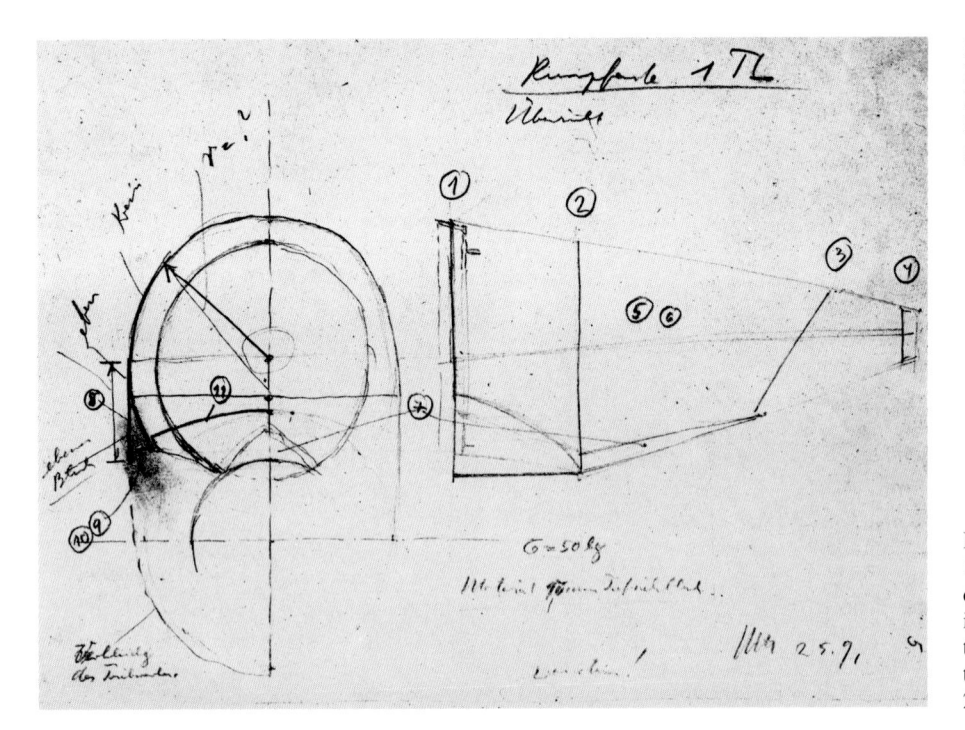

Design sketches by Professor Messerschmitt, dated 29 September 1944, showing the rear fuselage area of the single engine jet aircraft (P 1101).

Basic layout drawing of the P 1101 by Willy Messerschmitt, dated 15 October 1944. The tanks in the upper fuselage area were to have been rivet sealed, a practice already employed on the Me 261.

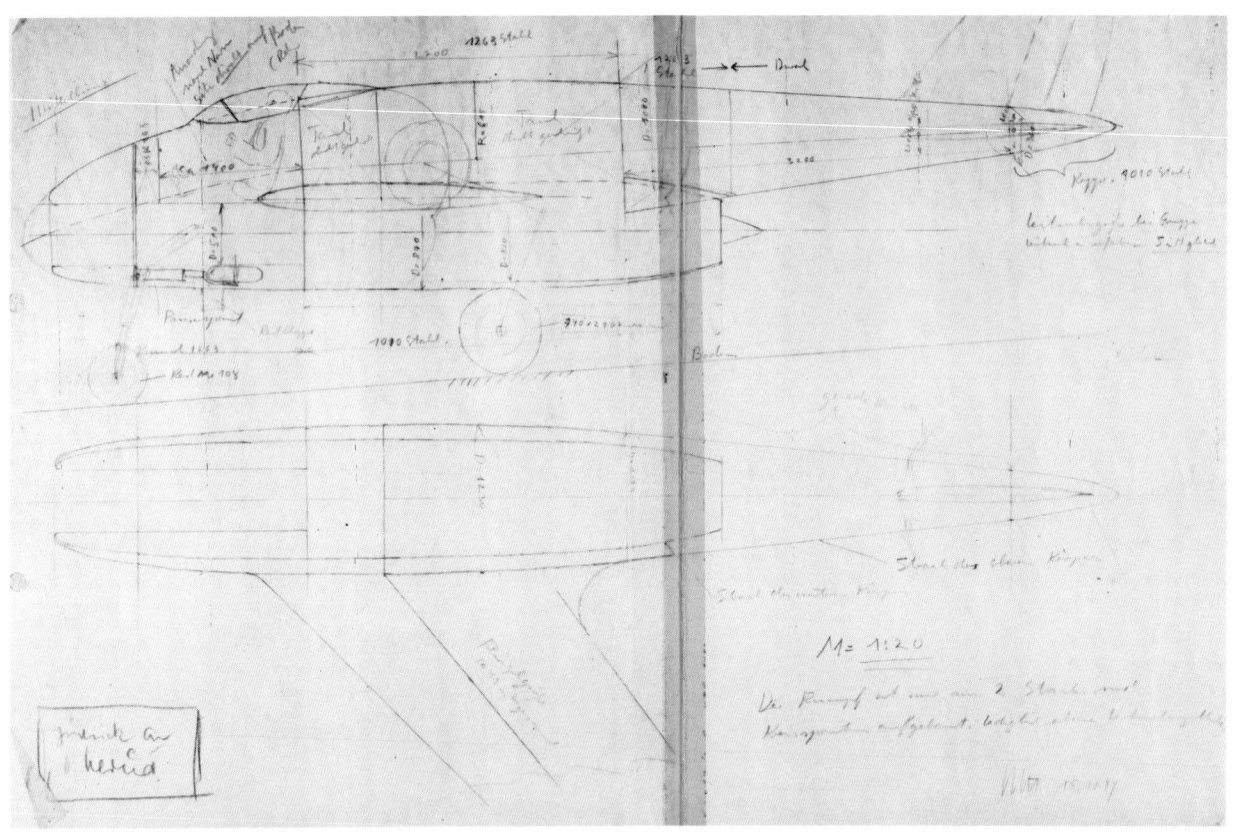

each) to the point where comparative studies could be made. The Messerschmitt P 1101 project, with the smallest wing area, turned out to have the lowest equipped weight, and with 1,000 km/h the second highest speed after the Heinkel project. The representatives from the individual companies tended to overestimate the flight performance of the competing designs, and there differences of up to 45 km/h. In the opinion of the experts from Heinkel and Focke-Wulf, the calculated speed level was too high, whereas the Messerschmitt engineers held the opposite viewpoint based on their experience with the Me 262.

One of the outcomes of the meeting was to expand the circle of involvement to include the companies of Junkers and Henschel, and another was Messerschmitt establishing a guideline for designing an *Ein-TL-Jäger* by reworking the P 1101 project yet again. Throughout the months of September and October Willy Messerschmitt was heavily involved in constructive details, which in turn were immediately passed on to the *Kobü* for reworking. One of his sketches from 15 September clearly shows the difference between the new airplane and the design from 30 August. It is also apparent that the layout was taken from an older project, the P 1092 from early 1943, with the exception being the much more pronounced 40% sweepback.

During an internal conference on 30 September Willy Messerschmitt decided to have an "experimental plane for studying performance and characteristics" built in order to research the unknown problems with the swept wing concept.[5] At the same time, however, development and planning of the military version was to continue unabated. The interim equipment plan from 2 November shows that several variants were in the works, i.e. clear-weather fighter, all-weather fighter, night-fighter, interceptor, and tactical reconnaissance aircraft. Featured aspects of the design included: RATO packs (2x 500 kp), pressurized cockpit, ejection seat, drop tanks, and different types of ordnance, e.g. bombs and air-to-air missiles (X-4 and Hs 298), as well as field conversion kits for fitting armor strong enough to withstand 12.7 mm shells from the front and 20 mm shells from the rear.

On 10 November the P 1101 project was turned over to the Ko*bü*[6] under the direction of Walter Rethel. The transfer also addressed the requirement for the experimental plane with regard to the military prototypes and series production. Adhering to the stipulation of utilizing as many parts and components form the Me 262 series as possible, the following list shows up just how radical a departure from the 30 August design this new plane was.

The fuselage consisted of an upper and lower cylindrical section joined together by a linking section. The central air inlet for the Jumo 004B engine now proposed (the He S 011 engine, not available until later, was required to be interchangeable) was located in the lower section.

The wings were based on the Me 262 wing and had a sweepback of 40%. The experimental plane was to have a wing which "would pivot at the spar join point in such a manner that a 35% and 45% sweepback would be possible

---

[5] (Mtt-AG, *Probü*), memorandum for record re: P 1101 equipment, Oberammergau, 2 Nov '44
[6] OFA, Hornung/Voigt, P 1101 project transfer, Oberammergau, 2 Nov '44

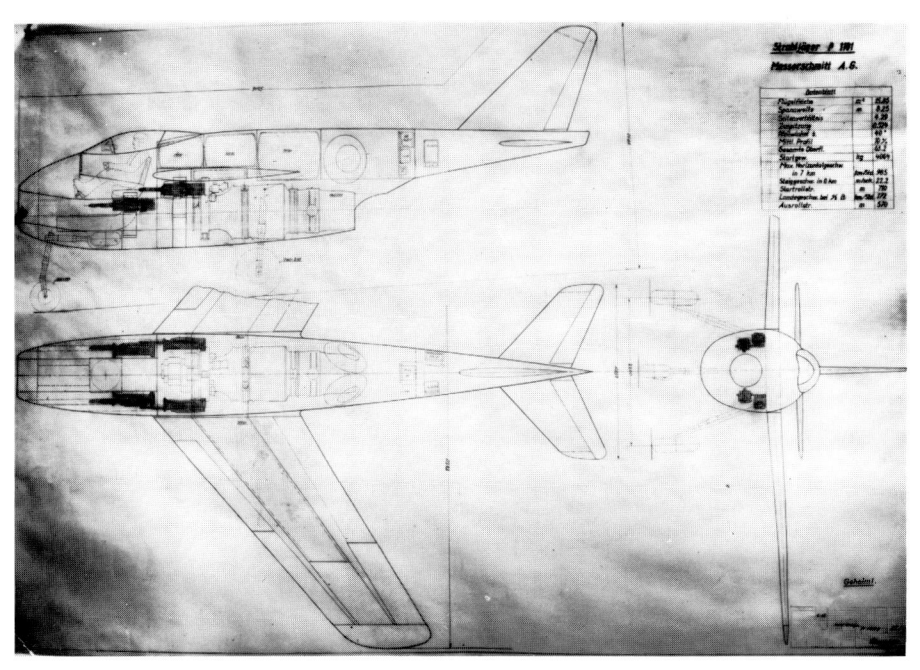

Three-view of the P 1101 from 22 February 1945, as submitted to the fighter design competition.

In order to measure the loss in thrust of the P 1101's approximately 2 meter long air intake, static tests were carried out in Lechfeld using a Me 262 with a 3 meter long tube attached to the front of the Jumo 004. Results showed that thrust loss remained constant at around three percent.

along the quarter-chord line. Dihedral must be able to fluctuate between +2% and -3%"

This was the first instance where the variable wing sweep and unique significance of the P 1101 was documented. Half a year before, the Messerschmitt design team had discussed the idea of varying the wing sweep in flight in reference to other P 1101 and P 1102 configurations. However, it wasn't until six years later that the concept was realized with the Bell X-5.

The empennage initially took on a normal configuration, with the tailplane of wooden construction. Over the next few months, six different configurations were reviewed using normal, butterfly, and T-tail layouts.

A model with a wingspan of nearly two meters was built to study rudder, leading edge slats, and flap effect in the DVL's wind tunnel at Adlershof.

P 1101V-1 Under Construction and as Pattern for Postwar Aircraft

Completion of the project transfer can also be considered the point at which construction of the P 1101V-1 began, with Willy Messerschmitt personally playing a key role. Assembly took place at the prototype construction department, which had relocated to Oberammergau, under the supervision of the reputable M*eister* Asam. Construction of the V-1 progressed rapidly over the next few months, and it was hoped that the first flight would take place at Lechfeld by March 1945. However, by early 1945 the material and transportation situations in the Reich had become so desperate that not even a single Jumo 004 engine could be diverted from production and delivered. As a result, on 29 April the Americans found the airplane 80% completed—and still without its engine.

Supported in his thinking by the RLM, Messerschmitt deserves much of the credit for the initiative in building the P 1101V-1. With it he had hoped to gain important data in the field of aeronautical research and become the winner of the competition as well. The decision to build should optimally have come before the end of the competition, which was attracting ever newer projects from the companies involved.

That Willy Messerschmitt himself did not consider the P 1101 as the optimal solution is evidenced by another rework of the design which led to the P 1106 in December of 1944. A variant with rocket propulsion, it was hoped that supersonic speeds could be attained. But even this interesting project did not satisfy his demands, for Messerschmitt withdrew the P 1106 at the decisive meeting of the development commission (Ent*wicklungskommission*) and the Chief of Air Armament (*Chef der Technischen Luftrüstung*) on 27 and 28 February 1945, instead substituting two new projects alongside the P 1101: the P 1110 and P 1111.[7] The competition produced no clear winners before the war's end, but it did generate recommendations for building prototypes of several of the projects which showed promise. His early decision to go ahead with the construction of the P 1101V-1 had given Willy Messerschmitt a marked advantage over the other competitors.

The P 1101 generated considerable interest among the victorious powers; in three instances it served as a direct pattern.

Similar were the cases for the majority of other projects which had been designed to the requirement; the numerous criteria they embodied with regard to jet aircraft were valid

---

[7] OFA, *Probü*, Hornung/Voigt, comparison of proposed *Ein-TL-Jäger* designs (preliminary work for the meeting of 27 and 28 Feb '45) TB no. 139/45 from 26 Feb '45

The P 1101V-1, about 80 percent complete, with a He S 011 engine mockup fitted, at a hangar in Oberammergau following capitulation.

Another photo of the P 1101V-1 with its wings swept back 40%. Wing components, engine covers and landing gear doors have not yet been fitted.

until well into the '70s. This one fighter requirement alone was of inestimable value to the Allies.

The Americans Arrive

On 29 April 1945 American troops occupied Oberammergau, where they found the nearly completed P 1101 prototype with a mockup of the He S 011 installed. The first American experts of the Aier Technical Intelligence Team, which had arrived on 7 May, very quickly discerned that what they had before them was a prototype of a high-speed fighter with swept wings, wings whose sweep could be variably controlled (albeit initially on the ground only). Among the first specialists of the Combined Advanced Field Team (CAFT) was Robert J. Woods, chief engineer and co-founder of Bell Aircraft Corporation. The French seized a portion of the P 1101 design blueprints and took them to Paris. The P 1101 prototype, 80% complete, was taken by the U.S. Air Force to Wright Field in Dayton for evaluation. The numerous reports on swept wings and high-speed aerodynamics were cataloged and translated and, after being published, were made available to the aviation industries of the victorious powers.

The P 1101 Becomes the Bell X-5

Robert Woods had such a strong interest in the P 1101 that the Bell company proposed to the Air Force an idea for developing and building 24 variable-geometry fighters based on the P 1101. But the Air Force rejected the proposal, claiming that the P 1101 had too little space for carrying the ordnance wanted or the fuel capacity for longer-ranged missions. In short, it was too small. In February 1949 Bell then proposed construction of an experimental plane using variable sweep—again based on the P 1101—and in July of that year was awarded a contract for building two prototypes under the designation X-5. Originally Bell had hoped to overhaul and modify the P 1101, but the airplane had suffered considerable damage in transit during the summer of 1948. The Jumo 004 jet engine was no longer reliable and would need to be replaced by a more powerful American motor. A test fit of a 21.8 kN/2225 kp Allison J 35-A-17 revealed that this engine could be used in its stead.

Bell borrowed both the idea and external shape of the P 1101 for its X-5 experimental aircraft, with only minor dimensional differences. The wings could move from 20% to 60% in about 20 seconds. On 20 June 1951 the first Bell X-5 took to the air, followed by the second on 10 December the same year. By the ninth flight it was possible to evaluate the entire sweep range. In the meantime, interest in the swept wing concept had grown considerably in the U.S., with NACA and the Wright Field Aircraft Laboratory becoming involved and organizing a comprehensive flight test program. The second X-5 crashed in 1953, and the first was retired in 1955 and can be seen today in the Air Force Museum in Dayton, Ohio.

The Bell X-5, built on the basis of the P 1101, was the first plane to fly with variable geometry and the ancestor of the Grumman XF-10 Jaguar naval fighter (first flight May 1952), the General Dynamics F-111 fighter-bomber (first flight December 1964), and also the multi-role Panavia Tornado (first flight 1974).

The P 1101 became a reality as the USA's Bell X-5. This 1951 photo shows the prototype with a wing sweep range of 20% to 60%.

Swing-Wing Types Throughout the World Designed on the Basis of German Research

The Swedish engineer Lars Brising, who later became the project director for the SAAB J 29, was able to make photocopies of German research reports on swept wings during a visit to Switzerland. Taking them home, he utilized the assistance of Behrbohm, a former Messerschmitt stress analyst and now a SAAB employee, to evaluate and apply the information therein. The result was the SAAB J 29 fighter, a design which relied heavily on the basic principles of the P 1101 and first flew on 1 September 1948. The wing sweep was only set at 25%, as any greater angle was felt to be too risky. The J 29 flew for many years, up until 1976, in fact, and over 600 of the type were built.

The first American design to employ sweepback was the North American F-86 Sabre (first flight 1 October 1947). Next to the B-47 this aircraft profited the most from Germany's wing sweep research. In the spring of 1945 North American project engineers got a glimpse of swept wing studies and Messerschmitt's project work and very quickly redrew their design to reflect a 35% sweep using the Me 262 profile of 11% thickness. According to these engineers, they estimated that they'd saved about three years of research time. Boeing's B-47 bomber project was redesigned with swept wings using documents and models found by Boeing developmental engineer George Schairer in Göttingen and Oberammergau, flying for the first time on 17 December 1947.

At the DVL in Berlin-Adlershof, Soviet engineers discovered documentation concerning swept wings, which became reflected in the first Soviet swept wing design, the Lavochkin La -160 research plane. This design, too, relied upon the basic concepts of the P 1101 and had a wing sweep of 35%. Its first flight took place on 24 June 1947. The experience gained was exploited in the MiG-15, a slightly later type which the USSR put into full scale production. The MiG-15's first flight occurred on 30 December 1947.

Shortly after the war Professor Kurt Tank and several engineers from Focke-Wulf went to Argentinia, where they developed the Pulqui fighter by making use of the Focke-Wulf Ta 183 project with 40% sweep. Its maiden flight took place on 16 June 1950.

For all of these initial swept wing designs it was the Messerschmitt P 1101 fighter—despite never having taken to the air itself—which was their precursor, an exemplary role model, and their stimulus.

Enzian Remote-Controlled Anti-Aircraft Missile

In 1940 Willy Messerschmitt passed on a developmental contract for a remote-controlled glide bomb to his Department "L" under Dr. Alexander Lippisch. The bomb was to effectively act as a flying torpedo. The firm of Siemens developed the control system. Designated the FG 04/M 09, the body was of wooden construction. After completion, the finished missile was tested at AVA Göttingen's wind tunnel, and on 29 April 1941 it was launched from its Bf 161V-1 mother ship and remotely guided by Heini Dittmar observing from the nose of the Bf 161. It was lost during this first test, but once found and repaired underwent a second test. This time the FG 04 was destroyed. Consequently, Dr. Lippisch began working on his own designs for radio-guided bombs, making use of the Me 163's fundamental shape. But these efforts also ran into snags at an early stage. Whereas these projects were of the air-to-ground variety, in 1943 the stepped-up pace of Allied bombing raids necessitated a rethinking of the whole idea—what was needed were ground-to-air missiles (anti-aircraft). Under the direction of Dr. Hermann Wurster, Department "L" came up with the FR 1 through FR 6 anti-aircraft missiles based on the principles behind the Me 163. These projects all made use of a constant chord wing and a cylindrical fuselage. In June 1943 the Messerschmitt AG was awarded the development contract for the Enzian ground-to-air missile. The RLM's department responsible for overseeing such programs, GL/Flak-E, established the most critical points of the design as follows:

- High subsonic speed
- Reliance on the Me 163 rocket fighter, i.e. tailless swept-wing design
- Wooden construction, suitable for mass production
- Walter 109-739 double-fuel rocket motor
- Four jettisonable solid-fuel takeoff booster rockets
- 500 kg warhead
- 20 km range
- Joystick control from ground

Developed and built at Messerschmitt by the Lippisch team in 1940/41, the FG 04/M stand-off glide bomb is seen here at the AVA's large wind tunnel in Göttingen. The missile was subsequently released from the Bf 161V-1 and remotely guided in two separate tests.

Launch of one of the first E-1 "Enzian" prototypes at Peenemünde in June 1944. It was powered by a Walter liquid fuel R I-203 rocket motor and four Rheinmetall-Borsig RATO boosters. The missile had a launch weight of about 1,800 kg, a span of 4 meters, and a length of 3.75 meters. Maximum velocity was 240 meters per second. One third of the approximately 40 Enzian missile launches were successful.

Design work on the Enzian missile (Gerät 2002) began in November 1943 under the supervision of Dr. Wurster. Part of his team took up residence at the Holzbau-Kissing GmbH company in Sonthofen, Allgäu. Dr. Wurster and his close colleagues, including Fliegerstabs*ingenieur* Peuter Nauschütz (aerodynamics), technical engineers Schnaller (stress analysis) and Bosch (design), *Oberleutnant* Kurth (flight mechanics) and Dr. Graf Thun and *Oberstingenieur* Dr. Thiel, occupied rooms at the Linderhof palace.

Construction of the first E-1 Enzian prototype got underway in January 1944 at Sonthofen. Except for the warhead housing and a few metal fittings, the entire missile was built using conventional aircraft wood construction. The design required the application of several new manufacturing techniques, and the wood manufacturing company's skills were quite valuable when it came to building pattern pieces and airframe parts. The innovative pressed wood method was used to make fuselage and wing halves pressed into shape from wood shavings and a binding agent. The halves would then be glued together. Output of the E-4 production version was expected to reach 5,000 per month.

Development of the Walter 109-739 rocket moter was delayed considerably. This engine had been specially designed for the Enzian missile and burned nitric acid as an oxidizer. With the delays, however, it was decided to fit the lower-powered Walter R I-203 (109-502) engine, which had an output of 14.7 kN/1,500 kp and a burn time of 45 seconds. The four Rheinmetall-Borsig takeoff booster rockets (R I-503) each provided 14.7 kN/1,500 kp of thrust for a four second burst. The Enzian anti-aircraft missile was fired from a 7 meter long rail fitted to an 88 mm gun mount.

The electro-hydraulically operated rudder made use of servo motors from the Siemens K 12 autopilot. Control signals were sent using the Kehl/Straßburg system.

The first test launch took place on 29 March 1944 at Peenemünde. There were many problems and several premature crashes in the test program (involving a total of 38 Enzians), particularly at first. Speeds of Mach 0.79 and altitudes of 7,000 meters were recorded. Relatively good results weren't achieved until flights 31 through 38, occurring in early 1945. The E-1 series underwent numerous modifications over the course of the test program, which lasted about a year. These changes in turn led to delays in the delivery of the test missiles.

At the same time, work was underway on follow-on developments, especially the E-2 version with integrated fuselage tank and a Walter 109-739 motor, and the E-3 version with spherical tank and Konrad rocket motor. Design work began on the large-scale production variant, the E-4, in early 1945. This model was to have incorporated the liquid fuel rocket engine developed by Dr. Konrad (TH Berlin), a motor which had an output of 19.6 kN/2,000 kp—climbing to 2,200 kp—and a burn time of 70 seconds. With a launch weight of 1,900 kg, the Enzian E-4 was expected to reach a velocity of Mach 0.9 and a ceiling of 16,000 m. Ultimately, a design was drawn up for a supersonic E-5 variant, a sharply tapered cruise missile with reduced payload.

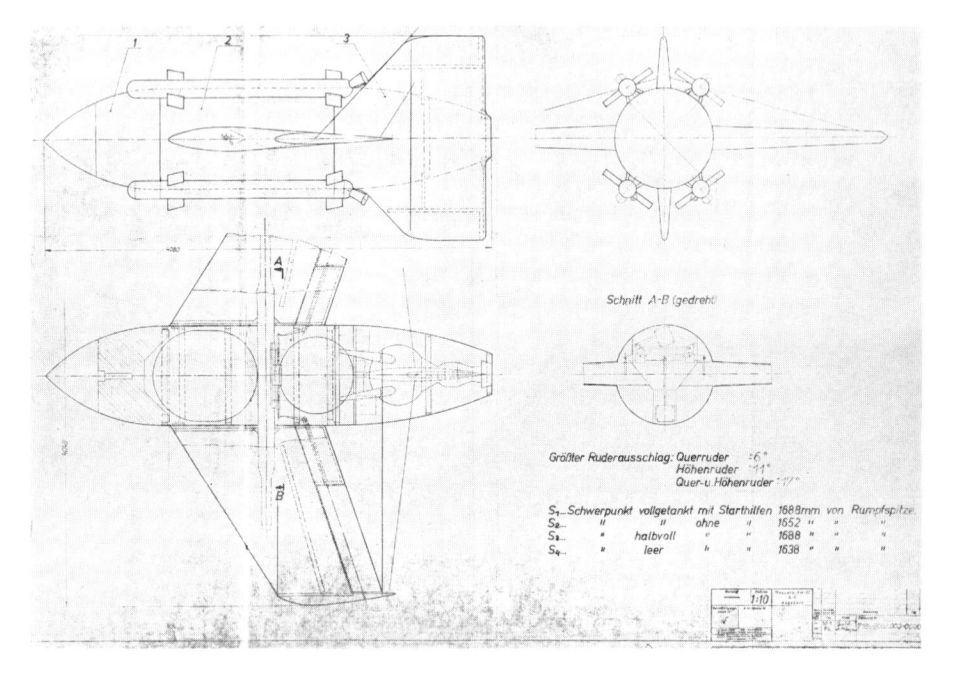

Three-view of the planned production model of the E-4 Enzian (Gerät 2002), from January 1945. It was to have been powered by the newly developed Konrad liquid fuel rocket engine having a burn time of 72 seconds. Maximum velocity was to have been 300 meters per second, with a ceiling of 16,000 meters and horizontal range of 25 km. Launch weight was 1,920 kg, span 4.05 m, and length 4.08 m.

In any event, Messerschmitt produced a total of just 60 Enzian anti-aircraft missiles, none of which were used operationally against bombers. After the war, British and American interrogation officers and techical experts were completely taken by surprise upon discovering this wooden missile, a weapon which had hitherto escaped their attention completely. The wooden construction method found no subsequent application, and the design has thus remained unique in the annals of aviation.

Right: Today only a few of the approximately 60 E-1 Enzian missiles still exist, such as this one in the Royal Australian Air Force Museum at Point Cook near Melbourne.

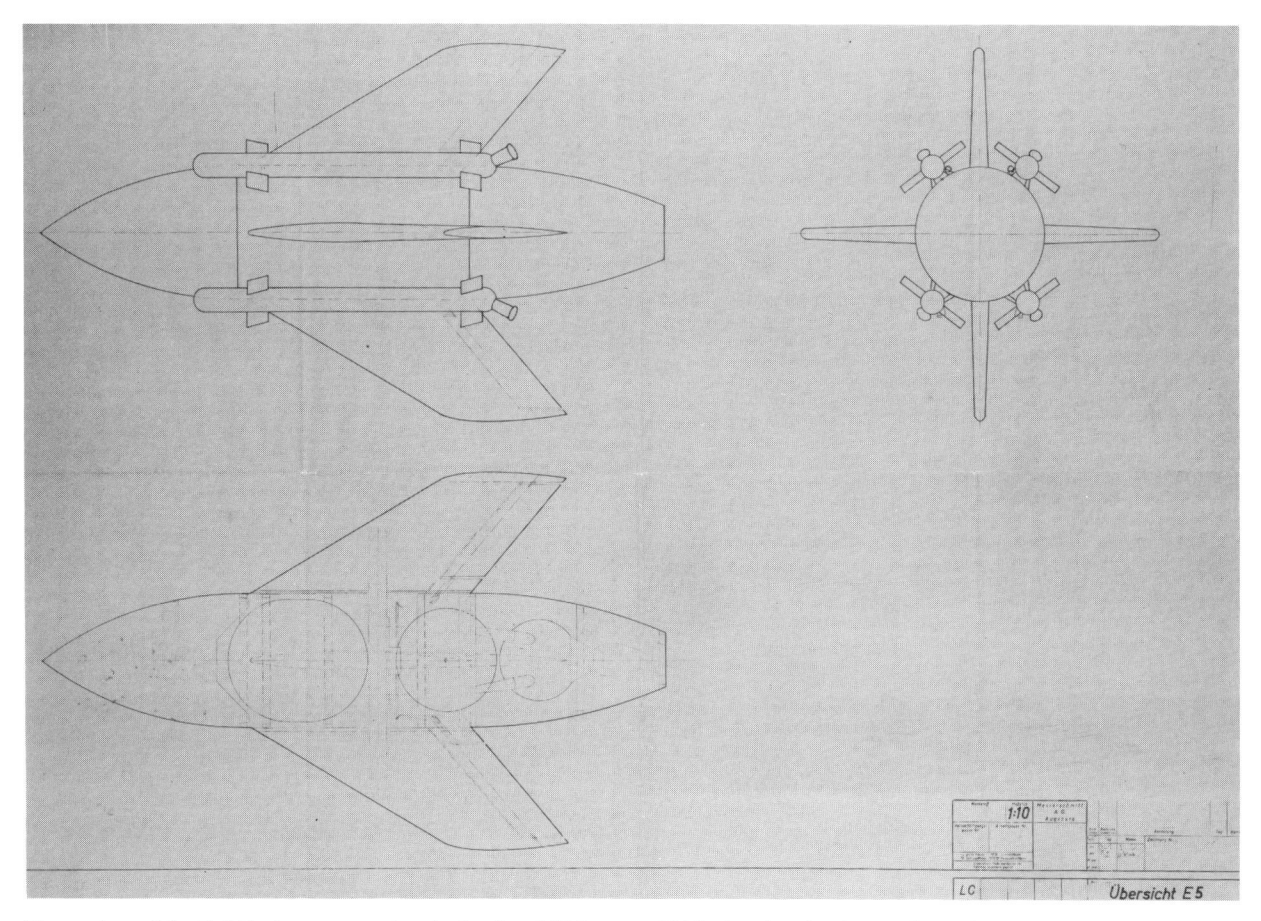

Three-view of the E-5 Enzian supersonic missile, dated 31 January 1945, showing the Konrad liquid fuel motor and fuel for a burn time of 54 seconds. It had a velocity of Mach 2, a launch weight of about 1,900 kg, a span of 2.45 m, and a length of 4.60 m.

Messerschmitt Projects From P 1000 To P 1112

The numbering of those projects tackled by the Probü had most likely been started by Kurt Tank as early as 1930 in the BfW's newly established development section, beginning with the number 1000. By 1932 the numbers P 1012 and P 1020 had been confirmed. Robert Lusser continued the list from 1933 onward, with the first verifiable number being P 1035 (= Bf 110). During the war, Hans Hornung maintained the listing. According to him, it ended in 1945 with the number P 1112. (The number P 1116 often seen found in publications is based on a typographical error and is in fact a variant of the P 1106.)

With the original list missing, the following contains all the P-numbers which can be verified through documentation. It does not include those many projects which were not given a number, noted down by the Americans during and after their interrogation of Messerschmitt employees after the capitulation and which later cropped up in various publications. Nor will the reader find projects from Department "L" (Lippisch, 1939-1943), which had their own numbering system. Those Lippisch products which were relevant to Messerschmitt aircraft are referenced, however. Those projects which did make it to production are referred to by their RLM numbers in their respective chapters and throughout the book generally. Technical details for many of the projects listed here are sketchy due to the fragmentary nature of the documents still available.

P 1012 - High-speed mailplane (1932)
P 1020 - High-speed airplane (1932, became He 72)
P 1035 - became Bf 110, Bf 161, Bf 162
P 1051 - became Bf 163
P 1053 - started as Bf 164, round-the-world plane (1935- 1937)
P 1059 - became Me 209
P 1060 - became Me 210
P 1061 - became Me 264
P 1062 - became Me 261
P 1065 - became Me 262 (also initially designated P 65)
P 1070 - Twin-jet fighter design (1940). Wingspan: 8.20 m, length 8 m, height 2.90 m, area=13 m2, take off weight =2,800 kg, much smaller than Me 262 of same period. The engines, located halfway out along the wings had a diameter of just 60 cm according to manufacturer. Armament was two machine guns.
P 1072 - Strategic bomber (1940). Several versions were drawn up: Ho 1 with 4x Jumo 223 (diesel), 62- ton takeoff weight with 2 t bombload, range 15,800 km, wingspan 40 m; Ho 2 with 2x DB 613, 47.5-ton takeoff weight, range 11,600 km, wingspan 36.7 m; Ho 3 with 2x DB 604 and 2x Jumo 223, 33-ton takeoff weight, range 10,000 km, wingspan 31.6 m; S 1 with 6x Jumo 223, 107.5-ton takeoff weight, wingspan 54 m, speed 560 km/h.
P 1073 - Study for a long-range mother plane (P 1073A) carrying 3 parasite fighters (P 1073B) for Atlantic/US operations (1940) with a range of 16,000 km. P 1073A: 8x Jumo 223 engines, empty equipped weight 48.6 tons, add'l weight 79 tons, including 6 ton bomb load and 4.8 tons for the three fighters. Takeoff weight 128 tons, speed 420 to 600 km/h, wingspan 63 m, wing area 330 m². P 1073B: 1x BMW 3304 jet engine, weight 1,620 kg, wingspan 4.4 m, wing sweep 35° (foldable), length 5.9 m, armament 2x MK 103/108. P 1073B (1944) new variant, brought up to He 162 standards and compared: length 8.65 m, wingspan 7.2 m, wing area 11.16 m², takeoff weight with BMW 003 turbojet and 475 kg fuel was 2,500 kg. Speed 790 to 835 km/h.
P 1075 - Long-range bomber. Follow-on development of P 1061 (Me 264) with six engines (1941)
P 1079 - Project study series by Rudolf Seitz for a high- speed bomber using pulsejet engines (spring/summer 1941). Several variants with swept wings and spans of 5 to 6 m. Powered by one or two 500- or 600 kg jets. Bomb payload up to 1,700 kg externally. Built as Me 328 with non-swept tapered wings (1942).
P 1085 - Ultra-long-range bomber and reconnaissance aircraft (1942/1943). Me 264 with six engines and having lengthened fuselage and enlarged wings. Engines: 6x BMW 801D, takeoff weight 75 t, range up to 17,000 km (see also Me 264).
P 1090 - Twin engine multi-purpose aircraft (modular system, 1942/1943) with 2x DB 603 or BMW 801D or 2x turbojet engines. Easily converted for a variety of roles. Variable wing area ranging from 28 m² to 36 m² by swapping out wing sections. Takeoff weight 9.7 to 11 t (see also chapter "*Aircraft Using the Modular Assembly Principle - The P 1090 Project*"
P 1091 - High-altitude fighter development in three stages, based on Bf 109/Me 209 (1943). Stage I with DB 605 and O$_2$ system, wingspan 13.20 m; Stage II with DB 605 and O$_2$ system, wingspan 21 m, Stage II with DB 603A and TK 15, wingspan 21 m (see also Bf 109).
P 1092 - Multi-role jet aircraft. As a result of Willy Messerschmitt's efforts toward commonality among aircraft types (see P 1090), in May 1943 such an example was explored using jet propulsion. The basic model had the slightly swept outer wing sections of the Me 262 with a span of 8.4 m and an area of 12 m²; length 9 m. The aircraft was therefore much smaller than the Me 262. The P 1092A turbojet version was similar to the P 20 project being developed by the Lippisch team at the same time (Me 163 with Jumo 004) and was planned as a fighter. Version B with rocket engine was an interceptor, Version C a high-speed bomber with 2x As 014 pulse jets (as Me 328), Version D a heavy fighter with two turbojet engines, Version E a nightfighter. Other applications included long-range fighter, torpedo bomber, dive bomber.
Hans Hornung first began working on the *Ein-TL-Jäger* program with second series, the P 1092/1-5, drawn up in June and July 1943. The variants differed primarily in the fuselage and wing areas. The basic design remained untouched and formed the basis for the P 1101 a year later (wingspan 9.4 m, length 8.10 m, area 12.7 m²).

P 1070
(1940)

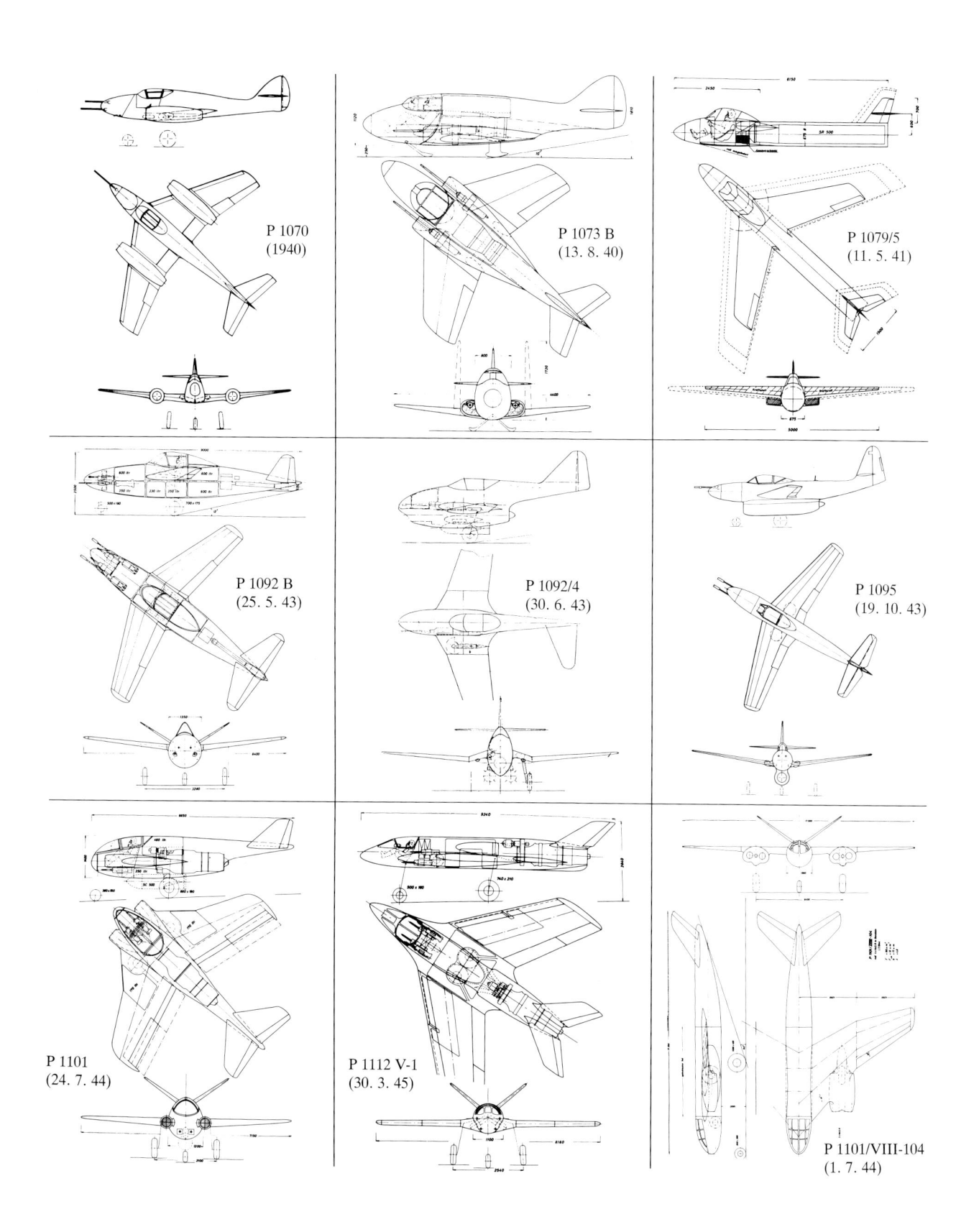

P 1073 B
(13. 8. 40)

P 1079/5
(11. 5. 41)

P 1092 B
(25. 5. 43)

P 1092/4
(30. 6. 43)

P 1095
(19. 10. 43)

P 1101
(24. 7. 44)

P 1112 V-1
(30. 3. 45)

P 1101/VIII-104
(1. 7. 44)

P 1095 - Single turbojet multi-purpose aircraft (1943). Searching for a role for the halted Me 328 project, Rudolf Seitz's team initially worked out several variants using the Me 328 on hand with a Jumo 004 engine, then explored other options as this study from 10/19/43 was to utilize 50% of the components from the Me 262 (wings, control surfaces, cockpit) and the Me 209 (empennage, landing gear). The Jumo 004 was to have given it a speed of 810 to 860 km/h. Having a takeoff weight of 3,620 kg, the aircraft could be employed as a figher (with 2x MK 103/108), a ground attack plane (5x MG/MK), a fighter bomber (+ 2x 250 kg bombs), a reconnaissance plane or a high-speed bomber. Wingspan 9.77 m, area 15.3 m^2.

P 1099 - Two-seat fighter, heavy fighter, night fighter based on Me 262 with completely new (larger) fuselage (early 1944). Engine: 2x Jumo 004C, later He S 011, empty equipped weight 5,364 kg, takeoff weight 10,784 kg, wingspan 12.61 m, length 12 m.

P 1100 - Two-seat bomber/high-speed bomber based on Me 262 like P 1099. Bomb payload of up to 2,500 kg in fuselage, defensive armament in barbettes possible both fore and aft. Comprehensive project working in early 1944 to mockup stage.

P 1101 - Jet aircraft project studies (summer 1944). Numerous designs with wing sweeps ranging from 30° to 45° were drawn up by the *Probü* in conjunction with these studies for a new generation of swept-wing jet aircraft. They included crescent wing designs, variable-sweep, and even scissor-wing concepts, plus VTOL combinations—all falling under the P 1101 project number.
An invitation to tender by the OKL for an E*in-TL-Jäger* (single-jet fighter) with specifications drawn up by the RLM's Technisches Amt led the *Probü* to produce its first design in late July 1944 (with cheek intakes). Messerschmitt presented another design to the Amt on 20 August, which generated much discussion among those participating in the competition. In September the final design was drawn up having a central intake (basically reworking the P 1092 project from the previous year). Beginning in the fall of that year, the project (retaining the P 1101 nomenclature) ran on three separate tracks: 1 - involvement in the competition and comparison with competing designs, 2 - based on Willy Messerschmitt's decision, construction of an experimental plane with modifiable wing sweep of up to 45°, 3 - preparation for construction of the fighter with combat equipment (production, with 4x MK 108, empty equipped weight of 2,594 kg, takeoff weight of 4,064 kg, wingspan 8.25 m, wing area 15.85 m^2, wing sweep 40°. Estimated speed would be 980 km/h (see also chapter on P 1101).

P 1102- Study for a jet bomber with three turbojet engines, two located on the forward fuselage and one in the tail. Variable sweep wings, in landing position = 27 m^2 (1944).

P 1103 - Rocket-powered parasite fighter (1944). Wings based on V1 missile with a span of 5.76 m and an area of 6.1 m^2, length 5.83 m, armament 1x MK 108 with 100 rounds, takeoff weight 1,600 kg, later also listed as 2,382 kg. Aircraft also later planned for rigid tow, e.g. behind an Me 262.

P 1104 - Rocket-powered interceptor (Walter HWK 109-509 A-2), probably similar to P 1103 with V1 wings. Wing span 6.2 m, length 5.48 m, takeoff weight 2,400 kg. Take offs (also horizontal) and landings on skid undercarriage. Willy Messerschmitt proposed the idea in October 1944 as an economical alternative to the Me 163 or Ba 349, as the type would only require 650 man-hours to build and 500 to 1,000 per month could be built.

P 1106 - *Ein-TL-Jäger*. Directly derived from P 1101 and briefly (December 1944) submitted to OKL for the fighter competition. A light fighter, it had an empty equipped weight of 2,370 kg and a takeoff weight of 3,788 kg. Wingspan was 6.65 m, length 9.19 m, wing sweep 40°. Armament included 4x MG/MK, with a potential external under wing payload of 2x 250 kg bombs or X-4 missiles. Increased range with blended-in fuselage plug for additional fuel.
Separate from the competition, the P 1106R variant was drawn up powered by two Walter rocket engines, each with 1,800 kg of thrust. This was theoretically to have given the type supersonic capabilities (wingspan 6.74 m, length 8.42 m, area 13 m^2).

P 1107 - Long-range jet bomber and reconnaissance aircraft (1945). Although by early 1945 the emphasis was on building fighters and there had been no winner yet declared in the emergency fighter program competition, a requirement went out for a strategic bomber. Willy Messerschmitt became intensely involved in work on this project, particularly since he had an eye toward using such an airplane as a commercial airliner after the war. The P 1107/II project from 24 February 1945 envisioned a mid-wing design with four He S-011 engines embedded in swept wings. Wingspan was 17.3 m, length 16.7 m, area 60 m^2, aspect ratio 17.3 m, takeoff weight 30 t, of which 15 t was fuel and 4 t was bombs. Calculated range was 5,800 km at a cruising speed of 900 km/h. The P 1107/I differed from the original by having a different engine layout in the wings and by a T-tail.

P 1108 - Strategic bomber (1945). Willy Messerschmitt submitted three proposals for the requirement, including the tailless P 1108/I and II with four turbojet engines (strongly influenced by the Lippisch team). Wingspan was 20 m, length 16.25 m. Another design also carried under the designation P 1108 was a butterfly tail concept with four jet engines buried in the wing trailing edge (span 19.80 m, length 18.20 m). It was this design in particular, not unlike the future de Havilland Comet, that Willy Messerschmitt hoped to be able to employ in the civil market. It was estimated that the aircraft could cover the distance from Frankfurt to New York in about six hours' non-stop flying.

P 1110 - *Ein-TL-Jäger*. Messerschmitt's main proposal for the OKL fighter competition of 1944. The narrow fuselage with oval cross section had its wings mounted low and swept 40°. The He S-011 engine was located in the aft part of the fuselage. There were two proposals for the

P 1102/5
(1944)

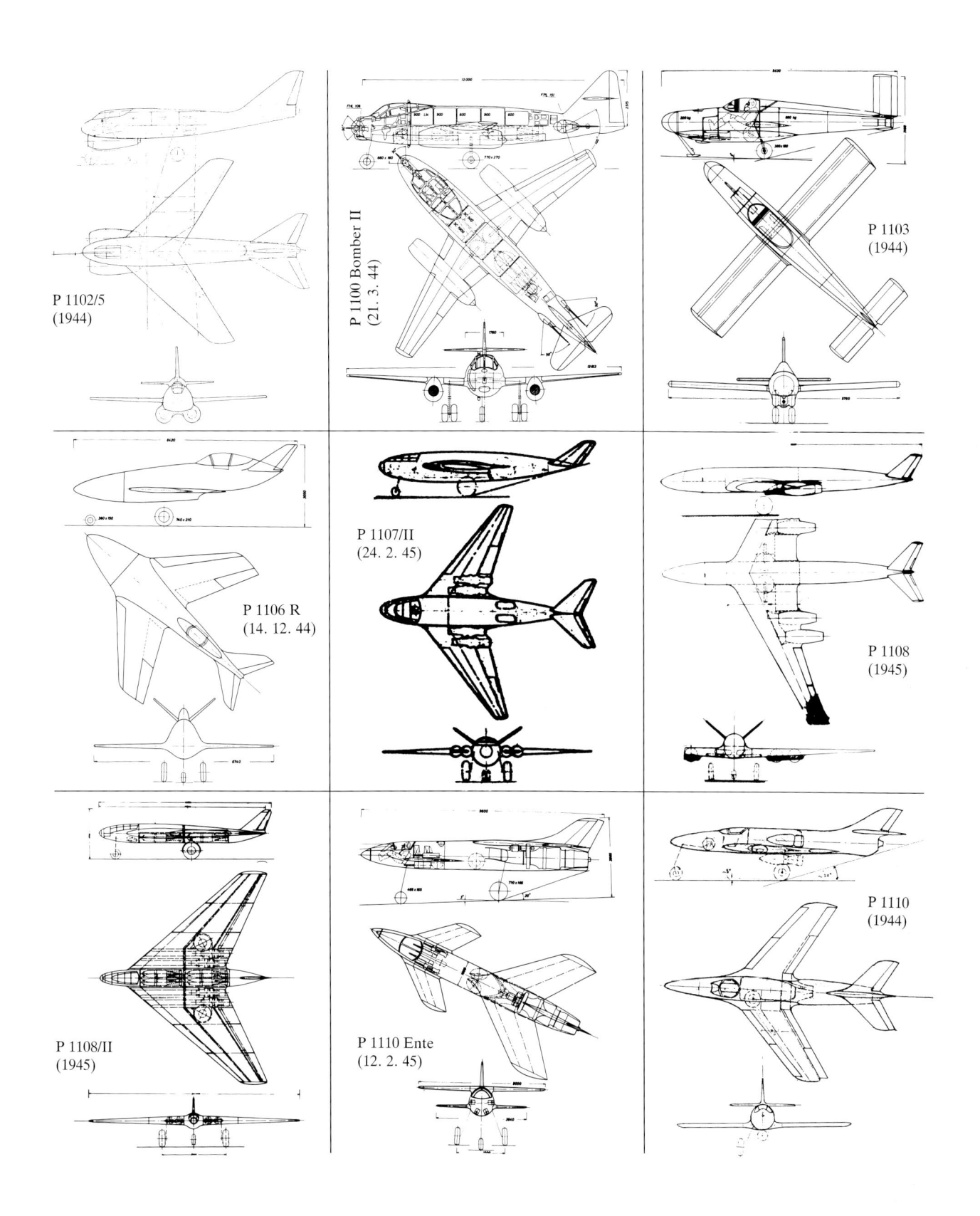

P 1100 Bomber II
(21. 3. 44)

P 1103
(1944)

P 1106 R
(14. 12. 44)

P 1107/II
(24. 2. 45)

P 1108
(1945)

P 1108/II
(1945)

P 1110 Ente
(12. 2. 45)

P 1110
(1944)

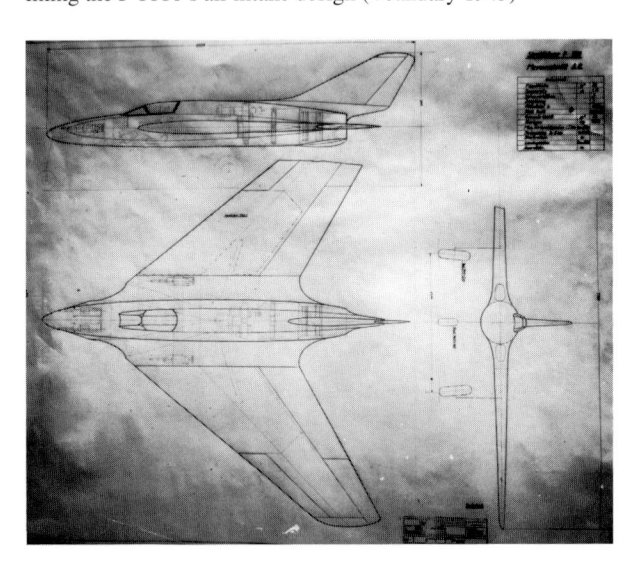

intake: 1) intakes on either side of the fuselage, 2) annular, around the entire aft fuselage section. This latter variant had a V tail. Main data for the competition included: takeoff weight 4,290 kg, wingspan 8.25 m, area 15.85 m², profile thickness 12%, maximum speed 1,000 km/h at 7,000 m, armament 3x MK 108. In February 1945 the P 1110 was redesigned with a canard wing layout, with the swept main wings having a span of 5 m and the swept canards a span of 3.54 m. The length was 9.6 m.

P 1111 - *Ein-TL-Jäger*, belatedly submitted (early 1945) to OKL for fighter competition. Alternative proposal to P 1110. Tailless aircraft with a span of 9.16 m and an area of 28 m². Wing sweep was 45°. Calculated speed was 995 km/h. Empty equipped weight was 2,740 kg, with a takeoff weight of 4,282 kg. Armament was 4x MK 108.

P 1112 - *Ein-TL-Jäger* (1945). Based on the P 1111, several designs with modified wings and a cockpit fully blended into the lines of the fuselage. Intensive studies progressed to mockup stage and combining the P 1110/I and II projects with the P 1112 (drawing shows P 1112V-1 from 3/30/1945. Wingspan 8.16 m, length 9.24 m, armament 4x MK 108.

Professor Willy Messerschmitt's instructions to the *Probü* outlining the P 1110's air intake design (4 January 1945)

Below left: The P 1111, second-to-last project before the war's end in 1945.
Below right: P 1101/XVIII-108: scissor-wing design in both monoplane and biplane configuration. Wingspan was 9.40 m, and length 12.05 m (11 July 1944).

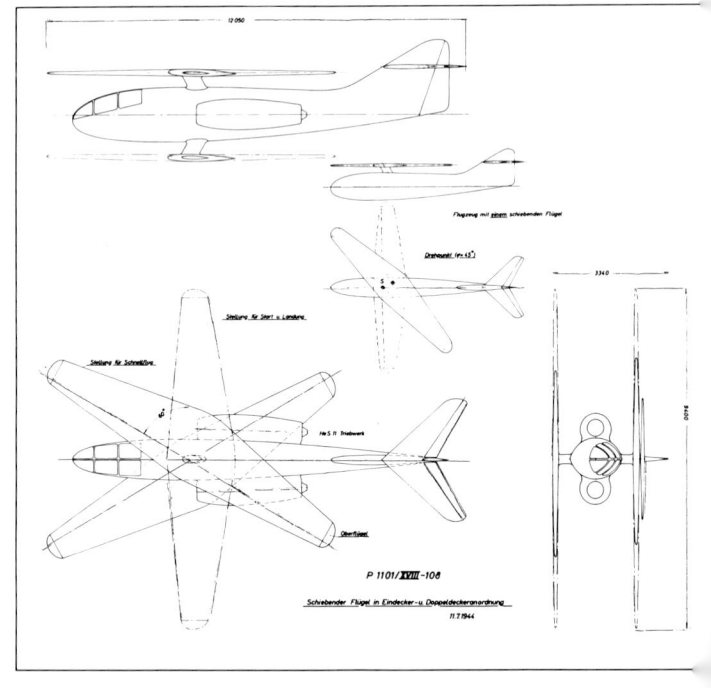

294

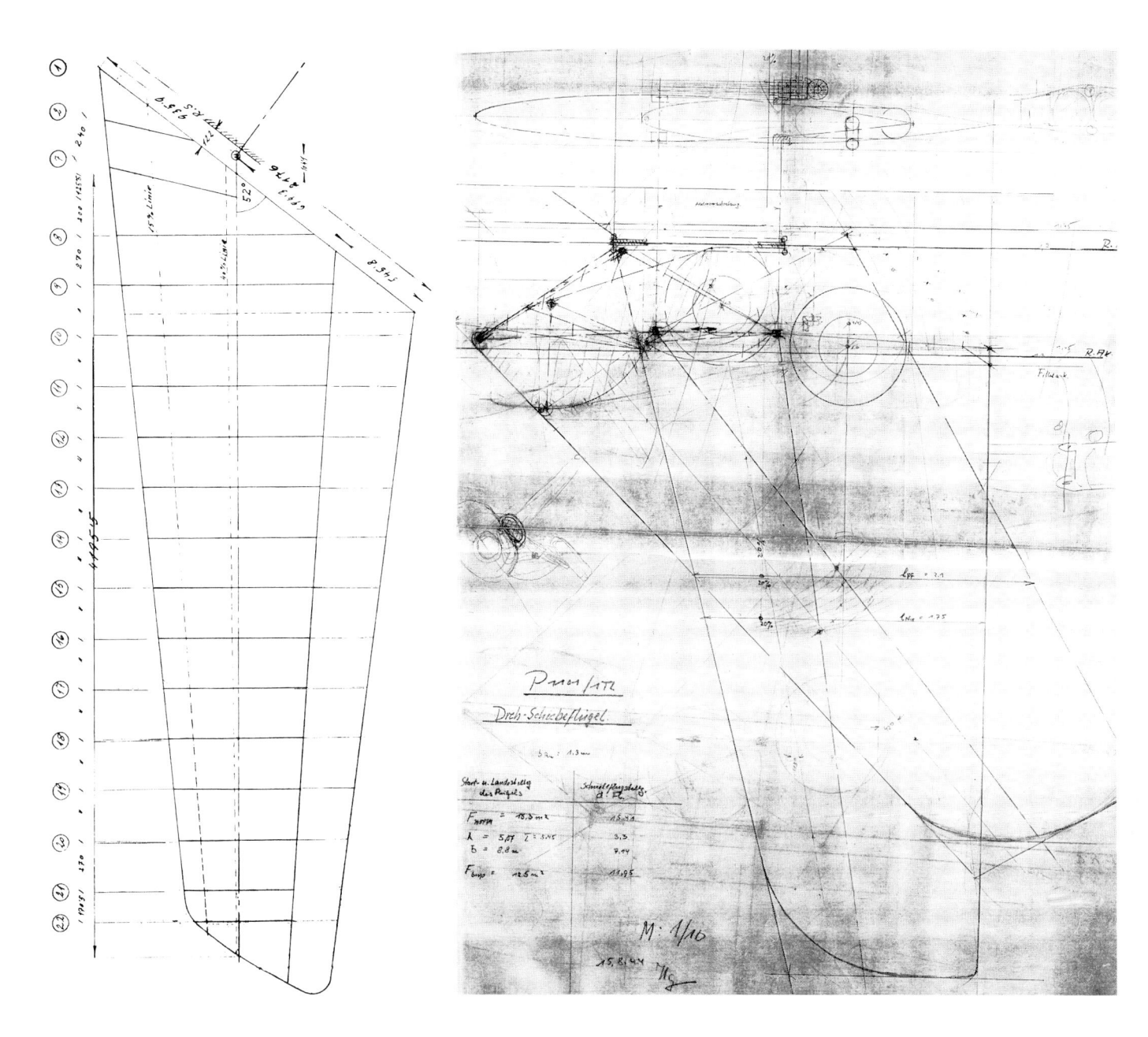

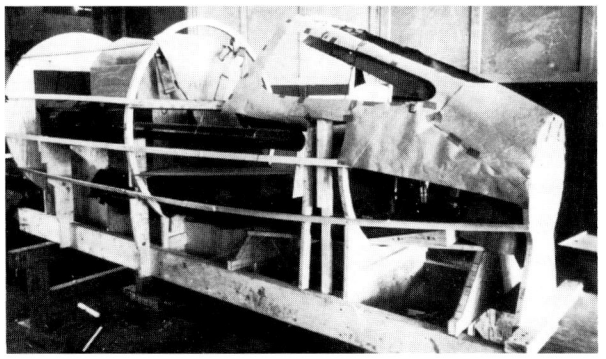

Above left: The concept of the Me 262's outer wing section became the basis for the P 1101's standard swept wing (drawing) and numerous other Messerschmitt projects. Even the Americans utilized it in the first swept wing fighter to go into production on a large scale, the F-86 (first flight 1947).

Above right: In the summer of 1944 studies were carried out on variable wing sweep at Mtt's *Probü*. Modifications were made and a patent was applied for a scissor-wing design. This shows a study by Hans Hornung for the P1101 turbojet fighter, dated 15 August 1944.

Left: Cockpit mockup for the P 1112 in Oberammergau (1945).

Projects with Official or Unofficial RLM Type Numbers

It was common practice to track project offers made to the RLM under the project numbers assigned by the companies. Once the RLM awarded a contract, the project was given a type number. Companies also requested specific numbers which were submitted to the RLM for approval. The following RLM or Messerschmitt request numbers remained projects:

Me 155 - Special carrier fighter based on the Bf 109G (1942)
Bf 164 - Aircraft for 'round-the-world flight (1935-1937). Number recycled for:
Me C 164 - (Me 308) Commercial airliner. Developed in 1941/1942 at Caudron
Bf 165 - Long-range bomber project (1937)
Me 265 - Apparently Lippisch P 10 (unconfirmed in currently available documents)
Me 308 - Commercial airliner (see Me C 164)
Me 310 - Me 210 with increased dimensions for high-altitude operations (1941/1942)
Me 329 - High-speed bomber project
Me 334 - Lippisch Me 163 project with DB 605 engine (1942/1943)
Me 364 - Unauthorized number for six-engined Me 264
Me 409 - High-altitude/carrier fighter based on Bf 109G (1942)
Me 609 - Cover designation for test-ready Me 262s (1944/1945)

Messerschmitt Propeller Developments (Me P)

In order to ensure his aircraft had optimal propellers, in 1936 Willy Messerschmitt started developing variable pitch airscrews in a separate department under the direction of Robert Prause. Although several types were developed, only the Me P 7 went into production.

On the orders of the RLM the development program was turned over to the VDM company in Frankfurt am Main in July 1942. Despite the fact that the Messerschmitt variable pitch and reverse pitch propellers were considered to be of high standards, VDM did not continue with production. Those propellers built included:

Me P 2 - Variable pitch for engines up to 200 hp/147 kW. Developed and tested in 1940. Production planned for Bücker Bü 181 etc., but never materialized.

Me P 7 - Variable pitch for engines between 200 and 600 hp/147 to 440kW. Testing began in September 1937 with Bf 108 aircraft. Produced and retro-fitted to Bf 108B from 1939 onward.

Me P 6 - Variable pitch for engines up to 1,400 hp/1,030kW. Three-blade airscrew with speed control for landing and dive braking. Development began in 1939. First flight on Me 210 on 15 May 1940. 500-hour evaluation concluded in November 1940, with 500 hours logged on Ju 87 by April 1941, as well. Me P 6 planned for several aircraft types, including Bf 109, Bv 138, Fw 200, but only about 90 were built.

Me P 8 - Variable pitch for engines up to 2,000 hp/1,470kW. Developed from 1940 onward for DB 603 and Jumo 213 engines. Development transferred to VDM in July 1942, with evaluation following in 1943. Not produced.

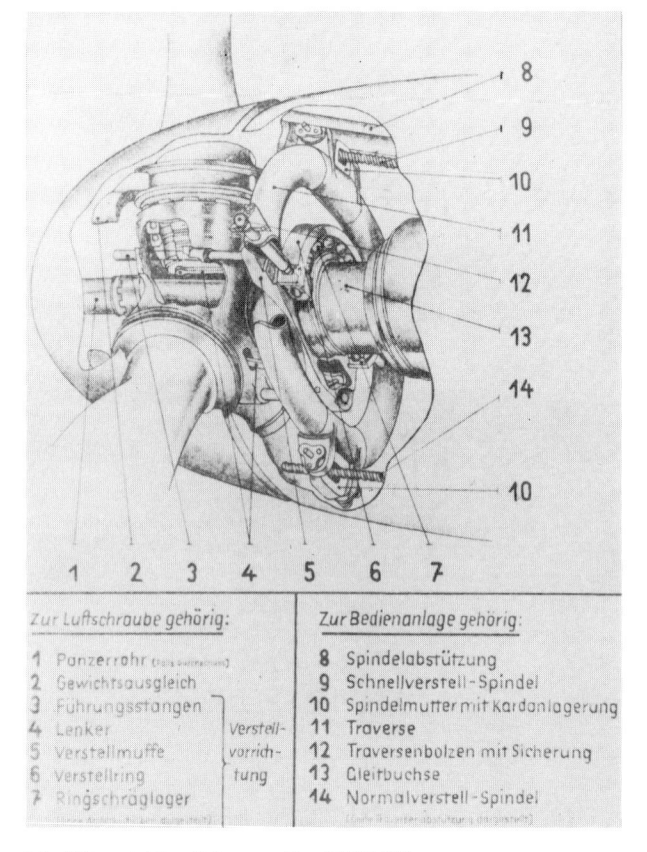

Me P 8 variable pitch propeller (1942/43).

Zur Luftschraube gehörig:	Zur Bedienanlage gehörig:
1 Panzerrohr	8 Spindelabstützung
2 Gewichtsausgleich	9 Schnellverstell-Spindel
3 Führungsstangen	10 Spindelmutter mit Kardanlagerung
4 Lenker	11 Traverse
5 Verstellmuffe	12 Traversenbolzen mit Sicherung
6 Verstellring	13 Gleitbuchse
7 Ringschräglager	14 Normalverstell-Spindel

Type Names

The company did not issue names for Messerschmitt aircraft. The only exception to this was the name "Taifun" (Typhoon) given to the Bf 108. After Elly Beinhorn had this painted on her first Bf 108, the moniker was applied to the entire series and made the airplane recognizable worldwide. For propaganda purposes, higher authorities instituted the name of "Jaguar" for the Bf 162 and "Gigant" (Giant) for the Me 321 and Me 323. The names "Komet" (Comet) for the Me 163 and "Schwalbe" (Swallow) or "Sturmvogel" (Stormbird) for the Me 262, often seen in writings, are not found in company records. These names were first used in Anglo-American publications after the war. The name "Enzian" (Gentian), applied to the Messerschmitt anti-aircraft missile, was a cover term.

The War's End

On the 8th of May 1945 began a new chapter in German history, but for Willy Messerschmitt personally and the Messerschmitt AG collectively it heralded a time of hardship and struggles. This difficult period is described on the following pages. From the years immediately following the capitulation we have only the memories of eyewitnesses, which have also provided the major source material for recalling the comments made by Willy Messerschmitt.

The Americans and French "Capture" the Project Bureau

In April 1945 the U.S. 7th Army under the command of Lieutenant General Alexander Patch was pushing out from Stuttgart in a southerly and southeasterly direction. On 29 April 1945, a Sunday, the Americans occupied Oberammergau. There, back in October of 1943, the entire Messerschmitt AG development program had been relocated in the Gebirgsjäger (mountain infantry) barracks, including the project and design offices, the stress analysis section, and the prototype development department. Disguised under the name of "*Oberbayerische Forschungsanstalt*" (Upper Bavarian Research Institute), the developmental group's location, function, and importance remained unknown to the Allies. This ignorance went on for an entire week until the Combined Advance Field Team got underway, and it wasn't until 17 May that the first members of the Air Technical Intelligence section from the U.S. Strategic Air Forces arrived in Oberammergau. They occupied Haus Osterbichl as their headquarters and immediately began methodically cataloging the materials found at the barracks.

14 days before the Americans arrived, the project department began packing up the most important documents "with optimistic expectations that the Americans would allow us—as Allies, so to speak—to continue working after the occupation," recalls Wolfgang Degel. Documents and files could therefore be found neatly sealed up in metal containers at four different hiding places, including the basement of Eberhard Stromeyer's fiancee in Wertach. Here, however, the French beat them to it. During the combined advance, the French 1st Army under General Jean de Lattre de Tassigny had pushed into the sector assigned to the Americans on several occasions and thus the documents first fell into the hands of French specialists. The material included production files for the P 1101 prototype as well as blueprints and documents for the P 1102 through P 1112 projects. The Americans didn't get them back until 9 June 1945.

Eight of Messerschmitt's employees remaining behind at Oberammergau were arrested after they appeared reluctant to cooperate with the Americans. However, they were released after just a week and were lodged in a hotel with the others. Later they could even go outside the hotel. The Americans hoped that by extending this generosity they would be more cooperative. In actuality, by this time the Messerschmitt employees were indeed displaying an increased willingness to discuss their former work, most certainly due to the fact that the war had ended in the interim. These interviews resulted in the Survey of Messerschmitt Factory and Functions Oberammergau, Germany, completed in February of 1946. The survey was conducted throughout the remainder of 1945 with workers from the company's other sites being brought in, while members of the interrogation team also traveled to other Messerschmitt plants.

The survey basically contains testimonies of leading managers covering the overall structure and operations of the whole company and, based on the results of interviews with senior engineers, detailed technical discussions of important Messerschmitt projects and programs.

Internment

Following the arrival of the Americans at Oberammergau, Willy Messerschmitt was at first detained in Murnau, briefly flown to London in a DC-3 for interrogation, then brought back to Murnau. From there he spent a long period in the "Bärenkeller Settlement" outside of Augsburg, then was shipped off to Ludwigsburg. The Ludwigsburg camp was overcrowded, and by a counting off process, every tenth person was sent to Heilbronn. One day Willy Messerschmitt's number came up. Of all the sites, Heilbronn was the worst. The internees spent day and night in a field under open skies in the rain and cold. These conditions, plus a lack of adequate food coupled with depression took Messerschmitt to the brink of death. An American guard found him and sheltered him in a barracks building, an action which in all likelihood saved his life. It wasn't until then that the Americans realized how important this particular internee was.

Messerschmitt's Nephew Arranges First Postwar Work

Elmar Messerschmitt, Willy Messerschmitt's nephew and grandson of the astronomer Prof. Dr. Johann Baptist Messerschmitt, was drafted into the military while still undergoing graduate studies. He served in the *Luftwaffe* and initially flew transports (including the Ju 52 and the Savoia-Marchetti SM 82) before flying the Fw 190 in ramming operations. In late 1945 he moved to Munich, became a buyer in Bavaria for a north German department store called "Waren aller Art," soon started his own business up and moved to Dachauer Straße 93-95.

On one of his business trips he learned that his uncle was to be interrogated as a witness for the accused at the Nuremberg trials. Along with Eugen Kogon, Rudolf Diels and others, Willy Messerschmitt was being "detained" at Nuremberg. His quarters were in a small room at a villa. Elmar visited him and learned of his desire to take up engineering work again as soon as possible after being released. At his uncle's request he made contact with Willy Messerschmitt's former work colleagues. Soon Messerschmitt received visits from Wolfgang Degel and Hans Hornung. Elmar Messerschmitt bought drafting tables from the trustee of the Messerschmitt AG at the time and set them up at the Dachauer Straße offices. In addition, he contacted Dr.-Ing. Karl-Otto von Faber-du-Faur, who had been an assistant to Willy Messerschmitt during the war years, and persuaded him to help. Faber-du-Faur lent his name to the engineering office, which was initially furnished and financed by Elmar; it was to form the nucleus for the later technical office. Elmar Messerschmitt also made contact with Rudolf Münemann on behalf of his uncle in order to lay the groundwork for financing his planned projects. Münemann was known as the inventor of the "revolving system" of credit—turning short-term borrowed funds into long-term secured credit. After the war he financed the projects of larger German companies and also showed an interest in Messerschmitt's new ideas.

Elmar Messerschmitt—at 25 years of age—also met with Bavaria's Minister of the Economy, Ludwig Erhard. He applied for a so-called wholesaler's license, a necessity for carrying out business transactions in postwar Germany. But Erhard, who agreed to Messerschmitt's petition, was

Willy Messerschmitt's first postwar office (second floor, above the car) at his nephew Elmar's at Dachauer Straße 93-95 in Munich. Elmar also chauffeured his uncle on his first business trips in his DKW convertible. Messerschmitt required few amenities while working; desk, drawing board and paper sufficed. And an ash tray needed to always be close at hand for the heavy smoker.

replaced just a week after the conversation, and it was a long time before approval was forthcoming from his successor.

All the while, Wilhelm Binz, Lorenz Bosch, and Hans Hornung had been busily engaged at the "Faber-du-Faur engineering office." Elmar Messerschmitt played the role of courier between Nuremberg and Munich, bringing sketches and instructions from his uncle along and handing them over to the engineering team.

Even while interned, Willy Messerschmitt remained active with many projects which reflected his creative genius. Among other things, he invented a wristwatch which had only three mechanical parts. At Nuremberg he finished up the design drawings for the watch. With the help of a Swiss observer in an American uniform, Messerschmitt's nephew was able to sell the idea to a Swiss watchmaking company for 5,000 francs. Messerschmitt's first "startup capital" after the war had just been made.

In mid-1947, following a year of "custody," Messerschmitt was set free from Nuremberg. At first he lived with his cousin Dora Storp, a daughter of the painter Pius Ferdinand Messerschmitt, at Fröhlichstraße 2 in Munich-Solln and set up shop in "his" office, already waiting for him at his nephew's on Dachauer Straße.

New Beginnings

Messerschmitt Seeks - and Finds - Alternatives to Building Aircraft

It has already been mentioned that during his internment period Willy Messerschmitt, clearly recognizing that aircraft construction in Germany would not be permitted anytime in the foreseeable future, began thinking how he could apply his unique skills and talents—and those of his former colleagues—to what was for him "unconventional" fields of work.

The scope and results of this thinking are remarkable. For various reasons not all of the ventures were successful, certainly not from a financial standpoint. In each and every instance, however, they showed how unconventional, novel, even better solutions could be found in areas where Messerschmitt and his colleagues employed their manner of thinking, techniques and methods of aircraft construction. In doing so, Messerschmitt demonstrated a feel for the market needs in the post-1945 era. That he wasn't able to push through some of his ideas and products was due to the fact that they were either prematurely conceived or directed at a market dominated by established, reputable companies which shut out "the new kid on the block" because he was unable to shoulder the requisite high investment costs.

Windmills for Bavaria's Energy Needs

Even a year before his release, Willy Messerschmitt was thinking about how he might alleviate Germany's acute energy crisis. He hit upon the idea of a wind-powered storage power station.

This project was to be accomplished conceptually and by the construction of a test facility in conjunction with "his" engineering department; Faber-du-Faur and Rudolf Münemann were to be the financial backers. He himself would be required to remain inconspicuous and follow the orders of the military government when working. His nephew Elmar carried out any "official" correspondence. Furthermore, in order to obtain a business license Elmar Messerschmitt had listed the Messerschmitt KG with the government in Upper Bavaria as "a wholesale company for industry and trade requirements, as a purchasing and production organization, company representative, and for technological exports." The license would give the ongoing activities of his uncle a legal standing in the business world.

Turning to the Bavarian government, Messerschmitt proposed setting up a windmill on the Herzogstand for driving a pump system to draw water up from the Kochelsee to the Walchensee power station. The windmill would drive the pump directly via Cardan shafts with appropriate intermediate bearings. His first hand drawings and figures revealed a wind-driven power generator with all the characteristics of a modern wind-driven power plant: variable pitch propellers for maintaining a constant rpm rate, automatic switch-off at wind speeds over 20 m/s, alloy construction

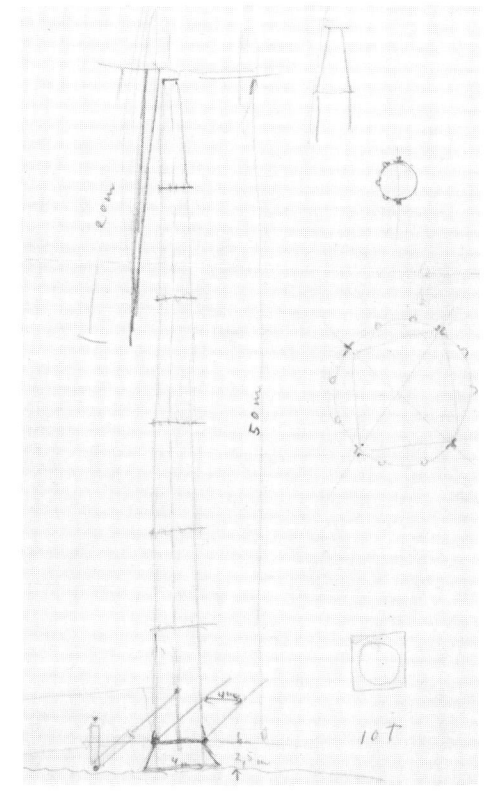

Willy Messerschmitt's hand-drawn sketch for his wind-driven power generator project, made while he was still at Nuremberg. The test system's tower was to be 50 meters in height and could be raised using a simple hydraulic lift mechanism.

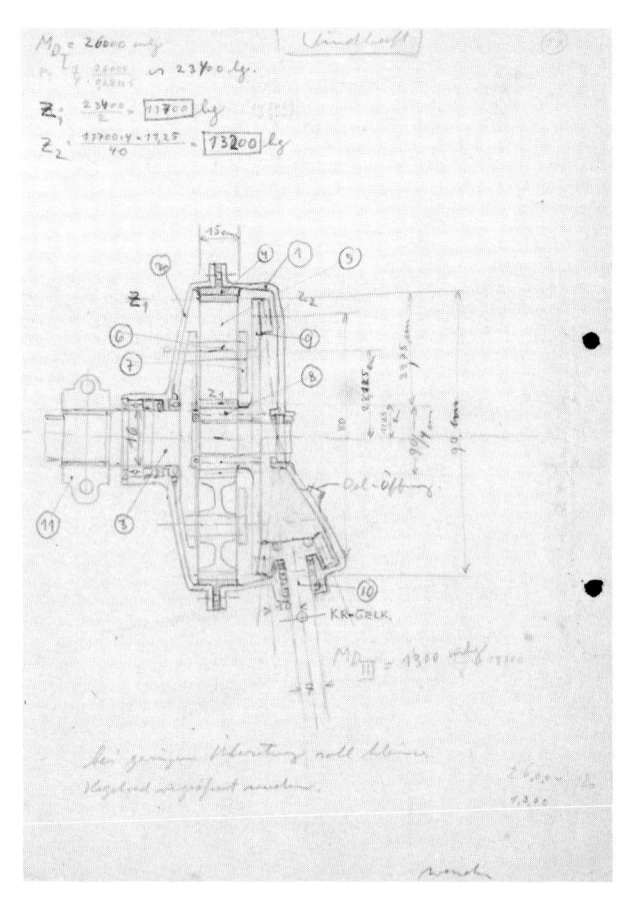

The main drive of Messerschmitt's wind-driven power generator, drawn by Messerschmitt himself. As was his habit, he simultaneously drew up a parts list (on the reverse of the sheet). This included: 1) rear housing cover, made of sheet metal, 2) front housing cover, sheet metal, 3) axis with arms, 4) fixed toothed gear ring, 5) planetary gear wheels, 6) bolts, 7) washers, 8) pinion, 9) large bevel gear wheel, constructed as a unit with 8, 10) small bevel gear wheel, and 11) propeller hub. Messerschmitt added the comment: "entire unit to be attached a pivot mounting on the housing ring, as is the weather vane."

of tower and rotor. A prototype station designed by him—with a 50 m tall tower and a two-blade rotor having a diameter of 20 m—was expected to reach a peak output of 2,600 kW

Messerschmitt began design work on the prototype system. This work was carried out under the auspices of his engineering office on Dachauer Straße, which by now had evolved into Neue Technik GmbH on Dachauer Straße. By the time lack of necessary funding forced cancelation of the project in 1948, it had progressed to the point where it would have been possible to begin on the working drawings.

Neue Technik GmbH Founded

In 1949 Messerschmitt's engineering office moved to Tölzer Straße 186, the former home of the Messerschmitt conglomerate's Eiso Schrauben GmbH. The rooms on the second floor facing the street were rented out to Rudolf Münemann, while those on the courtyard side now housed the "Technische Büro Prof. Messerschmitt" (Prof. Messerschmitt Technical Office). Münemann became Messerschmitt's first financial backer and was jointly involved in the business initiative of the Neue Technik GmbH, which provided the framework for Willy Messerschmitt's first postwar business ventures. It had been set up as a developmental company on 3 March 1948 by businessmen Peter Höllein and Karl-Otto von Faber-du-Faur, and attorney Konrad Merkel. Shortly thereafter, on 14 June 1948, there followed the establishment of the "Fertigungsgesellschaft Neue Technik GmbH" (FNT) by businessmen Konrad Merkel, Rudolf Münemann, and the Neue Technik GmbH. Managing directors were Merkel, Münemann, and Karl Lindner, who had been the senior manager of Messerschmitt's Regensburg plant up to 1945. Any business being set up at the time required the approval of the military government, and this was given without hesitation.

The FNT rented space from the Messerschmitt AG at the Augsburg plant, now partially rebuilt, for producing components for prefabricated housing. There were plans to merge the FNT with the Messerschmitt AG once property controls had been lifted. It therefore received financial support from the AG, and its newly established (July 1948) board of directors consulted the senior trustee with all matters affecting the AG. However, by the end of November 1949 the FNT was forced to stop issuing paychecks. A shortage of funds ruled out any judicial settlement. Subsequent bankruptcy proceedings ground to a halt due to lack of assets, and the FNT was stricken from the register of companies.

Innovative Prefab Homes

One of the most pressing concerns in Germany during the early postwar years was the housing crisis, a problem which stemmed from the wholesale destruction of Germany's cities. There seemed to be an urgent need to develop a faster and cheaper method of building houses than was currently being employed. Thus, nothing made more sense to an aircraft designer and businessman looking for useful (and permissable) work than to make houses from prefabricated components. With the support of a few rehired employees, Willy Messerschmitt set about converting his goals into reality.

The resulting "Messerschmitt Concept" was not a prefab house in today's sense of the term, but a method of construction using standard, universally applicable finished products (wall sections, door sections, window sections). It gave the architect the largest amount of space to play with, with no limitations on size and room dimensions. A particularly good feature of the concept was that it allowed the construction of multi-storey homes. Commonly found raw materials were used for building, including wood for the doors, steps and window frames.

The sections, or plates, consisted of steam-treated aeroconcrete, a material that had been tested in other countries for several decades. To produce these plates, Willy Messerschmitt obtained a license from a company in Hamburg. The plates were reinforced using diagonal cross-bracing and—as a second major component—were edged with galvanized steel. They were joined together by horizontal connecting rods and the gaps between the plates filled in with concrete afterwards. Floor beams were made of steel plating and tubing; ceiliing beams comprising steel tubing and flooring made of lightweight concrete plates rounded out the building program in production at Augsburg.

In October 1948 the city council of Munich decided to have a model house built using the Messerschmitt concept. This resulted in the construction of a three-storey building incorporating six small apartments on Bad-Schachener Straße, which followed a single-family house using the same principle in Munich-Solln and a four-family house in Burghausen a. d. Salzach. Munich's mayor at the time, Thomas Wimmer, took part in the house-raising ceremonies.

Earlier, when handing over the model home in Burghausen in early 1949, Willy Messerschmitt outlined the reasons which led him into the housing construction field: basically, it was the recognition that the construction industry was in need of a new breath of life through modern design and new technology production methods. "I was also thinking of the social problems which have cropped up in the world during the last few decades. As an engineer and scientist, I know that it's possible to create a healthy living environment using available materials and resources if those materials and resources are applied using economically sound principles."

In May of 1949 the Messerschmitt concept was given the stamp of approval by the senior building code authorities within the Bavarian State's Ministry of the Interior. It was estimated that a home built in Ramersdorf enjoyed a savings of 30 to 45 percent compared to the standard, "conventional" house style.

Following approval in other German states, by 1951 a further 200 housing units had been built in Munich, Burghausen, Stuttgart, Landshut, Schliersee, Obernburg am Main, Solingen, Frankfurg, Nuremberg, Mainz, Oberstaufen, in the Ruhr district and near Bremen. During

Die Messerschmitt-Bauart

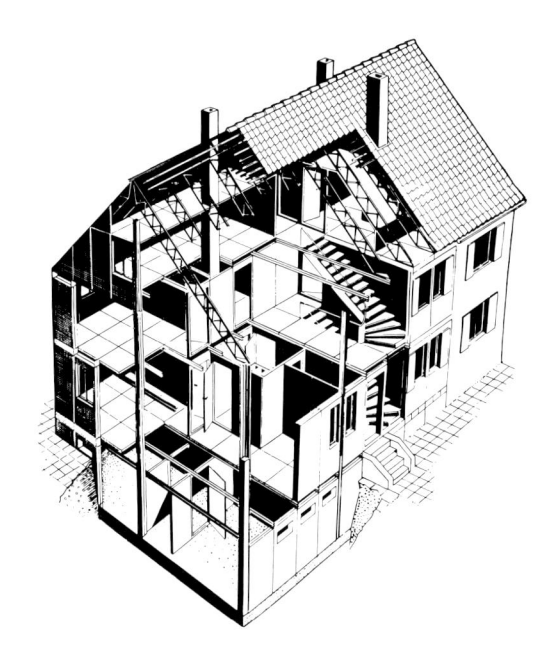

Fertigungsgesellschaft Neue Technik m. b. H. · München · Tölzerstraße 186

Layout of a four-family house based on the "Messerschmitt Concept." In addition to the numerous houses built in Germany, this concept formed the basis for proposals ranging from a large housing project in Israel, one in South America, and for farming houses in Australia.

the same period Messerschmitt produced steel roof trusses for about 300 apartments, industrial buildings, and warehouses.

Although the Messerschmitt concept was shown to have excellent insulation qualities and met structural codes, it seemed the urgently needed prefab approach to housing construction had been "a bit premature." It was impossible to shout down an established "competition," whose tried and true brickbuilding method hearkened back over a thousand years and who, despite a shortage of conventional buildings, utterly rejected this unconventional approach to building homes. Too, there were various technical problems. These included corrosion problems with the aeroconcrete/steel combination, for which there had been no data to fall back on. Ludwig Bölkow remembers warning Willy Messerschmitt about this problem beforehand. But Messerschmitt—not for the first time in his life—trusted in the approving officials who, in his opinion "should have known." Independent of this, it was not possible to generate sufficient sales from the components manufactured by

Assembling the components of the second floor of a house in Munich-Ramersdorf. A makeshift crane was adequate for lifting and positioning the pre-finished sections.

The Messerschmitt concept allowed for the construction of multi-story houses built from prefabricated standardized components.

Hardening ovens for the aeroconcrete segments seen in the casting plant at the Messerschmitt-Werk in Augsburg.

Neue Technik. Due to the season, production stopped in the fall of 1950 and did not resume again the following year.

Postwar Struggles to Keep Aviation Industry Alive - Messerschmitt's "Japanese Funds"

In the fall of 1948 the Neue Technik GmbH needed a loan to the tune of 150,000 DM. To resolve the matter, a meeting took place in Augsburg with the military government, with Rakan Kokothaki representing Messerschmitt's interests. During the course of the meeting, fundamental questions were brought up which not only shed light on Willy Messerschmitt's contractual affairs prior to the war and his company's struggle for survival, but on the entire postwar aviation industry as a whole.

It turned out that the loan could not be secured as a result of certain conditions the military government set for the trustees. In connection with the proceedings, the director of the property control department also questioned the right of Willy Messerschmitt's claim to 7.5 million DM from the so-called "Japanese funds." As a result of negotiations between the German government and the Japanese Ministry of War on 2 March 1944, in December of that year the Ministry of the Imperial Japanese Navy and the Messerschmitt AG signed an agreement for handing over the copyrights to the Me 163 and the Me 262, for which the Reich deposited the sum of 20 million RM into an account at the Bayerische Vereinsbank Augsburg. This account was carried on the books as a fiduciary account under "Inventor's Group" (Japan Deposit). An audit report from August 1946 shows the breakdown of this account as follows: Messerschmitt AG - 10 million RM, Dr. Lippisch as inventor - up to 2.5 million RM, Willy Messerschmitt as inventor - no less than 7.5 million RM. Messerschmitt's claim was used by the Bayerische Vereinsbank as collateral for a 5.3 million RM loan which he secured and drew out on 27 April 1945.

The military government took the view that the claim as outlined—documents no longer existed—was not proof to the contrary in view of the available (and submitted for review) agreement from 1938, whereby Willy Messerschmitt was paid a one-time lump sum of 6 million RM for handing over all his inventions and patents.

Kokothaki explained that the 1938 agreement had come about solely as a result of the demands of the Reichsluftfahrtministerium. At the time, the RLM demanded such an assurance "so that the armament program would not be jeopardized by any potential claims on Messerschmitt's part." However, the Messerschmitt AG's board of directors and executive board were united in their opinion that this agreement was only applicable to direct and indirect contracts for the Reich and/or the RLM, whereas

---

Willy Messerschmitt asked his nephew Elmo to continue working with him when he moved into the facilities on Tölzer Straße. However, Elmo Messerschmitt had other plans. He began studying at the Technische Universität Munich, and after earning his degree later made his own way in life as an inventor and businessman. He played a major role in the development and advancement of the screen printing process. In the words of Elmar Messerschmitt:

"As a businessman, my uncle had more of the 'big picture' in mind than the day-to-day business matters. In business matters he gave the appearance of not really quite knowing what to do. Only after asking advice from several people would he then make up his mind and follow a course of action. It was always a joy to have him occasionally call up for advice on one of 'his matters' even long after we'd gone our separate ways.

As a designer he obviously responded quite differently. With regard to his homebuilding activities, I once wanted him to get together with an expert who was quite familiar with aeroconcrete. 'Yes, quite right,' responded my uncle. 'But maybe later.' He wanted to approach the task without any preconceived notions. For his design of wind power generators, Messerschmitt's colleagues chided him for wanting to use torsion shafts for transferring the power output. 'If I always had such reservations, I probably would never have done anything at all,' said Willy Messerschmitt of the matter. 'Never have biases, Elmar,' he would always say to me—and trust yourself enough to pursue what even seems to be technologically absurd ideas.

the interested parties agreement of 1927 remained in full force for governing all other private domestic and foreign transactions.

At the time of the 1938 agreement's signing the RLM representative had expressed his opinion that it was entirely in keeping with the Ministry's wishes if the Messerschmitt AG were to specially recognize any special services rendered by Messerschmitt the man. The RLM, so it was said, had a specific interest in providing a special incentive for Messerschmitt's creative talents.

Even before the capitulation it was generally assumed that the "Japanese Funds" would be used for financing a startup of peacetime production for the Messerschmitt Works. Of the remaining 1.2 million DM following the conversion from Reichsmarks to Deutschmarks, 500,000 DM went to the Messerschmitt AG—of which 300,000 DM was funneled off to the Fertigungsgesellschaft Neue Technik GmbH. An additional 324,000 DM went directly to the FNT.

By the war's end, the Reich had accumulated an accounts payable tab amounting to hundreds of millions of marks, bills which in all likelihood would never be paid. On the other side of the spectrum were the subcontractors, whose claims remained in effect for their full amount. Given this situation, Germany's post-1945 aviation industry found itself fighting for survival, something which Willy Messerschmitt was often forced to point out. At a meeting of North Rhine-Westphalia's Arbeitsgemeinschaft für Forschung in Düsseldorf, a lecture on 9 January 1952 by Friedrich Seewald on "Aeronautics and its Effect on Technical Advances as a Whole" stimulated a discussion wherein Messerschmitt expressed his feelings thusly:

"Gentlemen, I'd like to broach a rather delicate subject which is probably more appropriate to the aviation industry than to aviation research. Economically, the industry has suffered serious damage because of the war and its consequences. My factories at Augsburg and Regensburg still have many high-value claims outstanding against the Reich. Up until the currency reform, our plants were in a financially sound state any way you look at it—this despite dismantling, despite the six year work prohibition, despite being plundered, and despite a trustee administration which has pawned off most of our material assets. Meaning that if the Reich had paid its debts, the company would have been able to pay off its creditors fully and happily and still have been quite sound financially.

However, in one stroke the currency reform turned matters completely upside down. The Reich's outstanding accounts were not revalued, while we were forced to accept a 1:10 conversion for company debts. It goes without saying that the situation has become untenable.

The currency reform law has provided a type of protection for the industry, including the aircraft industry, in the form of a §21 which contains a contract relief clause. By this contract relief, debts may be reduced following a legal review of the interests and situation of the creditor and the debtor. This law went into effect in 1948. Nevertheless, even today no judge has ever seriously considered applying the law where it is actually needed. Yet it is extremely important that the aviation industry is supported in its efforts to keep what little it still has and not to wipe out what's left of the aircraft industry by unjustifiably interpreting §21 of the reform law. If the Reich doesn't pay us for loans, loans which were forced upon us at the time in the interests of the Reich, then we are not in a position where we can pay the debts we incurred for the Reich. Up until 1942 the Reich had made down payments for its swelling number of contracts, then made incremental payments during the production phase with the final payment being made upon delivery of the product. However, because of financial liquidity from 1942 onward the Reich forced the industry to accept bank loans, meaning the Reich refused to make any payment until the products were delivered, and the banks were compelled to cover the industry's costs through these loans. The industry is now stuck on these loans, revalued to ten percent, but we're not being paid accordingly by the Reich or its successor, the Federation or the states...

The aviation industry has a considerable impact on the national economy, as well. You will certainly be interested to know that just a single plant, my factory at Augsburg, has done over 500 million goldmarks of foreign trade in the 25 years of its existence. It should be noted that it's not even the largest German aircraft factory, such as Junkers and Heinkel. You can therefore see that Germany's aviation industry has played a major role in international trade, as well. And it will need to play its role as a currency earner again in order to help our country, now suffering from shortages of raw materials and foodstuffs, to be able to feed itself."

The End of Augsburg

The war had a devastating effect on the factories of the Messerschmitt AG, particularly the parent facility at Augsburg, which suffered approximately 75 percent damage. The outer walls of the large works buildings were still standing, but interiors and machinery had for the most part been completely destroyed. When the Americans marched in, the military government immediately issued orders prohibiting the Messerschmitt AG from further production.

Horst Brausewaldt, whom Dr. L. S. Rothe hired as the company's press secretary in 1957, has thankfully annotated Messerschmitt's postwar history. With regard to the time immediately after the war, he writes:

"...Administration of the factories came under the authority of the American military government. Captain Neil was responsible for the Augsburg district, and exercised his duties from the military government's administration buildings at Gersthofen. Captain Neil originally appointed H. Mathy as trustee for our factories, later replacing him with chief trustee H. Brüning, who had come from the Höchste Farbwerke subsidiary at Gersthofen. Brüning fulfilled his responsibilities until being replaced by H. Menzel in early 1947. Menzel served for a year, then was succeeded by H. Steinbach, then by H. Schnause. Deputy chief trustee and technical director during Augsburg's reconstruction years was H. Belz.

Just four weeks after the arrival of the Americans, an eight-man team from the Messerschmitt work force began cleanup work at Plant IV under the supervision of H. Mösinger. They set up a small workshop in the washroom, fixed up some of the machines and then began shoring up the walls against collapse and generally making the facilities usable again. Shortly thereafter, the team began producing cooking pots, fittings for building and furniture, curlers and similar useful and needed items at the time. By and by, the work force grew to 40 men....

Working on behalf of the trustees, Mösinger sat inside the porterhouse of Plant IV on Haunstetter Straße and sold the freshly made cooking pots and curlers to passers-by, where they found a ready market.

Before 1945 was out, efforts were even being made to collect what remained of machine tools and materials, which for the most part was scattered about the outlying areas of Augsburg at dispersed manufacturing sites....From Oberammergau, Großaitinge, Ettringen, Stätzling, Memmingen, Leipheim and many other dispersed sites, tools, machinery and materials began wending their way back to the parent factories at Augsburg....

In 1947 the trustees received permission to sign lease agreements with companies wanting to set up their own small production shops on the site in order to make use of the manufacturing facilities that had been somewhat restored. As a result, the Messerschmitt AG was afforded the opportunity to meet its responsibilities with the income from these leases. For the Messerschmitt Works, however, the general ban on production remained in effect.

In Plant II the Cerwinsky Maschinenbau company, later renamed the Tepag Textilmaschinen, set up shop. In Plant IV the company of Schweda & Tröger settled in. At first they used the 5,000 ton press to make milk cans—initially under the name of "Messerschmitt" even—and produced eating utensils and milking equipment. The firm of Korndorfer, a producer of textile machinery, also established itself in Plant IV. By the way, up until 1948 the former Ungarische Waggon- und Maschinenbau Budapest operated out of Plant I. This company had formerly license built the Me 109 in Hungary, pulling back to Deggendorf during the general retreat before settling down in Augsburg. The Froitzheim Fahrzeugbau worked out of the Seibenbrunn dispersed facility before eventually moving into Plant III.

In 1947 all surviving storage and air-raid bunkers on site were blown up on the orders of the military government. This was followed by the stockpile of confiscated machinery being shipped off to meet the Allies' reparations demands. All reparations goods from Swabia were collected together in Plant I, from there to be sent to those countries claiming compensation.

Operating under difficult circumstances at a time of general collapse, a small group of former Messerschmitt employees with a firm belief in the future gradually began the work of reconstruction....The materials needed for rebuilding were quite often obtained through the barter system. For example, sheet metal, screws, tubes, and electronic components were exchanged for urgently needed T-beams, cement and bricks."

Messerschmitt Falls Under Trust Company

The seriousness of the situation for the Messerschmitt AG at the time is revealed in a handout prepared by the company's trustees in 1946:

Lately questions have been mounting by our former creditors regarding the fate of blocked old accounts receivable.

This gives us occasion to point out that we still remain under property control. As we are considered an armaments company, the relaxing of restrictions since their implementation do not apply to us. Furthermore, we are legally prohibited from meeting those debts incurred before 9/7/1945 (date of effect for trusteeship). For this reason, to avoid unnecessary effort and expense we are asking our creditors to refrain from petitions and from taking legal action against us. In this context, we would like to point out that any enforcement based on a writ of execution is not authorized without special approval from our controlling agent, the Bavarian State Office for Administration of Assets and Reparations, Augsburg, Haunstetter Straße 19. Given the situation as outlined above, such approval is generally not forthcoming. We are also asking our creditors to understand that we insist on non-preferential treatment for each creditor in all circumstances.

Our assets for the most part consist of claims against the legal successors to the Reich. It must therefore be expected that even once property control restrictions have been lifted there can be no immediate thought of debt settlement. It goes without saying that the asset situation is crucially dependent on the fate of these accounts receivable and the yet to be determined ruling affecting them.

Problems with material deliveries during the last months of the war and the ceasefire have made it impossible to fulfill many outstanding contracts. The question remains as to who bears responsibility for the damages resulting from these circumstances (higher authority and legal action by the occupying forces). Nevertheless, some of our suppliers have billed us for the cost of work stoppage, i.e. wages and materials. To prevent any misunderstandings, we would like to make it expressly clear that we do not recognize such claims...

...On the other hand, we are meeting our financial obligations incurred after 9/7/45 in a timely manner. All of our business partners may therefore expect to be paid for these new claims on time...

We regret that we cannot provide our business associates with better news regarding outstanding old claims at present, and remain most

Sincerely,
Messerschmitt AG
Trustee:
(signed) Schnause

Holdings of Messerschmitt AG in 1950:

Messerschmitt GmbH, Regensburg	135,500 DM
Leichtbau GmbH, Regensburg	150,000 DM
Uher & Co, Munich	350,000 DM
Eiso-Schrauben GmbH, Munich	250,000 DM
Schwäbische Formholz GmbH, Augsburg (SFG)	102,000 DM
Gemeinnützige Wohnungsgesellschaft GmbH, Augsburg (GEWOG)	12,500 DM

Dismantling Destroys Capital

Horst Brausewaldt continues:

"...The Messerschmitt AG also carried out dismantling projects for the Office of Reparations within the Bavarian Ministry of Economy in Munich, management work for the tenant companies, and set up a truck repair workshop. The monetary reform whittled away a considerable chunk of the company's financial resources. In order to meet the expenses of ongoing jobs, it became increasingly necessary to rent out plant facilities, as it was no longer possible to sell off capital in the form of materiel and facilities....At the end of 1949 the workforce encompassed 76 wage earners and 62 salaried employees.

Machinery which was to have been shipped out in 1948/49 to meet reparation obligations never materialized, and was ordered by the military government to be sold on the open market just prior to lifting the control over the company's assets. As a result, the AG was left with only somewhat antiquaited and well-worn equipment. Notable in this context is the fact that in September 1949 intervention on the part of Bavaria's prime minister failed to prevent the 5,000 ton press from falling victim to the cutter's torch on the orders of General Clay.

Effective 25 November 1949 the Messerschmitt AG was released from property control and Siegfried Keller and Konrad Merkel were made members of the executive board. On 15 July 1948 the board of directors had consisted of Baroness Lilly von Michel-Raulino, Konrad Merkel, Edgar Huth and Rudolf Münemann. Willy Messerschmitt was elected to the board in place of Huth in 1949.

With the lifting of the controls, the company first took stock of its reduced value material assets, then stopped their continued reduction by halting any further sell-offs, and began working on laying the foundation for industrial production. It was also necessary to deflect the intentions of a few former creditors with regard to the company's remaining assets, and the company underwent a general reorganization....

In addition to truck repairs, by 1950 the manufacturing program included producing aeroconcrete sections and roofing joists for home construction, plus making radio housing components and industrial cabinetmaking....

The first measure taken by the newly elected board of directors to balance the books occurred in early 1950. Although it originally involved arranging a grace period for any outstanding debts, subsequent agreements and settlements resulted in many of these either being reduced considerably or written off altogether.

By 1950 all the company's assets had been sorted out and unused facilities were being rented, and by 1951 production of a home sewing machine got underway, a step which was to herald a rebirth in industrial activity."

Sewing Machines Bring Initial Success

After the annexation of Austria prior to the war, the Messerschmitt AG set up a branch factory at Kematen near Innsbruck for manufacturing and supplying aircraft parts. With the rebirth of the Republic of Austria, however, this factory was placed under a trusteeship and thus was cut off from the main Messerschmitt works.

In 1950 Willy Messerschmitt learned that the Messerschmitt Ges.mbH Tirol was building a home sewing machine and selling it under the name of Messerschmitt. Looking for something to keep his Augsburg works busy, he felt that the product—unaffected by the mandatory restrictions then in force—was interesting enough to pursue negotiations with the managers at Kematen. In the end, he was provided with a set of blueprint documents and immediately began thinking about production, sales, and follow-on developments, while also meticulously working out calculations and design features.

With the blueprints reworked by Kurt Brandt, work got underway in Augsburg for starting up production on the "Kematener Modell." This was a problematic task indeed, for the requisite manufacturing area had to be rebuilt and the assembly equipment set up. Much of what had survived from the machine shop dated back to 1940. As it turned out, it was not possible to achieve the necessary precision in machine tooling; the measuring equipment simply wasn't available. An "optimal finish" in the spirit of Willy Messerschmitt seemed to be an unattainable goal. Indeed, even in later years manufacturing quality wasn't able to keep pace with improvements in design and often gave rise to complaints.

An event in the foreign market led to a court decision resulting in the Kematen Works being forbidden from marketing its sewing machines under the name of Messerschmitt, while for its part the Messerschmitt AG was forced to stop production. Nevertheless, operating wholly independently of the Kematen company the Messerschmitt AG ultimately succeeded in selling 6,200 KL-51 machines, a modest success in a difficult, hitherto uncharted market dominated by well-established competitors.

Willy Messerschmitt and Kurt Brandt then plunged headlong into the development of a new sewing machine. Messerschmitt's focus, as always, was geared toward simplifying the mechanical aspects as much as possible. His efforts resulted in the KL 52 zigzag machine, equipped with the "magnet gripper," a jam-preventing device which had been invented by Brandt and later patented.

This new technology was also applied to the KL 53 (straight-stitch machine with gripper at right-angle) and KL 54 (automatic zigzag machine). All sewing machines were available as portable and tabletop models, with foot pedal (manual), electric, or combination power.

Messerschmitt and Brandt invented an automatic decorative stitch function for the KL 54, whereby depending on the desired stitch, a selected button controlled the needle's lateral movement by means of a plate cam package.

A worldwide network of distributors was organized. Sales naturally ran into repeated problems with an innovative, well-designed product suffering from manufacturing flaws. Ultimately, this doomed the sewing machine, for the internal market was too small and the export market was never able to be fully exploited.

A license agreement with the Italian company Filotecnica Salmoiraghis S.p.A. in Milan failed, as did the American Singer company's acquisition of the copyright for the automatic machine. In 1959 the Pfaff company rejected an offer to take over sewing machine production and manufacturing equipment, an understandable action since Messerschmitt had never been able to garner more than 0.5 percent of the German market. This despite the fact that his competitors recognized that "an unconventional approach has been taken with the Messerschmitt sewing machines, which is attributable to the fact that wrestling with relevant problems was a group of unbiased engineers who came from the aviation world and not from within the industry." Despite the setbacks, the patent for the magnet gripper was eventually sold to the Singer Company.

Selling the sewing machines increasingly became more difficult. Established manufacturers had their contractually obligated core of dealers, and their cost structure was of an entirely different magnitude due to the volume of machines they produced. Thus, it would have required the investment of considerable financial resources in order to produce the machines more economically and in greater numbers. Air-

Willy Messerschmitt with the three basic models of his sewing machines.

Messerschmitt sewing machine production in 1951 at Plant II in Augsburg. The photo shows workers testing the machines. Even today, over thirty years later, replacement parts for these apparently reliable machines are much in demand in Augsburg.

craft production was getting underway in 1957, and as this branch would require an infusion of a major portion of the company's resources, it seemed easier to kill the production of sewing machines instead. In 1959 the dealers were informed that Messerschmitt would no longer be producing its sewing machines.

In a report from late 1957, Siegfried Keller reported that "the sewing machines enabled us to build up a solid core of experienced workers, to train them on building jigs, to improve the company's available machine equipment and thereby ensure the company would be ready when the time came to restart aircraft production."

All told, about 20,500 sewing machines were manufactured.

By the end of 1951 the Messerschmitt AG numbered 197 wage earners and 66 salaried employees. In November Konrad Merkel, who had been assigned to the executive board, returned to the board of directors. Hans Strauch took his place on the executive board, and banker Georg Eidenschink was also called to the board of directors.

In 1952 the sewing machine took top priority within the scope of the Messerschmitt AG's manufacturing program. In spite of the fact that Messerschmitt had succeeded in establishing a foothold—and being noticed—in the brutally competitive sewing machine market, it was apparent that sewing machine production alone would not keep the plant working at capacity. Thus the search for new products continued. In the meantime, shares in the Gemeinnützige Wohnungsgesellschaft (GEWOG) were sold off, and 95 percent of the shares in the newly established Messerschmitt GmbH Rheinland acquired.

First Postwar Aeronautical Inroads

The Messerschmitt GmbH Rheinland had been set up in Essen as a mailbox manufacturing company. It served Willy Messerschmitt as a negotiating and contract partner within the state government of North Rhine-Westphalia. Since 1949 Leo Brandt had been actively campaigning in Düsseldorf for the advancement of research and technology. Brandt later became state secretary in the Ministry of Economy and Trade for North Rhine-Westphalia and, from 1953 onward, the managing director of the commendable "Arbeitsgemeinschaft für Forschung" research consortium founded by the state's prime minister at the time, Karl Arnold. In June 1953 Willy Messerschmitt applied for funding to support research and development of a "hydraulic turbojet engine for stationary facilities and various types of vehicles." Messerschmitt stated that he had resumed work on a project begun by Dr.-Ing Johann Endres, an assistant professor at the Technische Üniversität in Munich, for an hydraulic jet propulsion system, and that these efforts "had led to results which may have an impact on future developments in the field of maritime propulsion." Endres' earlier endeavor had been carried out in cooperation with the Navy High Command and professors Föttinger and Triebnigg at the Technische Üniversität at Charlottenburg, Berlin.

Indeed, Endres had been working for Willy Messerschmitt on this very matter since 1951. The concept of hydraulic jet propulsion lies in the combination of a dual-effect opposed free-piston diesel engine coupled directly to an opposed-piston jet pump, whereby the combustion force generated in the engine was transmitted directly to the water column by means of the jet pump, producing thrust in accordance with the law of momentum. Drive efficiency was said to be higher than with standard ship propulsion systems.

An initial contract was awarded, and Endres continued working for the Messerschmitt GmbH Rheinland until 1958. Looking back, it isn't hard to see that—some two years before Germany had even regained control of its own airspace—a farsighted contractor encouraging a dedicated supplier to conduct research again (albeit under "camouflaged terminology") for the first time since 1945 would also bear fruit in the field of aeronautics at some future point, as well.

In early 1953 the Messerschmitt AG took over management of Prof. Messerschmitt's technical department as the "Konstruktionsbüro München," or Munich Design Bureau. It bore the personnel expenses since it would later be the main beneficiary of the Munich department's results. Willy Messerschmitt remained responsible for management and tasking. Manufacturing of wood products, which had primarily consisted of radio housing components for the Schwäbische Formholz GmbH (SFG), ceased as of the year's end. By this time the Messerschmitt AG had a total of 1,750 employees, including 675 at the Regensburger Stahl- und Metallbau (RSM), 530 at SFG, and a further 198 at the Augsburg plant. Interests in the Uher company were sold off, with profits going toward acquiring the remaining 49 percent of the SFG for a total takeover.

Tram line 4 still has a stop called simply "Messerschmitt" in front of MBB's Plant IV in Augsburg.

New Concepts for Engine and Automobile Construction

The financial reforms in 1948 heralded Germany's economic renaissance after the Second World War. It soon became apparent that one of the personal, immediate goals of its people included motorization, however modest the scale. Willy Messerschmitt had already dealt with the topics of "cars" and "engines" while at Heilbronn, creating his first designs in conjunction with the Faber-du-Faur engineering division. Searching for work for the Augsburg and Regensburg factories, Messerschmitt saw in the automobile a product that required enough technical expertise to keep his former aircraft designers busy, not to mention the challenge every engineer faces when he's challenged to apply "classical" knowledge to a new, entirely different, technical field.

P 511 Mid-Size Automobile

Ultimately, the automobile is a product which is closely related to the aircraft with regard to light construction technology; on the one hand meaning application of available know-how would be possible and, on the other, new advancements in related technology could be achieved.

A car was to be created with the following characteristics:

- in the international mid-size category in view of performance and space, but with a one-liter engine.
- individually spring-mounted wheels and a clutchless, automatic hydrostatic transmission, which dispensed with the need for a differential, providing greater driving and operating comfort.
- simple construction and ease of maintenance.
- good driving performance, which meant a low empty weight in view of the modest engine size. This in turn was possible by utilizing a self-supporting lightweight body incorporating technology familiar to aircraft construction.

In order to make assembly easier and simplify repairs, Messerschmitt came up with the idea of breaking down the body into a torque and bending resistant center section and a front and rear chassis section (comprised of two strut assemblies linked by a cross member) attached to the center section at four points. The cross member of the rear section would also support the drive system with engine and having the gearbox and differential gear flanged to it. The large front and rear hoods could be removed. By eliminating the fenders the body could be utilized to its outer contours, resulting in an internal width of 1.45 m. This meant that the vehicle could accommodate up to three people both in the front and back seats.

Messerschmitt selected a chassis with telescoping struts, which had proved so reliable in aircraft (and motorcycle) design. He anticipated a good suspension ride with these, with good handling as a result of the relatively smooth motion both in the direction of travel and crosswise to it.

He hoped that the car's lightweight engine would later evolve into an aircraft engine. As a result, he envisioned a

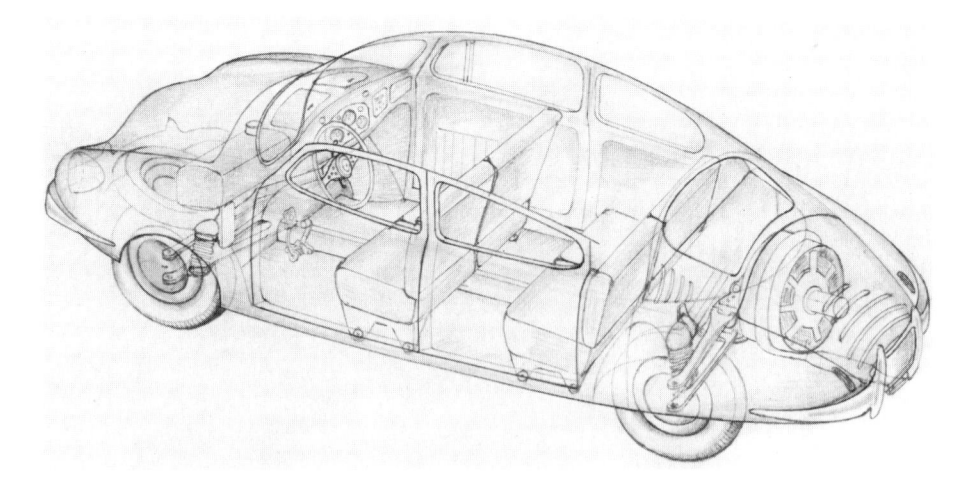

See-thru view of the P 511 shows the characteristic arrangement of the engine.

Center section of the P 511 with hood and front section removed to show the front wheel assembly.

motor fitted with a rotary slide valve in place of a governor, for which he had data from the initial concept testing carried out during the war. The engine was to be air-cooled and be of a radial design—a small-diameter approach which the designers had so much experience with from working with aircraft that the developmental risks were considered extremely low. Dr.-Ing. Otto Steigenberger was hired as an advisor. During the war he had worked on a similar project for Willy Messerschmitt, after having played a major role in the development of the As 10C aircraft engine as director of Argus Motoren GmbH's development plant.

Work began on the engine in 1950. At first, Willy Messerschmitt tried to locate a suitable transmission as well as accessories, something which proved to be quite difficult at the time. In the end he decided to develop the transmission at the Tölzer Straße offices himself as well. It was to be a hydrostatic transmission based on the principle of vane-type pumps with a two-piece slide functioning as the motor, thus producing a differential effect. This dispensed with the need for a separate differential gearbox. Siegfried Lubinski, the inventor of the Lubinski converter, managed the test program for the transmission. Development continued on into the mid-fifties. By this time a reduced-scale testbench model had reached an efficiency of 85 percent, meaning that the predicted 90 percent plus efficiency rating for a larger transmission seemed achievable.

In June of 1951 a Swiss businessman by the name of Reitz paid a visit to Messerschmitt in Munich in order to learn more details about the planned automobile. As a result of this visit, Reitz brought together an Italian financial consortium and in September 1951 contracted for the development of a prototype car. The main components were to be finished by 1 July 1952.

On 26 March 1952 the first run of the single-cylinder testbench engine and its slide valve gear was completed, and testing began on the hydrostatic transmission on 5 May. A contract for building the body had been issued to the Spohn company in Ravensburg back in February, albeit with the stipulation that work would only begin if there were no further anticipated changes to the engine and chassis. Spohn used this time to draw up a counterdesign for the body's external shape. In place of the streamlined rear section, he proposed a design with a slightly stepped appearance (with more interior height for the backseat passengers) and made further suggestions to the body design based on experience. For the most part the lower portion of the body remained untouched. Based on this, Messerschmitt made additional changes to the blueprints and some modifications to assembly procedures. In August 1952 the Spohn company received the reworked drawings and began construction of the body.

In September of 1952, Messerschmitt determined that "tests with the hydrostatic transmission (on a provisional test bench) show that this development is still a long way from success. As a result, it is being shelved in favor of obtaining a simple semi-automatic transmission."

In the meantime, evaluation of the single-cylinder test engine continued. Construction of the five-cylinder radial engine was put on hold because of problems with the sealing and lubrication of the slide valve, as well as the mounting and lubrication of the high-speed vertical shaft driving the valve. The single-cylinder engine attained the requisite performance of 6.6 kW/9hp on 24 October 1952.

It was soon discovered that the first test automobile wouldn't be completed until the beginning of 1953 at the earliest. In the meantime, construction of the five-cylinder engine had begun, and this completed its first run-up on 16 April 1953. The contractors (including the Reitz brothers) had become impatient, and in view of this it was decided to prevent further delays by ordering a replacement engine from Porsche in late 1952, which was fitted on a temporary basis.

On 19 May 1953, nearly a year and a half after the contract had been awarded, it was time: using its own transmission and a Porsche engine, the P 511 automobile was driven for the first time along the former airfield grounds between Works II and III. Wolfgang Degel recalls that "during subsequent test drives, including cross-country, high temperatures caused the rubber boots for the small hydraulically operated couplings in the transmission to give us problems. We'd thought of alternative solutions, but weren't able to apply them, and had to use available spare boots.

Despite the delays, the contractors were still interested and willing to continue their support of development. On 3 June 1953 they watched a demonstration of the car, also conducted at the former airfield. There followed several test drives with distances ranging from 11 km to 200 km (Munich). These short "day trips" were plagued by sealing problems which kept cropping up with the hydraulic shifting components inside the transmission.

Initial run-up of the improved five-cylinder engine took place on 22 February 1954, and in May of that year reached an output of 34kW/45hp at 5,500 rpm. On 8 July 1954 the P 511 was driven for the first time with its intended engine. Nevertheless, by this time it was becoming apparent that unless the contractors were to come up with more funds,

Angled view of Messerschmitt's five-cylinder four-stroke engine shows the arrangement of the components. The design of the fan gives it an almost turbine-like appearance. The carburetor is at the left front. Total volume was 1 liter (200 cm per cylinder). Maximum rpm was 6,000 min^{-1}, with a compression ratio of 7.5:1. Constant power output was, in the initial developmental stage, 26 kW/35 hp at 5,000 rpm.

Main components of the five cylinder DVL-style engine designed by Otto Steigenberger: on the left is the upper section of the cylinder head with stroke chamber, while the stroke baseplate with rotary plate and drive pinion is on the right. In the front is a gasket and sealing ring.

further work on the project would have to be terminated. "Basically, with regard to the layout and design established for our automobile, our approach is still quite advanced in comparison with the current state of technology," wrote Wolfgang Degel, director of the design department, in November of 1953. "Even with the Italians pulling out, we feel it therefore expedient to continue working." Further financing of the project on a wide scale never materialized, however. Available funding slowly dried up, and it was accepted that the program's goals hadn't been achieved with one prototype alone. Thus ended this chapter, one which had begun with such enthusiasm.

Willy Messerschmitt (left) and Wolfgang Degel discuss details of the P 511's engine design in September 1953.

Automobile and engine development ceased forever in late 1957. Looking back, it is obvious that an outsider wouldn't have stood a chance in gaining a foothold in the world's automobile marketplace and its relatively high investment costs. Nevertheless, many of the conceptual ideas of Messerschmitt and his colleagues later found application in the manufacturing of automobiles and engines.

The P 511 had an empty weight of 780 kg and a capacity of 450 kg, giving it a total weight of 1,230 kg. The car's fuel consumption rate was 8 liters per 100 km and it had a maximum speed of 120 km/h. The length of the P 511 was 4.47 m, with a width of 1.66 m, a height of 1.48, and a ground clearance of 0.23 m.

In December 1953 the American industrialist E. P. Aghnides visited the Messerschmitt offices and presented Wolfgang Degel with his idea for a single-axle amphibious vehicle. Aghnides' design called for a vehicle with just two semi-spherical wheels, an arrangement which he held a patent on. According to Aghnides, the vehicle would have required no external support for a stable drive and would, in addition to having pleasant driving characteristics, be able to easily traverse soft terrain and even water. A gyroscope would be used to keep the car stable, shifting the cabin and engine, or just the engine itself, as needed. The project proved to be unfeasible. Wolfgang Degel remembers that "it took a lot of nerve and work to convince a non-technical person and stubborn inventor that his ideas were not practical. This didn't change even though he was paying us a lot of money ($20,000) for this information." The "Rhino" study, as it was called, remained nothing more than a novelty for the technical department.

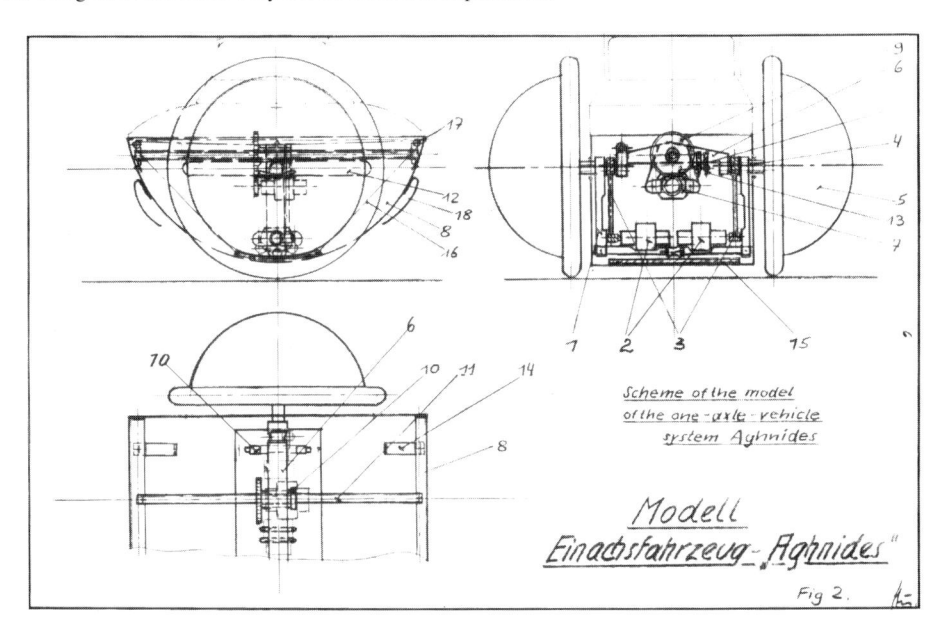

Messerschmitt's Bubble Cars in Full-Scale Production

Willy Messerschmitt first became noted for his bubble car, developed by Fritz Fend in Rosenheim, in 1952. During the war, Fend had worked as a technical engineer at the Rechlin Test Center, becoming involved with landing gear studies, among other things. His meticulous data obtained from landings and takeoffs by the Me 262 eventually led to the eradication of the undercarriage problems on the type. It was also during this time that he became acquainted with Willy Messerschmitt. After 1946 he built manual- and motor-powered vehicles for handicapped people (Fend "Flitzer"), as well as motorized delivery scooters, some of which even went into limited production.

Fend was looking for larger production facilities, while at the same time Willy Messerschmitt was searching for ways of utilizing the capacity of his newly established subsidiary, Regensburger Stahl- und Metallbau GmbH (RSM) to its fullest. The RSM had risen from the ashes of Messerschmitt AG's destroyed Regensburg Works and was otherwise busy with building railroad cars, steel, bridges, and machinery; it had initially begun with contracts for repairing railroad cars.

Once both partners had settled on an agreement, Fend began working on the prototype of a two-seat bubble car in the summer of 1952. Before giving his nod to production, Messerschmitt had stipulated that the bubble car was to be a two-seater and that Fend's design blueprints were to be made production ready at the technical department in Munich.

With its 6.6 kW/9 hp single-cylinder, two-stroke engine, the KR 175 "K*abinenroller*" bubble car reached a maximum speed of 90 km/h and had a fuel consumption rate of 2.5 liters per 100 km. Despite this, it could carry 150 kg. Nearly 15,000 of the first models were built.

Fend's basic thinking was to create a covered motorcycle or scooter, which would provide protection against the weather and greater stability as a result of its double tracked wheels. This addressed the market potential of scooter and motorcycle drivers, who were now presented with a reasonably priced and much more comfortable alternative in the form of a bubble car costing 2,375 DM at the time.

The agreed upon shape was in many respects a compromise of ideas: the track was to be as wide as possible, but not so wide as to prevent it from finding easy parking/storage. In addition, in light of manufacturing costs the top had to be designed as small as possible while keeping within minimum required interior dimensions. Whereas the trend in the automobile industry was to fully utilize the track width for occupant space, with the bubble car a compromise had to be made in this area as well with regard to drag and engine power, a consideration which led to the tandem seat arrangement.

Messerschmitt was able to personally drive the KR 175 prototype before Christmas 1952. At the same time the RSM began full-scale production of the type based on the blueprints supplied by Fend. In March 1953 the KR 175 was shown to the public for the first time at the Geneva Automobile Salon. It soon became apparent that the available drawings and lists were inadequate for optimum full-scale production. As a result, Willy Messerschmitt sent Wilhelm Binz to Regensburg to sort things out; he was later followed by other employees from the technical department.

Despite numerous complaints (some of which led to modifications to the various subassemblies), vehicle production continued on. This despite the fact that Willy Messerschmitt had recommended stopping production until all deficiencies had been rectified.

Those employees which had been despatched were able to be recalled back to Munich in September 1953, where they continued to work on improvements in design for the RSM. These included developing, building, and testing a new forward chassis in conjunction with Fend and Messerschmitt.

In July 1954 Fend unveiled the design for an improved KR 200 in Rosenheim. This differed in design from the KR 175 by having an improved canopy with a double curved windscreen and elastic engine mounts, as well as a Cardan drive to the rear shaft.

Willy Messerschmitt took the view that putting this new type into production was the best way to get rid of all defects encountered up to that point. He sent Wilhelm Lanhammer, later to become managing director of the Manching plant, and a team of four design engineers to Regensburg to carry out improvements in the design. They remained there until the fall of 1956. The Messerschmitt AG and the RSM jointly worked out the areas to be reworked, and the technical department was assigned the responsibility of making these changes. Important points of consideration included cost saving measures, e.g. making the "tub" out of sheet steel instead of steel tubing, as well as methods for making assembly easier.

During the course of these efforts, Fritz Fend successfully pushed to have complete assembly blueprints made up, and an employee from the technical department was sent to oversee work on the prototype.

According to Reinhold Ziegler, about 60,000 KR 200 models were sold—nearly 12,000 in 1955 alone. This was solid proof that the car had been tailor-made to meet the needs of a postwar Germany, where the trend toward larger, "proper" automobiles was already well underway.

At the same time he was involved with the bubble car program, Willy Messerschmitt was hard at work ironing out details in design and manufacturing—despite spending most of his time since 1951 on renewing his work in aviation in Spain. Reports and correspondence from this period hint at a certain—perhaps natural—tension between Fritz Fend, the one who had the original idea, and the more production savvy engineer Willy Messerschmitt. Even from Spain he would send numerous hand drawn sketches along with recommendations and improvements.

It should be noted that in Willy Messerschmitt's mind there was no question Fritz Fend deserved most of the credit for the bubble car. This is borne out in many of his writings. He drove a bubble car himself and expressed concern for ensuring chaplain Paul Adenauer's bubble car was well cared for in a letter to Hans Strauch, the mercantile board member of Messerschmitt AG at the time. On the other hand, he had no reservations about expressing his view that, "in light of the financial crisis, contests and world records only

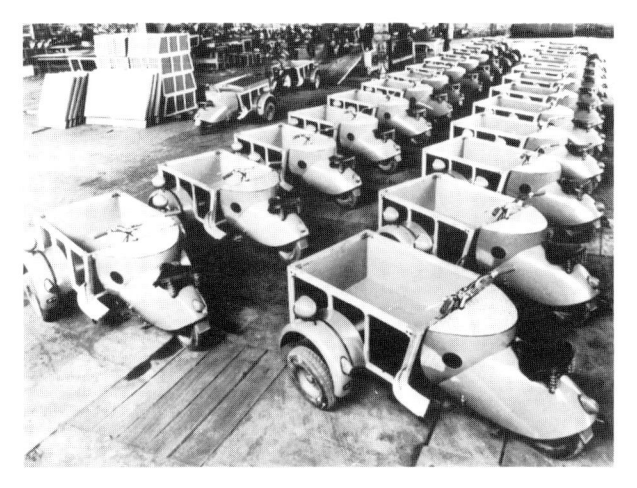

In addition to the *Kabinenroller* mini-cars, Fritz Fend also developed two models of diminutive three-wheeled transport vehicles for the RSM in 1953 and 1954. These tax exempt "delivery mopeds" required neither a driver's license nor a vehicle registration to operate, and were powered by either a 50 cubic centimeter or 100 cubic centimeter Sachs engine. These scooters went into production under the name "Mokuli" (from the German Mot*or*-*Kuli*).

cost money; they certainly don't bring in contracts." He was referring to the events of 29 and 30 August 1955 at the Hockenheim Ring, where automobile journalist Helmut Werner Bönsch and Fritz Fend himself, among others, set 25 world records. Nevertheless, the 24-hour non-stop driving did result in some valuable information being obtained with regard to linkage, suspension, sound damping, power train, four-speed transmission, and tires.

With a credit loss of 300,000 DM for a delivery of bubble cars to the U.S. (a hurricane buried the delivered bubble cars under water and made them unfit for sale), the RSM found itself facing serious liquidation problems. The result was to stop production of the bubble car and liquidate the RSM in an out-of-court settlement. The company site and its manufacturing equipment were acquired and later sold by the Bavarian Office of Structural Financing to cover the credit it had given to the Messerschmitt AG.

In early 1957 Fritz Fend established the Fahrzeug- und Maschinenbau GmbH Regensburg (FMR) together with Valentin Knott, the well-known industrialist and engineer from Eggstätt. Fend thus assumed management of the Messerschmitt bubble car's production himself. At the International Automobile Exhibition in September 1957 Fend exhibited a follow-on development to the bubble car principle alongside the Mokuli, the KR 200 and the KR 201 roadster (all still marketed under the Messerschmitt name). It was the "Tiger" sportscar—later renamed the FMR Tg-500 for trademark purposes. This was a four-wheeled car with tandem seating and a top speed of 130 km/h.

Fend developed the folding-top KR 201 "Messerschmitt-Roadster" from the KR 200, the successor to the KR 175. Having a sticker price at the time of DM2395, it came factory equipped with an interior which was luxurious even by today's standards. Average fuel consumption was 3.5 liters per 100 km, with a top speed of 100 km/h.

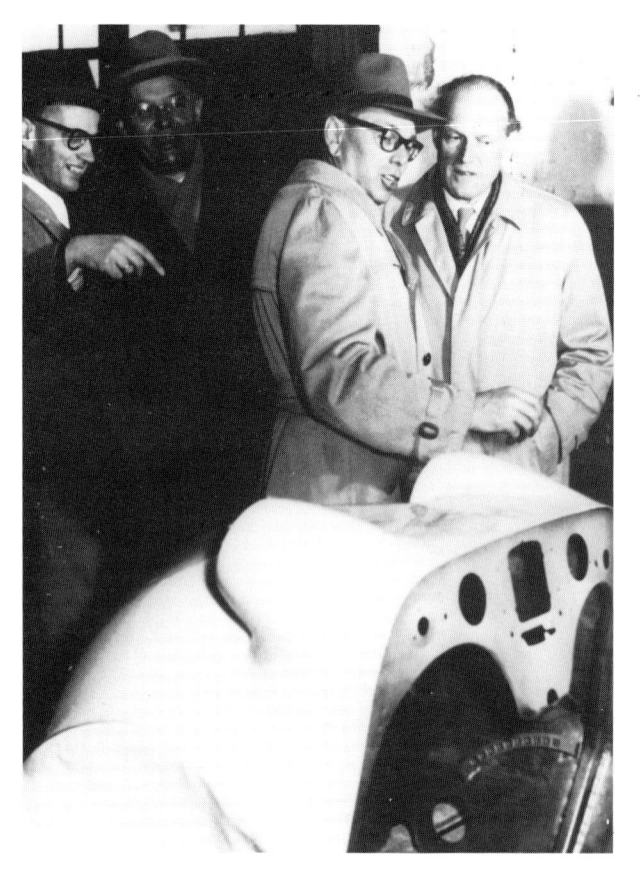

This vehicle was both the high point and the end point in the bubble car's developmental lifecycle. Customers' demands began to change, difficulties increased in finding sub-contractors to supply parts for fewer and fewer automobiles. Bubble car production ended in 1964, after nine years and nearly 75,000 of five different models had been built.

Willy Messerschmitt hadn't been involved with the bubble car for years, yet as in earlier times his name remained connected with this product of his day. There are Messerschmitt clubs throughout the world today, and his vehicles are bought and sold at collectors' prices.

Willy Messerschmitt inspects the record-breaking KR 200.

KR 200 record-setter in September 1955 at the Hockenheimring track.

Vespa Built Under License - Contacts with Alfa-Romeo

Vespa motor scooters began rolling off the assembly line at the Augsburg factory just one year after talks between Messerschmitt and Dr. Rinaldo Piaggio got underway in 1954. License manufacturing plans called for the Messerschmitt AG to pick up roughly half of Vespa's total production. The Vespa-Messerschmitt GmbH, founded by Messerschmitt AG and Piaggio on 18 January 1955, initially obligated itself to manufacture up to 8,000 vehicles per year. Later there was even talk of increasing production up to 1,250 scooters per month. Furthermore, it was planned to develop a joint operation between the Vespa-Messerschmitt GmbH (scooters) and the RSM (bubble cars). Obviously, the intended goals were never attained. However, in 1955 3,500 motor scooters left the factory, followed by 5,200 in 1956 and 6,000 in 1957. Independent manufacture only took place on a limited scale (e.g. wheel forks).

The Vespa-Messerschmitt GmbH was at first only able to sell a fraction of its production due to the fact that the former authorized dealer in Germany had been given permission to supply the scooters only through September 1955. After further startup problems—which incidentally were not of Messerschmitt's makings—Piaggio insisted on a change of contract. This would have resulted in Messerschmitt receiving only a percentage of the profits in exchange for the use of his name. In addition, he intended that the Messerschmitt AG would assume ten percent of the manufacturing funding for Vespa-Messerschmitt GmbH. The board of directors rejected these proposals and, as a result, in late 1955 the Messerschmitt AG reduced its former equal partnership in the Vespa-Messerschmitt GmbH to just 10 percent. The remaining 40 percent went to the Swiss industrialist Martial Fréne. The Messerschmitt AG eventually cut off all ties in late 1957.

Looking for opportunities to expand production, Willy Messerschmitt also made contact with the Italian automobile manufacturer Alfa Romeo in 1954. Wolfgang Degel made a business trip to Turin, where he obtained files for the Model 1900. Alfa Romeo was to have supplied engine, transmission, chassis, and the parts for the body. Assembly of the body, final assemby, and installation of accoutrements was to have taken place at Messerschmitt. These intentions, however, were never pursued beyond the initial stages, as it was felt that such a license production arrangement would have been unprofitable.

Workers at Augsburg pose with the "front" of the plaster model of the Messerschmitt K 106 mini car. Hilmar Stumm is second from left.

Minicars Follow Bubble Cars

While production of the bubble car was still underway, Willy Messerschmitt was already thinking of expanding his product line into a "minicar family" in order to meet the anticipated changes in the market. The trend from bubble car to minicar was already making itself known, was ultimately the reason for the demise of the bubble car and, albeit only temporarily, the impetus behind the rise of the minicar's popularity in Germany.

Detailed studies had already been made with regard to the most appropriate approach to parallel production with and/or replacement of the bubble car. One of the most pressing needs was to get rid of the bubble car's tandem seat arrangement, which was coming under increased criticism from customers as it made communication between driver and passengers difficult.

With this in mind, as well as problems with load distribution, asymmetrical loading, and ease of entry moved Willy Messerschmitt to draw up a small four-wheeled automobile which would accommodate two adults and two children or large suitcases, or tolerable space in the back seat to even accommodate two adults for short distances.

The goal was to remain in the economical middle class of "scooter-mobiles" and avoid developing a "full-grown" car, like the Volkswagen, for example.

In order to beat the competition—BMW Isetta, Goggomobil, Lloyd—in the performance category, it was Willy Messerschmitt's opinion that they would need to achieve 13kW/18hp, something which would be possible with a 400cm2 two-stroke engine.

Messerschmitt's K 106 mini car. Weighing just 370 kg empty, it could carry 280 kg, giving it a maximum weight of 650 kg. The vehicle was to have had a cruising speed of 95 km/h and a fuel consumption rate of 5.5 liters per 100 km.

In 1955 work began on the K 106, as the new minicar came to be known. Its layout as a two-door sedan with rear motor and Willy Messerschmitt's patented self-supporting body with removable engine and chassis segments—as on the larger P 511—in the main followed his ideas of light-weight design and simple assembly. By the year's end the blueprints were virtually complete, and construction of the prototype began at Werk II in Augsburg. The search for a suitable motor among German motorcycle firms was unsuccessful. As a result, Dr. Otto Steigenberger was given a contract for the design of a 400cm2 engine. An interim, or stopgap solution was found by using a Fichtel & Sachs 200cm² two-stroke single-cylinder engine, which was installed prior to testing of the frame and chassis in early 1956. The car exhibited quite good driving characteristics and acceptable power even with this "underpowered" version.

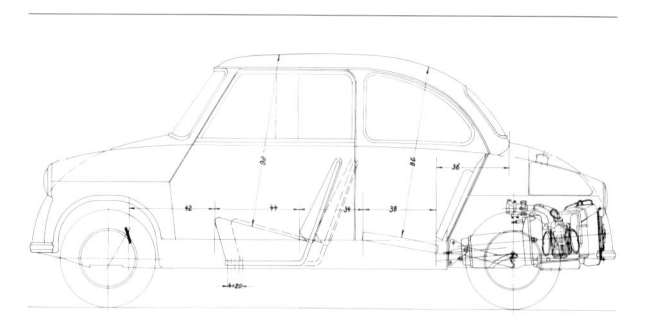

Cutaway of the K 106 showing the location of the 400 cm² Messerschmitt engine.

Further variants of the K 106 were looked at, drawn up and calculated—with a keen eye towards the American market and the competition in that country. These models included:

- K 107 (400cm2 engine, straight cut doorframe, door hinged at the front)
- K 108 (800cm² engine, 125 km/h maximum speed, sporty styling, longer nose, shorter rear, cabin smaller than K 106, only bucket seat in the back)
- K 109 (400cm² engine, 105 km/h, convertible, only bucket seat in the back)
- K 110 (600cm2 engine, 125 km/h, sports coupe, straight lower window edge)
- K 111 (600cm² engine, 125 km/h, sports coupe, curved lower window edge, straight cut doorframe)

With the re-emergence of aircraft production stretching the financial resources of the company to its fullest, it was not possible to pursue the K 106's development with the hoped for intensity. In August 1958 Willy Messerschmitt attempted to sell the developmental files, including those of the K 106 prototype, in the U.S. for a one-time license sum of $200,000, along with follow-on parts licenses. A letter dated 12/23/58 from the Messerschmitt AG to General (ret.) Dogchel, an interested party in America, summarizes the situation at the time quite well:

"...we developed the K 106. Unlike the minicars which had been appearing on the market, we wanted to achieve more space, better driving performance, and a more pleasing shape. Those minicars coming out in Germany today are following the same pattern that we'd envisioned back then. Unfortunately, as a plant which was in complete shambles because of the war, we lacked the funding to see the development program through to its end and enter into production. With aircraft construction resuming, our limited facilities meant that we could no longer pursue further automobile construction...

...The calculations for the 400cm2 engine have been completed and work has begun on construction of details. However, we lack the majority of construction drawings. Nor could a prototype engine be built...

...Testing (of the vehicle) has not gone beyond its early stages....Sufficient manufacturing plans have been drawn up. The vehicle has been calculated as carefully as possible, and investment plans have also been prepared..."

Willy Messerschmitt later offered Fritz Fend the opportunity to take over production development by his company. Fend was agreeable to this offer, but only if the American contractors noted above would come up with the funding for it—something which never materialized.

A "*Kurvenleger*" prevented the occupants from slipping out of their seats when going around turns. In 1956, the Messerschmitt studies department became involved in the further development of the MOBO concept built in the United States (this frontal view shows the chassis "banked" in a left turn).

Technical Office Keeps Busy With A Kurvenleger

In 1956 Dr. Ludwig Vogel, who had previously negotiated with Willy Messerschmitt in the U.S. for assembling the K 106 in that country, now approached him with the MOBO project. Vogel represented the interests of design engineer Louis de Monge and the company of Etienne Boegner and was searching for a company which would take over continued development of the project. De Monge had designed a four-wheeled vehicle which could "lean" into curves in much the same manner as a motorcycle. This so-called "*Kurvenleger*," or "curve-leaning" principle had been around since the '20s and, although it had advanced beyond the testing stage, the idea had yet to be applied to a production vehicle.

The "*Kurvenleger*" had many advantages, including a better ride for the occupants since there was no unpleasant leaning to one side or sliding along the seat in a turn. The hotshot powerslide was also under threat, as the increased wheel camber on a "*Kurvenleger*" produced high lateral load forces even without a high angle of incidence. Turning remained normal, and there was less danger of sliding.

The pivoting wheels also prevented the vehicle from tipping over even when making tighter turns.

Although recognizing the positive qualities of the "*Kurvenleger*," the Messerschmitt studies department came to the conclusion that the MOBO concept offered no significant advantages over conventional layouts, particularly since vehicles were starting to come off the drawing boards designed low enough so that they generally slid before tipping over. Furthermore, the "*Kurvenleger*" required a relatively great deal of effort to be put into its design and construction. And finally, the common two- to three-seat width of a vehicle which had become standard for car design established limits for what was—in theory at least—the ideal turn. There was no perceptible need for continued refinements in design, such as that incorporated into a Citroen car (hydraulically operated wheel extensions for counterbalancing the tendency to pull to the outside, caused by the centrifugal force in a turn). Therefore, for Messerschmitt and his department the "*Kurvenleger*" project became nothing more than a minor diversion, albeit an interesting one.

Messerschmitt In Spain

This critical chapter in the personal career of Willy Messerschmitt—a forerunner of his subsequent, renewed industrial involvement with aviation in the Federal Republic of Germany—is largely based on the preliminary work of Johann B. Kaiser, who meticulously analyzed and weighed relevant correspondence material obtained from available files. His "Chronologie Spanien" serves as source material for much of the information in this chapter. Another important source is "From Fabric to Titanium" by Jesus Salas Larrazabal and published by CASA in 1983, which is probably the first work to properly do justice to the history of Spain's aviation industry.

The close ties between Willy Messerschmitt and the Spanish aviation industry, particularly with La Hispano Aviacion S.A. (HASA), were important on two counts; not only did it provide Messerschmitt with the first opportunity since 1945 to pursue independent aeronautical developments, but as a follow-on to the HA 200 piston-engined trainer also led to the first ever jet airplane for Spain, the HA 200 trainer and, as a result of a license production contract, to Egypt's first jet as well, the HA 300. The HA 200 was Messerschmitt's sixth jet aircraft development, following on the heels of the Me 163, Me 262, Me 263, Me 328, and the P 1101. Two types, the HA 100 and the HA 200, went from drawing board to first flight in under five years. These accomplishments, achieved with but a small German leadership team and a group of qualified HASA employees yet again reveals Messerschmitt's typical approach method. From the viewpoint at the time, as well as with today's hindsight, it is only too understandable how an airplane builder could eagerly and passionately seize upon the opportunities presented to him in Spain and later in Egypt and become so involved in the aviation industries within those two countries.

Aircraft Development Again

The opportunity came at a time when Willy Messerschmitt and the Messerschmitt AG were working to branch out into other areas of development and production as a result of a ban on aircraft development in the Federal Republic. Messerschmitt's former chief stress analyst, Julius Krauß, received a request from the Me 109 license holder, Hispano Aviacion in Seville, to act as an advisor to the company on further development of the Me 109. The request had come through the former manager of the license production department in Regensburg, Frederico Prasthofer, who had remained in Spain after the war as a liasion technician. Following the Spanish Civil War, Hispano Aviacion had maintained the Messerschmitt fighters formerly of the Condor Legion on behalf of the Spanish Ministry of Aviation, later obtaining the license for producing the Me 109G, and after 1945 developed and manufactured the "Spanish" Me 109 (HA 1109 in various subtypes). However, the tides of war prevented the German engines from being delivered for the more than 200 Spanish airframes built during the war. As a result, the Spanish initially decided to install a Hispano-Suiza engine, then settled upon the Rolls Royce Merlin powerplant which had proven its worth in the Vickers Supermarine Spitfire, among other types. Questions arose regarding the airframe's ability to survive the marriage with the new powerplant, and these matters prompted HASA to open up communications channels with its former licensor.

A request from HASA was made through Guillermo F. Mallet, the former Messerschmitt representative in Madrid, and in March of 1951 Willy Messerschmitt found himself traveling to Spain for the first time since World War II, accompanied by Baroness Lilly von Michel-Raulino, his future wife. During the course of his visit he inspected several aircraft factories and took the opportunity to establish ties with leading figures in the Spanish aviation community.

In April 1951 Messerschmitt flew to South Africa in order to explore the possibility of developing aircraft in that country. This appeared to have such little chance of success that he did not pursue this course of action any further. On the other hand, events were proceeding apace in Spain. At first working independently of Messerschmitt, Julius Krauß was supervising the continued development of the Spanish Me 109 at Hispano Aviacion. A second visit to Spain occurred in July 1951, with even more visits to key industrial sites and, in the interim, negotiations for the establishment of a development division together with Hispano Aviacion, headquartered in Seville. These negotiations were carried out by Dr. Hans Heinrich Ritter von Srbik, representing Messerschmitt's interest on behalf of his company "Internationale Treuhand & Verwaltungsgesellschaft München."

In his "Spanish Memorandum" of 17 July 1951 Messerschmitt assessed the capabilities and potential of the Spanish aviation industry, possibly a bit subjectively, as being quite favorable, but pointed to gaps in production of semifinished materials, in powerplants and in the area of aircraft systems. He drew up task proposals, and by doing so hoped to be of use to the Spanish Ministry of Aviation with the establishment of a joint developmental department with HASA. These proposals included the design of fighters and the construction of an aircraft through the full scale production stage, the design of a "modular plane" (transport, commuter plane, with potential for conversion to bomber) with two or four engines—also to achieve production status, the development of a jet engine, follow-on development of a 750 hp engine, subsonic wind tunnel tests with INTA (Instituto Nacional de Technica Aeronautica), as well as researching a remote controlled, airplane-like missile for air defense. This multifarious portfolio of development programs—only some of which were ever contracted for—expanded even further with the involvement in a requirement for trainers issued by the Spanish Ministry of Aviation in June 1951.

The "Spanish Memorandum" was the basis for the agreement between Messerschmitt and HASA, and offered the foundation for Ministry of Aviation contracts for Hispano Aviacion. By the way, the ministry placed considerable value on the approval of these events by official agencies within the Federal Republic of Germany. Consequently, in August 1951 Willy Messerschmitt and Georg Eidenschink presented their intentions to Hans Speidl and Adolf Heusinger, advisors in the German chancellor's office, and in Amt Blank to Oberst Count Kielmannsegg, among others. Although these departments did indeed give their nod of approval as a result, they also expressed their intention not to bear any responsibility. It was recommended that Messerschmitt maintain regular contact with the Bonn agencies.

Close Cooperation With Hispano Aviacion

The first agreement between Hispano Aviacion's general director, Don Pedro Arritio, and Willy Messerschmitt was signed on 26 October 1951 in the office of Dr. von Srbik in Munich. It called for the establishment of a "Prof. Messerschmitt Developmental Department" (Oficina Tehnica Prof. Messerschmitt) at HASA in Seville, beginning on 1 January 1952.

Supervised by Julius Krauß, the first development team arrived on 3 and 4 January 1952. It consisted of Hans Hornung, Lorenz Bosch and Georg Ebner. August Hoffmann arrived about six months later. With his arrival, from the old Messerschmitt team Krauß now had a team chief for projects, design, stress analysis, and systems.

The tiny Messerschmitt team first began working on Hispano Aviacion contracts for a trainer initially powered by a 330kW/450 hp engine (later the HA 100) and on the specifications for a jet trainer (later to become the HA 200 Saeta, or Arrow).

Messerschmitt spent almost the entire month of February in Seville and, as Mallet wrote, enjoyed "being able to devote his time to developing airplanes once again, instead of having to deal with endless business and legal stuff back in Munich," fields in which he was not as naturally gifted.

HA 100 Piston-Engine Trainer

Messerschmitt had begun thinking about a piston-engined trainer starting in late 1951, so that within the first half of 1952 rapid progress had already been made on the prototype design. Although Messerschmitt only worked part time in Spain personally, he nevertheless drew up all the important design details from Munich. By July of 1952 he was able to project the first flight of the HA 100 for March/April 1953.

However, development work evolved at a slower pace than anticipated. Acquisition of engines, propellers, radiators, landing gear parts, instruments, even standard parts, semifinished materials, and insulation proved problematic. Even the construction of frames for static and dynamic testing ran into difficulties in the factory unaccustomed to building modern airplanes and led to additional unexpected delays.

On the other hand, by mid-1952 development of the jet trainer (HA 200) was moving along quite well, and even the jet fighter (later the HA 300) began taking shape in the minds of Messerschmitt's engineers. It was decided to utilize the Turboméca Marboré II (for the HA 200)—to be delivered sometime that year—and there were discussions with Oerlikon regarding the armament for the HA 300.

An attempt to hire German engineers who had been working for several years at CASA in Spain and wanted to transfer to Messerschmitt failed due to heavy governmental involvement in drafting up an agreement. Spanish aviation companies were subsequently prohibited from soliciting employees from each other. His team was in urgent need of beefing up, and as a result Messerschmitt was forced to provide the numbers from his Munich department and through new hirings.

Messerschmitt was personally in Spain again in June/July 1952. He was chiefly involved in design work in Seville and also carried out discussions in Madrid with Chief of Staff General Vigon, dicussing among other things the performance and armament of the HA 300. During this time, he learned from a messenger dispatched from the Amt Blank that discussions had taken place between Spain and the Federal Republic regarding cooperation between the two countries on matters of airborne weapons systems. At Spain's request, Messerschmitt was to represent that country in Bonn as a kind of "technical development aide." Even Minister of the Army Munoz Grande asked him to convey his personal requests to Minister Blank. These primarily addressed technical, organizational, and personnel problems within the Spanish armament industry.

At the same time, Messerschmitt and Hispano Aviacion concluded another agreement, which went into effect in August following approval from the Spanish government. This agreement effectively tied Messerschmitt to Spain until at least the end of 1954.

In view of America's initial activities as part of the first "airbase treaty" (building airfields, a naval port, and pipelines, as well as subsequent maintenance and repair contracts from the U.S. Air Force for the benefit of Spain's aviation industry), it appeared that stretching his portfolio to include a jet fighter was a tenuous proposition even then. Nevertheless, Messerschmitt and his Munich team continued working diligently on this project despite the lack of a contract.

As before, the focus of the team's development work remained concentrated on the HA 100. Messerschmitt went to Seville in September 1952 for this express purpose. Afterwards he had all critical drawings sent to Munich prior to clearance being given for prototype construction, and finally authorized an increase in personnel for the Seville team starting at the beginning of 1953. First to arrive were Ferdinand Miller and Gert Wischhöfer. Lorenz Bosch, the designer, left in February for a job in the U.S. and was replaced by Alfred Nerud. Max Blumm (flight handling) joined the team in August, and on 1 September Gero Madelung, Messerschmitt's nephew, began working. When it became known that Julius Krauß would leave at the year's end to take over the newly established department of aeronautical engineering at the Technische Universität in Munich, Madelung was appointed as his successor to become Prof. Messerschmitt's "permanent representative." The office grew to ten with the assignment of the aero-elastics engineer Erich Schmidt on October 1st, 1953.

By this time the design of the HA 100 was virtually complete. It was initially to have been powered by ENMASA's 330kW/450 hp Elizalde Sirio engine. Assembly of the prototype was delayed, however, by the need to incorporate critical changes resulting from the durability tests, such as strengthening the nosewheel. Flight certification for the engine and delivery of parts from abroad, such as landing gear wheels from England, also ran into delays. In the end, ENMASA's older and heavier Elizalde Beta B4 (555 kW/755hp) had to be installed in the first prototype.

The Beta motor stemmed from a development by the American company of Wright, built during the Spanish Civil War under license in the Soviet Union and sent to Barcelon for assembly. Despite these setbacks, the HA 100E took to the air for the first time on December 10th, 1953, just 26 months after the contract was signed.

1953 also saw considerable progress made on the HA 200 jet trainer. By mid-year the rough draft for the initial version had been completed with two Turboméca Marboré engines, each outputting 3.7 kN/380kp. Design work began from this time on. At about the same time Willy Messerschmitt instructed his team in Seville to research alternative configuration variations for the HA 300 jet fighter in preparation for the expected development contract.

Feelers extending outward from Munich at the beginning of 1953 to the Eidgenössische Flugzeugwerke Emmen and its high-speed wind tunnel in Switzerland led to a HASA contract with that company later that year, once HASA itself had been covered by an appropriate government contract.

Messerschmitt was on the move quite a bit during 1953. In March, April, and June he was at work in Seville or Madrid. On 15 May he gave a speech in Düsseldorf at the Rhein-Ruhr Club on "German Potential in the Field of Aeronautics." In July he visited the Paris Airshow, something he hadn't done since the '30s, and in September he was at the Farnborough Airshow near London. He then went back to Seville to await the date for the first flight of his HA 100 (which didn't occur at this time). In October he returned to Munich and from there flew to the U.S. There he met with his brother-in-law, Prof. Georg Madelung, renewed old acquaintances and established new ones. By the end of October he was back in Seville in order to personally secure the HA 100's maiden flight before the year's end.

For the development team in Seville, the Beta B4 powered HA 100V-1 (HA 100E) was soon put on the back burner, so to speak. As a result of the Elizalde engine's unsatisfactory operating characteristics, the decision was made to use the 588 kW/800 hp Wright Cyclone 7 (R-1300). The V-2 (HA 100F) was finished by late 1954. It flew for the first time in February 1955. The second HA 100E with B4 engine took to the air for the first time on 3 May 1956. When this motor proved to be unreliable as well, consideration was again given to modifying the airframe to accept the Sirio engine. This never happened, and there weren't enough funds to purchase the R-1300 in quantity. The planned production of 40 aircraft was halted. Where possible, parts were set aside for later use on the HA 200.

HA 100 Triana - First Postwar Airplane

The first airplane designed and built by Willy Messerschmitt in Spain was the HA 100, created with the support of the Hispano Aviacion design bureau. This basic and advanced trainer was laid out using an airframe having a load factor of 7 so that two different roles could be met: a basic trainer with a 330kW/450 hp radial engine and an advanced trainer with a 590kW/800 hp engine. The two variants differed externally only in the engine assembly, i.e. the nose section ahead of the firewall.

The all-metal low-wing design had two seats in tandem beneath a single panoramic canopy made of plexiglas. The single-spar, two-piece wing was fitted with Messerschmitt's trademark split flaps, operated manually by simple kinematics, plus a planned dive brake mounted as a separate kit on the fuselage underside. The fuselage had an oval cross-section and was of standard construction, meaning load bearing and "developable" skin (the Spanish didn't have a stretching and drawing press at the time) with formers and stringers. The non-developable engine housing and the wing join had been carefully "hammered out" by hand. Perhaps because of this, the HA 100 was meaningfully given the name "Triana." Triana was the suburb of Seville where the headquarters of Hispano Aviacion were located; it was also the old quarter where the gypsies had once had their smithies. The plane's landing gear retracted hydraulically. The main gear, mounted into the wing, retracted into the fuselage, with the nose gear (separate from the engine supports) retracting rearward into the fuselage.

Messerschmitt had paid particular attention to ease of maintenance. For example, the engine housing consisted of four sections, each comprising a top and bottom section bolted onto the fuselage. As a result, the engine housing was fitted snugly to the airframe, thus preventing the windscreen from getting dirty. The side sections were hinged at the top.

In addition to pilot training the HA 100 could be used for preliminary fighter training. Accordingly, two fixed machine guns and a gunsight could be fitted, along with a gun camera. For fighter-bomber training, the plane could also be fitted with pylons for carrying up to four 50kg bombs. The type even had the potential for providing reconnaissance training by fitting a photocamera in the fuselage.

Type	HA 100[1]	HA 100
Engine	ENMASA Elizalde "Sirio" 330kW/450 hp	ENMASA Elizalde "Beta" 550kW/750hp
Crew	1+1	1+1
Length (m)	8.5	8.25
Height (m)	3.0	3.0
Wingspan (m)	10.4	10.4
Wing area (m^2)	17.0	17.0
Empty equipped (kg)	1497	1743
Add'l load (kg)	420	700
Takeoff weight (kg)	1917	2443
Max. speed (km/h)	387	478[1]
Cruise speed (km/h)	330[1]	408[2]
Rate of climb (m/s)	7.3	10.5
Service ceiling (m)	7000	9100
Range (km)	650[1]	1200[2]
Takeoff to 15 m (m)	470	530
Landing fm 15 m (m)	470	500
Notes	[1] estimated data. This version not built	[1] at 3000 m, 426 at sea lvl [2] at 3000 m

The HA 100 running up its engine just prior to its maiden flight.

Prof. Georg Madelung, the father of Gero Madelung, wrote a letter to Hans Hornung on 17 June 1955 in which he stated that he was revamping the descriptions for the Me 100 and Me 200 and translating them into English. Two paragraphs are worth reprinting here, for not only do they provide a glimpse of the spirit of the times, but also reemphasize the work methodology of Willy Messerschmitt:

"I see that the mention of the Me 109 as well as competing aircraft designs is to be dropped. The Me 109 was pride of place for the company and brought renown to the name of Messerschmitt which can never be erased. Nevertheless, that was twenty years ago. Perhaps it's better not to give our competitors the opportunity to say that Messerschmitt is still trying to peddle twenty-year-old, long-since withered laurels. On the other hand, the Me 109 was far ahead of its day and in many respects even now a model example. There can be no denying the fact that the Me 100 has fundamentally been cut from the same block as the Me 109. The "law of unfair competition" prevents mentioning the competition (North American T-28), even though in fact there's nothing unfair about it at all."

I agree with you in that the data provided in the brochure is enough to whet the appetite of those interested in the airplane. They can then ask about any further details. But the brochure should also dispel any prejudices which are most likely still in circulation. Especially the idea that Messerschmitt aircraft aren't designed strong enough. I therefore feel we need to make sure the customers fully understand that the great savings in weight has been achieved through entirely normal means, namely by not only sharply reducing the surface area through combining several functions into a single component, but also the conscientious workmanship which goes into every component—even those seemingly trivial ones which many a chief designer leaves to his assistants. In the public's eye, the fact that Messerschmitt personally has been involved in every facet of the design can't be emphasized enough. I would therefore like to begin the brochure with words to the effect that both aircraft have been personally designed and worked through by Willy Messerschmitt down to the last detail."

Comparing the HA 100 with the T-28A

A data comparison between the North American T-28A—one of the most commonly flown airplanes in the world at the time—and the HA 100 reveals the latter's much smaller dimensions and resulting lower empty equipped weight with a higher additional load, despite the fact that both types were designed for the same mission. The HA 100 is superior to the American trainer in nearly every category. However, the HA 100 suffered from the handicap of having an engine running at too great rpms. As a result, the propeller was of necessity optimized for higher speeds, which in turn led to the only comparable category being that of takeoff performance. Estimated manufacturing costs for the HA 100 were lower as a direct result of its lower empty equipped weight and correspondingly reduced material requirements. In addition, the HA 100 was planned from the outset to accommodate larger and heavier engines. Flight handling was flawless, and the control forces were considered to be pleasant and well balanced.

Type		HA 100	T-28A
Engine		ENMASA Elizalde "Beta" 550kW/750 hp	Wright R-1300 590kW/800 hp
Crew		1+1	1+1
Length	(m)	8.87	9.76
Height	(m)	3.05	3.86
Wingspan	(m)	10.4	12.23
Wing area	(m²)	17.0	24.9
Empty equipped	(kg)	1743	2320
Add'l load	(kg)	900	748
Takeoff weight	(kg)	2643	3068
Max. speed	(km/h)	478[1]	435[1]
Service ceiling	(m)	9500	8200
Takeoff to 15 m	(m)	450	455
Landing fm 15 m	(m)	450	412
Notes		[1] at 2900 m	[1] at 1800 m

The upward opening engine covers gave easy maintenance access to the HA 100's engine.

Above: M 17 from 1925, the oldest existing Messerschmitt aircraft, seen here in the Augsburg plant. Today the airplane is housed in the Deutsches Museum in Munich.
Left: BFW brochure from 1927 showing its first product, the U 12 Flamingo.
Right: Drawing 29 from the Messerschmitt Flugzeugbau, Bamberg, from 1927. It shows a twin-engined M 17 on a typical blueprint from the 'twenties.

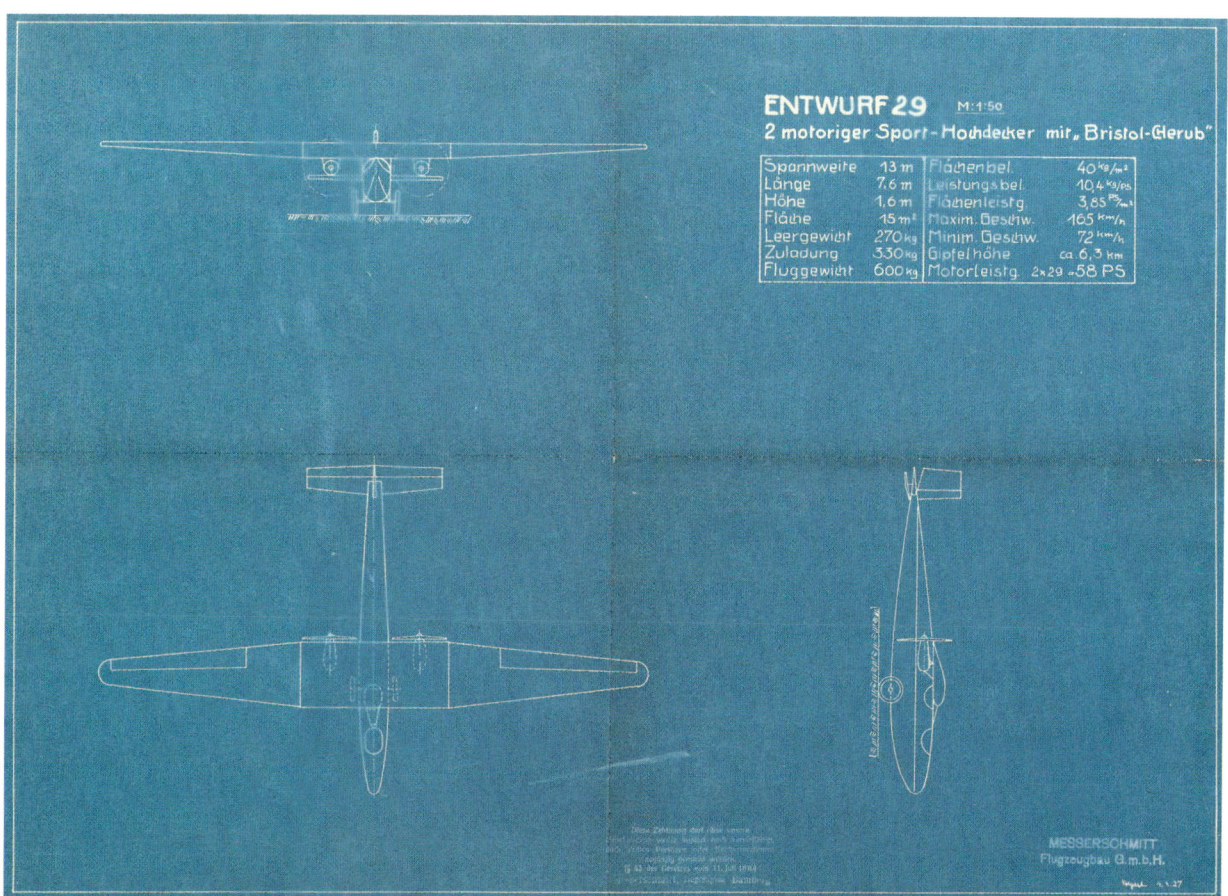

Bayerische Flugzeugwerke A.G. Augsburg
Metall-Verkehrsflugzeuge Konstruktion Messerschmitt
Avions entièrement en métal pour voyage et commerce

BFW

ILA 1928 BERLIN

BAYERISCHE FLUGZEUGWERK

EXPOSITION INTERNATIONALE AÉRONAU
INTERNATIONALE LUFTFAHRTAUSSTEL
INTERNATIONAL AIRCRAFT EXHIB

En publiant cette brochure, nous voudrions offrir un souvenir de l'exposition des Bayerische Flugzeugwerke aux visiteurs de l'ILA. En même temps, elle pourra donner après coup une idée de notre Stand à ceux qui n'ont pu voir l'Exposition.

Mit dieser Schrift beabsichtigen wir, den Besuchern der ILA eine Erinnerung an die Ausstellung der Bayerischen Flugzeugwerke in die Hand zu geben. Den vielen, die die Ausstellung nicht besuchen konnten, soll sie ein Bild unseres Standes vermitteln.

Through the issue o chure we propose tors to the Show with recollection of the E by the Bayerischen To the many who we the Aero Show, let be a survey of

Brochure for the M 18 showing the company's new logo, from 1928.

The 1928 ILA exhibiting the original M 18, 20, 21, and 23.

A Messerschmitt Bf 108B-1 company aircraft in a civilian scheme, during a photo flight over the Alps (1939)

A modern photo showing MBB's Me 108D (EF+PT) with Argus variable-pitch propeller and a military paint scheme from the 'forties.

MBB's Me 109G-6, FM+BB, born from a Spanish 109 fuselage and a DB 605 license-built engine (first flight 23 April 1982).

MBB's two company planes, the Me 108D and Me 109G, preparing to take off for an open house show.
Below: In 1939 and 1940 Switzerland received 80 and Yugoslavia 73 Bf 109E-3 fighters. These two photos show some of the aircraft prior to being delivered.

Fuselage and final assembly of Bf 109s at the Regensburg works (1940).

The sole original Bf 109G-2 flying today is this aircraft, Werknr. 10639, built by Erla in Leipzig in 1942 and flown operationally in North Africa. It is owned by the RAF Museum, Hendon, UK. After restoration, it first flew on 17 March 1991.

Baugruppen-Benennung	B-1	C-1	C-2	C-3	C-4	C-5	C-6	C-7		D-0	D-1	D-2	D-3	D-4		E-1	E-2	E-3	E-4
Fertiggruppen Nr. für Rumpfwerk	-	F10	F11	F10	F11	-	-	F19		F16	-	F17	-	-		F18	F12	-	-
Fertiggruppen Nr. für Leitwerk	-	F30	F31	F30	F31	-	-	F31		F31	-	F31	-	-		F32	F32	-	-
Fertiggruppen Nr. für Tragwerk links	-	F50	F52	F50	F52	-	-	F52		F54	-	F54	-	-		F56	F56	-	-
Fertiggruppen Nr. für Tragwerk rechts	-	F51	F53	F51	F53	-	-	F53		F55	-	F55	-	-		F57	F57	-	-
Fertiggruppen Nr. für Triebwerk links	-	F60	F62	F60	F62	-	-	F62		F62	-	F62	-	-		F64	F64	-	-
Fertiggruppen Nr. für Triebwerk rechts	-	F61	F63	F61	F63	-	-	F63		F63	-	F63	-	-		F65	F65	-	-
Rumpf	10	13	13	13	13	-	-	13		19	-	19	-	-		19	19	-	-
Rumpfanbauteile	11	14	14	14	14	-	-	14		101	-	101	-	-		101	101	-	-
Verkleidung a.R.f. Zusatzbehälter	-	-	-	-	-	-	-	-		17	-	17	-	-		17	-	-	-
Rumpfeinrichtung	12	15	15	15	15	-	-	15		15	-	15	-	-		15	15	-	-
Schutzscheibeneinbau	-	103	-	103	-	-	-	103		103	-	103	-	-		103	103	-	-
Platteneinbau	-	102	102	102	102	-	-	102		102	-	102	-	-		102	102	-	-
Schlauchbooteinbau	-	-	-	-	-	-	-	-		16	-	16	-	-		16	-	-	-
Einziehfahrwerk	24	27	27	27	27	-	-	27		27	-	27	-	-		207	207	-	-
Bremsanlage	22	28	28	28	28	-	-	28		28	-	28	-	-		208	208	-	-
Sporn	20	29	29	29	29	-	-	209		209	-	209	-	-		209	209	-	-
Querruder links	30	305	305	305	305	-	-	305		305	-	305	-	-		305	305	-	-
Querruder rechts	31	306	306	306	306	-	-	306		306	-	306	-	-		306	306	-	-
Höhenflosse	302	302	302	302	302	-	-	302		302	-	302	-	-		302	302	-	-
Höhenruder links	303	303	303	303	303	-	-	303		303	-	303	-	-		303	303	-	-
Höhenruder rechts	304	304	304	304	304	-	-	304		304	-	304	-	-		304	304	-	-
Seitenflosse links	161.35	161.35	161.35	161.35	161.35	-	-	161.35		161.35	-	161.35	-	-		161.35	161.35	-	-
Seitenflosse rechts	161.36	161.36	161.36	161.36	161.36	-	-	161.36		161.36	-	161.36	-	-		161.36	161.36	-	-
Seitenruder links	161.37	161.37	161.37	161.37	161.37	-	-	161.37		161.37	-	161.37	-	-		161.37	161.37	-	-
Seitenruder rechts	161.38	161.38	161.38	161.38	161.38	-	-	161.38		161.38	-	161.38	-	-		161.38	161.38	-	-
Handsteuerung im Rumpf	400	400	400	400	400	-	-	400		400	-	400	-	-		400	400	-	-
Handsteuerung im Flügel	45	45	45	45	45	-	-	45		45	-	45	-	-		45	45	-	-
Fußsteuerung im Rumpf	41	401	401	401	401	-	-	401		401	-	401	-	-		407	407	-	-
Fußsteuerung im Leitwerk	406	406	406	406	406	-	-	406		406	-	406	-	-		406	406	-	-
Klappen- u. Flossenverstellung i. Rumpf	42	402	402	402	402	-	-	402		402	-	402	-	-		402	402	-	-
Vorflügelkupplung	47	47	47	47	47	-	-	47		47	-	47	-	-		47	47	-	-
Flügel links	508	58	58	58	58	-	-	58		58	-	58	-	-		58	58	-	-
Flgl.Einb.f. Zusatzbehälteranlage l.	-	-	-	-	-	-	-	-		526	-	526	-	-		528	528	-	-
Flügel rechts	509	59	59	59	59	-	-	59		59	-	59	-	-		59	59	-	-
Flgl.Einb.f. Zusatzbehälteranlage r.	-	-	-	-	-	-	-	-		527	-	527	-	-		529	529	-	-
Triebwerkverkldg. i. Flügel links	510	502	502	502	502	-	-	502		502	-	502	-	-		502	502	-	-
Triebwerkverkldg. i. Flügel rechts	511	503	503	503	503	-	-	503		503	-	503	-	-		503	503	-	-
Vorflügel links	54	54	54	54	54	-	-	54		54	-	54	-	-		54	54	-	-
Vorflügel rechts	55	55	55	55	55	-	-	55		55	-	55	-	-		55	55	-	-
Klappe links	52	522	522	522	522	-	-	522		522	-	522	-	-		522	522	-	-
Klappe rechts	53	523	523	523	523	-	-	523		523	-	523	-	-		523	523	-	-
Triebwerksverkleidung linke Gondel	61	63	63	63	63	-	-	63		63	-	63	-	-		63	63	-	-
Triebwerksverkleidung rechte Gondel	62	64	64	64	64	-	-	64		64	-	64	-	-		64	64	-	-
Triebwerksgerüst	60	162.63	162.63	162.63	162.63	-	-	162.63		162.63	-	162.63	-	-		162.63	162.63	-	-
Auswechselbares Triebwerk links	-	606	606	606	606	-	-	606		606	-	606	-	-		606	606	-	-
Auswechselbares Triebwerk rechts	-	607	607	607	607	-	-	607		607	-	607	-	-		607	607	-	-
Flugmotor links	713	730	730	730	730	-	-	730		730	-	730	-	-		730	730	-	-
Flugmotor rechts	714	731	731	731	731	-	-	731		731	-	731	-	-		731	731	-	-
Triebwerksbedienungsgestänge	745	741	741	741	741	-	-	741		741	-	741	-	-		741	741	-	-
Drahtzüge	-	742	742	742	742	-	-	742		742	-	742	-	-		742	742	-	-
Kraftstoffleitung mit Armaturen	716	746	746	746	746	-	-	746		746	-	746	-	-		746	746	-	-
Schmierstoff-Ltg. linker Motor	717	747	747	747	747	-	-	747		747	-	747	-	-		747	747	-	-
Schmierstoff-Ltg. rechter Motor	718	748	748	748	748	-	-	748		748	-	748	-	-		748	748	-	-
Zus.Kraft-u.Schmierst.Anlg.i.Rumpf u.Flgl.	-	-	-	-	-	-	-	-		756	-	756	-	-		756	756	-	-
Zus.Kraft-u.Schmierst.Anlg.unt.dem Rumpf	-	-	-	-	-	-	-	-		759	-	759	-	-		759	-	-	-
Kaltstartanlage	755	753	753	753	753	-	-	753		753	-	753	-	-		753	753	-	-
Kühler	71	715	715	715	715	-	-	715		715	-	715	-	-		715	715	-	-
Kühlwasserleitung l.Motor	721	728	728	728	728	-	-	728		728	-	728	-	-		728	728	-	-
Kühlwasserleitung r.Motor	712	729	729	729	729	-	-	729		729	-	729	-	-		729	729	-	-
Abwerfbare Behälter	-	-	-	-	-	-	-	-		758	-	758	-	-		758	758	-	-
Kabinenheizanlage	-	-	-	-	-	-	-	-		-	-	-	-	-		754	754	-	-
Triebwerksbehälter	82	84	84	84	84	-	-	84		84	-	84	-	-		84	84	-	-
Betriebsgeräte	904	919	919	919	919	-	-	919		919	-	919	-	-		944	944	-	-
Luftschraubenautomatik A-Motor	-	981	981	981	981	-	-	981		981	-	981	-	-		981	981	-	-
Luftschraubenautomatik N-Motor	-	989	989	989	989	-	-	989		989	-	989	-	-		989	989	-	-
Tankpumpabsicherung	-	945	945	945	945	-	-	945		-	-	-	-	-		-	-	-	-
Meßleitungen	905	941	941	941	941	-	-	941		941	-	941	-	-		941	941	-	-
Navigationsgeräte	910	943	943	943	943	-	-	943		943	-	943	-	-		943	943	-	-
Bordnetzgeräte	902	-	-	-	-	-	-	-		-	-	-	-	-		-	-	-	-
Rumpfeinbauteile f. Abwurfwaffen unt.d.Rpf.	-	-	-	-	-	-	-	948		-	-	948	-	-		974	974	-	-
Notwurfeinbau im Rumpf	-	-	-	-	-	-	-	950		-	-	950	-	-		950	950	-	-
Abwurfwaffeneinbau unter dem Rumpf	-	-	-	-	-	-	-	949		-	-	949	-	-		949	949	-	-
Hydraulische Fernbetätigung	92	920	920	920	920	-	-	920		920	-	920	-	-		920	920	-	-
Kurssteuerung und Patinanlage	915	-	-	-	-	-	-	-		-	-	-	-	-		-	-	-	-
Zusatzsauerstoffanlage	-	-	-	-	-	-	-	-		980	-	980	-	-		980	980	-	-
Elt-Ausrüstung	-	925	-	925	-	-	-	-		-	-	-	-	-		978	978	-	-
Elt.Anlage f. Luftschraubenautomatik	-	971	972	971	972	-	-	972		972	-	972	-	-		-	-	-	-
FT-Anlage	912	912	-	912	-	-	-	-		-	-	-	-	-		982	982	-	-
Elt-Ausrüstung mit FuG X Einbau	-	-	965	-	965	-	-	965		965	-	965	-	-		-	-	-	-
FuG 25 - Einbau	-	973	973	973	973	-	-	973		973	-	973	-	-		973	973	-	-
Elektrische Leitungen	913	-	-	-	-	-	-	-		-	-	-	-	-		-	-	-	-
Elektrische Kühlklappenbetätigung	958	-	-	-	-	-	-	-		-	-	-	-	-		-	-	-	-
Elektrische Anlage für starre Bewaffnung	959	-	-	-	-	-	-	-		-	-	-	-	-		-	-	-	-
Elt.Ausrüstung für Zus.-Kraftstoffanlage	-	-	-	-	-	-	-	-		984	-	984	-	-		-	-	-	-
Elt.Anl.f. Abwurfwaffen unter d.Rumpf	-	-	-	-	-	-	-	947		-	-	947	-	-		947	947	-	-
4 MG i. Rumpfspitze	908	908	908	908	908	-	-	908		908	-	908	-	-		908	908	-	-
MG FF i. Rumpfboden	918	918	918	-	-	-	-	-		-	-	-	-	-		-	-	-	-
MG FF-M i. Rumpfboden	-	-	-	918	918	-	-	918		918	-	918	-	-		918	918	-	-
ESK 2000-Einbau	916	916	916	916	916	-	-	916		916	-	916	-	-		916	916	-	-
Bewegliche Schußwaffe	986	986	986	986	986	-	-	986		986	-	986	-	-		986	986	-	-
Abwurfwaffeneinbau	-	-	-	-	-	-	-	-		-	-	-	-	-		977	977	-	-

B-1 u. übernommen für C-1 C-2 C-3 C-4 C-7 D-0 D-2 E-1 E-2
C-1 u. übernommen für C-2 C-3 C-4 C-7 D-0 D-2 E-1 E-2
C-2 u. übernommen für C-4 C-7 D-0 D-2
D-0 u. übernommen für D-2 E-1 E-2
E-2

A Bf 110C(KD+TM) with a test scheme for night fighting operations (1941).

Left: This Bf 110B-E (1938-1940) shows both new and inherited components as well as their use on the Bf 161 (control surfaces) as of 1940.

A Me 410A-1(GF+TM) converted from Me 210A-1, Werknr. 10022 (1943).

Remains of the record breaking Me 209V-1, Werknr. 1185, D-INJR, currently housed in a museum in Krakow, Poland. During the war the aircraft had been kept in storage in the eastern part of the Reich. In the lower photo can be seen the water reservoir for the aircraft's water cooled radiator.

Fritz Wendel prior to the first flight of the Me 262V-3 at Leipheim on 18 July 1942.

Opposite: A production Me 262A-1, Werknr. 110604, Red 1, of JG 7 with pilot Helmut Lennartz at the controls, prior to takeoff at Lechfeld (1945).

The Me 163B-1 restored by Messerschmitt at Augsburg in 1965, shortly before its handover to the Deutsches Museum in Munich.

333

Regensburg's Städtische Galerie put on an exhibition 11 January to 2 February 1986 under the motto "*Die flotten Flitzer aus Regensburg*" (roughly translated: "The Sporty Speedsters from Regensburg"). The exhibit included a nearly complete showing of Fritz Fend's life work, including the major variants of the Messerschmitt *Kabinenroller*. Above left is the KR 200, below left is a version of the "Mokuli" (with a Fend "*Flitzer*" in the background, while below right is the record-setting KR 200.
An advertisement for the Messerschmitt KR 200 *Kabinenroller* during its heyday (above right).

Willy Messerschmitt's first postwar development, the HA 100, during engine run-up shortly before its maiden flight.

A one-of-a-kind picture: in Spain, a HA 100 overflies license-built Me 109s and a Ju 52.

Production HA 200 planes at the San Pablo airfield in Seville, prior to delivery.

An Egyptian HA 200, "El Kahera.200," at the Silver Hill depot of the Smithsonian Institution's National Air and Space Museum in Washington, DC.

336

The HA 300V-1, as presented to Werner Blasel in Helwan prior to returning to Germany. Among other refinements, the aircraft has been newly repainted. The negatives have been retouched by the company's security officer.

The HA 300V-1 was returned to Germany lacking its main instrument panel. The right and left consoles, however, are generally in their original state.

Model of the Me P 2030 "Rotorjet" project in hover configuration. When cruising, the rotors would fold up and retract.

Willy Messerschmitt as painter—and how he was portrayed by other artists. Above left is a water color he painted in 1947 showing the Raulino House in Bamberg. On the right is Messerschmitt's portrait painted by Professor Paul Mathias Padua.

On the occasion of his 80th birthday, Willy Messerschmitt sat for the Munich portrait artist Günter Rittner.

The pinnacle of development at MBB's Manching plant, founded by Messerschmitt in 1961, was the Tornado European multi-role combat aircraft program. The plant has been equipped with the most modern equipment for final assembly, flight testing and care of high performance aircraft. This photo shows a Tornado belonging to the Marineflieger (Naval Aviation) overflying MBB's facilities.

Following the HA 100's maiden flight on 10 December 1953, pilot Raffael Lorenzo Vellido is clearly pleased with the way the airplane performed.

This amateur photo shows a rare "group picture," taken during the time Willy Messerschmitt was working in and for Spain. The HA 100 overflies Me 109s and a Junkers Ju 52 still in service with the Spanish Air Force.

HA 200: First Post-1945 Jet Development

The activities of the design department increasingly became focused on the HA 200. Messerschmitt had issued the challenge to make the most use of components from the HA 100. For saving time and money, he'd been keenly interested in the "modular concept" idea from the outset. He was subsequently able to convincingly argue its effectiveness in the coming months and years.

In actual fact, the HA 200 differed externally from the HA 100 basically through the replacement of the latter's radial engine and propeller with a smooth forward fuselage having quite pleasing lines. Two air intakes in the front belied the plane's new powerplants in the form of two small jet engines. Exhaust was carried along inside the wing roots—hardly noticeable from the outside—and exited out the wing trailing edge. The HA 200 was equipped with two Turboméca Marboré II turbojets, each with an output of 3.9 kN/400kp of static thrust.

Along with France's Fouga Magister, the type was quickly becoming a serious competitor in the international marketplace—particularly in the Federal Republic of Germany. Its maiden flight on 23 July 1952 gave it a considerable time advantage, as well. Messerschmitt hoped to shorten this even further and be able to produce his HA 200 "cheaper and better." For the first time performance comparison tables were drawn up and a design description prepared, so that in discussions with German officials "we'd have something to show them." At the year's end, it was decided to allow Dr. E. W. Pleines, a former DVL test pilot, to take the HA 200 up on its maiden flight. German mechanics under the direction of Hubert Bauer were to be brought to Seville for final assembly and testing.

Wind tunnel testing had not yet been completed, and this was carried out by INTA (Instituto Nacional de Tehnica Aeronautico) at Torrejon near Madrid, albeit after the first flight had already taken place. The results fortunately showed that only a minor "cleanup" of the design was needed.

Despite the first flight of the HA 200 still pending, by mid-1954 almost two-thirds of the German development

Assembly of the first HA 200 prototype at La Hispano Aviacion S.A. in San Jacinto (Triana) at Seville.

team was working on the HA 300: with project development, stress analysis, wind tunnel data, systems studies, and preparations for construction. HASA's contract awarded to the Eidgenössische Flugzeugwerke Emmen ensured that the extensive high-speed evaluation program was financially covered in the short term, whereas the Messerschmitt developmental team continued to work along without a contract.

Around this time the first fruits of his Spanish involvement began to ripen for Messerschmitt personally: Hispano Aviacion wanted to extend his contract through 1955 and offered him a 20 percent interest in the company as part of an increase in capital. This was to have resulted in the form of material deliveries from Germany. A letter from Mallet dated 6/14/1954 contains indications of a joint German-Spanish-American effort planned by the Ministry of Aviation, to be headed by HASA. Lockheed would be the American partner, while Willy Messerschmitt would head a combined HASA-CASA-Lockheed development department in Madrid. In this context, Lockheed would also have acquired holdings in HASA. In addition, Lockheed was showing a keen interest in the HA 300. However, the idea never materialized. Nathan "Nate" Price, a senior engineer from Lockheed, and his wife Genevieve ("Gen") spent three months in Seville, much of the time in the pleasant company of Willy Messerschmitt.

In July 1954 Messerschmitt was named as a member of the board of directors at Hispano Aviacion. Messerschmitt-Bölkow GmbH's holdings later grew out of this initial involvement with Spain's aviation industry, and even today Messerschmitt-Bölkow-Blohm GmbH retains shares in Spain. In honor of his services rendered to Spain's aviation program the Spanish government awarded Willy Messerschmitt the Gran Cruz de Merito Aeronautico (Grand Cross of Meritorious Aeronautical Service) on June 25, 1954. In a private discussion the Minister of Aviation asked Willy Messerschmitt to address the current matters facing aviation in Spain. This he did in a letter dated 5 July 1954.

Although matters did not exactly turn out as he had recommended at the time, this request for his opinion alone shows the trust and the high esteem in which the Spanish government held Willy Messerschmitt. Although the Messerschmitt-Lockheed-HASA-CASA never materialized, this probably had little to do with what was at the time quite disturbing attempts by his "old rival" Ernst Heinkel to do business with the Spanish through CASA (the holder of the license for the He 111). Rather, it was most likely due to pressure from the American government as well as Messerschmitt's repeatedly expressed desire to keep his hands free for the time when Germany would be permitted to build planes again.

In 1955 the Federal Republic achieved another step on its road to sovereignty when it reacquired limited airspace control and the option of establishing its own aviation and defense industry. Messerschmitt supervised the expansion of personnel in Spain in order to accelerate development of the HA 300 jet fighter. At the same time, he agressively pursued flight readiness for the HA 200 jet trainer and its subsequent flight certification in Germany so that he could then offer it to the Luftwaffe. Plans were drawn up and calculations made for prototype testing and production in Germany. Cost comparisons showed a 5 to 6 ratio in favor of the Me 200 (as it had been christened) compared to license building the Fouga Magister. Despite the fact that the decision makers in Bonn had been given comprehensive information from early on and senior officials from the Ministry of Defense had personally been briefed in Spain on Messerschmitt's developments at HASA, in the end the HA 200/Me 200 didn't stand a chance.

The decision had virtually been made by the time Juan Valiente had flown it on its maiden flight on 12 August 1955. In early September, Messerschmitt informed Spain that Turboméca and the French government would not sup-

Eleven production HA 200A planes lined up in front of the tower at Seville's San Pablo airport, awaiting delivery.

342

The first prototype of the HA 200 as flown during a goodwill tour through Europe in the spring of 1956. The nose carries the flags of those countries in which the plane was demonstrated: Luxembourg, Germany, Austria, and Switzerland. The pilot was Juan Valiente.

ply Germany with the Marboré II engines for the Me 200 (the Marboré was the engine of choice for both the Magister and the HA 200).

Along with other German companies he was forced to accept building the Magister under license in the Federal Republic. The agreement also bound him to drop the development of his own jet trainer for the Federal Republic and other NATO countries for a period of four years.

There were most likely several reasons behind the Magister's selection over the HA 200. For one thing, it was at a much more advanced stage in its development—the first pre-production aircraft had flown for the first time by 7 July 1954. In addition, there was definitely the political desire to cultivate the newly acquired, good neighborly relations with France. Another significant factor perhaps lies in the certain polemic on the part of the media against the exposed "airplane builder for the Third Reich," something which would not have made the Bonn government's decision in favor of the Me 200 an easy one.

Despite the disappointment in now having to shoulder the development costs for the HA 100 and HA 200 themselves, the Spanish were still willing to extend the contract with Messerschmitt through 1955. Willy Messerschmitt also remained "persona gratissima" (as Mallet aptly put it in a letter dated 11/18/1955) in the eyes of Spain's Ministry of Aviation—in spite of occasional "interference tactics" from his old rival Ernst Heinkel.

Willy Messerschmitt may have been able to take some comfort in the fact that his joint venture company Flugzeug-Union-Süd (FUS) GmbH, founded in 1956 with Heinkel Flugzeugbau GmbH for the purpose of license building the Fouga Magister, was given the lead management for the type, with a significant portion of the work going to the Messerschmitt AG's Augsburg assembly operations, and that it was his company in Munich-Riem which carried out final assembly and the systems checkout flights.

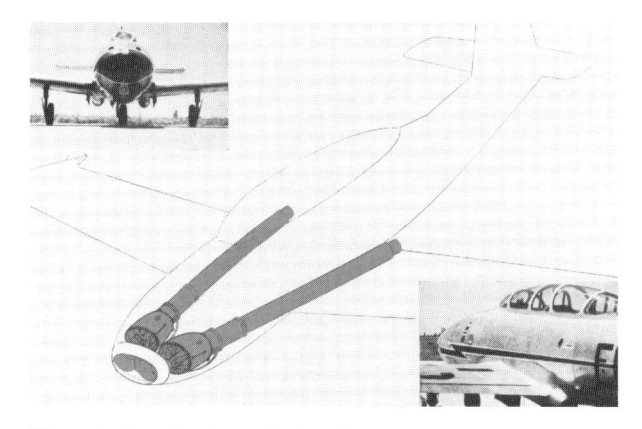

The typical engine layout in the plane's nose, showing the injector cooling for the jet pipe developed by Helmut Langfelder.

343

HA 200 Saeta - First Jet Aircraft after 1945

The HA 200 Saeta ("Arrow") was designed as a basic and advanced trainer, but was also capable of filling the weapons familiarization and reconnaissance roles with equal ease. Armament consisted of two 7.7 mm Breda machine guns in the nose. In addition, it could carry four Oerlikon 80 mm rockets and a gun camera, with the optional fitting of a photo camera, as well.

Along with the HA 300, the HA 200 was in all likelihood the last aircraft which fully embodied Willy Messerschmitt's typical thinking and work methodology. But not just in its conception. Here Messerschmitt was yet again afforded the opportunity to get involved at the designer level, applying his own unique style, as witnessed by the layout of the aircraft and many of its details. As always, he seemed unable to dispense with drawing up parts lists along with the blueprints. He felt that only in this manner would it be possible to come to grips with the real "meat" of a design task by minimizing the labor demands in the assembly stage, among other things.

His "modular concept," i.e. the extensive use of components from the HA 100, led to an unusual arrangement for the HA 200's engines—in the nose ahead of the pilot. Optimization calculations showed that the loss caused by the long exhaust pipes was offset by the minimal loss of intake flow with the forward-set engines. Of course, the jet pipes had to be cooled where they ran by the fuel tanks. Helmut Langfelder developed what was then a novel ejector solution, which drew the cool air to be circulated around the jet pipes through the bleed air doors in the nose and mixed it together with the exhaust gases aft. Lanfelder also supervised the testing of this system at INTA. The engine arrangement on the HA 200 had another, totally different, advantage: the HA 200 was virtually immune to foreign object damage during landings and takeoffs.

Another typical feature was the "bump" in the middle of the wing root, clearly visible on photos of the HA 200. It covered Messerschmitt's "bone" ("El Hueso"), which connected the wing's main spar with the fuselage former by going around the jet pipe. Messerschmitt consciously chose the "bump," a more optimal solution in terms of weight and drag than a potentially more elegant fairing along the entire wing root join which would have increased surface area. Undercarriage was initially offered from other countries, but in the end Wilhelm Binz and Ferdinand Miller developed the system using a Messerschmitt design.

Technical ground was also broken with the development of the pressurized cockpit fitted into the pre-production and production models. A "swimming pool" in the courtyard of the HASA factory served as the test center for pressure testing the cockpit, which was submerged and subjected to high-pressure blasts of water.

The first prototype of the HA 200 (designated in the Spanish Air Force as the XE.14-1) took to the air on its maiden flight on 12 August 1955, with the second prototype (XE.14-2) flying on 11 January 1957. It later flew as part of the Paris Air Show program at Le Bourget. Initially, the Air Ministry placed an order for a batch of ten pre-production models. In 1959 there followed a contract for thirty HA 200A production planes. Five of the pre-production aircraft were sold to the United Arab Republic in 1959, these being delivered in 1960 after being converted to HA 200B standards. Flight testing of the five Spanish airplanes got underway in 1961 and ended in 1965. In 1963 the Spanish Ministry of Aviation contracted for an additional 55 Saetas under the designation HA 200D, with HASA completing delivery of these in 1967.

X-15 Pilot Appraises the HA 200

Major Robert "Bob" White of the American Flight Test Center at Edwards Air Force Base in California had the opportunity to evaluate the HA 200 in the air in early 1963. In a detailed test report from April 1963 White, who among other planes had been the test pilot for the North American X-15 hypersonic rocketplane, assessed the HA 200 to be "a pleasant and easy-to-fly airplane which enjoys inherently good handling characteristics throughout the entire flight envelope." His only criticisms were leveled against some of the avionics and engine systems, and he recommended that these be corrected during subsequent flight testing. White observed that "the HA 200 has the potential to become an outstanding trainer for transitioning to fighter jets, as well as for navigation, instrument, and weapons training."

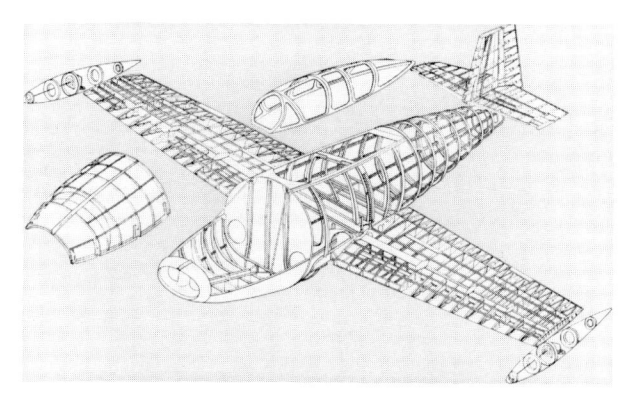

Type	HA 200	HA 200E	Type	HA 210	HA 57
Engine	Turboméca Marboré II 2x3.9 kN/400kp = 7.84kN static thrust	Turboméca Marboré VI 2x4.7 kN/480kp = 9.4kN static thrust	Engine	Turboméca Marboré VI 2x4.7 kN/480kp = 9.4kN static thrust	Turboméca Marboré VI 2x4.7 kN/480kp = 9.4kN static thrust
Crew	1+1	1+1	Crew	1+1	1
Length (m)	8.93	8.97	Length (m)	8.97	8.97
Height (m)	2.85	2.85	Height (m)	2.87	3.26
Wingspan (m)	11.03	10.93	Wingspan (m)	10.9	11.17
Wing area (m^2)	17.4	17.4	Wing area (m^2)	17.4	17.4
Empty equipped (kg)		2020	Empty equipped (kg)	2100	
Fuel (kg)		1090	Add'l load (kg)	1600	
Crew (kg)		180	Takeoff weight (kg)	3700	3400
Payload (kg)		310			
Add'l load (kg)		1580	Max. speed (km/h)		690[1]
Takeoff weight (kg)	2636[1]	3600	Climb rate (m/s)		16.0
Max. speed (km/h)	640[2]	690[1]	Range (km)		1600[2]
Cruise speed (km/h)		579[2]			
Climb rate (m/s)	13.6		Takeoff to 15 m (m)		962
			Landing fm 15 m (m)		950
Service ceiling (m)	12000	13000	Takeoff run (m)		675
Range (km)	620[3]	15000	Landing rollout (m)		528
Takeoff to 15 m (m)	715	1080	Notes		[1] at 9000 m [2] without reserves
Landing fm 15 m (m)		700			
Takeoff run (m)	490	800			
Landing rollout (m)		400			
Notes		[1] 3350kg max [2] at 8000 m [3] 1480km at a takeoff wt of 3000kg			[1] at 7000 m [2] at 6000 m

HA 200B with an externally mounted HS 804 cannon under the forward fuselage. The Spanish later based their single-seat ground-attack HA 57 on this variant.

This see-thru drawing shows a typical Messerschmitt design approach. Willy Messerschmitt designed the so-called "bone" himself; its purpose was to join the main wing spar to the fuselage former by circumventing the jet pipe.

HA 200 in Comparison with the Fouga Magister

Type	HA (Me) 200[1]	CM-170[1]
Engine	Turboméca Marboré II	Turboméca Marboré II
	2x3.9 kN/400kp	2x3.9 kN/400kp
	= 7.84kN static thrust	= 7.84kN static thrust
Crew	1+1	1+1
Length (m)	8.88	10.05
Height (m)	3.00	2.74
Wingspan (m)	10.40	11.27
Wing area (m^2)	17.40	17.20
Empty equipped (kg)	1694	2155
Fuel (kg)	993	800
Crew (kg)	168	168
Payload (kg)	311	43
Add'l load (kg)	1472	1011
Takeoff weight (kg)	3166	3166
Max. speed (km/h)	721[2,4]	715[3,4]
Climb rate (m/s)	20.9	17.0
Service ceiling (m)	13000	12000
Range (km)	1075[5]	925[5]
Takeoff to 15 m (m)	540	800
Landing fm 15 m	778	967
Takeoff run (m)	373	555
Landing rollout (m)	430	527

Notes:
[1] performance based on equal (overloaded) takeoff weights
[2] performance at average takeoff weight of 2350kg (same mission)
[3] performance at average takeoff weight of 2850kg (same mission)
[4] at 9000 m
[5] at 9000 m

HA 300 Supersonic Fighter Developed

The development program for the Spanish Air Force took top priority at first. Messerschmitt was able to beef up his team; by the end of 1955 there were 18 employees in Spain. In addition to the previously mentioned personnel there were Josef Kempter, Helmut Langfelder, Max Schäffer, Karl Wassermann, Rudolf Leistner, Wilhelm Geduldig, Fritz Klotz, Joachim Puffert, and Paul Ruden. In order to speed up the acquisition of equipment, accessories, materials, and standard parts he was given virtually a blanket import license and, upon providing cost justification, permission to develop specialized sub-programs in Germany provided that Spain did not have the capability. Finally, the Spanish Ministry of Aviation quickly settled on the type of engine to be acquired for the HA 300.

In the interim, the HA 300's project layout had come under heavy discussion. Ernst Heinkel had offered to develop a "near-supersonic" figher with 19.6 kN/2,000kp of thrust and a takeoff weight of 2,000kg for Spain. Development would take place in Germany, but CASA would build the first prototype; spin-up to full-scale production was expected to take no more than three years. Messerschmitt pointed out that such a plane was no longer on par with the latest state of technology. Indeed, he had been on the verge of carrying out the first flight of a similarly designed plane (P 1101) back in 1945. Based on the advances in engine development in Britain and the U.S., Messerschmitt nevertheless felt that it would be worthwhile to rethink the HA 300—laid out at the time with a takeoff weight of 5,500kg, 34kN/3,500kp thrust and a speed of 1,400km/h—and possibly come up with a much lighter design. An offer was accordingly made to Spain's Ministry of Aviation towards the end of the year.

From mid-1956 onward the Hispano Aviacion worked on establishing contact with Egypt, something which was supported and encouraged by the Spanish government. Initially these led to contracts for acquiring the Spanish Me 109, followed later by the HA 100 and HA 200. With the Egyptian government's interest in the HA 300's development, it was entirely possibile that Egypt would be willing to share the developmental costs.

Despite the German government's decision not to purchase the HA 200 (Me 200), Messerschmitt continued to push for the type's certification so that it could be demonstrated in Germany and Austria and so that plans could go forward with a commuter plane variant. Following what was a rather involved ferry flight at the time, the HA 200 was demonstrated for the first time at Düsseldorf's 1956 National Open House, an event running from 1 to 4 June and sponsored by the North Rhine-Westphalia Flying Club. The aircraft had been test flown beforehand in Seville by Heinrich Beauvais, the former top pilot at the Rechlin Test Center during the war. As a result, the aircraft's flight envelope was expanded to include speeds of up to Mach 0.8 (in a shallow glide). Afterwards the aircraft was demonstrated in Vienna, then at the end of June in Munich-Riem before being flown back to Spain. Great interest in a commuter version of the HA 200 type was not only expressed by those in the flying circle, but also from the business world both at

The HA 300 configuration from 13 April 1956 showing the forward swept wing-root intakes, designed for speeds up to Mach 2.0. In 1958 "normal" cheek intakes were eventually chosen following extensive discussion between Gero Madelung, Helmut Langfelder and Professor Klaus Oswatitsch of the Rheinisch-Westfälisch Technische Hochschule (RWTH) in Aachen. Messerschmitt explored the possibility of having the Lear company develop the intakes, as they had been responsible for the forward swept intakes on Republic's F-105.

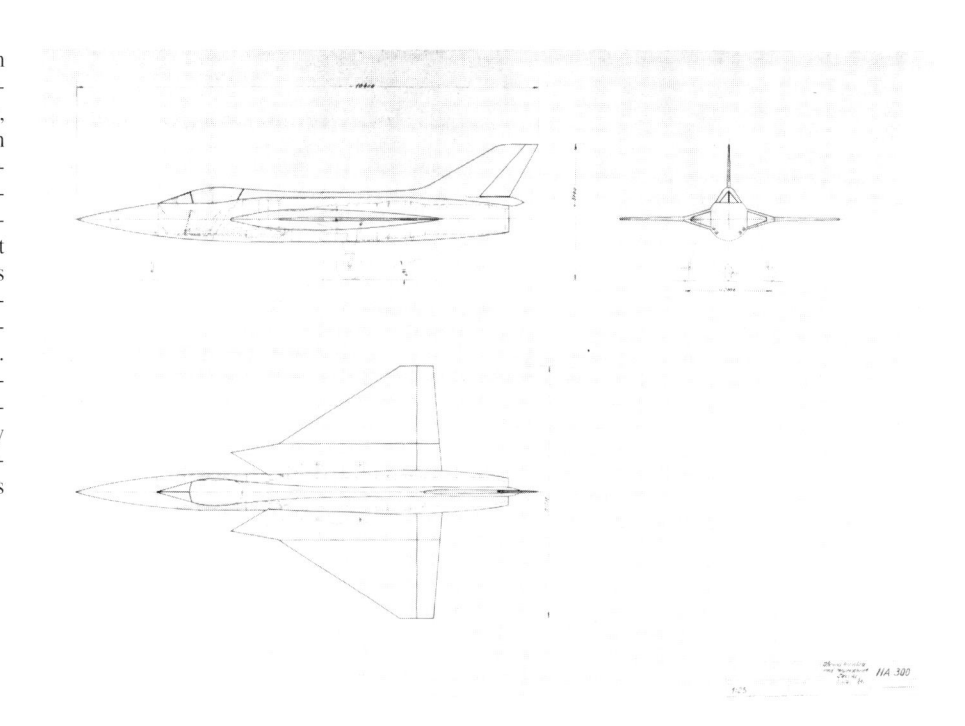

home and abroad. None of this interest, however, generated enough impetus for any agreement which would serve as the basis for a production order or even a follow-on development as a business commuter plane. Work on the INTA certification testing for the HA 200 was completed in the latter half of 1956, as was flight testing for the V-2 (with wingtip tanks).

By the end of 1956 the "Oficina Tehnica" team had been joined by Wilhelm Binz, Gottfried Haase, Fritz Liebmann, Paul Klages as design supervisor, Herbert Schnabel, Reinhold Wiesner, and Martha Blocher. By this point the department's primary focus was on the HA 300.

The Spanish government's shortage of funds and foreign currency slowly began to have their effect on the paychecks of those members of the Messerschmitt team employed in Spain. More seriously affected, however, were the development and assembly plans for the two prototypes of the HA 300, which had now been firmly decided upon. An example of this was the decision to use the test framework built to evaluate the HA 200's static vibration (still ongoing) for the HA 300, as well. The first noticeable delays in the partially assembled airframe components for the first prototype of the HA 300 were also being felt. Wind tunnel testing in Grainau and Emmen had to be broken off for a time or stopped altogether. Acquisition of materials, finishing machinery, etc. for assembling the prototype were either deferred or dropped completely.

Inspection reports, letters, and minutes from meetings, however, indicate that Willy Messerschmitt ignored these

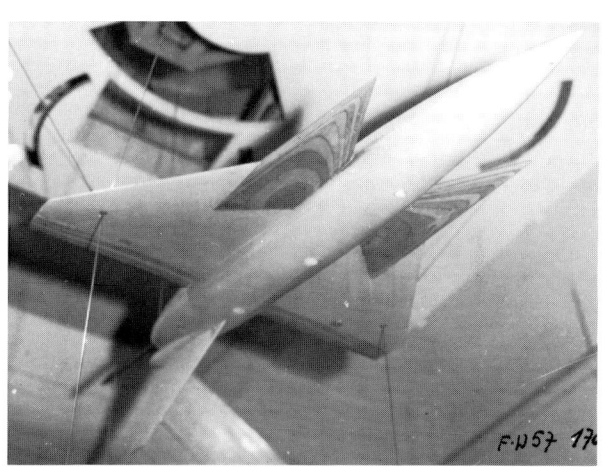

A 1/6th scale model of the HA 300S glider provided aerodynamic measurements in the Eidgenössiche Flugzeugwerke Emmen's small subsonic wind tunnel in 1957. The photo shows the model being tested in the following configuration: fuselage + wings + large rudder + forward swept wing-root intakes. This research was conducted primarily to determine the effectiveness of the elevators, both with and minus ground effect, as well as their impact on the aircraft's low-speed stability.

problems himself, and that Gero Madelung and all other managers in Seville and Munich diligently pressed forward with the development of the HA 300. Equally resolute were the efforts to procure usable wind tunnel data as the basis

347

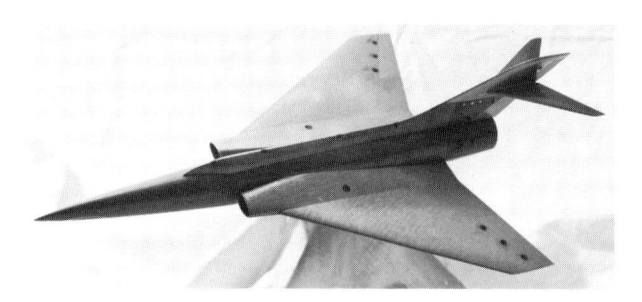

HA 300 variant with reduced span wings and the elevators centered on the vertical stabilizer. The wind tunnel model was evaluated in August 1955 at the Eidgenössische Flugzeugwerk Emmen's facilities.

for calculating the dimensions of the wings, empennage, ailerons, and flaps.

With the appointment of Dr. Leonidas Rothe as chairman of the board in 1957, the Munich developmental team, i.e. Prof. Messerschmitt's technical department, broke away from the Messerschmitt AG. As an independent company, its first job was a study contract for Germany's Ministry of Defense. Some of the "Oficina Tehnica" team, which as part of the agreement spent two months of each year in Germany (taking their vacation during this period), now began working on a limited basis for this study contract, and as a result were suspended from the contract with the Spanish developmental department. Thus began the dismantling of Messerschmitt's developmental team in Spain. By the year's end only ten employees were left.

Correspondence between Munich and Seville/Madrid (Mallet) grew less and less frequent, another sign of the diminishing priority the Spanish were placing on a cooperative venture with Willy Messerschmitt. In addition to Spain's critical financial situation, contracts from America began playing an increasing role in the aircraft works' operations, as evidenced by Hispano Aviacion accepting a contract for repair and assembly of the Lockheed T-33 for the Spanish Air Force.

It was during this period that development of the HA 300 suffered another major blow. It turned out that Bristol did not want to be solely responsible for putting the 12 SR afterburning version of its Orpheus engine into production. For the Spanish, partial funding of development was out of the question.

Prototype construction was canceled against this backdrop in 1959—at a time when design work on the airframe and systems had been completed and the major portion of systems had already been contracted for.

As early as 1956, there were plans in the works for a 1:1 scale wooden flying model glider of the HA 300 to evaluate its unique delta configuration. HASA passed the contract for building the aircraft, designated HA 300P, on to the AISA company. Willy Messerschmitt and his team made improvements to the assembly blueprints and thereby assisted in accelerating the work. With Pedro Santa Cruz Barcelo at the controls, the HA 300P took off on the first of its two flights on 25 June 1959 from Seville's San Pablo airfield.

Messerschmitt's urgency had not been unfounded. The Egyptian government had made the successful first flight of the HA 300P as a major prerequisite to signing a license contract for the HA 300, as well as for the HA 200. Even the direct acquisition of ten Saetas was depending on this flight. Negotiations begun in 1959 concluded in 1960. By now virtually complete, the entire HA 300 project and all associated documentation were sold to the United Arab Republic. In doing so, Spain recouped all the developmental costs for the HA 300 up to that point and earned a not insubstantial profit through the sale of the HA 200. By this action Hispano Aviacion was able to hold onto its position as an aircraft development and manufacturing company, even if the Spanish Ministry of Aviation was unable to compensate it for its losses suffered because of the HA 300 contract—something which led to serious financial burdens.

Collapse of Spain's aircraft production capabilities was only a matter of time at this point. In Janaury 1966 La

During its two flights on 25 June 1959, the glider version of the HA 300 (with Pedro Santa Cruz at the controls) was towed by a CASA 2111, the Spanish license-built version of the He 111. This doubtless made for another unique "group photo"—an airplane by Willy Messerschmitt in conjunction with one from Ernst Heinkel!

Hispano-Suiza became insolvent. The Spanish government felt the solution to its economic problems would be for the state holding company of INI (Instituto Nacional de Industria) to immediately take Hispano-Suiza's shares in HASA and later pass them on to another aviation company. This occurred in February 1967. German stockholders held 27.5 percent of HASA's stocks with the expectation that the Spanish government would guarantee the company contracts totaling 12.5 billion pesetas. Despite initial promises by aviation minister General Julio Salvador, these views were obviously quite unrealistic. It was therefore hoped that a French company or even CASA would get involved. In October 1967 INI acquired half of Hispano-Suiza's shares in HASA, thereby increasing its interests in the company to 53 percent. The other half went to CASA. The original 27.5 percent interest held by German shareholders in HASA was dropped to 1.5 percent each for Willy Messerschmitt and Messerschmitt-Bölkow-Blohm GmbH in CASA. For Messerschmitt-Bölkow-Blohm's part, Willy Messerschmitt was elected as a honorary member of the board of directors along with assessor Josef Fuchshuber and Herbert Hellmann. With the death of Mallet, Hellmann had assumed the responsibility of representing Messerschmitt's interest in Spain. Fuchshuber remained at his post, following subsequent developments in the Spanish aviation industry until retiring from CASA's board of directors in February 1989. Messerschmitt-Bölkow-Blohm GmbH had bought up Messerschmitt's shares even before his death. MBB's member of the board from 1976 to the writing of this book has been Luis M. Ecenarro Balparda, with Peter Maximovic also serving from 1988 to 1989.

Shortly after this first step in redirecting Spain's aeronautical industry HASA was awarded a contract for delivering 25 HA 200E Super Saeta airplanes which had been under negotiation since May of 1967. Although this meant that Hispano Aviacion's developmental capabilities would be employed to the fullest, its production facilities were still not operating. The contract did, however, provide an effective means of bridging the gap to solvency. HASA was also given the contract for refitting 55 HA 200D aircraft with weapons. In doing so, the E.14E (Air Force designation for the HA 200) trainer became the C.10B combat aircraft. 21 were converted in 1968, with the remaining 34 following in 1969.

In 1968 Fernando Orduna Gomez, formerly INI's representative on HASA's board of directors, became president of Hispano Aviacion. He was followed a short time later by Emilio Gonzales, who soon was seeking discussions with Willy Messerschmitt and Ludwig Bölkow. As a result, through the influence of Messerschmitt-Bölkow GmbH Gonzales was able to improve his company's precarious financial standing by acquiring the rights to produce the wing center sections for CASA's C.212 transport plane. These center sections had been developed by the Hamburger Flugzeugbau GmbH (which, incidentally, had joined with Messerschmitt and Bölkow to form the Messerschmitt-Bölkow-Blohm GmbH in 1969) as part of a contractual arrangement with CASA. HASA later licensebuilt MBB's piston-engined Flamingo trainer, developed as the SIAT 223 in Donauwörth, of which the Flugzeug-Union Süd (FUS) was able to successfully market a total of 46.

Gifted Designer

Gero Madelung recalls that "in Seville, Willy Messerschmitt lived in the same room at the elegant Hotel Alphonso Trecce, with a view overlooking the famous tobacco factory from the opera Carmen (today the university library). It was during my four years of involvement there that I had the most contact with him, as I did with other members of the development group who provided me with many of the following anecdotes.

I wasn't able to visit with him as often when he later moved to his abode of El Velerin in Estepona. From 1958 onward he spent up to six months out of the year there. In addition to his brilliant creativity he also had a saturnine air about him—this lesser known side of Messerschmitt is hinted at in Claudio Bravo's portrait of him. Generally speaking, he did not naturally approach people, but was all the more affable and congenial when approached himself. An evening at the opera—especially when Richard Wagner was on the program—was a particular delight for him. Dora Stoop, his socialite cousin, would organize these outings. Given a bit of wine, Messerschmitt could also become quite merry in a game of skat, and despite the most intense game was usually the first to pay up at the end. His sense of humor was apparent even in these cases, laughing heartily even as he was losing at such get-togethers.

From Seville he treasured the weekend excursions into the mountains, to the beaches and to visit architectural attractions. He was a fast driver—more than once I found myself in fear for my life. Later on he would let me do most of the driving, but except for one occasion along the 'Messerschmitt Curve,' as Fritz H. Hentzen christened the Augsburg-Munich stretch, he never had an accident. During those days Théodore von Kármán would regularly visit Seville in the autumn and sometimes took the opportunity to call on Messerschmitt. Once, when I was chauffeuring the two of them, von Kármán said to me: 'Young man, if you want to become famous very quickly you should take out a lamppost about now.'

At home, his wife Lilly exercised a strict regimen. Smoking and alcohol were kept to a minimum, and everywhere he turned it seemed there were rules. But he resigned himself—often with a smile—and even seemed to enjoy this facet of her affection. Lilly's death in 1973 affected him deeply. At El Velerin Willy Messerschmitt was well looked after by the

This portrait of Willy Messerschmitt, painted in Spain in 1964 by Claudio Bravo, shows a sensitive, somewhat reserved "artist"—characteristics not altogether inapplicable to the brilliant designer.

tempermental Traudel Berger, who for many years had been a manageress. While living in the Mauerkircher Straße in Munich his stepdaughter-in-law from Bamberg, Karin Stromeyer (Hilmar's wife), took care of him. Despite leading a not altogether healthy lifestyle, his health was generally good. At El Velerin he enjoyed swimming on a regular basis.

Even as a trainee, I was able to experience Messerschmitt's creative engineering activities in an intense way. It was obvious that he had mastered the art of 'product design' at all levels, from configuration blueprints through sub-assemblies down to the individual components. Often drawn to scale, his design sketches were punctuated with explanations, crossed out calculations and parts lists and bear witness to this. Calculations were usually made with a small pocket slide rule, although sometimes these were done off the top of his head. If he wasn't working at his drawing board, he would reach for any scrap of paper: letter envelopes, the back of an old pack of cigarettes, the edge of a newspaper—or once while at the opera even a shirt cuff. Messerschmitt expected his engineers to apply their scientific knowledge to his sketches, yet he was willing to discuss well-founded alternative approaches with the patience of a saint. I was often amazed by this aspect of his. He would question the engineers in detail when it came to loads and dimensions, and would force into submission any hold-outs not based upon well-founded ideas. The brilliant and dedicated stress analyst Schorsch Ebner was frequently brought in to settle matters.

At the time, there was no such expression as 'value engineering.' Calling upon his vast wealth of experience, Willy Messerschmitt would always intuitively focus on the 'buildable' design. In light of this, a poor design or a cavalier approach to weight and space (such as a flanged edge outside of specifications) would invariably incur his wrath—which without exception had a telling influence. I personally experienced this with my first design of the HA 100F's mounts for the Wright engine. My design called for six attachment points on the firewall. In Messerschmitt's opinion this was much too laborious. His counterdesign sufficed with just four points ("You won't get by with more than four points.") and quite elegantly integrated the nose gear housing, as well. He was so dissatisfied with the wing/fuselage fairing a HASA engineer had designed for the HA 200 that in just one afternoon he single-handedly drew up the two-meter long section to full scale on his large drafting table. In spite of Messerschmitt's design focus on weight, he was ever mindful of details which might potentially lead to accidents. Such as improper assembly going unnoticed.

One of his main concerns was economy of operations; he was constantly striving to recycle designs and components. Thus his strict instructions for using as many components as possible from the HA 100 in the design of the HA 200. This was one of the reasons—if not the only one—for the HA 200's unusual engine layout. He would accept second-hand materials, but only if they had first been thoroughly evaluated as to their repair potential. As a young worker, fellow employee Hilmar Stumm learned this first-hand when he cheekily suggested that an ill-fitting engine mount be "thrown away." Messerschmitt was so enraged by this that he fired the young man on the spot, but once his anger had blown over, the man was rehired and became one of his most valued employees.

Work had to move rapidly along, as well. The example of the Me 321 transport glider (from drawing board to first flight in just three and a half months for a behemoth with a wingspan of 55 m!) amazes the experts even today. Meinhard, a senior foreman, told me that during his time at Bamberg—when Messerschmitt would sometimes have his sheet metal foreman make airplane fittings, the designer had once gotten him out of bed at 10 o'clock at night asking him to make a quick change to a part....

No matter the pace, Willy Messerschmitt seldom felt compelled to give his nod to a design he wasn't convinced of. I first witnessed this on the main spar/wing/fuselage for the HA 200, whose shape was in actuality quite a complicated affair due to the fact that the engine exhaust tube had to be circumvented while at the same time avoiding as much as possible a thickening of the wing root. Blueprints for the wing and center fuselage section were virtually complete and the design was ready to be approved for assembly, yet Messerschmitt was still troubled by this 'keystone' of the entire airframe. In the end, he drew up an unusually refined welded steel section ("El Hueso" = "The Bone")—and without the need for later modification.

Nor was he reluctant to develop aircraft systems himself if it was to his advantage. This especially applied to landing gear and hydraulics, and for the HA 200 (due to a shortage of funds) wheels and brakes plus the actuating arms for the undercarriage and flaps. His wealth of ideas are marked by the fact that all these components were able to be produced with the minimum expenditure conceivable."

Messerschmitt Reaches Pinnacle of Fighter Development in Egypt

Guillermo F. Mallet, Willy Messerschmitt's former Spanish representative, recognized the export potential of both the HA 200 and the planned HA 300. After the war, Mallet had become the representative with the Swiss company of Oerlikon-Bührle. It turned out that he was successful in broaching the matter with his colleague Hssan Sayed Kamil, the general representative for Oerlikon-Bührle in the Near East. Kamil, who had earned his degree at the Eidgenössische Technische Hochschule in Zurich and was considered to be something of a financial wizard and industrialist, immediately recognized the role he could play as a middleman. He soon was involved in discussions with the Egyptian government and, when these took a more concrete shape, Kamil established the MECO company (MEchanical COrporation, headquartered in Zurich) for the purpose of establishing formal ties between Messerschmitt and the Egyptian government. MECO then became the contractual partner for the Egyptians in subsequent negotiations.

The agreements obtained by MECO called for assistance in expanding Egypt's own aviation industry to include obtaining all the necessary machinery and facilities and acquiring the needed workforce. In wrapping up the license manufacturing agreement for the HA 200 and HA 300, the Egyptian government also contracted with Messerschmitt personally. This contract obligated him to maintain regular supervision of the work at Helwan.

Willy Messerschmitt soon realized that he would have to appoint a proper representative who would be able to watch over the team. Initially, he found his "chief of staff" in the person of Fritz Hentzen and later in a technical engineer by the name of Schönbaumfeld, a former Junkers director and subsequent general manager of the Andritz machinery company in Austria. From November 1967 on, he was able to get Kurt Tank, who had just finished his developmental work in India and brought with him "a good air of mutual respect," as von Srbik put it. Given the time frame of his arrival, Tank managed what proved to be the program's "final push."

The background behind Egypt's interest in gaining German cooperation was the country's desire for independence in equipping its air forces. In delivering spare parts for the Egyptian Air Force's MiGs, the Soviets had attached certain political conditions which President Gamal Abdel Nasser was unwilling to meet. One of the means by which he made the Soviets aware of this was by calling home his brother, who was studying at the aeronautical academy in Moscow.

Dr. Hans-Heinrich Ritter von Srbik noted the numerous, involved discussions which he and Messerschmitt took part in at the Egyptian president's villa. These meetings not only concerned themselves with the aviation industry, but also focused on fundamental structural matters and the shakeup then going on in Egypt. Nasser saw an independent aviation industry not just as an opportunity for greater independence in equipping his armed forces, but as a significant industrial-political tool in his country's growth. Nasser was concerned that, given the continuing political unrest, he would not have time to follow through on long term industrial programs of this kind.

These concerns later proved to be well-founded. It should not be overlooked that the Egyptians had asked Ernst Heinkel back in 1952 (when General Nagwib held the political reins) to construct a single-engined fighter-bomber and its engine in the country. As Ferdinand Brandner wrote, for whatever reason "these ties were severed two years later, however." In spite of this setback the Egyptians still seemed willing to work with the Germans.

It was against this background that the Spanish were able to sell the manufacturing license for the HA 200 and the developmental blueprints for the HA 300 to the United Arab Republic in 1959/1960. The deal also included delivery of all subassemblies and components prepared for the HA 300 thus far, enabling the assembly of three prototypes, plus the engine-less HA 300P glider, ten HA 200 aircraft, and a complete set of jigs and tooling for their construction.

Saeta Becomes Al Kahira

The Spanish Air Force permitted Hispano Aviacion to pull five Saetas from the pre-series production line for the purpose of converting them to HA 200B planes for Egypt. The HA 200B was fitted with a Hispano-Suiza HS-804 20 mm cannon located on the underfuselage to the right of the nose gear door and a Ferranti gyroscopic gunsight. The forward fuel tank's capacity was reduced in order to locate the cannon's ammunition as close to the center of gravity as possible.

The HA 200B pre-production aircraft took to the air for the first time in early 1960. Transports ferried them to Cairo's Helwan Factory #36 in April of that year. There, HASA test pilot and technical engineer Francisco Esteva Salom was tasked with flying the airplanes and familiarizing the pilots of the UAR with their handling characteristics. This was done under extreme pressure, for President Nasser had demanded that the planes were to fly in the 7 July 1960 military parade commemorating the eighth anniversary of the revolution which had brought him to power and subsequently resulted in him becoming president of Egypt in 1954. Esteva was able to complete his difficult task in time and led the Saeta's formation fly-by himself. The Egyptians revealed the aircraft to the public under its new designation of "Al Kahira" (the name for Cairo, meaning "The Conqueror").

The five remaining aircraft were delivered in 1961, with production tooling following in late 1961 and 1962. To assist in spinning up production, HASA alone sent over a hundred employees to Helwan, where they worked under the direction of Francisco Esteva Salom. An additional 30 complete undercarriage sets, plus 60 hydraulic systems and other subassemblies and components were also sent.

Engine Troubles Delay the HA 300

Deliveries for the HA 300 encompassed overall design documentation for the aircraft, parts breakdowns and blueprints for fuselage, wings, empennage, cockpit, intakes, fuel tank size and layout, engines, climate control, electro-hydraulic systems and armament, plus all subassemblies and equipment parts hitherto produced. Even the HA 300P "glider version" was part of the package, as well as a mockup of the forward fuselage with cockpit and engine intakes.

The first prototype of the HA 300 took off from Cairo on its maiden flight on 7 March 1964, with the second one taking to the air in 1965. Both prototypes were equipped with the British Bristol Orpheus 703 engine, whose installation arrangement had been worked out together by Wilhelm Geduldig and L. Salas. The Bristol Orpheus 12 with afterburner had been planned for the third prototype, a boost which would have given the HA 300 a capability of Mach 2. Bristol had, however, subsequently decided against producing this version. But an offer was made to Egyptian officials to continue working on the engine if the Egyptians would shoulder all costs. Price and availability were unacceptable for the Egyptians, who then turned to a Soviet-designed engine. These plans were later shelved, however.

HA 300 in Helwan, prior to its first flight on 7 March 1964.

HA 300 being prepared for its flight, during its maiden flight (with landing gear extended), and landing for the first time on 7 March 1964 at Helwan. Verdict of the Indian test pilot: "The plane is quite easy to take off, fly and land in, better than the (Folland) Gnat and the (Hawker) Hunter."

Following these two false starts in an attempt to obtain a powerful engine for the HA 300, Willy Messerschmitt found a solution by offering to develop an indigenous product in this area as well, something which the Egyptians had originally hoped to avoid. At first, Messerschmitt was asked by the Egyptians to develop the HA 300's powerplant himself, but refused the offer by pointing out that he was not a builder of powerplants. However, he and Dr. von Srbik soon came up with the idea of soliciting help from Ferdinand Brandner.

During the war, Brandner had worked at Junkers on several engine designs, including the most powerful German engine of the day, the Jumo 222. Taken to the Soviet Union with the rest of the Junkers team after the war, there he developed the most powerful turboprop in the world, the Kuznetsov NK-12 m rated at 8,800kW/12,000 hp. In April of 1960 Messerschmitt invited Brandner to a meeting in Munich and there introduced him to Hasan Sayed Kamil. At the time, Brandner was chairman of the board of directors at Austrian Airlines (AUA), but problems in the company and the attractive offer which the Egyptians made

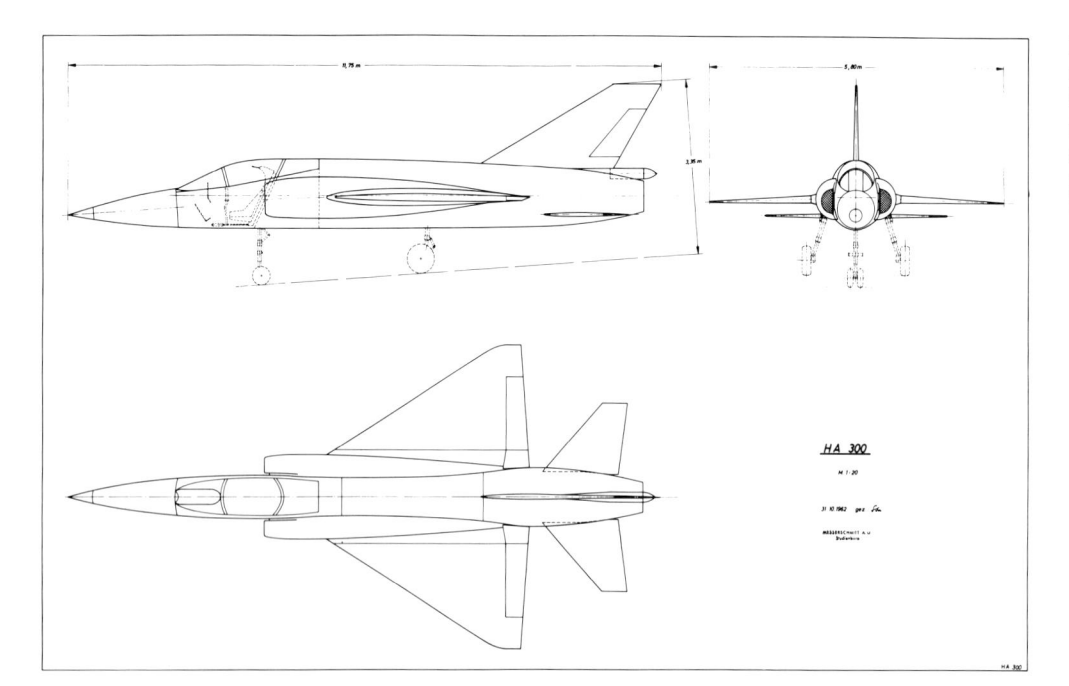

HA 300 in its configuration from 31 October 1962; this remained basically the same as for the first prototype (maiden flight in 1964).

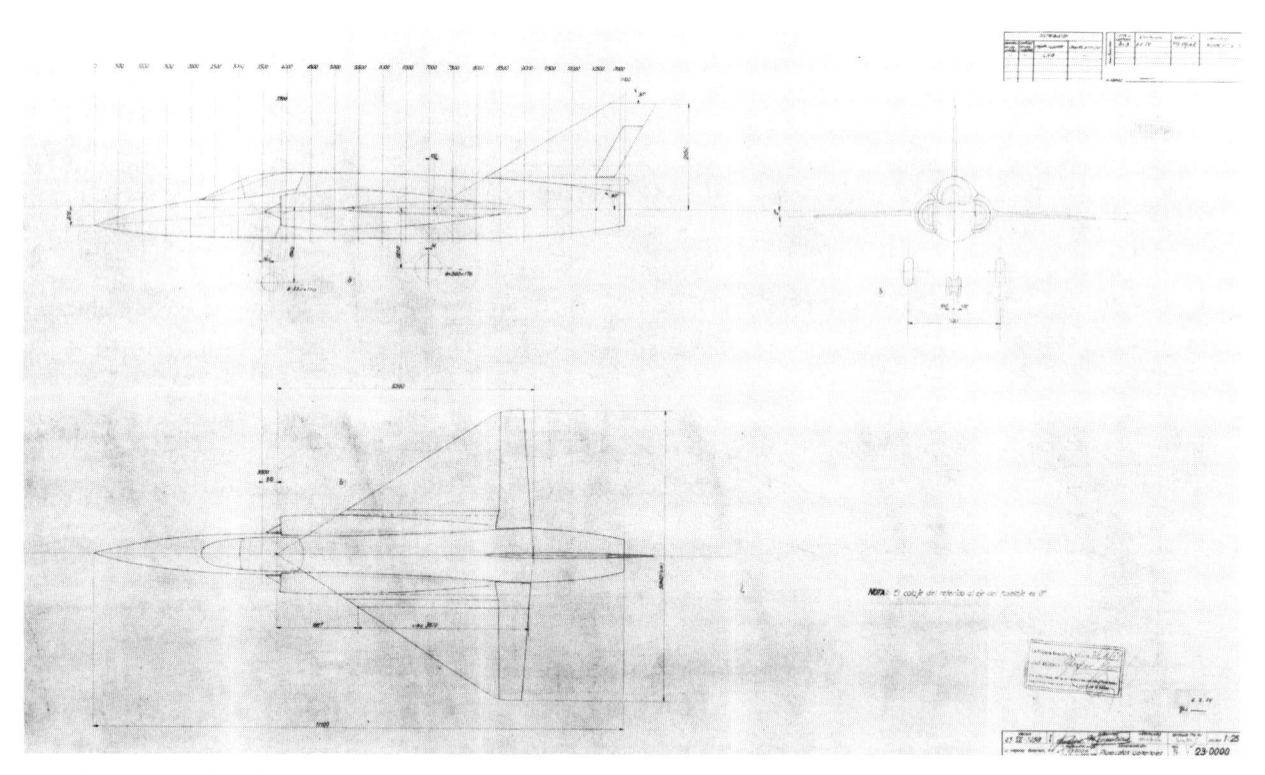

A HASA blueprint from 29 December 1958 shows the layout of the HA 300 at the time—now with cheek intakes and still lacking horizontal stabilizers. The aircraft's wingspan had increased to 5.84 meters and the length to 11.1 meters.

A front view of the HA 300 shows the aircraft's delta wing design, cheek intakes, and the low position of the elevators to good effect.

meant there was little hesitation in accepting. For him, too, this was the first opportunity after 1945 to begin work on an indigenous engine development program. Brandner was excited by the opportunity to develop an ultra-modern engine for (as he put it at the time) "the smallest, most modern jet fighter in the world." In addition, like Messerschmitt he felt himself to be "first and foremost a designer, and not a manager." And like Messerschmitt, he set himself to the task of "stamping out an aviation industry on a sand dune from the ground up." Kindling for the project was an aircraft factory built by the British at Helwan near Cairo during the war, which would need to be expanded and fitted with the proper machinery.

The HA 300 wasn't expected to be the only recipient of Brandner's proposed E 300 (sometimes designated HE 300) turbojet, rated at 33.3kN/3,400kp static thrust without afterburner and 49 kN/5,000kp with afterburner. Another candidate was the HF 24 Marut fighter, which Kurt Tank had developed for the Indian government. Indeed, an agreement to this effect between the governments of Egypt and India had already been signed on 29 September 1964.

In the meantime, the Messerschmitt team was hard at work building the third prototype of the HA 300. In order to accommodate the E 300, the airframe had to be modified considerably. Brander later felt that Messerschmitt may have underestimated the time involved in getting the first engine ready for flight testing. This may indeed have been the case. What is known is that he was unable to come to grips with the serious difficulties in getting parts (such as equipment fittings) and the political intrigues within Egypt's organizations. These "hiccups" by default led to delays in meeting deadlines.

As was standard practice with flight testing a new engine, during the winter of 1966/67 the E 300 was first tested in another airframe, a HF 24 which had been disassembled in Bangalore and sent to Helwan for this purpose. After reassembly, it flew in Egyptian skies for the first time powered by two Bristol Orpheus 703 engines. Had the E 300 progressed in its development to operational maturity, it would have undoubtedly made a capable fighter of the HF 24 as well. The Egyptian officials later informed the Indians that the engine would not be made available to them. In the interim, India had begun license production of the MiG-21 independent of this program, meaning that the marriage of the HF 24 with the E 300 would most likely have met with marketing failure in any case. All the more reason for the teams of Messerschmitt and Brandner to look forward to the first flight of the HA 300 with the E 300 powerplant.

In June of 1967 the third war in the Middle East between Egypt and Israel broke out. The "Six-Day War" left the Egyptian Army crushed and its air force practically destroyed on the ground. However, this event only interrupted work for a short time. Both teams, with Kurt Tank now managing the program for Messerschmitt, soon were awarded new contracts and continued to work at a feverish pace. Brandner recalls: "Together with Professor Tank, we concentrated our efforts on demonstrating the new HA 300's supersonic capability to the public by the summer of 1969. In doing so, we still harbored the hope that the Egyptians would be able to sell the type to an interested country and thereby at least recoup their developmental costs." However, this was not meant to be. Both teams were sent back home, the reason most likely being Egypt's new political and economic ties with the Soviet Union.

In the end, two HA 300 airframes with Orpheus engines were left in Egypt along with another two aircraft lacking powerplants. At least one of these aircraft was put into storage at Helwan. It was shown to Brandner during a visit in 1975, and editors from the American journal "Aviation Week and Space Technology" also saw it in 1981. The first prototype was offered for sale to Messerschmitt-Bölkow-Blohm GmbH in 1990, and a MBB delegation under Werner Blasel had the opportunity to view and photograph the well preserved HA 300V-1 in October of that year. At the time of this writing, the aircraft is at MBB's Manching works, where it is undergoing restoration to eventually become part of an exhibit at the Deutsches Museum.

The HA 300 is a testament to the capabilities of a small, enthusiastic nucleus of workers under superior management, operating under unfavorable external conditions. In a relatively short period of time, Messerschmitt and Brandner were able to develop an ultramodern supersonic aircraft and

Despite a lack of drawings for the wing, at Helwan Ferdinand Brandner's team modified an Antonov An-10 to fly as a testbed for the E 300 jet engine. The port inboard Ivchenkov AI-20K turboprop (2,942 kW/4,000 hp) was swapped out for the E 300 with an appropriately modified engine mount. Here Ferdinand Brander protects his ears against the engine's noise prior to the initial flight, on which the Egyptian pilot insisted Brandner accompany him.

a suitable powerplant. Without a doubt, the HA 300 can be considered the penultimate product of Willy Messerschmitt's fighter development career.

Nor was the political playing field an easy one to negotiate. Messerschmitt's involvement in Egypt came under heavy criticism from both the German and international press in the early '60s, criticism which was most likely unwarranted in light of today's knowledge of the military and political balance in the Middle East at the time.

HA 300 - Lightest Supersonic Fighter in the World

The HA 300 was laid out as a light, single-seat, single-jet supersonic fighter with a delta wing design. Its role was primarily to have been high altitude interception with secondary roles of a fighter-bomber, ground attack, and reconnaissance.

Initial work began in mid-1953 on what was then known as the P 300 project. Due to a lack of documentation, project design originally was made using a hypothetical powerplant. Helmut Langfelder addressed the matter in a letter from Seville to Max Schäffer in Munich, dated 13 May 1954:

"Dear Mr. Schäffer,

I am writing today to request your assistance in helping us with our work on the engine intakes for the P 300. As we here in Spain have neither the engine blueprints nor the ability to acquire them in the foreseeable future, we've decided to draw up a hypothetical engine in order to begin working on preliminary studies. Prof. Ruden, who has been given the responsibility of working out the intakes, is also of the opinion that this is a viable alternative in view of the absence of proper drawings. The results can then be compared with the 'ATAR,' whose data may be accessible to Prof. Ruden."

Data later became available for the most modern engines at the time, such as the Rolls Royce Avon and Armstrong Siddeley Sapphire. In light of the knowledge of maximum velocity aerodynamics at the time, it was at first thought that supersonic speeds would only be attainable in level flight using these powerful engines. Studies soon revealed, however, that a combination of newer, lighter engines such as the Bristol Orpheus and aerodynamic principles such as the application of area rule would easily provide the same results.

In the autumn of 1955 this knowledge, coupled with a certain concern about the aircraft's costs, led from a design originally weighing between 5,500 and 6,000kg to a light fighter with a takeoff weight of just 3,000 to 3,500kg.

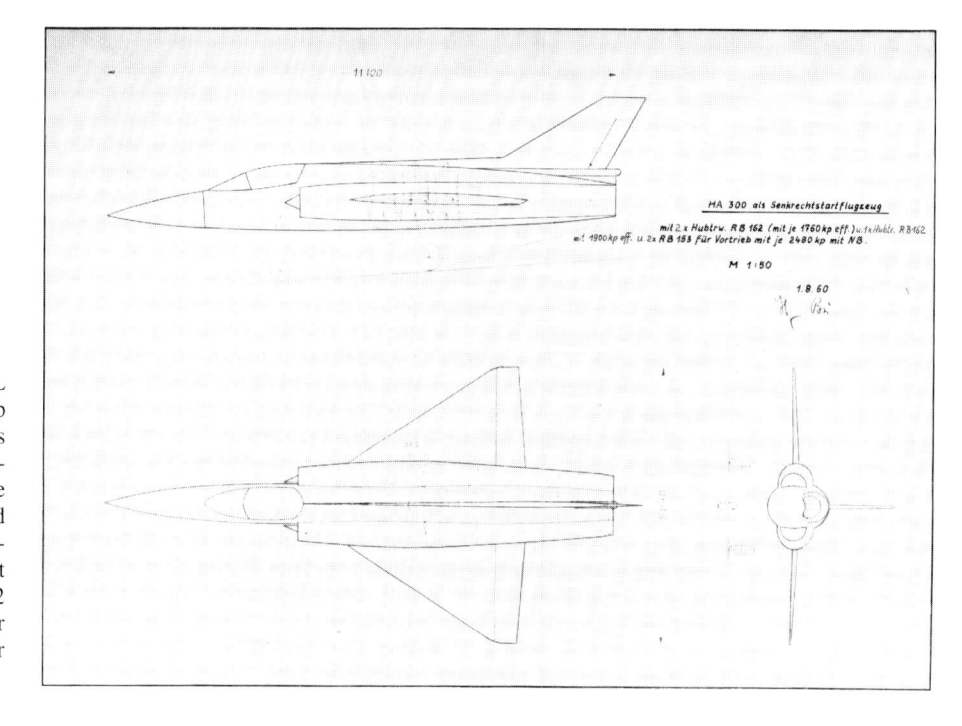

Messerschmitt also had a VTOL version of the HA 300 drawn up using most of the components from a standard HA 300. "Performance possibly better than the VJ-101C" was how he presented the concept. In addition to the layout shown here, another variant was planned having two RB 162 lift engines and two RB 153 (for forward propulsion and also for VTOL using thrust vectoring).

With the signing of the agreement between Spain and Egypt in late 1959, Max Blümm—working in close cooperation with Willy Messerschmitt—modified the pure delta wing design and drew up a tailplane. Using a 1:3 model built by INTA in Madrid and subsequently modified in Emmen, Switzerland, the pure delta design and the modified design with tailplane underwent subsonic wind tunnel testing at the Eidgenössische *Flugzeugwerke* in Emmen in September of 1960. Transsonic and supersonic testing took place at Bedford, England, during the summer of 1961. Max Blümm recalled that "according to Mr. Hills, the director at Bedford, (the design) had the best model results in the history of the facility. In actual flight, the pilot flying with the Orpheus 703 engine which powered the two prototypes was not aware when he broke the sound barrier in horizontal flight."

In a letter dated 6/11/1962 written from Cairo to Hans Hornung, Willy Messerschmitt offers an interesting comparison between the HA 300 and the MiG-21:

"Dear Mr. Hornung,

Regarding the testing contract for the 300, I would like to ask you to plan the systems as well as the aerodynamics. This is a day fighter and thus not suited for IFR landings. I had the opportunity to see the MiG-21; it had a radar dish of no more than 300 mm diameter. We in the West have got to find similar systems. It also had only a single 30 mm cannon with 60 rounds and the ability to carry guided missiles like the Sidewinder (two of them). In place of missiles, it could carry two 500 lb bombs or unguided rocket pods, as we already knew. A teardrop-shaped drop tank with a capacity of 480 liters is carried under the fuselage. The MiG-21's total fuel capacity is estimated at between 2,500 and 3,000kg, a wing area of 24 m, and a takeoff weight of 8 tons (without drop tank or bombs). We could carry about 300 to 350 more liters of fuel in the fuselage than originally planned. This can occur under the following conditions: 1) that the intake can simultaneously serve as a fuel tank wall—which is permissible, and 2) by lengthening the tank forward up to the nosewheel. For the weapon we'll have to make a bulge for housing the ammunition (inside the tank area). We'll also have to make another bump to accommodate the barrel (the MiG-21 has loads of bulges).

We need to have the following options:
1) pure fighter (interceptor) designed for high altitude flight exclusively, optimal takeoff and climb both with and without afterburner.
2) same, with brief Mach 2 capability.
3) reconnaissance, operating at maximum altitude with simplified optics carried in underbelly pod (no faster than Mach 0.9)
4) ground attack, capable of carrying rocket pods or bombs (low to medium penetration altitude), return flight can occur at 11,000 m if profitable, outbound leg also possible at high altitudes (all at about Mach 0.9).

Please have Stach provide you with advice on:
a) the most modern systems, i.e. the latest equipment which is not yet available, but will be by the time the first plane goes into production three years hence (keep the equipment designations generic).
b) currently available equipment.

Using these recommendations we'll be able to design the final layout within the fuselage.

Sincerely yours,
Willy Messerschmitt

Test Flying the HA 300V-2

Under the title of "Egyptian General Aero Organization/Aircraft Factories Helwan/Factory 36/Technical Staff," Fritz Schäffler wrote to Willy Messerschmitt on 3/16/1967. The letter reflects the team's heavy involvement in Egypt, even if it only reveals a single detail problem.

"Dear Professor,
On 15 March 1967 we conducted another vibration flutter test flight with the HA 300V-2. As we discussed in the last conversation with you, we fired off all destablizing thruster jets at an altitude of 9,000 m and a speed of Mach 0.95 and achieved a velocity of up to Mach 0.95 at 12,000 m. As on the previous flight, the vibrations noted earlier were not apparent. According to the pilot, the aircraft flew calmly and felt normal. At a speed of Mach 0.95, a bit of noseheaviness (Mach effect) was noted at both altitudes (9,000 m and 12,000 m). In the next flight, we will fly the same profile, but at Mach 0.97, i.e. at 9,000 m with jets affecting all axes and a speed run of up to Mach 0.97 at 12,000 m.

The reason for this letter, however, is not so much the results of this vibration testing, but to report the break of a vibration absorber in the nose gear which occurred upon landing after the flight.

Weather conditions were not exactly favorable. On takeoff and landing we had a sidewind equivalent of about eight knots and a tailwind running at five knots. As we had a barrier net set up at the northern edge of the runway—due to the prevailing north winds—we always takeoff and land in a northerly direction, even when there's a slight tailwind.

According to the recorder, the pilot touched down normally with his main gear at a speed of 250km/h, followed seconds later by the nose gear. Vibrations were immediately felt just after the wheel made contact with the ground, as had been noted earlier when the nosewheel began shimmying. These vibrations increased when braking (normal shimmy) and when rollout speed reduced even further through the use of the braking parachute. The nose gear's steering arm was broken. The break seems to be due to force, not to wear. However, the matter will be researched further when the materials are examined. It is not yet clear whether the break was due to shimmying (which is most likely the case) or whether the break occurred when the braking cylinder was activated. External inspection of the braking cylinder shows that it had been properly filled with oil, with both filler screws on either side fully extended. The locking mechanism on the nose gear was open during landing, as is proper, so that the nosewheel was free to rotate 45 degrees.

We do not intend to disassemble the shock strut cylinder just yet; instead, it will be fitted to a replacement nose gear we've got in storage and we will swap out the V2's nosewheel with this one. The operation of the shock strut will then be tested on the aircraft, and then we'll take the cylinder apart. If at that point we don't find any defects then we will no longer be able to duplicate the state of this cylinder. In this manner we hope to establish the cause of the shimmy and have the aircraft flying again after the weekend break."

HA 300 flight profile using the planned E 300 powerplant.

A creative break in the office at Herzog-Heinrich-Straße. Left to right: Hans Hornung, Willy Messerschmitt, Gero Madelung, Wolfgang Degel.

Signs of Renewed Aircraft Development

In addition to the dominant "non-aircraft projects," key words such as "aircraft construction (general matters)" and "Spain (info re Me 109) began appearing on the Munich Messerschmitt office's cost planning documents in 1951. The costs were listed as 2,400 DM and 711.89 DM, respectively. Given the hourly pay within the technical department at the time, these sums indicate that one employee spent about six days per month working on "aircraft construction." Given the fact that Willy Messerschmitt went to Spain twice and South Africa once that year, this expenditure seems surprisingly low—particularly as that year also saw Julius Krauß assume a role as an advisor to Hispano Aviacion for changes to the "Spanish" version of the Me 109. On the other hand, Messerschmitt's memorandum of 17 July 1951 to the Spanish Ministry of Aviation hinted at various other aeronautica projects being worked on by his Munich development team.

Aviation Development Begins Anew in 1955

Ten years after the end of the Second World War, the signs finally seemed to be favorable in the Federal Republic of Germany for renewed development in the field of aviation. On 9 May 1955 Germany entered NATO following the London and Paris conferences resulting from the Paris Accords of 23 October 1954. At the same time it joined the Brussels Pact, which thus expanded to become the West European Union (WEU) and led to the Federal Republic obtaining its sovereignty on 5 May 1955.

Within the civilian sector, the capability for a national aviation development program shifted over to the ministries of economy and trade, a move which at first made it virtually impossible for the industry to establish effective business partnerships for developmental and production contracts. In comparison, the military sector had been subordinated to the chancellor's office as "Dien*ststelle Blank*" until coming under a newly formed (7 June 1955) Ministry of Defense, giving it clear-cut lines of responsibility from the outset.

The Messerschmitt AG—whose chairmanship rotated from Karl Thalau through E. Stromeyer to Hubert Bauer as acting member and technical director—kept its head above water financially by manufacturing parts for Vespa bubble cars, producing shock absorbers for cars, and building machine jigs.

As early as September 1951, at the same time he was beginning negotiations for developmental work in Spain, Willy Messerschmitt made initial contact with the "Blank" department for the purpose of garnering the support of the chancellor's office for such activities—something which Spain's Ministry of Aviation was desirous of gaining, as well. Shortly after these first steps had been taken, the highest levels within Spain's military became keenly interested in establishing cooperative programs in the areas of industry and defense technology. Messerschmitt was asked in 1952 to establish the necessary contacts in Bonn. For his part, these discussions didn't take concrete shape until early 1955. As mentioned in the chapter on his activities in Spain, Messerschmitt's focus was on offering the HA 200 to the Federal Republic as part of re-equipping the new *Luftwaffe*.

A concentrated effort was made from the beginning to gain German approval for the Me 200, as the aircraft was designated in the marketing campaign in Germany. After receiving an offer, *"Dienststelle Blank"* gave assurances to Willy Messerschmitt in April 1955 that his Me 200 would be taken under consideration once it had successfully completed its first flight. At the time, it was obvious that France's Fouga Magister jet trainer, remarkably similar to the Me 200 in performance characteristics, was the leading contender from a political standpoint. Even semi-official visits by leading German minstry officials to Hispano Aviacion in Madrid and Seville didn't change matters much. By the time the HA 200 completed its maiden flight on 10 August 1955, Germany's newly established Ministry of Defense had already decided in favor of the acquisition of the French jet trainer.

Thus the intense efforts to market the Me 200 in the Federal Republic of Germany did not initially meet with the success originally anticipated. However, these efforts proved to be more of an impetus for the defense ministry to build upon the limited developmental resources which had either survived or become newly established in Germany.

Licenses, Programs, Cooperative Efforts

In 1957, the German government began issuing its first contracts in the area of aircraft manufacturing. Initially, the Messerschmitt AG was awarded a maintenance and overhaul contract for the North American T-6 Harvard Mk IV, introduced into the Luftwaffe's inventory as a trainer. There followed a similar contract the next year for the Lockheed T-33, an American jet-powered trainer. An alternative locale for these programs had to be found since the Augsburg airfield was still being occupied by a U.S. Air Force unit. The airport at Munich-Riem offered possibilities, where the Flughafen GmbH was willing to lease a 4,000 m² hangar and the use of the runway. A total of 117 Harvard Mk IVs were overhauled at Riem between October 1957 and August 1959. Initial deliveries of the Fouga Magister began in 1958, followed in February of 1959 by the overhaul of no less than 691 Lockheed T-33s—first at Riem, then at Manching until the program ended in mid-1972. The move to Manching had begun back in May of 1961. Step by step, aircraft and equipment made their way to the new location and the Munich-Riem facilities were abandoned to new owners. These maintenance contracts formed the cornerstone for a number of development and production programs—some of which later turned into joint ventures—for the "airplane builders" at Messerschmitt AG, then Messerschmitt-Bölkow GmbH, and finally Messerschmitt-Bölkow-Blohm GmbH. Thus, programs and mergers overlapped one another, a not uncommon occurrence in a field which saw developmental times lasting ten or more years, production runs lasting just as long, and "life expectancies" involving planning and financing issues of up to thirty years or longer. The first critical step on the way to a new beginning, however, was made with these license manufacturing programs, which in subsequent stages grew significantly to encompass responsibility for entire weapons systems within Germany's new Luftwaffe.

Overhaul and maintenance of the American T-6 Harvard Mk IV at Messerschmitt's Munich-Riem plant.

Finishing and fitting out of a Fouga Magister fuselage at the Augsburg factory.

Fouga Magister License-Built

The Ministry of Defense had awarded the Ernst Heinkel Fahrzeugbau GmbH and the Messerschmitt AG the contract for license building the Fouga Magister CM 170R as early as August of 1956. Messerschmitt AG's Augsburg Works assumed responsibility for assembling the fuselage and fittings, while Riem was tasked with final assembly and flight checkout, and Heinkel's Speyer plant produced the aircraft's wings, empennage, flaps, ailerons, and nose.

By combining an equal amount of shares, Heinkel and Messerschmitt had set up the Flugzeugunion Süd (FUS), headquartered in Munich, for the dual purpose of fulfilling the contract and at the same time consolidating the aircraft manufacturing resources in southern Germany—analagous to the concentrated resources in the north. The FUS acted in a fiduciary capacity between the publicly recognized contractor and the two companies involved. Stipulation for the award of the contract was solvency and an acceptable financial standing, something which the Messerschmitt AG wasn't able to achieve until after its restructuring in early 1957. As part of this capital reorganization, the Messerschmitt AG was released from its obligation to the Regensburger Stahl- und Metallbau in the sum of 5.8 million DM of federally guaranteed credit. In exchange, the Messerschmitt AG handed its business shares over to the Messerschmitt GmbH in Regensburg. As this reorganization did not occur until early 1957, production didn't begin until after that time.

The Air Fouga company delivered the first license production documents in December 1956, the last in September 1957. The drawings and tables first had to be translated, then the materials for potential German suppliers needed to be converted over. As Hubert Bauer, the technical director of the Messerschmitt AG, wrote at the time, the choice of suitable materials initially posed a problem, as standardization and labeling of aircraft materials had to be accomplished from the ground up following the many years of dormancy in the German aviation world. The two German companies involved suddenly found themselves confronted with the fact that manufacturing tolerances were much lower than had initially been stated. This particularly applied to the precision required in the sheet metal parts, the entire airframe, and the riveting. The number of swappable components, i.e. those parts and subassemblies required to be interchangeable with each other, was roughly the same on the Fouga Magister as it had been for the Messerschmitt fighters in the Second World War.

Suitable production facilities had to be created for what was to be the first production contract awarded to Germany's aviation industry after the war. The managing directors of Messerschmitt AG at Augsburg accordingly put every effort into rebuilding several areas of Wer*k II*. Two hangars were rebuilt in *Werk IV*, one of these being lengthened by moving over a hangar framework from another site. A cafeteria popped up, then a new, albeit modest office building; the old, generally undamaged administration center had previously been leased out long-term.

As impetus for the program, the French company provided components for 22 aircraft. Messerschmitt AG delivered the first fully assembled aircraft to Riem in July of 1958, with the first indigenously assembled aircraft arriving in April of 1959. The military received the last of a total of 194 aircraft produced in March 1961. There followed maintenance and overhaul of 285 Fouga Magisters altogether—first in Riem, then in Manching. The overhaul program ended in 1970.

License Production of Fiat G.91R3 Ground Attack Plane

In the fall of 1958, initial discussions got underway for the license manufacture of the Fiat G.91R3 ground attack plane, which NATO had called for back in 1953. These discussions involved the Bund*esamt für Wehrtechnik und Beschaffung* (BWB, or Department of Defense Technology and Acquisition) on the one side, and the companies of Dornier, Heinkel, and Messerschmitt on the other.

In 1955, Fiat Aviazione had emerged as the winner of the NATO competition for a light ground attack plane with its G.91—effectively a "small scale" version of North

After structural assembly of the fuselage forward and rear sections, the Augsburg plant undertook the fitting out of these two major subassemblies for the Fiat G.91R3 (below) and G.91T3 (right).

American's F-86 Sabre (whose aerodynamic profile, in turn, was based on the Messerschmitt P 1101). All NATO countries were therefore encouraged to either buy or license build this aircraft. However, it was only the Federal Republic of Germany which actually stepped up to the plate.

The three companies mentioned earlier formed the "G.91" coalition and in July 1959 were awarded a manufacturing contract. As part of the program Messerschmitt AG was responsible for producing the forward fuselage with cockpit, hydraulic and mechanical fittings, and the aft section. Dornier assembled the aircraft. Messerschmitt AG delivered the first "spin-up components" (ten sets of individual parts) to Dornier in February 1961, with the first indigenously produced subassemblies going out in May of that year and the last of a total of 294 being supplied by February of 1965. An additional 22 G.91T3 trainers were produced in 1970/71 as part of a follow-on license production program.

F-104G Starfighter Most Significant License Program

Also in 1958, initial discussions got underway for the license manufacture of an interceptor for the German Luftwaffe. This aircraft was expected to replace both the obsolescent F-84 Thunderstreak and the F-86 Sabre. Final contenders were the Dassault Mirage IIIA, the Grumman F11F Super Tiger and the Lockheed F-104A Starfighter. A studies group, comprised of Hubert Bauer and Karl Frydag from Messerschmitt and Fritz Walter of Heinkel, among others, submitted a preliminary report to defense minister Franz Josef Strauß in July of 1958. Discussions regarding license production within the BWB and involving the southern German industry began in late 1958. Decision to license build the F-104 was reached late in 1959 after the idea was submitted to the budget committee of the *Bundestag*. Following Germany's decision in favor of the F-104, several other countries also jumped on the bandwagon. The actual career of this aircraft began at this point, which at the time resulted in a multinational license production program whose scope was without precedent. The European Starfighter production alone employed at one time over 100,000 personnel, and in western Europe involved four joint venture enterprises, the companies of Fokker, Hamburger Flugzeugbau (HFB) and Vereinigte Flugtechnische Werke (VFW) in "Group North," Avion Fairey and SABCA in "Group West," Messerschmitt, Dornier, Heinkel and SIAT in "Group South," and Fiat, Aerfer, Aermacchi, SIAI-Marchetti and Piaggio in Italy.

As part of "Group South," Messerschmitt produced the forward fuselage section (minus cockpit), mated the fuse-

Completed fuselage sections (minus aft portion) of F-104G jets. The parts bay is in the background.

lage center section supplied by Dornier with the forward section, fitted these components out and assumed responsibility for final assembly including engine fitting, plus initial flight testing and subsequent release to the contracting agent. This corresponded to 38 percent of the aircraft's total production.

Manufacturing began at Augsburg in early 1960. In the fall of that year, Messerschmitt workers from Riem began reassembling 27 F-104Fs at the airbase at Nörvenich, where they were assembled, checked out in flight, and handed over to the contractor (Riem was not able to be used for checkout flights).

The first F-104G was flown on 25 July 1961 at Manching, with the last aircraft being delivered in late 1965. As part of this program, an initial 210 components for the F-104G were manufactured in Augsburg. In March 1964 the defense ministry decided to acquire an additional 32 TF-104G trainers. Here, too, the Messerschmitt AG undertook assembly of the fuselage center sections and the fitting of equipment, as well as carrying out the necessary

F-104G modification overhauling at the Manching works.

functionality testing for the forward fuselage sections supplied by Lockheed. A follow-on program in 1967/68 saw Messerschmitt AG supply components for a further 23 TF-104Gs, plus carry out final assembly and flight testing of the aircraft. In addition to the 27 reassembled F-104Fs, a further 88 F-104Gs and 42 TF-104Gs were built from sub-assemblies supplied by the U.S. The remainder (210 F-104Gs and 23 TF-104Gs) were assembled from components license built in Germany. A new-build program in 1971/72—with MBB as the primary contractee—embraced another 50 TF-104G airframes. All in all, Riem and Manching produced and delivered a total of 440 Starfighters. Overhaul of this airborne weapons system began in 1963. Up until the aircraft's retirement, the Manching plant overhauled no less than 3,319 Starfighters from both the air force of Germany as well as that of Canada.

In 1966 Entwicklungsring Süd GmbH was given design responsibility, and in 1967 weapons systems management control for the F-104G. This not only provided Germany's entire aerospace industry—including Messerschmitt AG—with much needed "know how" in the production of ultramodern, complex aircraft, but also with critical experience in evaluating solutions to reliability problems. By rebuilding its plants, acquiring facilities and equipment for the production and maintenance of combat aircraft, and bringing together a core group of technical experts, Messerschmitt AG was able to create a position for itself as a competitor in future projects of the same magnitude as the Starfighter program.

Parts for the Bell UH-1D

Rolf Besser wrote in 1982: "In the U.S., developmental work began on a Bell 205 with increased fuselage dimensions starting in 1960. The YUH-1D prototype took to the air on its maiden flight on 16 August 1961. The U.S. Army began taking deliveries of production UH-1D Iroquois in August of 1963. The basic model was designed to accommodate 14 soldiers and their equipment or six stretchers and a medic and was powered by the 810kW/1,100 hp Lycoming T 53 L 11 turbine used in the Bell Model 204. In 1967 the Lycoming T 53 L 13 was introduced on the UH-1H variant, having an output of 1,030kW/1,400 hp.

The same year saw license production of 343 UH-1Ds for the Luftwaffe and the Marineflieger begin in the Federal Republic—with Dornier as primary contractor after beating out IGLR in the competition." Ten helicopters had previously been reassembled at Dornier.

Within the scope of this license building program, which also involved Heinkel and SIAT, Messerschmitt AG was responsible for manufacturing the cockpit and seats, the fuel and oil systems, and eventually took over installation of electronic systems and half of the final assembly. Altogether 303 helicopters were produced by the joint venture company.

Fuselages for the Sikorsky CH-53G

The Ministry of Defense evaluated the Sikorsky CH-53 medium lift helicopter in 1966 and 1967. In 1968 the Bundestag's budget committee agreed to a license-building contract. In 1970 Sikorsky supplied the first manufacturing blueprints, and in May of 1972 the first components for a total of 92 helicopters were delivered. As part of this program, Messerschmitt-Bölkow- Blohm GmbH produced the forward body and center sections (with the exception of the nose section) at its Augsburg plant, as well as the main rotor and engine housing at Donauwörth. Final components were completed in late 1972.

Transall C-160 Center Wing Sections

In 1964 talks began with the goal of involving southern Germany's aviation industry in the Franco-German Transall program, then being run by the companies of Hamburger Flugzeugbau, Nord Aviation, and the Vereinigte Flugtechnische Werke. Nord Aviation offered up its center wing production, which subsequently were manufactured at the Augsburg and Donauwörth plants. Between July 1965 and mid-1971 135 center wing sections were delivered for final assembly. MBB assumed production of the entire fuselage, the main gear doors, and all control surfaces for the second Transall production series, begun in October of 1976. Production ceased in 1986 with the delivery of 35 aircraft to France and Indonesia.

CH-53G forward and center fuselage section at Augsburg prior to delivery.

Gleaning "Know-how" from the F-4 Phantom II

To fulfill a contract won by establishing suitable deadlines, quality, and cost, MBB built approximately 7,600 components (including complete aft fuselage sections, outer wing sections, and moving slats) for about 500 of the U.S. Air Force's F-4 Phantoms. The company was given overall system responsibility for the RF-4E reconnaissance aircraft and the F-4F tactical fighters of the German *Luftwaffe*, meaning it was responsible for managing the weapons system, design program, technical development and overseeing the logistical support of the program. Manching saw the maintenance and overhaul program start up in mid-1972, which by late 1991 had delivered over 1,400 Phantoms following a complete technical overhaul. In later years, this number included those RF-4E and F-4F aircraft of the *Luftwaffe* which underwent an extended-life upgrade.

Here in the Augsburg works the landing flap is being fitted as part of the Transall program, which saw Messerschmitt initially building wing center sections.

Elevators (in foreground) and rear section assemblies for the McDonnel-Douglas F-4 Phantom at Augsburg.

Modification overhaul for the F-4 Phantom II at the Manching works.

The CASA C-101, most modern trainer in the Spanish Air Force's inventory, flying with its HA 200 predecessor (left).

Casa C-101

MBB became involved in the prototype phase of the Spanish Air Force's C-101 trainer program as a subcontractor to CASA and was responsible for designing, assembling, and fitting out the aft fuselage sections for the four prototypes and additional two airframes used for stress testing.

Tornado European Combat Aircraft

Willy Messerschmitt followed the early stages of the European Tornado program with interest. The program stemmed from the "NKF" concept, and partners included MBB (42.5%), British Aerospace (42.5%), and Aeritalia (15%) working under the management company of Panavia

In July 1971 senior company officials invited Willy Messerschmitt to visit the Tornado (at the time MRCA) component assembly plant in Augsburg. Left to right: Gero Madelung, Volker von Tein, Oskar Friedrich, H. Fiebrich, Dr. Wolfgang Herbst, Max Dronsek, Willy Messerschmitt, Günter Gans, and H. Sebald.

Group photo taken on the occasion of a celebration in Augsburg; left to right: Hubert Bauer, Willi Stör, Fritz Wendel, and Willy Messerschmitt.

Aircraft GmbH. Developmental work began in 1970, and on 14 August 1974 the first Tornado took to the skies at Manching. Full-scale production was initiated in July of 1976. The Augsburg factory produced 929 fuselage center sections for lots 1 through 7, while by the end of 1991 the Manching plant had delivered 357 aircraft for its German customers (*Luftwaffe* and *Marineflieger*).

Average test day at the MBB Manching plant. On the left is a Tornado testbed, on the right a production aircraft during its acceptance phase.

VJ 101:
World's First Supersonic VTOL Plane

The development, construction, testing and technology of what was to become the first significant independent German aircraft project after the Second World War, developed by an industry without the advantage of prior years of research work and, in many technical aspects, charting completely unexplored territory, has been given more than adequate coverage by other sources. However, since this aircraft project had a significant impact on Willy Messerschmitt and his company from a technical standpoint and, as will be seen later, from an industrial-political perspective, a condensed version of the history behind Germany's first VTOL project is presented here.

Project Competition for Interceptor

The "project competition" for an interceptor which Franz Josef Strauß (appointed as defense minister in October of 1956) issued can be considered the first tangible manifestation of contractor and industry's efforts to secure a new contract for a German-developed product. From this point onward, Prof. Messerschmitt's technical department (which from mid-1956 on mainly constituted employees returning from the Spanish group) began working on numerous preliminary studies for fighters with the intention of outlining the guidelines for the competition, as well as providing the defense ministry with suitable proposals and encouragement. It therefore comes as no surprise to learn that, from June to October 1956, Hans Hornung oversaw several turbojet-powered fighter projects, even including a VTOL tail-seater.

On 2 November 1956 the Ministry of Defense formally called for an interceptor project competition. At first, there was no requirement for a vertical/short take-off and landing (V/STOL) capability. However, recognition of the need for an airborne weapons system's survivability in the event that the Federal Republic's runways were rendered unusable eventually led to an additional V/STOL requirement. The Ministry of Defense accordingly reissued the request for tender to include this requirement on 3 December 1957. As a result, the design teams at Bölkow Entwicklungen KG in Stuttgart-Echterdingen, Ernst Heinkel Flugzeugbau GmbH in Speyer and at Messerschmitt AG all began working on design studies.

Bölkow's Erich Haberkorn drew up an unmanned guided weapons system based on the experimental VTOL ring-wing C 450 Coleopter by SNECMA, a project which he'd worked on earlier. This project was later dropped, as it did not fit the requirements of the contractor.

At Heinkel, Siegfried Günter developed an airplane which envisaged four pivoting engines, one each mounted at the wingtips and at the ends of rather large canard surfaces. The vectoring powerplants were expected to provide the necessary thrust for both lift as well as forward flight.

Following numerous studies, Willy Messerschmitt decided to submit a V/STOL design by Hans Hornung based upon the P 1227/1 project, which saw the four engines mounted inside the fuselage around the center of gravity. Thrust was provided by directing the jet flow downward, while forward flight resulted when the exhaust was directed horizontally out the aft fuselage.

As the Ministry of Defense had made clear that it didn't wish to issue several developmental contracts for the project simultaneously, on 11 June 1958 the companies involved agreed—after initial discussions in 1957—to join forces and

Engaged in heated discussion, sometime during the '60s: (left to right) Willy Messerschmitt, Josef Kempter, Hans Hornung.

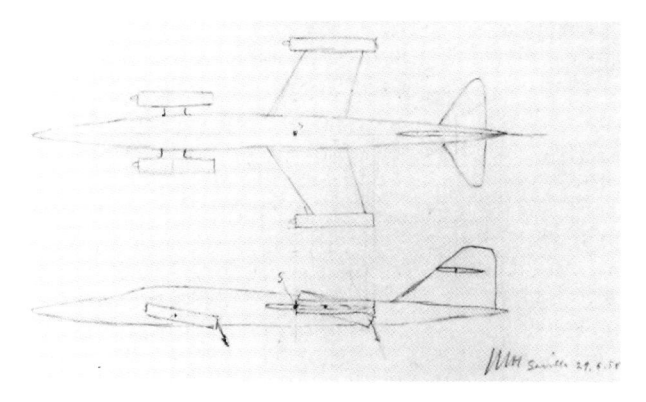

This hand sketch from Seville, dated 29 May 1958, shows that from early on Willy Messerschmitt had been thinking of a four-engined C variant of the VJ 101. He felt that the aircraft could be made smaller and lighter and therefore would have virtually the same performance with four engines as that of the "A" project. A lift engine could be fitted as needed at the aircraft's center of gravity point in order to attain the same performance as the F-104G with external tanks and ordnance.

establish a cooperative venture along the lines of the Fouga Magister, G-91 and F-104 programs. A major consideration behind this decision was clearly the recognition that such a project was beyond the technical and personnel capabilities of any one company at the time, nor did the ministry give much indication that the winning company would be the sole recipient of a production contract. On a side note, the driving force within the Ministry of Defense was Dr. Theodor Benecke (the senior editor of this book series when published in the original German), director of the air and space department within the defense technology branch.

Bölkow Entwicklungen KG, Ernst Heinkel Flugzeugbau GmbH, and Messerschmitt AG accordingly tendered their offer to the Ministry of Defense to accept the developmental contract on 4 November 1958. At a meeting in Munich on 23 December Defense Minister Franz Josef Strauß called upon the three companies to bring their developmental teams together and establish the joint venture company. This took place on 23 February 1959. The joint venture was given the name Entwicklu*ngsring Süd* (EWR, or Development Ring South) and quickly moved into offices within the library at the Deutsches Museum in Munich.

The EWR's governing body was created as a "standing committee." It consisted of a chairman of the board from each of the three companies. The firm of Heinkel sent Prof. Karl Thalau, Messerschmitt AG provided Dr. L. S. Rothe, while Ludwig Bölkow represented his company's interests personally. At first, the committee met only sporadically, then every two weeks, later on a monthly basis, and eventually quarterly. Its "instruments" included secretaries, business leaders, and technical directors brought together into a "technical committee."

The "standing committee" made all the fundamental decisions in matters affecting the interests of the businessmen it represented, in the type and scope of tasks (contracts) and approaches to tasking (planning), and in the company's representation vis-a-vis the contractors. The joint venture company's inner organization was also covered by this.

The "technical committee" consisted of a technically qualified director and deputy director from each of the companies involved. The "standing committee" gave this committee its marching orders and met roughly once a week. At first, Willy Messerschmitt acted as chairman. He later delegated his responsibilities to Gero Madelung, while Ludwig Bölkow appointed Dr. Otto Pabst, and Prof. Karl Thalau named Karl Schwärzler to represent their interests.

A visit by Robert Lusser to the Ministry of Defense in the summer of 1959 revealed that the Ministry—after it originally said it was ready to finance two different prototypes developed parallel to each other—was now insisting that only one type would be built as an experimental aircraft. None of the rival project groups was willing to back away from its own design. A compromise seemed out of the question. There followed months of technical discussions between the team of Ernst Heinkel Flugzeugbau GmbH, whose design was designated VJ 101A, and that of the Messerschmitt AG and its VJ 101B. Both designs made use of six small turbojet engines. After more heated discussions between the project groups the Heinkel group backed off from its canard design in favor of rear mounted control surfaces, while the Messerschmitt group agreed to nacelled

In Augsburg, a full scale mockup of the VJ 101B (P1227) was constructed in 1959 which simulated thrust control in the three axes. Although the mockup, mounted on a frame so that it could move, lacked a powerplant, it was fitted with compressed air control jets on the wingtips and under the fore and aft fuselage. The control proved to be quite effective.

engines on the condition that, for safety reasons, two engines would be housed in each pod. Thus was born the joint solution "C," as proposed by Gero Madelung.

In a meeting of the standing and technical committees on 22 September 1959 the vote was unanimously in favor of this joint concept. The basis for this decision was a carefully prepared point-for-point evaluation system, which had been discussed in detail beforehand with Messerschmitt, Heinkel, and Dr. Otto Pabst (representing Bölkow). This saw all three projects—A, B, C—matched side-by-side with four engines providing forward thrust and two providing lift thrust. Following a detailed review of the evaluation table and with other evaluation factors added in, the C project was awarded the highest point value.

Willy Messerschmitt presented the results of his own research, showing that Project C was also possible with just four engines. In the ensuing discussion, he repeatedly emphasized that exploration should be made of the "smallest possible" design for the aircraft before any determination of solution C's layout was made.

Ludwig Bölkow subsequently proposed that version C should be given priority and that the matter of the number of engines could be optimized at a later date. The committee agreed to this proposal.

Two Prototypes Built

The VJ 101C was a mid-wing design with unbroken slightly swept wings having a pivoting pod accommodating two turbojets on each tip behind the center of gravity, plus two more lift engines in the fuselage body ahead of the center of gravity. The advantages of this approach included:

- the entire output of all engines was available for vertical takeoff
- afterburner thrust could also be utilized for vertical flight
- no loss of thrust as with a thrust diverting system, and circumvention of interference problems with a canard lay out
- simple transition from hover to horizontal flight and back due to adjustable thrust vectoring (pivoting engines)
- control in VTOL flight through thrust modulation, dispensing with the need for widely separated control jets (required for a fuselage layout)

Once data for the VJ 101C had been obtained through extensive testing in both sub- and supersonic wind tunnels, the defense ministry decided to initially contract for two experimental aircraft, the VJ 101C-X1 with four forward propulsion non-afterburning engines and two lift engines for speeds below Mach 1, and the VJ 101C-X2 with four afterburning forward propulsion engines and two lift engines for supersonic flight. Both aircraft were also expected to be able to land and take off vertically.

Proof positive of the thrust modulation control concept and a key step in the progress toward a high-performance VTOL aircraft was the development, design, and construction of a "see-saw" contraption and a hover frame. The see-saw, which was basically a platform pivoting on a fixed horizontal axis, would be used to test the smooth function of engine and throttle control using thrust modulation.

After experimentation using a tripod, research in all six axes of freedom as well as the majority of critical systems for the experimental craft was carried out using the frame in over 200 hover flights, some of these being conducted in extremely poor weather conditions.

Construction of the VJ 101C-XI proceeded apace while the testing of the hover frame continued. The tripod mount enabled dynamic evaluation and optimization of onboard

Free flight demonstration of the VJ 101C hover frame on 9 July 1962 at Manching. Among the spectators were (left to right) Willy Messerschmitt, Inspector of the Luftwaffe General Josef Kammhuber, BWB president Dr. Theodore Benecke, Germany's defense minister Franz Josef Strauß, and Ludwig Bölkow.

369

systems to be carried out on the ground. With certain limitations, therefore, the aircraft, securely mounted to a column, could be tested as it moved around its center of gravity in its roll, pitch, and yaw axes. Partially or fully covering three blast vents in the tarmac permitted study of ground effects upon the aircraft.

Although the aircraft was basically created in EWR's joint design department under the direction of Karl Schwärzler, actual construction was assigned to the manufacturing sections of Heinkel in Speyer and Messerschmitt in Augsburg. Speyer produced the wings and control surfaces, with Augsburg manufacturing the fuselage and undercarriage. Rolls Royce supplied the engine pods together with the powerplants themselves. Final assembly took place in a special hangar at EWR's own test assembly center in Manching.

As early as the beginning of 1960 Willy Messerschmitt had offered to assume responsibility for construction of the undercarriage with his own employees. He was able to push this idea through over the objections of the technical committee—Robert Lusser was particularly opposed to the idea because airframe manufacturers generally obtained their landing gear systems from specialist undercarriage producers, as was predominantly the case in the U.S. The Messerschmitt AG made a fixed price offer for the undercarriage of the experimental planes manufactured to the same specifications as established undercarriage manufacturers such as Messier, Hispano Suiza, and Bendix. The contract included two complete sets for the two prototypes, plus half of a main undercarriage and a nose gear system for structural soundness and dynamic evaluation. Messerschmitt explained that he would be personally involved in the design, to which end he dispatched Wilhelm Binz and Ferdinand Miller to EWR. The undercarriage subsequently passed the evaluation phase with flying colors, even when the X1 once accidentally landed with its brakes applied. The successful design led to Messerschmitt being called upon to provide the undercarriage for the VJ 101D in 1962. Again, he assigned Wilhelm Binz with the design work.

Even after the "C" concept was agreed upon, a design which Willy Messerschmitt had specifically approved of, he continued in his quest for improvement. This first took the form of a "B" experimental model, which the defense ministry had originally agreed to in June of 1959. Willy Messerschmitt had indicated that this aircraft could be developed relatively easily using components of the Ha (Me) 300 then being produced in Spain. The Me 300 would have been able to provide the forward fuselage and cockpit, the wings and ailerons, as well as the complete vertical stabilizer and undercarriage, leaving only the aft fuselage behind the cockpit and the engine layout to be designed from scratch.

Willy Messerschmitt devoted much attention to the VJ 101C's landing gear. After successfully promoting his own design, he gave clear suggestions to Wilhelm Binz regarding their construction. These recommendations give insight into his typical manner of thinking.

VJ 101C-X1 in a hover, with the intake cover for the lift engine and forward gondola intakes on the swivel engines opened.

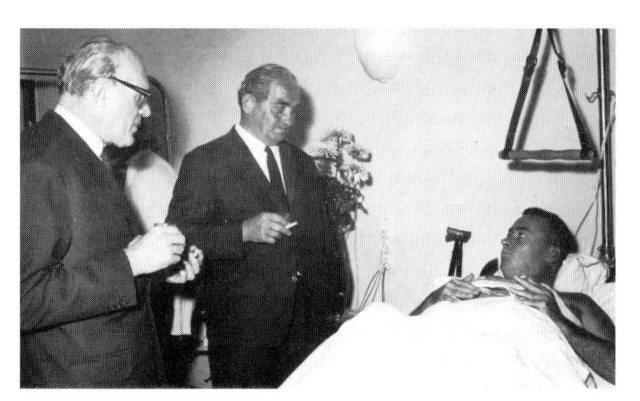

During a hospital visit, Willy Messerschmitt and Hubert Bauer share a cigarette with chief test pilot George Bright following his crash in the VJ 101C-X1.

Messerschmitt offered to undertake construction of this version in Spain, with first flights being carried out under the direction of the EWR. He felt that he could have the airplane available for initial flight testing within a year, meaning that the program would not interfere with that of the "C" version. This optimistic estimation was most likely due to his experiences during the war years, where many aircraft designs first took to the air within months of being designed (although admittedly, the majority of these were much less complex than the VJ 101).

The EWR rejected these ideas—initially using the excuse that it did not have the capacity. However, Messerschmitt continued in his efforts to prove the superiority of his concept. In March of 1961 he even went so far as to ask the committees to give him full authority to design and build an experimental aircraft "according to his formula." This met with strong opposition within the EWR and was rejected, for without a doubt it went against the spirit of cooperation being cultivated within the joint venture program.

On 10 April 1963 the X1 hovered in free flight for the first time. The first complete takeoff and landing transitions took place on 8 October 1963—after a previous total air time of just three hours. On 29 July 1964 the X1 became the first German aircraft and the first VTOL design in the world to attain supersonic speed (Mach 1.04). Until it was lost in an accident on 14 September 1964 due to improper assembly of the control system following maintenance, the X1 carried out 132 test flights, 65 of which were free flights.

Der Spiegel magazine's cover story on 15 January 1964 highlighted "Aircraft Builder Messerschmitt."

Lilly and Willy Messerschmitt in July 1965 at the International Transport Exhibition (IVA) in Munich. Messerschmitt has just been awarded the "Golden Windrose" neckchain for his pioneering work in aviation.

VJ 101C-X2 Tests STOL Capabilities

Flight testing continued with the first hover flights of VJ 101C-X2 on 12 June 1965. In comparison with the X1, the X2 had a higher takeoff weight by 7,700 kg due to the four afterburning RB 145R engines—which also necessitated a greater fuel quantity.

"While efflux recirculation problems with the X1 in hover mode were manageable," recalls Wolfgang Degel, "they were much more serious with the X2's afterburning engines, particularly when the aircraft was not held precisely above the exhaust deflector while hovering at a specified minimum altitude. Both close-proximity and residual recirculation led to a drop in thrust effectiveness and engine surging. The result was a hard landing with damage to the engines and a collapsed undercarriage. Thorough preliminary testing on the tripod and on the runway resulted in the development of a new takeoff method which was successfully tested. Similar to landings in afterburner, the pods were rotated to 70%, causing the plane to roll forward slightly prior to takeoff. Before takeoff, therefore, the aircraft kept ahead of the superheated exhaust—minimizing ground erosion and airframe heating and permitting a smooth takeoff. Test results indicated that takeoffs were also possible within the area affected by the superheated exhaust. This was successfully demonstrated in November 1969. The afterburners were lit before the aircraft began rolling. The run-up to the point of lift-off was three meters, with the 15 meter hurdle being cleared at 40 meters. This effectively proved that landings and takeoffs were possible without the need for prior, specialized ground preparations. Takeoffs and landings were improved even further through the development of a controller/regulator being undertaken simultaneously by EWR."

Engine and attitude control tests with the VJ 101C-X2 on the telescoping column at Manching.

This rolling vertical takeoff and landing method offered considerable increase in payload for operational aircraft. Although in much modified form, the Royal Navy basically employed the same concept when it introduced aircraft carriers having an inclined runway, or "ski jump," enabling the Harrier VTOL plane to make rolling takeoffs.

From a repair standpoint, it was not possible for the X2 in afterburner to make vertical landings upon unprepared concrete surfaces. However, with the engine pods angled at 70% to 75%, such landings with afterburner were indeed possible, with touchdown occuring within 46 meters from an altitude of 15 meters. These tests showed that such landings avoided wear and tear on ground surfaces, engine damage and overheating of the airframe.

The decision to end flight testing came in June of 1971. After 325 test flights, on June 8th, 1972, the X2 "flew" its final mission—suspended from the hook of a CH-53G from the Manching test center to MBB in Ottobrunn, eventually being handed over to the Deutsches Museum.

There the VJ 101C-X2 stands as a symbol for the level of aviation technology once again achieved by the Federal Republic of Germany following ten years of inactivity in this area, demonstrating that Germany's industry was not only on par with the rest of the world, but in fact was leading the way.

Looking back, it should be noted how little funding was actually necessary for Germany to reach the pinnacle of aviation technology—developmental costs for the VJ 101 program amounted to approximately 250 million DM, roughly the same as was allocated at the time for the introduction of a new production model of the VW Beetle. The consequences of this can be seen in the subsequent desirability of Germany's aviation and space industry in forming new partnerships. The roots of this success may perhaps be also found in the modest perseverence of those working in the VJ 101 program, those employees who approached their tasks with an enthusiastic and motivated spirit.

VJ 101D Follow-On Development

While the VJ 101C was fully directed toward meeting the requirements for an interceptor laid down in 1957, NATO requirement AC-169 from the summer of 1961 called for fundamental changes to the specifications originally established for the VJ 101C: an all-weather fighter-bomber in place of the interceptor variant; operational VTOL capability; modified requirements in regard to the minimum combat radius; and a lo-lo-lo profile (i.e. low-level flight for the entire mission). The combat radius now required for the new VTOL fighter-bomber could have been achieved by simply fitting additional lift engines into the VJ 101C—with a significant increase in airframe dimensions. Furthermore, for the fighter-bomber's low level operations it would seem necessary to install the newly developed and much more economical Rolls Royce/MAN RB 153/61 double flow engines. As these engines were larger, particularly in view of their diameter, it would no longer be possible to fit them into outboard pods as on the VJ 101C. With the backing of Willy Messerschmitt, a design evolved (similar to his first VTOL project, the P 1127/1 from 1958/59) which called for all engines to be housed within the fuselage. These were to include two RB 153/61 thrust vectoring forward propulsion engines in the rear airframe and five RB 123/23 lift engines behind the cockpit. Control of the aircraft in its pitch axis through thrust modulation was carried over from the VJ 101C.

Design work on the VJ 101D, as the fighter-bomber project was now known, began in late 1961. The first two prototypes were expected to be ready by mid-1964. However, the project subsequently became subject to several changes at the request of the contractor. For example, it was hoped that agreements could be reached between officials and manufacturers for the aircraft to serve as a testbed for the latest electronic systems available in the mid-1960s and their possible application. These included: radio systems, navigation computers, automated weapons delivery systems, air data computers, fire control systems, and infra-red search and track systems.

NATO, however, dropped the idea of such a project, a measure taken in conjunction with a fundamental change in its defense doctrine. This step left many foreign companies high and dry which had enthusiastically been working to meet the same requirement as EWR. The new doctrine of flexible response called for an increase in the combat payload to the level which ruled out vertical takeoff altogether. There was still the need for reduced dependency on long runways, however, i.e. the capability for short takeoffs was still much in demand. Use of variable geometry wings coupled with the thrust vectoring technology of the VTOL programs offered promising hopes. Following the cancellation of design work on the VJ 101D, external circumstances led to new studies being carried out under the designation of VJ 101E.

Willy Messerschmitt Plays A Minor Role

"Unlike the HA 100, HA 200 and HA 300 projects and the 'Rotor-jet' program from 1969, Willy Messerschmitt played a relatively minor role in the design work of the VJ 101 (with the exception of the landing gear system)," remembers Gero Madelung. "At the time, he seemed to be focused on other matters. Nor did he seem suited to the complexity of managing a consortium development program personally; he much rather preferred to be in direct contact at the work level with the designers, the aerodynamics engineers, the statisticians and the assembly workers. Sigfried Günter, Hans Hornung and I were the ones mainly involved in developing out the extremely complex and demanding overall design program. At the time Erich Haberkorn was still working with the unmanned interceptor idea.

We were soon in agreement that the best course of action lay with relatively small powerplants (in the J-85 class), which provided the optimal solution in view of performance to weight as well as integration possibilities. High-speed forward flight called for four such engines with afterburner, with more possibly needed for vertical flight. That so many engines would seriously be considered for what amounted to a relatively small fighter seems outrageous today, but at the time even foreign companies were making such proposals for their projects. On the other hand, the main parties did not see eye to eye when it came to integrating the engines with the airframe. Hornung was adamantly opposed to interfering with the aerodynamic lines of the plane and wanted to incorporate the powerplant fully within the fuselage, while Günter's priorities lay with maximizing the installed thrust (including afterburner) for vertical takeoff.

Willy Messerschmitt leaned more towards Hornung's philosophy, and he probably felt that by applying his ultralight design principles he could offset the inherent deficit of lift thrust—a goal that I didn't consider realistic in this case. Günter's design, with vector engines mounted on the canard foreplanes as well, seemed to be quite a risky proposition to most of his colleagues in the EWR. Not to mention the fact that we simply didn't have the time to fix this problem through extensive wind tunnel testing beforehand. Therefore, I ultimately recommended that we look at a design layout which was eventually accepted as the VJ 101C. There were risks enough with this proposal—particularly with the control system using thrust modulation and the engine vector system—but these were able to be worked out in stages. Tests later revealed that there was more than enough thrust for vertical take-off.

Even then, Messerschmitt regarded the suitability of jet engines for producing lift with skepticism, mainly due to their poor energy efficiency. Instead, helicopter rotors (as with the later Rotor-jet project) were what he had in mind. But even he was unable to find a practical configuration with this approach given the fact that the requirement called for a supersonic fighter. His proposal, the "C" solution making use of just four engines, ran into problems with application, as no suitable concept for pitch control in hover flight could be worked out.

Always difficult from an objective standpoint, the selection of a design was greatly influenced by EWR's temporary technical director, Robert Lusser. He was vehemently in favor of a rapid focus on establishing a concept. Without him, it would not have been possible to even get the single prototype we did to the flight testing stage. This viewpoint is borne out when, with Franz Josef Strauß leaving the post as defense minister, the Ministry of Defense quickly lost interest in continued development of the program."

German-American Weapons System

Newly established relations between Boeing in the U.S. and the Bölkow Entwicklungen GmbH almost immediately led to a joint AVS (Advanced Vertical Take-off and Landing System) study, building upon Boeing's work in the area of variable wing geometry and the farsighted V/STOL developments by EWR with its VJ 101 program. The fruits of this study resulted in the governments of both countries signing an agreement to begin preliminary work for the joint development of an advanced V/STOL weapons system. Consequentially, the U.S. Air Force assigned Fairchild-Hiller to be the American partner to EWR. From 1965 to 1967 both partners worked out the system parameters and definition for the AVS/A 400 together in Munich.

The most critical feature of the new weapons system was to be its adaptability to any changes in tactical doctrine and military requirements, something which had hitherto been unheard of in the military aviation world. The AVS was laid out as an airplane, whose operational and cost effectiveness was to be as impervious as possible to changes in military thinking. It was to be a two-seater, with variable geometry wings, two forward propulsion engines and, depending on configuration, would also be fitted with lift engines. Change in the priorities of the two governments eventually prevented the joint development of the system. Nevertheless, the Entwicklungsring Süd was able to gain much experience with the ultra-modern management techniques employed by America's Department of Defense, in dealing with the U.S. Air Force, in managing international programs, and along with this was able to broaden its technological base.

With the scrapping of the AVS program in 1967, the EWR applied this technological and organizational "know-how" to a new German program called the "Neues Kampfflugzeug" (NKF, or Next-Generation Combat Aircraft). Conceptual and definition phases for the NKF were worked out with the Vereinigte Flugtechnische Werke (VFW) in Bremen and the Bölkow GmbH in Ottobrunn during the first half of 1968, until later that same year when the venture crystallized into the trilateral Panavia MRCA (Multi Role Combat Aircraft) program, a project which would ultimately develop into the largest European aircraft acquisition program after the Second World War.

Looking back, it seems the Messerschmitt development team's ten-year mutual work with its colleagues in the EWR was a rollercoaster ride that eventually blossomed into international recognition in the once exclusive field of high performance aircraft. The maiden flight of the Panavia Tornado prototype in August 1974 at Manching—Willy Messerschmitt's 75th year—was a clear indication that, after nearly twenty years of struggling following the postwar disbandment of the original team, Messerschmitt had finally reached the pinnacle of international aviation technology.

The Road to Merger

The Entwicklungsring Süd, officially established as a consortium on the 23rd of February, 1959, initially consisted of 160 employees who had been detached from the three companies involved. This team had inherited the rather difficult task of developing the first post-war German combat aircraft and, in so doing, breach hitherto unexplored technical frontiers. Just four years later, in 1963, the group set a technological milestone with the initiation of the flight test program for the VJ 101C-X1 experimental aircraft.

In 1964 the Heinkel Flugzeugbau GmbH unexpectedly withdrew many of its managerial staff members from the EWR and reassigned them to a cooperative venture with VFW to work on the VTOL transport. The Bölkow Entwicklungen KG and the Messerschmitt AG therefore decided to pull their people out of the Entwicklungsring Süd as well, whose numbers dropped from 1,400 to 1,100 personnel. By concentrating all its resources, the EWR was able to continue with its labor intensive programs. Its personnel base slowly grew, reaching its zenith in early 1969 with over 2,000 employees. In addition to other projects, in 1961 the EWR had assumed logistical and construction responsibility for the F-104G Starfighter weapons system. This was followed in 1966 by the assumption of the program's design responsibility, and in 1967 by responsiblity for the F-104G weapons system management. Modification and overhaul of the aircraft itself were accomplished by Messerschmitt AG at its Manching facility.

The consortium was dissolved on 30 June 1965. On 6 July 1965 its legal successors became the Entwicklungsring Süd GmbH with partneres Bölkow GmbH (25%), Messerschmitt AG (50%), and Siebelwerke-ATG GmbH (25%). Walther H. Stromeyer became chairman of the board of directors, vice-chairman was Dr. Bernhard Weinhardt. Members included Hubert Bauer, Ludwig Bölkow, Dr. Hans-Otto Riedel, and Dr. Hans-Heinrich Ritter von Srbik. Hans Empacher, Kurt Lauser, Gero Madelung, and Dr. Otto Pabst were named as managing directors.

This action of 1965 was a major step in drawing those two business partners, Willy Messerschmitt and Ludwig Bölkow, closer toward the great integration movement of Germany's aviation industry.

Messerschmitt's Legacy: The "Rotor-Jet" Heliplane

In January 1966 Willy Messerschmitt and his studies department began wrestling with the problems of designing a vertical takeoff and landing aircraft. He set himself the task of turning a powerplant designed for high-speed horizontal flight into something which would also be suitable for vertical takeoffs. Furthermore, the airplane fitted with such engines would need to be just as fast as comparable aircraft taking off normally. The result of intensive research was a winged airplane which made use of rotor blade systems outboard on both the port and starboard wings for taking off and landing. After takeoff, these rotors would fold up and retract into gondola-like housings as the aircraft transitioned to horizontal flight.

In the forword to his conceptual description for such a "convertiplane," Messerschmitt wrote in October of 1967:

"Our future is based on progress, on studying new discoveries and gathering experience by evaluating those findings and creating ready-to-use products.

In the field of aeronautics, it is the vertical takeoff concept which leads to the future and offers new possibilities of critical import.

We need to follow all possible roads which may lead to this goal, and I believe that the path chosen here via the convertiplane is the safe one and will lead to success in the near future.

We know that every high-speed aircraft today is equipped with sufficient power that, when adjusted logically, would enable it to take off and land vertically, even if one engine were to fail. The greater an airplane's airspeed, the higher the power output of its engine, and therefore smaller the rotor diameter can be.

We must now find a way to sensibly transfer this engine power to the vertical lift rotors and enable the rotor blades to fold and retract. Our foremost and most urgent task is currently to establish a foundation for this.

This is a conceptual design worked out for high subsonic flight. It should be mentioned, however, that the powerplant configuration is also suited for supersonic flight."

Willy Messerschmitt, October 1967

Above all, the nature of the task drawn up by the studies department outlined the design's planned market:

"These types of aircraft will have a decisive impact and change future military thinking within the Army, Air Force, and Navy. They will also clear new paths in the civilian sector, particularly in the areas of airport design and the major problem of airport approaches. The required operations area would be but a fraction of that necessary for today's facilities. Airports could be built in the immediate vicinity of cities, noise pollution would drop, and the safety factor would rise as there would no longer be a need for high takeoff and landing speeds (most airplane accidents occur during takeoff and landing).

Not only do vertical takeoff and landing planes offer the possibility of considerable financial savings for airports, the availability of such aircraft may even be a prerequisite if we are to see further development of commercial air or the construction of new airports altogether."

Without assessing this approach from today's perspective, it undoubtedly called for the birth of a new type of aircraft rather than simply improving upon the operationally unique and reliable helicopter design. Messerschmitt therefore developed the concept of a high-powered winged

A private meeting. Left to right: Karl Thalau, Adolf Baeumler, Willy Messerschmitt, and Kurt Tank.

aircraft with the capability of landing and taking off like a helicopter.

The initial project for a test plane, begun in 1966 under the designation Me (P) 508, called for two General Electric T58 shaft turbines, each outputting 1,360 kW (1,850 hp). This power would be transferred to the rotors during "helicopter flight" via a collective and transfer gearbox. During "winged flight" (i.e. with rotors retracted), the aircraft would be propelled by a forward thrust powerplant in the form of a shrouded turbofan in the aft fuselage. Takeoff and landing transition would be managed by continuous interchange of power between rotors and fan. The novelty of this convertiplane lay not with the ability to fly at high speeds and the transition to vertical flight (something which any helicopter could do), but rather its ability to shut down and restart the unloaded rotors during flight, as well as storing and folding the rotor blades themselves.

In March 1967 the development department estimated the cost of building an experimental aircraft at DM30 million—a sum which in hindsight seems to be underestimated by a power of ten. Development time was projected at ten years, with the consideration that such a project would involve a series of technical problems whose solutions were by no means guaranteed. In order to keep the risks to a minimum, it was decided to proceed in five stages (something which, at the time seemed to be optimistic, but in principle was realistic):

- a rotor system mounted at an aircraft's fuselage center of gravity (a Mitsubishi Mu-2 commuter plane was considered)
- construction of a small testbed from steel tubing and appropriate skinning with the capability of showing transition
- construction of initial "convertiplane" with high subsonic speed and capability for demonstrating "the transformation"
- construction of a prototype
- production ("no earlier than 1975")

Competing Developers Reveal Advantages of Messerschmitt's Concept

A look at the V/STOL projects under consideration or being seriously studied at the time reveals a surfeit of proposed technical solutions. In several instances these are remarkably similar to Messerschmitt's ideas. A report by the studies department in August 1967 shows that Messerschmitt's concept of a "Rotor-jet" was no more exotic than those designs of the "big boys," particularly those of the American competitors. For example, numbered

"Combination of helicopter with standard airplane, with lifting blades able to fold and retract at high speeds" was how Willy Messerschmitt described the "Rotorjet" in his patent application at Estepona on January 1st, 1963. Above is a model of an early version, dated 1966/67, showing the takeoff and landing configuration with rotor blades extended and the cruise turbines located on either side of the fuselage. These also drove the rotors via a central distributor gearbox as well as a propulsion fan in the rear fuselage. This rear fan was to have made use of a boundary layer intake. The picture below is that of the Me P 408 with the propulsion configuration envisioned from 1968 onward. Gas generators driving the rotor turbines and cruise fans via tip turbines. When cruising, the rotors are folded and retracted.

among these were a Sikorsky project for a high-speed helicopter with coaxial fixed blade rotors and pusher-type propeller (cruise speed maximum 550 km/h and range less than 1,000 km), a composite helicopter with stub wings and supplemental forward propulsion such as Lockheed's 879 project (with limited cruising speed and range caused by the additional drag induced by its rotors), Hughes Aircraft Corporation's proposed "Hot Cycle Rotor Plane" with a takeoff weight of 20,600 kg, with the rotor blades located at the tips of the triangular-shaped wings. During hover flight

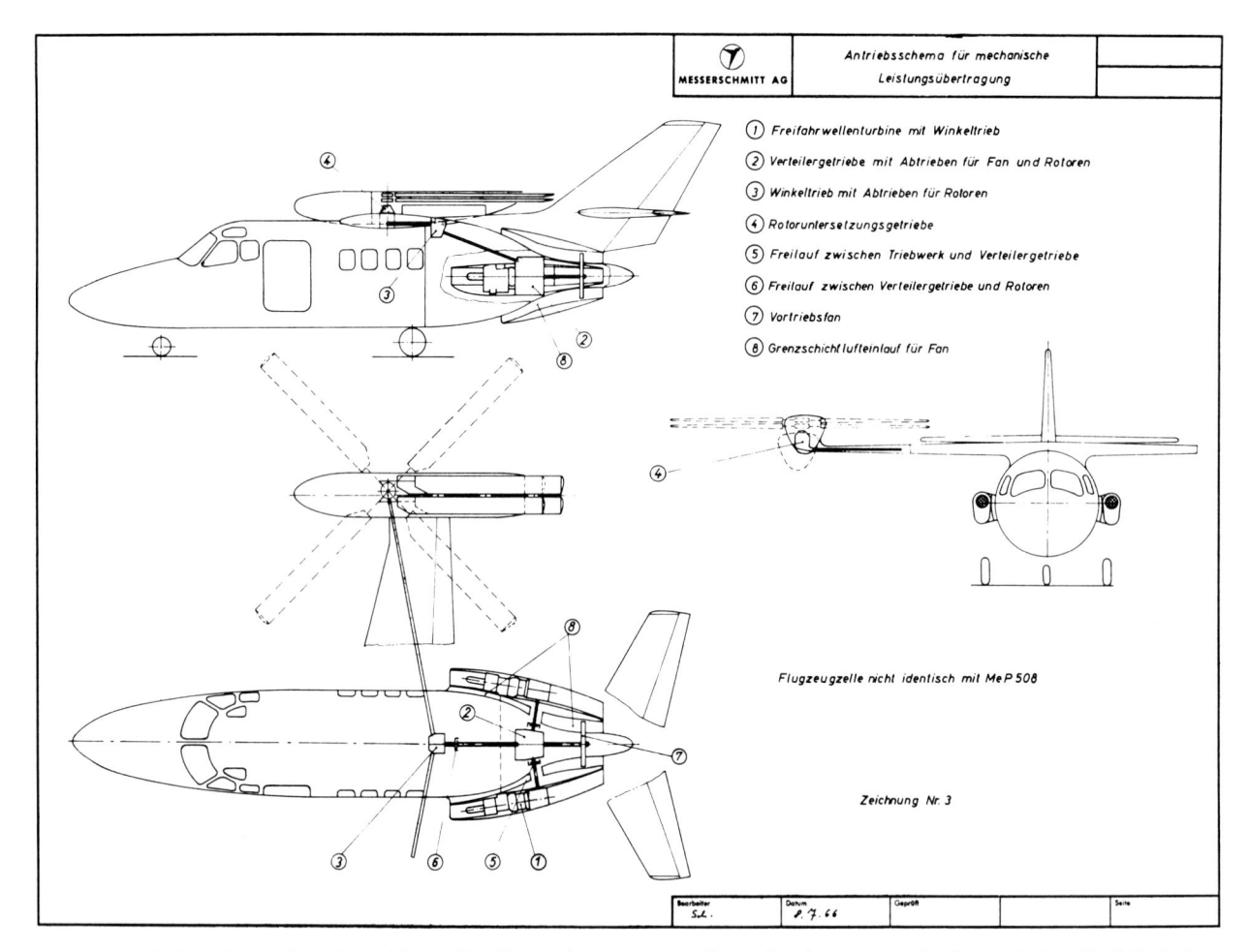

The Rotorjet in its early configuration with rear fan. The engine power was diverted to the rotors on landing and takeoff, while during forward flight they powered the turbofan propulsion system in the aft section of the fuselage. Up until the point where a safe transitional altitude was reached, i.e. at the beginning of transitioning from hover to horizontal flight, the rotors were fed all the power from the engines. Once at altitude, however, some of the power no longer required by the rotors diverted to the fan in the aft fuselage, thus causing forward momentum. As speed increased, power to the rotors was gradually reduced and taken over by the turbofan, until at the end of transition the aft fan was receiving the entire output from the turbines. The rotors were then brought to a standstill and retracted. Transiting from horizontal flight to landing occurred in the opposite order.

the rotor blades and wings together would be driven by a hot gas reaction process. A proposal from Boeing-Vertol called for a four-jet cargo plane with a takeoff weight of 36,600 kg and a single-blade engine on each of the wing tips which, during cruise flight, would be braked and form the outer wing sections (providing a gull-wing shape). Another Sikorsky project, with a takeoff weight of 20,400 kg, envisioned a central three-blade rotor above the fuselage with an anti-torque rotor on the vertical stabilizer. The main rotor would be stored in the upper fuselage, and the anti-torque rotor would retract into a teardrop shaped housing on the stabilizer.

Among the VTOL aircraft under development at the time, only the tilt rotor/tilt wing concept resulted in any kind of experimental aircraft, including the Hiller-Ryan XC-142A with a takeoff weight of 17,000 kg, the Canadair Cl-84 with a takeoff weight of 5,534 kg, the Bell X-22A with a takeoff weight of 6,800 kg, and the Curtiss Wright X-19A with a takeoff weight of 6,197 kg. There were additional projects, such as the VFW VC 400, the Bell D-266, and the Westland We 02. The Ryan XV-5-A, powered by a "liftfan," flew as a prototype, as did the Dornier Do 31 as the first jet-powered VTOl transport, as well as the Hawker Siddely Kestrel (later Harrier) fighter prototype fitted with thrust vectoring nozzles for its jet engine.

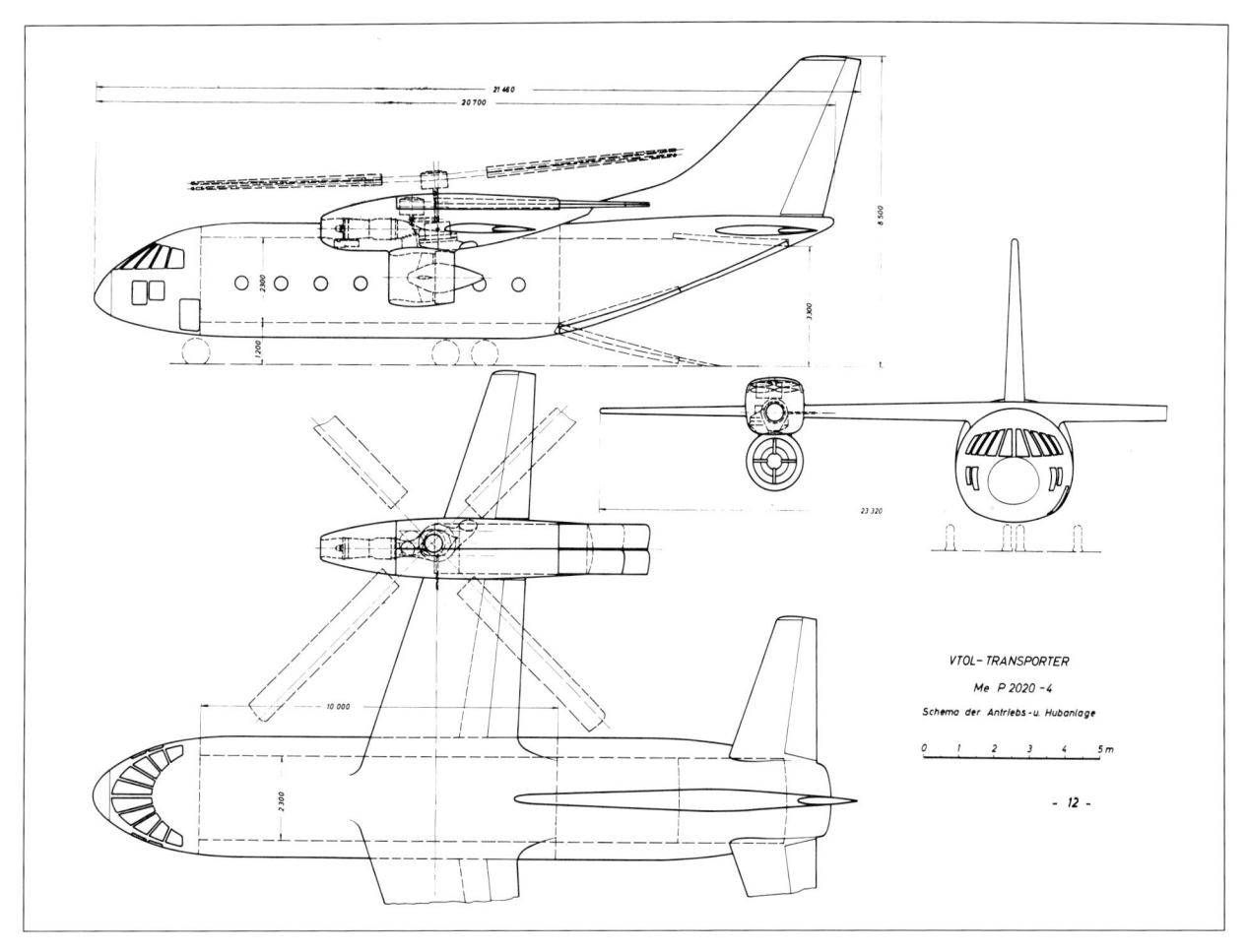

At the end of 1969, the company drew up the Me P 2040 aircraft project for a requirement "outlining a V/STOL transport airplane for military and civilian applications" issued jointly by Germany's Ministry of Defense and Deutsche Lufthansa. Using the same configuration, having the same VTOL takeoff weight and with a cruising speed of 840 km/h for both types, as a military transport the design was expected to be able to lift 10,000 kg vertically into the air and carry this load a distance of 800 km; the civilian version was to be able to carry 96 passengers, virtually the same capacity as the Boeing 737. Driving both the lift and the propulsion systems were four General Electric GE 1 gas generators located in twin gondolas under the wings. The generators were to run at a constant rate, their heated gases being diverted directly to the rotor turbines and turbofans.

Willy Messerschmitt and his studies department, in carefully researching the problems and comparing the different systems, came to the conclusion that a convertiplane design such as the Rotor-jet embodied many advantages in comparison with other VTOL systems without inheriting any of their flaws.

After working out the dimensions for the Me (P) 408 project and the test model, a rotor test stand was built at Augsburg in 1967. Parallel to this was the creation of a model rotor with a diameter of two meters.

Flight loading components were evaluated with a model blade in a wind tunnel of the Deutsche Versuchsanstalt für Luftfahrt (DVL) at Porz/Wahn. At EWR in Munich, vibration studies (natural frequencies and vibration behavior) were undertaken starting in 1968. Static tests using a spinning three-blade rotor began at Augsburg in the spring of 1968. The Messerschmitt test assembly center in Augsburg had built the rotor blades in cooperation with the Waggon- und Maschinenbau AG (WMD) in Donauwörth.

In June of 1968 new parts had to be produced due to a fractured bolt in the rotor hub, resulting in a simplified, non-spinning four-blade rotor hub and several dynamically similar blades of varying stiffness. The initial series of tests in the wind tunnel of the Eidgenössische Flugzeugwerke in Emmen proved the static and dynamic stability of the blades in the critical "frozen" (i.e. non-spinning) stage. In order to

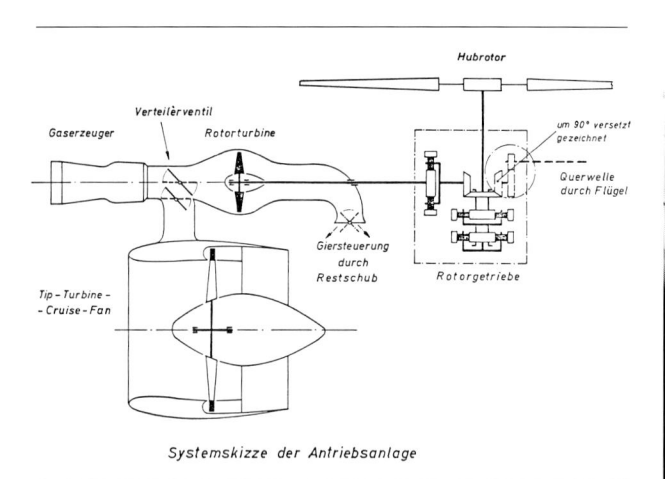

Rotorjet propulsion system as employed from 1968 onward: two Astafan IV single-shaft dual flow turbojet engines, with variable-pitch fan blades and a drive shaft for optional powering of the rotors. The two rotors were joined by a wind shaft in order to maintain symmetry in the event of a single engine failing during the transition phase or when cruising.

Static and dynamic stability of the Rotorjet's blades was measured in the wind tunnel at the Eidgenössische Flugzeugwerke in Emmen.

avoid a possible break due to oscillation flutter in the wind tunnel, the blades with the least amount of stiffness were first mounted on a car and tested on the runway at the Manching works. These tests revealed that the blades were not subjet to flutter at speeds even exceeding 110 km/h—the point at which the airflow completely separated. Given a speed scale of 1:3, this meant that it was safe to assume no flutter oscillation would set in at a relative wind velocity of 330 km/h.

This first series of wind tunnel tests using the four-blade rotor simulated flight conditions which would be experienced in the final transition phase, with the rotors locked and folding and the aircraft traveling at higher speeds. Flight loading was measured using pressure sensors set in 156 holes drilled in a rotor blade.

Further static tests resumed in late 1968/early 1969 once the three-bladed rotor had been repaired and improved. The second set of tests in the spring of 1969, simulating seven different airspeeds, focused primarily on the transition phase from hover flight to the beginning of horizontal/forward flight. By dropping the rotor's rpms in stages from 2,300/min to 0, it was possible to measure the rotor's stoppage at the estimated transition velocity of 250 km/h. The test's favorable results generally matched up with the data calculated beforehand.

The multiplicity of the work involved gives a clear indication of the complexity of the Rotor-jet project, a complexity which the studies department had identified from the program's outset. However, MBB-internal and other records from the time imply that there were yet many theoretical and experimental matters to be resolved—particularly those involving dynamic and aeroelastic questions.

For example, due to technical evaluation reasons, up to this point the tests involved evaluating rotor and model wings separately. In order to provide accurate data for the

Design work on the Rotorjet. In May of 1973 Willy Messerschmitt discusses details of the transmission gearing with Adolf Teufel in the studies department office in Arabellastraße.

Starttransition

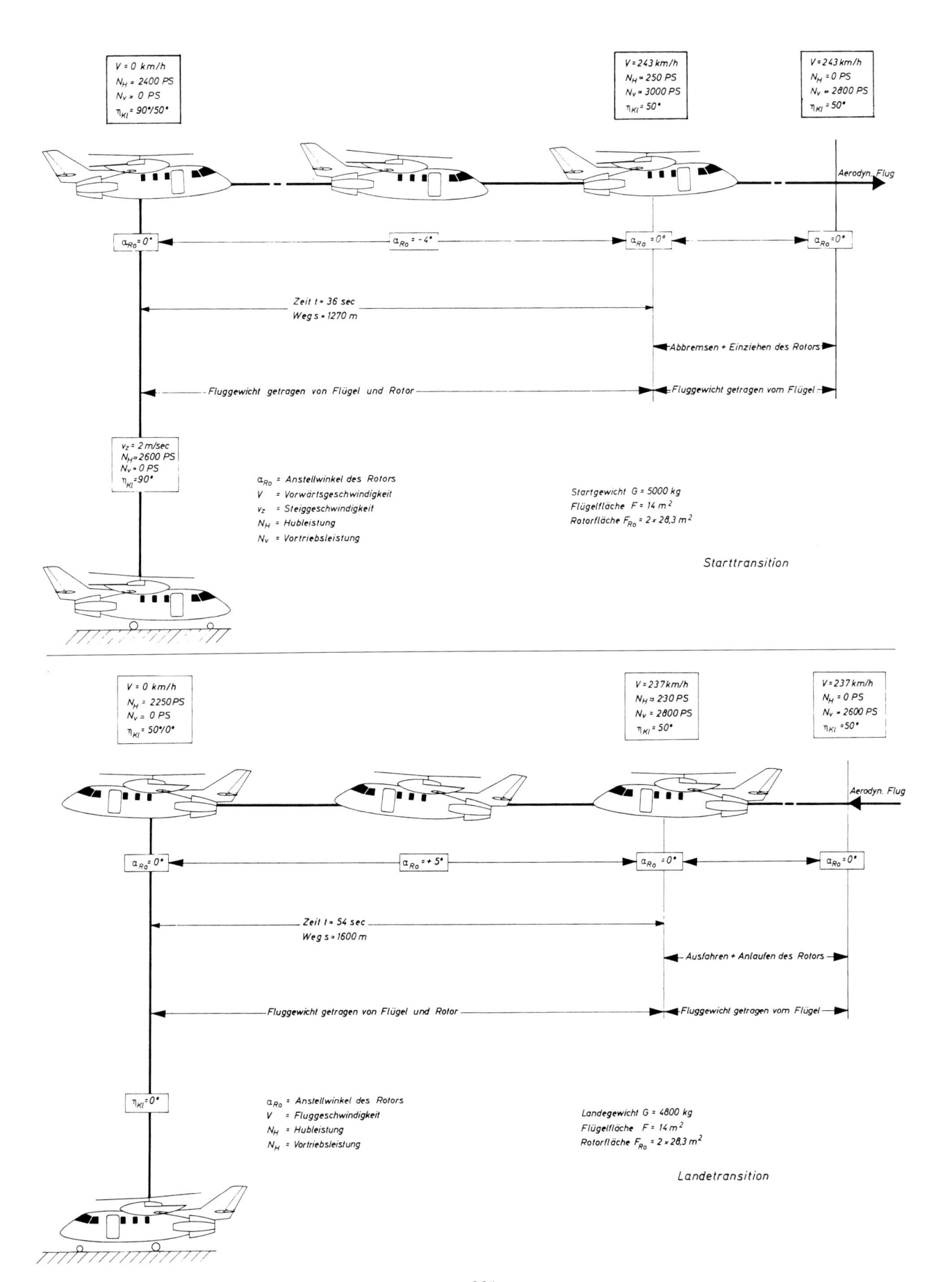

$V = 0$ km/h
$N_H = 2400$ PS
$N_V = 0$ PS
$\eta_{Kl} = 90°/50°$

$V = 243$ km/h
$N_H = 250$ PS
$N_V = 3000$ PS
$\eta_{Kl} = 50°$

$V = 243$ km/h
$N_H = 0$ PS
$N_V = 2800$ PS
$\eta_{Kl} = 50°$

Aerodyn. Flug

$\alpha_{Ro} = 0°$ $\alpha_{Ro} = -4°$ $\alpha_{Ro} = 0°$ $\alpha_{Ro} = 0°$

Zeit $t = 36$ sec
Weg $s = 1270$ m

Abbremsen + Einziehen des Rotors

Fluggewicht getragen von Flügel und Rotor — Fluggewicht getragen vom Flügel

$v_z = 2$ m/sec
$N_H = 2600$ PS
$N_V = 0$ PS
$\eta_{Kl} = 90°$

α_{Ro} = Anstellwinkel des Rotors
V = Vorwärtsgeschwindigkeit
v_z = Steiggeschwindigkeit
N_H = Hubleistung
N_V = Vortriebsleistung

Startgewicht $G = 5000$ kg
Flügelfläche $F = 14$ m^2
Rotorfläche $F_{Ro} = 2 \times 28{,}3$ m^2

Starttransition

Landetransition

$V = 0$ km/h
$N_H = 2250$ PS
$N_V = 0$ PS
$\eta_{Kl} = 50°/0°$

$V = 237$ km/h
$N_H = 230$ PS
$N_V = 2800$ PS
$\eta_{Kl} = 50°$

$V = 237$ km/h
$N_H = 0$ PS
$N_V = 2600$ PS
$\eta_{Kl} = 50°$

Aerodyn. Flug

$\alpha_{Ro} = 0°$ $\alpha_{Ro} = +5°$ $\alpha_{Ro} = 0°$ $\alpha_{Ro} = 0°$

Zeit $t = 54$ sec
Weg $s = 1600$ m

Ausfahren + Anlaufen des Rotors

Fluggewicht getragen von Flügel und Rotor — Fluggewicht getragen vom Flügel

$\eta_{Kl} = 0°$

α_{Ro} = Anstellwinkel des Rotors
V = Fluggeschwindigkeit
N_H = Hubleistung
N_H = Vortriebsleistung

Landegewicht $G = 4800$ kg
Flügelfläche $F = 14$ m^2
Rotorfläche $F_{Ro} = 2 \times 28{,}3$ m^2

Landetransition

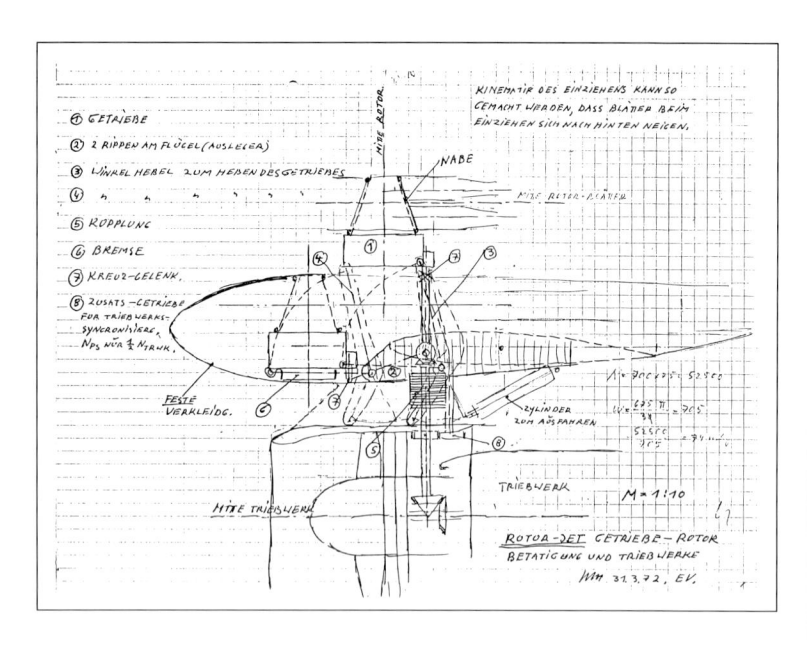

A sampling of Professor Messerschmitt's hand sketches pertaining to the Rotorjet. During the '70s, it was in this manner that Willy Messerschmitt presented virtually the entire Rotorjet concept, including the airframe, transmission, and rotor-folding system. These proposals were translated by his colleages into standard blueprints in the studies department. As in the '30s and '40s, Messerschmitt regularly dropped in on the design department and discussed problems with the technical engineers. He was impatient when it came to preparing blueprints, although he often had submitted his hand drawings only a few days prior. Inconsistencies with his drawings were always noticed immediately.

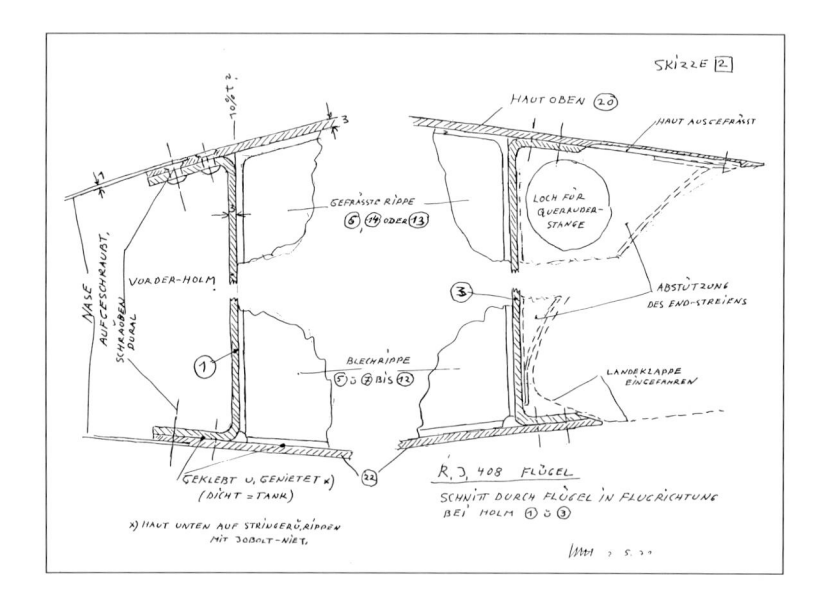

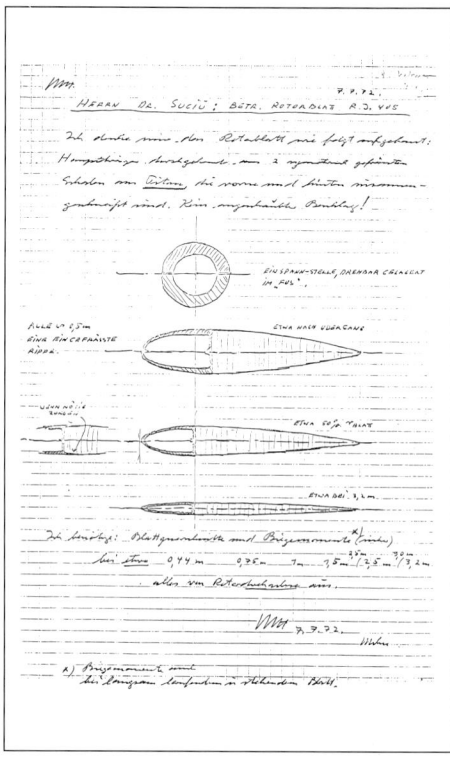

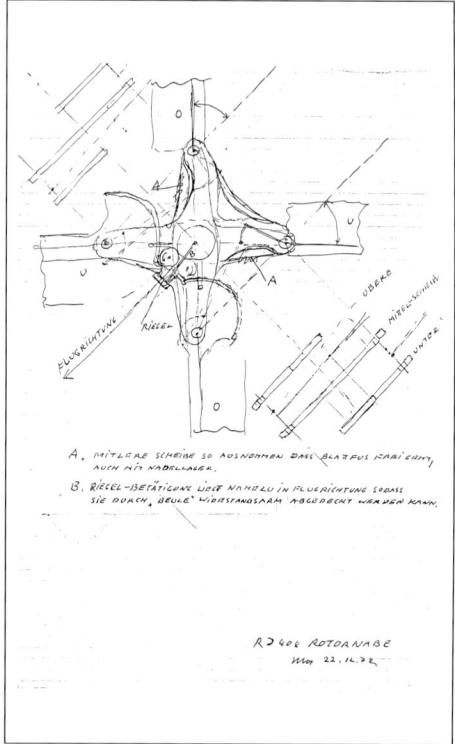

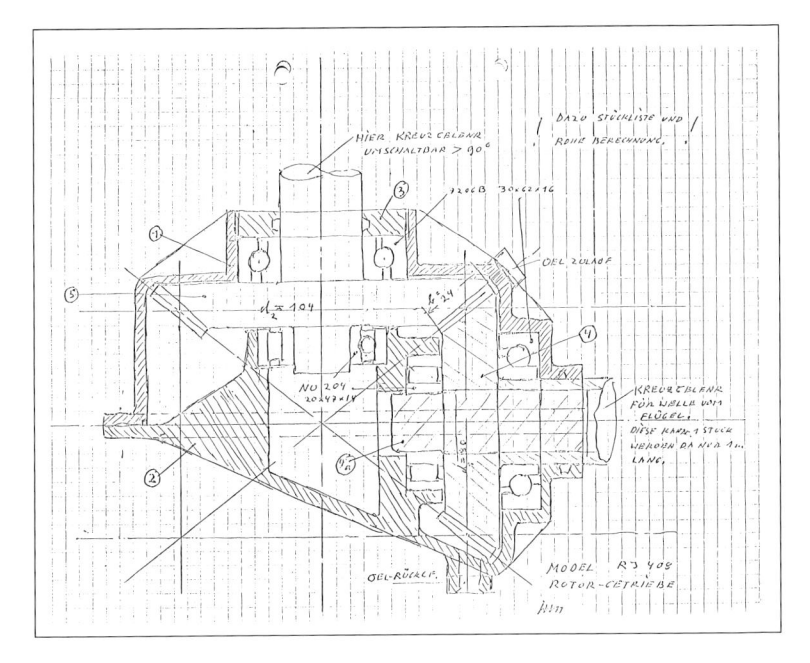

Representing German aviation pioneers, Willy Messerschmitt spoke at a celebration on 14 December 1973 in Washington in commemoration of the 70th anniversary of the Wright Brothers' first flight. Messerschmitt (the photo shows him during his speech, expressing gratitude and praise for the Wright Brothers' efforts) was presented an honorary certificate by the vice president of the Aero Club of Washington, Mr. Ford. In addition to Willy Messerschmitt, representatives from England, France, and Italy were also invited.

whole system, however, both rotor as well as wings needed to be dynamically true-to-life and both components would have to be elasto-mechanically linked to each other. Therefore, from spring 1971 onward the studies department concentrated its efforts on aero-elastic and constructive problems with the intent of building a model not only aerodymanically, but dynamically true-to-life, for subsequent use in static and wind tunnel evaluations. With new combination powerplants becoming available, the propulsion system would need to be revised as well.

During the next few years it was Dr. Titus Suciu who oversaw the project, but his team increasingly came into conflict with the management at MBB with regard to the immense costs being expended on the studies department, staffed with up to 25 persons. At a time when the company was able to turn little, if any, profit, even paychecks totaling in the millions within a single year played a critical role.

Despite this, both Ludwig Bölkow and the numerous experts were in agreement on at least one point: the technical originality and "feasability" of the Rotor-jet project was basically never in dispute. What was in doubt, however, was the marketability of the project in view of the fundamentally higher operating costs of VTOL aircraft and the question of whether a civilian or military requirement for such a convertiplane would be recognized in the near future. Aviation history has borne out the view of those who eyed the marketability of the Rotor-jet project with skepticism. With one exception (the Sikorsky X-72 "X-wing" from 1984), every other experimental model and even the single

Seen here in February 1968 with an American host and Titus Suciu (center), Willy Messerschmitt is brought up to date on the developmental work being undertaken by U.S. companies in the area of fixed and retractable rotors, as well as combination airplane-helicopter designs. In the co-pilot's seat, Messerschmitt experienced a flight in the Lockheed 286 fixed-rotor helicopter first hand.

prototype development of a transport convertiplane made use of tilt rotors or tilt wings. As of this writing, not one has found a market.

Looking back, it is obvious that the personnel and financial resources which Messerschmitt and his studies department had at their disposal fell far short of the mark compared with the resources available for competing designs in other countries, especially in the U.S. In the end, the year-long tug-of-war between Messerschmitt—firmly convinced of the correctness of his ideas—and a MBB business management under pressure from other things besides just matters technical ultimately came to nothing. For Willy Messerschmitt, the cancellation of a project to which he had dedicated his remaining creative years as an engineer was a bitter pill to swallow.

Projects

The following overview encompasses the postwar designs generated by Prof. Willy Messerschmitt's studies department. Not only do they reveal the extraordinary breadth of work and the plethora of new ideas, they also show how concepts proven in the past were applied in the post-1945 world. Messerschmitt did not hesitate in using both conceptual as well as complex layouts—ultimately trusting in his talent for creating products geared toward ease of manufacture.

The selection of data and facts sometimes appears contradictory, yet this is a result of the varying states of the designs' progress. Among the documents there may have only been a sketch or a page with rough calculations; on the other hand, a project may have been worked out down to the last detail. Brochures would have been printed, models displayed at aviation shows, and paintings made throughout the world. And some of these projects even made it to the prototype stage. Many of the designs began with design sketches made personally by Willy Messerschmitt. At that point Hans Hornung stepped in, working on the design until it reached the point of actual construction (assuming it made it this far)—then Messerschmitt once again became heavily involved. Thus, these projects are also a reflection of Hornung's creativity and his enormous influence.

Looking back, it can be said that the projects emanating from the studies department met the demands of the day fully, but because of the lack of business potential and poor national market conditions the projects more often than not never made it past the design stage. It was obvious that individual companies within the still fragmented German aviation and space industry were simply not able to bring financial power to bear against the leading competition in the world market, which at the time primarily came from the Americans. A welcome exception to this rule didn't come about until the advent of the VTOL development, sponsored by a new defense ministry. This program eventually led to partnerships within the German industry and established the foundations for joint venture programs which are the mainstay of the European air and space industry today.

The numbering system for the projects does not follow a standard pattern. The list begins with the "-1xxx" series, which are a natural continuation of the project numbers set at the end of the war, then concludes with the "1xx" series. For purposes of easy searching the designations have taken precedence over chronological order in drawing up the list.

Those aircraft concepts which actually made it to the test and prototype stage and/or even entered production include the HA 100 trainer, the HA 200 jet trainer (produced in Spain and Egypt), the HA 300P experimental glider, and the HA 300 fighter.

Among these designs, the P 1211 interceptor (for the 1957 fighter competition) proved to be a most visionary concept, having a canard layout and thrust-reversing engine. The Swedish Saab Viggen fighter-bomber (first flight 1967) was developed along the same lines as the P 1211 from 1958 to 1961 (probably with no knowledge of the Messerschmitt project). One can speculate that, with an unrelenting, aggressive defense-oriented approach by the Bonn government, a similar level as that achieved by the industrious Swedes might have been attained within the space of ten years.

It seems odd that Messerschmitt failed to successfully develop a single civil aircraft project in the postwar period. He was hesitant to become involved with any project which might endanger his company, a reluctance stemming mainly from the overindebtedness Messerschmitt's company found itself in as a result of the war. Even the Me 308 project, a small business jet design approved by the company's board of directors for prototype development, was not supported by Messerschmitt. At the time, Bonn criticized Willy Messerschmitt's conservative business approach. The fate which later befell the companies of HFB and VFW—resulting from their independent, labor-intensive civil projects—revealed the extraordinary risks such undertakings caused in the economic playing field of the Federal Republic of Germany.

	Brief Description	**Year of Development**		**Brief Description**	**Year of Development**
RJP 1005	V/STOL ground attack with two Turboméca Astafan W, each with 1325 kW (1800 hp), based on Rotor-jet concept	1973	P1200	long-range airliner with four piston engines, based on Me 264 (oval fuselage, tapered wings, twin rudders, nose gear); so-called "modular design", concept later refined several times; first mentioned in "Spanish Memorandum"; 50 to 92	1951
P1118	four-seat touring plane based on Bf 108(qv)				

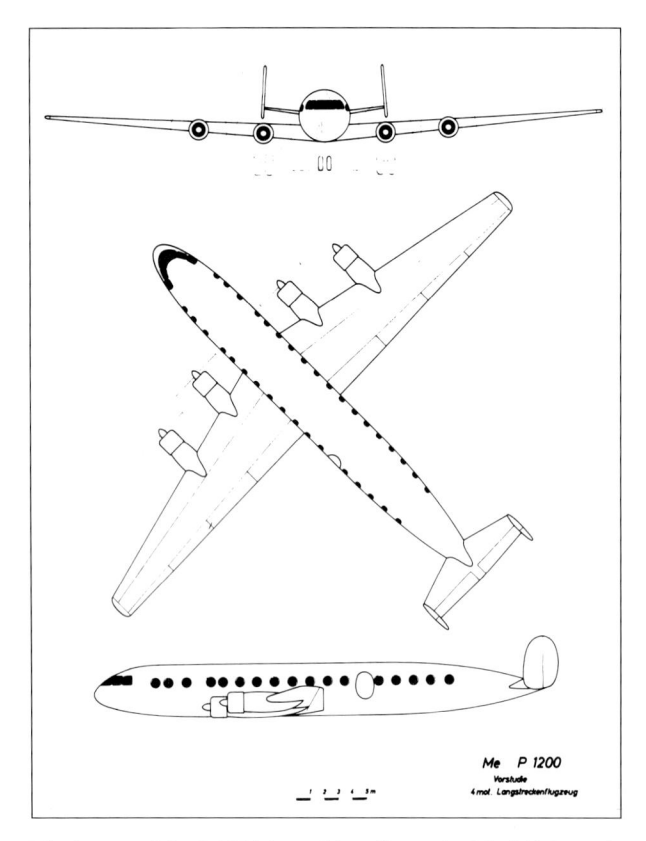

The layout of the P 1200 leaned heavily on the Me 264. Length was 32.1 m, wingspan 38.4 m, and area 155 m².

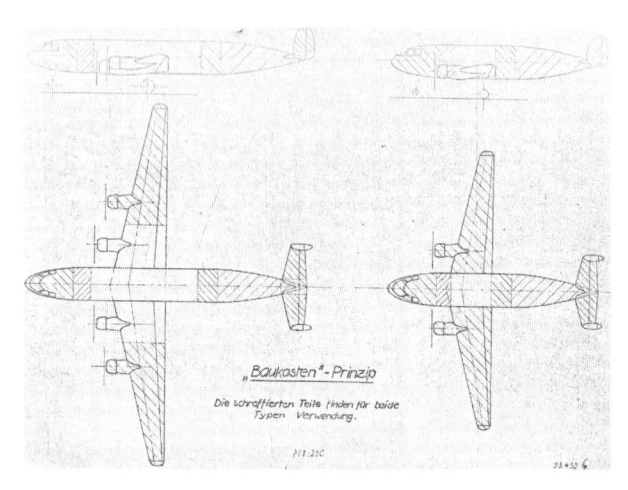

Drawings emphasizing the "component assembly principle" with wing and fuselage variations applied to an example of the four-engined (P 1200) and twin-engined (P 1201) passenger plane. Shadowed areas are common to both types.

	Brief Description	**Year of Development**
	passengers; P1201 derived using identical components (nose section with cockpit, tail with empennage, outer wing sections with engines)	
P1201	medium/short-haul airliner with two piston engines and seating for 30 passengers	1952
P1202	heavy lift transport with six piston engines, each outputting 1320 to 1550kW (1800 to 2110 hp); carried 20000kg of cargo or up to 220 passengers; capacity was 300 m2; takeoff weight 60000kg; design drew on Me 323 (cantilever high-wing, intermediate deck, nose gear with box housing, cargo hold dimensioned for standard gauge); capable of operating from remote airstrips (landing run 350 m); range 2600km	1953
P1203	basic transport with four piston engines, each having 590kW (800 hp) power; 5000kg capacity with a 20000kg takeoff weight; steel tube fabric covered fuselage similar in layout to Me 323 (but half size); 1600km range	1953
P1205	airliner with 2-4 turboprop engines, 12 passengers, 4800km range; pressurized cabin; aspect ratio of approx. 10 to balance range and cruising requirements with engine output; jet-powered variant also studied; planned joint venture with US-based Lear, which would have taken over development costs and fitting of components in the USA; achieving cruising speed of 800km/h meant a high propeller efficiency and keeping landing speed to 160km/h; boundary layer blower system required; no propeller capable of delivering such speeds available.	1953/1955
P1206	light supersonic trainer and light fighter; three variants with same basic airframe: two-seat trainer capable of supersonic speed in level flight without afterburner, two-seat trainer with Mach 1.2 capability with afterburner, supersonic fighter	1956

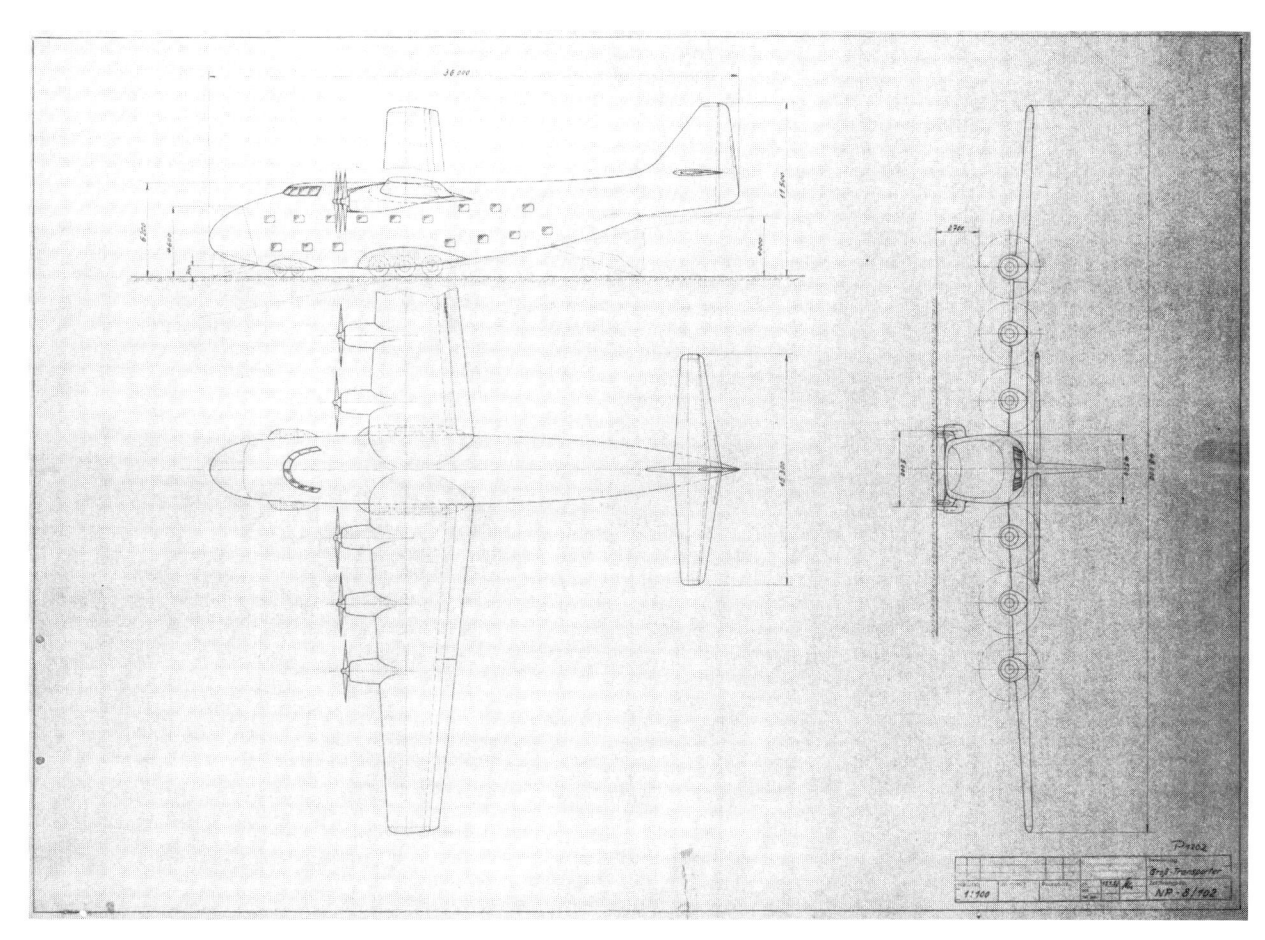

P 1202 heavy lift transport project with a capacity for 20,000 kg, based on the Me 323. Length was 36 m, wingspan 48 m, and area 255 m². Cruising speed was 331 km/h at 3,000 m.

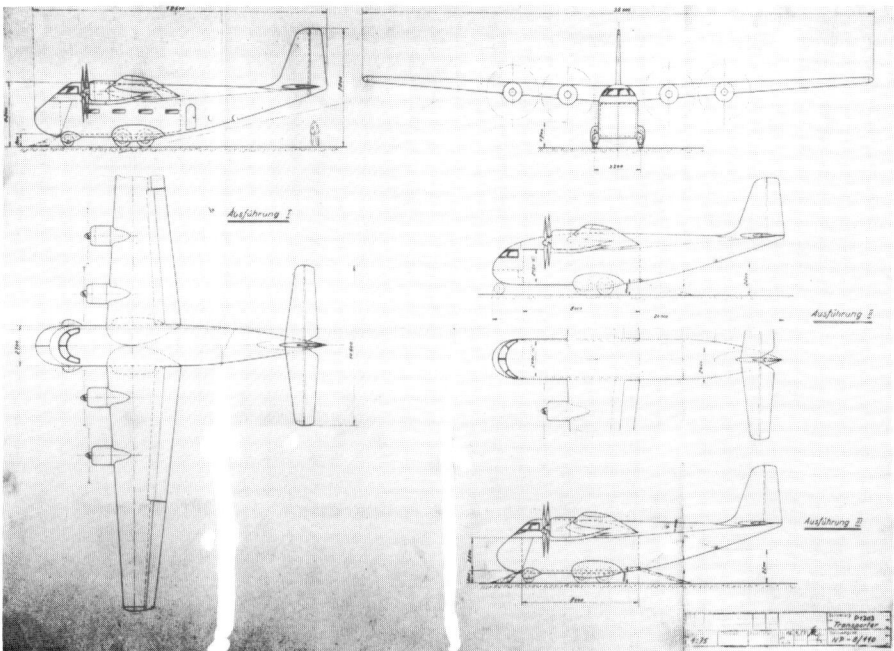

P 1203 basic transport project. Length was 20 m, wingspan 35 m, and area 125 m². Cruising speed was 305 km/h at 3,000 m.

	Brief Description	Year of Development
	(Mach 1.65) with 1000km range; area rule design, separating nose section in place of ejection seat, tapered wings.	
P 1207	Mach fighter as P 1206, but with two BMW 8004 turbojets; two pilots seated side-by-side	1956
P 1208 Tr	basic trainer with single BMW 8004 turbojet; wings and landing gear as well as empennage were to have been inherited with minor changes from Me 200	1956
P 1208 R	6-7 seat commuter plane with both military and civilian applications, powered by single BMW 8004 turbojet; same airframe as basic trainer with minor fuselage modifications	1956
P 1210	high-altitude interceptor with two afterburning turbojet engines and rocket propulsion; max. endurance 30 minutes; altitude up to 28,000 m	1956
P "BMW"	fighter with three turbojet engines; tapered wings with anhedral; "area rule" fuselage	1956
P 1211	Mach 2+ fighter/interceptor based on competition requirements issued by Defense Ministry on 11/15/56; canard layout, i.e. control surfaces forward of wing; negative sweep intakes in forward fuselage; mid-wing design with minimal sweep located far aft; boundary layer blower system; single afterburning turbojet, two liquid-fuel rockets on either side of fuselage; in addition to undercarriage designed for standard takeoff profiles, studies were also made of VTOL designs.	1957
P 1212	commuter/business/courier plane for up to ten persons including crew; two turbojet engines.	1957
P 1213	fighter-interceptor with single turbojet and two rocket engines; Mach 2.8 max. speed; altitude 34000 m; VTOL version with thrust vectoring	1957
P 1214	canard interceptor with tapered wings; two turbojets with thrust vectoring; two vector rockets located at fore end of wing tip tanks; twin V tail with high-set horizontal stabilizers.	1957

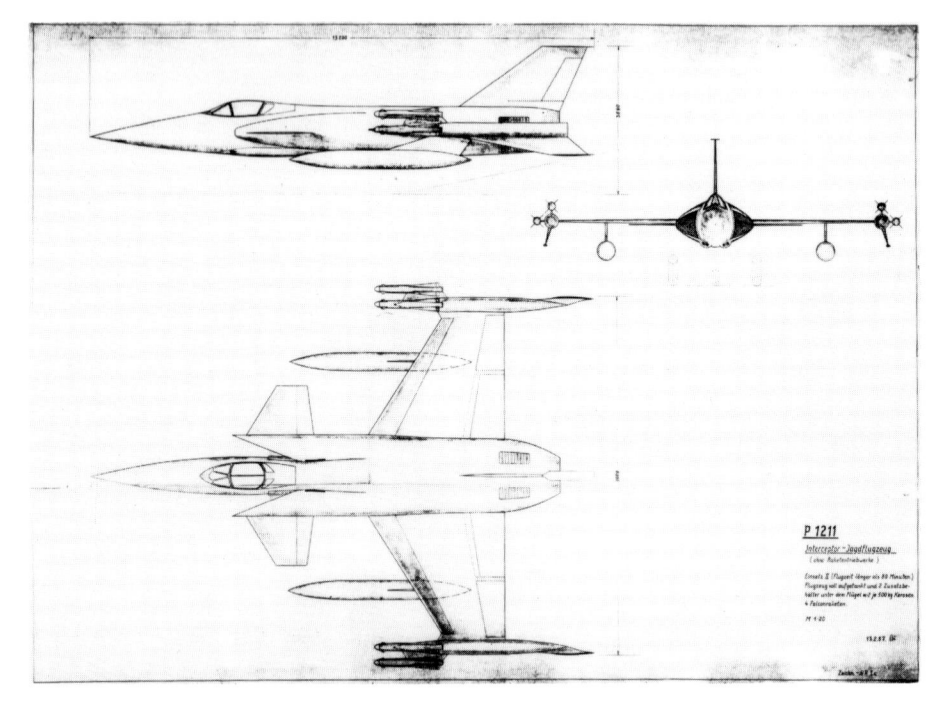

P 1211 interceptor project variant without rocket engine. Under the wings are two auxiliary fuel tanks. Armament was to have been four Falcon air-to-air missiles carried on the wingtips. Takeoff weight was between 6,800 kg and 7,800 kg. Maximum speed was Mach 2.8, service ceiling 34,000 m.

	Brief Description	**Year of Development**
P1215	commuter plane with 4-6 seats; two turbojet engines. F variant based on Fouga Magister; M version based on Me 200; P 1215 m later redesignated P 1216.	1957
P 1216	commuter plane with 4-6 seats; two turbojet engines. Goal: high-speed commuter with range encompassing European destinations, filling gap between twin-engine propeller commuters and twin-engine commercial airliners and approaching performance of twin-jet commercial airliners; in accordance with mission, as many components as possible were to have been utilized from Me 200 (Ha 200); variant with eight seats and extremly short takeoff and landing runs thanks to blowing air over outer wing flaps (see also Me 230)	1957
Me XI-21	VTOL interceptor; layed out as tailsitter with folding takeoff/landing frame	1957
P 1217	VTOL heavy lift cargo plane with six turboprop engines and rotating wings; 12000kg capacity; four engines in wings, two on nose mounted control surfaces	1957
Me XB1-0001	variant of Me X1-21 for conventional takeoff and landing	1957
P 1218	experimental VTOL aircraft to explore most cost-effective method of building a VTOL combat plane for Entwicklungsring Süd; dimensions as for planned strike fighter but with 1/4 of takeoff weight (1700kg); numerous variants, e.g. one with two afterburning turbojets and two vectoring rocket motors on forward section of wing tip tanks, one with rotating dual engines on fuselage sidewalls, one with thrust vectoring nozzles, one with a downward directed propeller in rear fuselage for control in hover flight; ground	1957/1958

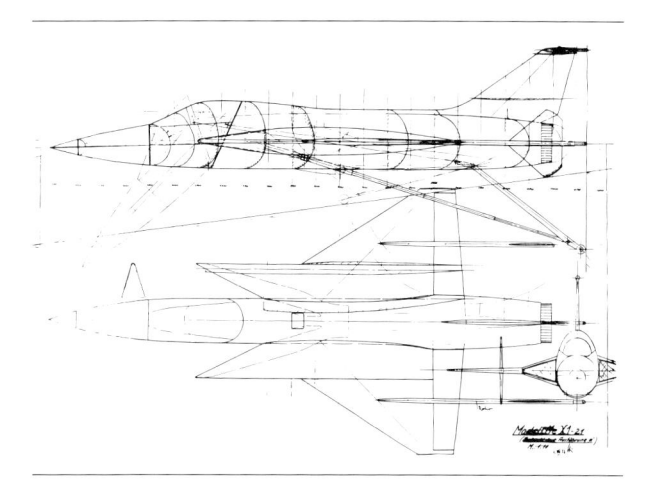

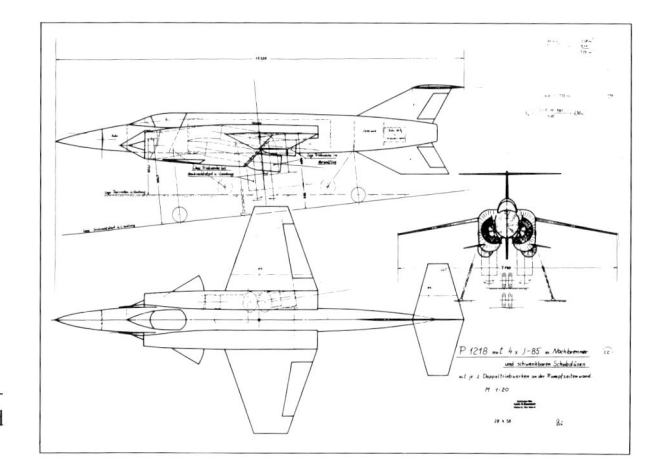

P 1218 with two twin engines on the fuselage sidewalls and vector thrust nozzles. Weight was around 2,800 kg, maximum speed Mach 1.7 at 11,000 meters.

The Me X1-21 was conceived as a taildragger design for vertical takeoffs and landings with a retractable launch and recovery strut.

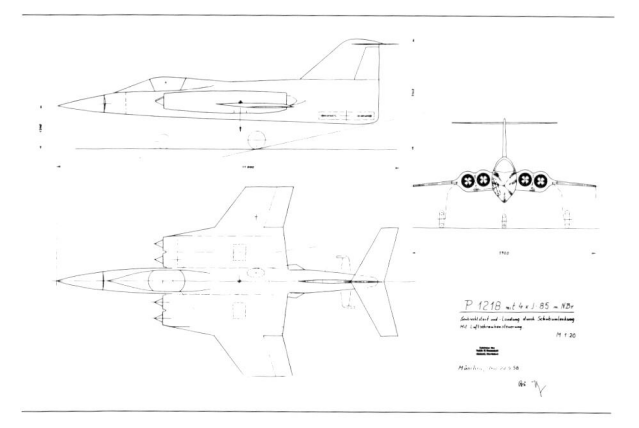

P 1218 variant with control supplemented by vertically arranged airscrews in the aft fuselage.

	Brief Description	Year of Development		Brief Description	Year of Development
	attack version as replacement for Fiat G.91: two thrust engines, one lift engine; takeoff weight between 2750 and 4200kg; Mach 1.7 at 11000 m altitude		P 1222	5xJ-85 engines; no further information known	1958
			P 1223	STOL commuter and courier plane with two 715 kW(975hp) turboprop engines; STOL effect achieved by adjusting wing and engine angle-of-attack in relation to the horizontal while simultaneously setting Fowler flaps to maximum; increased moments offset by a variable control propeller above vertical stabilizer or by blowing hot air out the back or over elevators	1958
P1219	front-line transport with four turboprop engines and 10000kg capacity, designed to specifications laid out by Luftwaffe Operations Staff; design based on experience with Me 323 and Me 264; high-wing design with relatively low wing loading and high aspect ratio (10); dual split flaps along entire length of wing; additional lift systems unnecessary, but examined; all-through fuselage: hinged nose doors, folding load ramp at end of cargo hold, which terminates in upward folding doors; dimensioned for standard gauge rails; crew located above nose doors; length 26.4 m, width 42.5 m, wing area 178 m2, cruising speed 400km/h near sea level	1957			
			P 1224	STOL airliner with four 715 kW(975hp) turboprop engines and variable sweep wings; 14000kg takeoff weight, 2850kg capacity (30 passengers); design as for P 1223	1958
			P 1225	VTOL supersonic ground attack plane; designed to same requirements as Fiat G.91 to include load and combat systems; four non-afterburning RB 153 turbojets, two for thrust, two for lift; 5340kg takeoff weight, 610kg load, max. speed Mach 1.8 at 11000 m, range 1200km	1959
P 1220	medium range cargo plane; civilian variant of P 1219; passenger/cargo version has hold split by additional deck to accommodate cargo in lower area and passengers in upper; climate control system planned; four turboprop engines; up to 12000kg capacity	1958	XB1-0001 P 1226	delta-wing turbojet powered fighter VTOL ground attack plane operating in the high subsonic region, up to Mach 0.9; also to same combat load requirements as Fiat G.91; as with P 1225, VTOL control via thrust vectoring; Bristol BE 53 lift/thrust (vector jets) engines; 5960kg takeoff weight, 600kg load capacity	1959 1959
P 1221	STOL transport with two turboprop and two dual cycle turbojet engines; background: Ministry of the Economy encouraged filling future needs of commercial airlines, charter companies and other parties with German aircraft: "In the opinion of the airlines there is still no true replacement for the reliable 33-seat DC-3"..."Short takeoff transports and airliners would be worthwhile developments in the near future"; use of blown flaps, i.e. blowing compressor air from second cycle of the two supplemental double flow turbojets over wing trailing edge; two turboprops, each rated at 1100kW(1500 hp) cruise; 2850kg capacity	1958	P 1227	VTOL fighter-bomber with four afterburning turbojets and thrust vectoring, plus two add'l non-afterburning turbojets providing lift (RB 153); basic design as for VJ 101; two variants; 8840kg takeoff weight, 230kg load (variant 1)	1959
			P 1228	VTOL fighter-bomber; four turbojet thrust vectoring engines in two twin wing nacelles, two lift engines in fuselage; variants 1-5 with various fuel loads and wing designs; 11350kg takeoff weight; 910kg payload (variant 5)	1959

	Brief Description	Year of Development		Brief Description	Year of Development
P 1229	VTOL fighter-bomber; four turbojet thrust vectoring engines in two twin wing nacelles, two lift engines in fuselage	1959	Me P 2040	V/STOL short-range jet; designed according to request for tender from Defense Ministry and Deutsche Lufthansa for "general requirements for military and civilian applications"; designed to carry same payload as Boeing 737 (9500kg); see "Rotor-jet" chapter	
P 1286	all-weather fighter with 2x twin afterburning J-85 turbojets in pivoting nacelles on wing tips, plus J-85 non-afterburning lift engine behind cockpit	1958	Me P 2050	military V/STOL transport based on Rotor-jet principle	1970
Me P 2010	V/STOL jet based on Rotor-jet principle		HA-56	four-seat commuter plane with two turbojets each having 6.86 kN(700kp) static thrust; major use of HA 200 components; 3800kg takeoff weight; 620km/h cruise speed, range 2800km	1960
Me P 2020-4	V/STOL transport with a payload of up to 6000kg or 55 passengers, based on Rotor-jet principle; 1200km range	1968	HA-57	single-seat fighter-bomber, derived from HA 200; max. takeoff weight (with wingtip tanks) 3400kg, 690km/h cruise speed; range 1600km (at 9000 m altitude)	1960
			Me 100	trainer based on HA 100, fitted with piston engine delivering between 330kW(450 hp) and 550kW(750 hp), or with two Turboméca Marboré II; propeller engine on HA 100 trainer can be modified to jet trainer by replacing engine assembly with turbojet engines; jet engines located in forward nose section together with nose gear and weaponry, replacing the piston nose sectionentire airframe including undercarriage acquired from HA 100 without major changes; takeoff weight varied between 2072 and 2655 kg	1953/1955

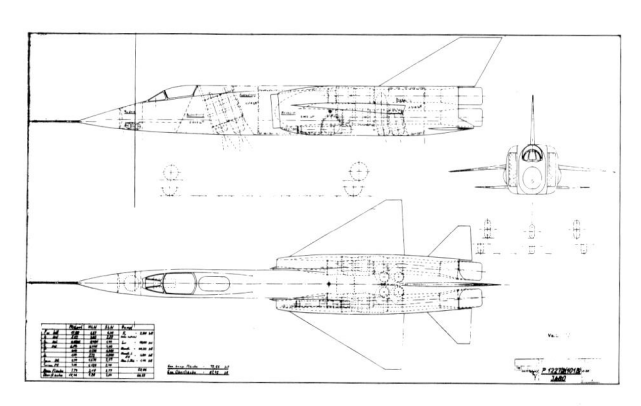

Variant 1 of the P 1227 VTOL project (101B). Wing area was 17 m². Powered by a total of six RB 153 turbojet engines.

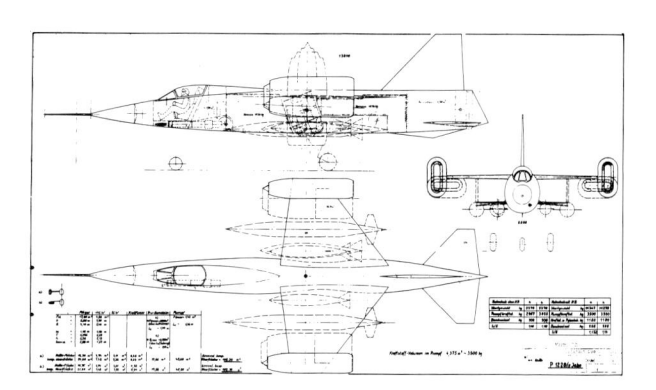

			SP 100	VTOL ground attack plane with GE X 353-7B/C plus two non-afterburning J-85 engines; delta configuration; 5360kg takeoff weight	1959
			SP101	VTOL turbojet ground attack plane with non-afterburning Rolls-Royce RB 153 engine providing lift; 4470kg takeoff weight; SP 101-1 variant with 4300kg takeoff weight; delta wing configuration	1959

P 1228 VTOL fighter-bomber project, variant 5. Length was 15.5 m, wingspan 6.6 m, area 17 m². Thrust to weight ratio without afterburner was 1.1 (at a takeoff weight of 9,370 kg), and 1.13 with afterburner (at a weight of 11,350 kg).

	Brief Description	**Year of Development**		**Brief Description**	**Year of Development**
Me 108D	"Super Taifun"; follow-on VTOL development of original Me 108 based on recommendation by Gero Madelung; one RB 108 engine rated at 12.25 kN (1250kp) to provide lift and one 110kW(150 hp) to 184kW(250 hp) piston engine; banking control achieved by blowing air over ailerons; elevator and rudder control using standard control surfaces in propeller wash	1958	SP 120-4	twin-jet commuter plane with swivel exhausts, for 8-12 persons (including crew); variant SP 120-5	1960/1961
			SP 120-8	airliner with four turbojet engines with swivel exhausts for 32 persons (including crew); variants include: SP 120-9 without swivel exhausts for 34 persons, SP 120-9-1 with three turbojets (lacking swivel exhausts), SP 120-9-2 with two turbojets and no swivel exhausts	1961
Me 108F	new Taifun project; attempt to bring German sportplane construction up to par with strong American and French competition; "new" design by Reinhold Ficht, with improvements to old Me 108 designed by Willy Messerschmitt himself; founding of the "Taifun GmbH" company by businessmen Mano Ziegler, Willy Messerschmitt, Dr. Hans-Heinrich von Srbik, Reinhold Ficht; Ficht planned production to take place at Weiden plant; all businessmen backed off from idea within the year, and project was abandoned	1975	SP 121	commuter plane for 8 persons (including crew) with two Turboméca Astazou driving single propeller; wings, control surfaces, and undercarriage taken from HA 200; SP 121-1 variant	1961
			SP 122	business commuter for 7 persons (including crew) with LFS 226 "propulsion fan engine" in aft section derived from two General Eletric GE T-58 turbojet engines; pressurized cabin; Mach 0.8 cruise speed; 2500km range; takeoff and landing roll just 400 m; variants SP 122-1/2/3; SP 122-4 single turbojet; SP 122-5/6/7/8 for 6-8 persons, two turbojet engines	1961/1962
Me 108W	four-seat piston-powered commuter plane based on "old" Me 108; developmental cost calculation for version with nose gear and new cockpit using many components of HA 100	1952			
SP 110	VTOL turbojet ground attack plane with two afterburning Rolls-Royce RB 153 engines providing lift; delta configuration; takeoff weights between 7200 and 8100kg (variants 1/2/3/4)	1959	SP 123	also Me P 123, later Me 308 Jet Taifun; trainer and commuter plane for 4-6 persons (including crew) and powered by 1-2 engines of various types; numerous variants (SP 123-1 thru -9) with blown flaps, fuselage boundary layer suction, wings of varying sweepback angles, straight trailing edge, forward swept (see Me 308)	1962/1963
Me 120	trainer with two turbojets, each outputting 3.9 kN(400kp) static thrust; tandem cockpit; 2500kg takeoff weight; cruise speed 530km/h; range 1250km (at 6000 m); predecessor of Me 200	1953	SP 124	also Me 124; V/STOL multi-role aircraft for civilian and military applications; two Turboméca Aubisque, each rated at 6.86 kN(700kp) static thrust as well as blown flaps and elevators (WM-1 variant); 2900kg takeoff weight (military trainer version), 4300kg (light bomber); variant WM-2 powered by two General Electric GE CJ 610-2B outputting 10.2 kN(1045 kp); Me P 124-5: two Turboméca Marboré II rated at 3.9	1962/1963
SP 120	twin-jet commuter aircraft with seating for 6-8 persons (including crew); variants 1/2/3	1960			

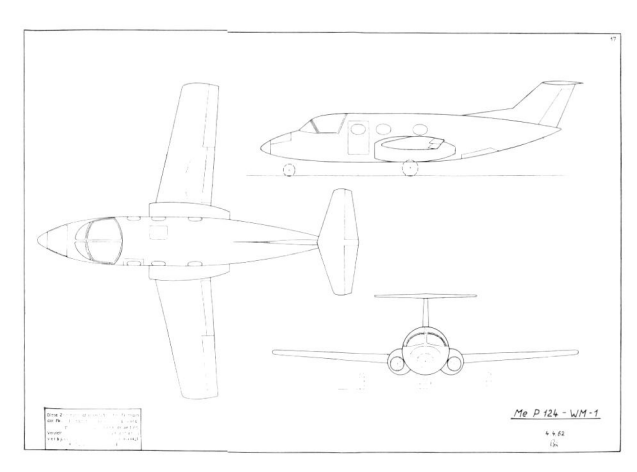

Me P 124-WM-1 multi-purpose STOL aircraft envisioned as a commuter transport. Length was 11 m, wingspan 10.4 m, wing area 17.4 m². Air was blown over outer wing flaps and elevators, giving it a takeoff run of less than 300 m and a landing roll under 200 m.

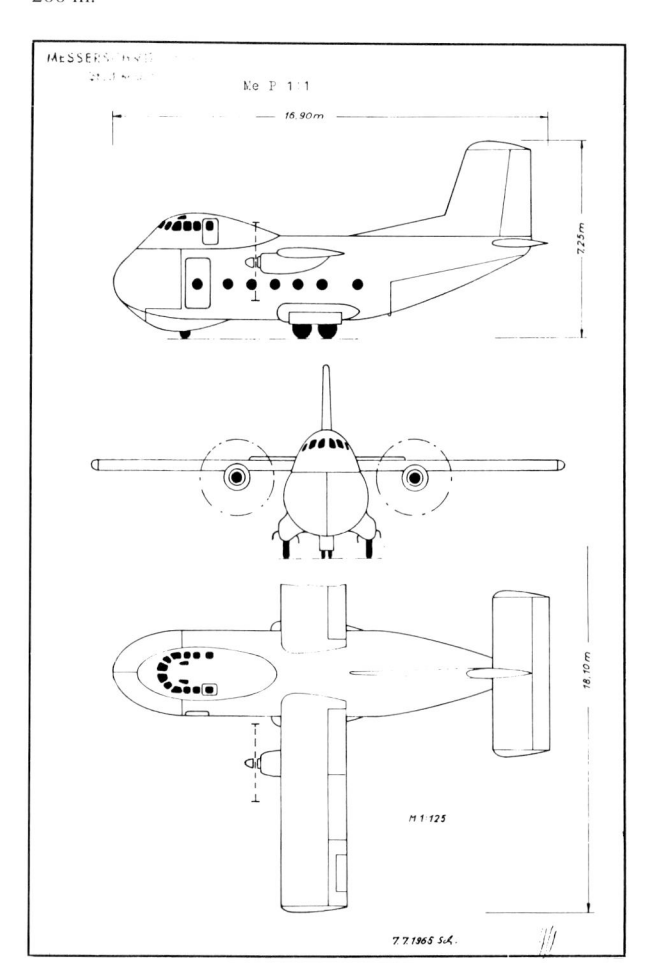

	Brief Description	Year of Development
	kN(400kp) each or Daimler-Benz DB TL 6 at 3.4kN(350kp) each (see also Me 208 Jet Taifun); takeoff weight 2210kg; Me 124-8: two Turboméca Aubisque, w/o blown flaps/elevators; takeoff weight 3260kg	
SP 130	VTOL interceptor/fighter-bomber/ground attack plane; one afterburning turbojet for thrust, one non-afterburning turbojet for lift	1961
SP 132	VTOL interceptor/fighter-bomber/ground attack plane; one afterburning turbojet for thrust, one non-afterburning turbojet for lift	1961
SP 140	transport with 2-3 Rolls-Royce RB 154 or 1-2 RB 168-1	1961
SP 141	basic transport with two turboprop engines; 6000kg takeoff weight; 3 m version with add'l aft turboprop engine driving pusher propeller	1961
Me P 141	basic transport in 3.5t payload class; based on construction/operational experience with Me 323; particularly suited for countries with limited road and rail systems; braced high-wing design with two turboprop engines, each rated at 735 kW(1000 hp), e.g. Turboméca Bastan, General Electric T-58, Daimler-Benz PTL 6; front and rear doors can be raised for simultaneous on- and off-loading, fixed gear, same profile for both wings and control surfaces; 7800kg takeoff weight; project exhibited at 1963 Paris Aérosalon; changes included pressurized fuselage, more powerful engines (Rolls-Royce Dart MK 503E, each rated at 882 kW (1200kW)), increasing takeoff weight to 9200kg; sales efforts focused on Portugal; planning of joint project with Portuguese gov't:	1962/1965

Left: Later version of the Me P 141 basic transport (7 July 1965) with pressurized fuselage and lacking struts. Wingspan 18.1 m, length 16.9 m, wing area 44 m². Takeoff to a height of 15 meters consumed a length of 590 m, while landing from an altitude of 15 m required 600 m. Cruising speed was 450 km/h.

	Brief Description	Year of Development
	development of 3 prototypes, followed by production of 100 aircraft at a rate of 4 per month; assembly planned for Beja; despite approval of Portuguese, project failed due to lack of funding; variants included SP 141 k with two Turboméca Astazou, SP 141-2 with two Turboméca Astazou and 5650kg takeoff weight, SP 141-3 m with three Turboméca Astazou (two tractor, one pusher prop); 7500kg takeoff weight	
Me P 142	basic transport, 4.5 t class, follow-on development of Me P 141 with two Lycoming PLF 1B-2 turbojet engines, each outputting 23kN(2350kp) static thrust; 11850kg takeoff weight; 4500kg capacity of cargo or 37 passengers	1965
Me P 143	transport with four double flow turbojets and blown flaps	1965
SP 150-1	trainer and commuter plane for up to eight persons (including crew); single turbojet engine	1961
Me P 151	commuter plane for 4-6 persons (including crew); single Turboméca Astazou X 470kW(640 hp) turboprop; takeoff weight 2700kg; cabin dimensions same as Me 308	1964
Me P 155	combat transport with 1.3t capacity (12 fully equipped soldiers), powered by two ZT 53 Lycoming KHD turbojets of 9.5 kN(970kp) thrust each, pressurized fuselage with oval cross-section; loading ramp beneath rear fuselage section; STOL performance through blowing mixed stream from the two double flow engines over trailing edge of landing flaps and ailerons; undercarriage fitted with low-pressure tires; takeoff weight 5310kg	1965
Me SP 160/ P160	short-range commercial airliner (replacement for Vickers Viscount 810 and Convair 440) of varying configurations (Me P 160-1 through -18) for 40-66 passengers; powered by 4-6 turbojets; example	1962/1964

Short-range commuter project, the Me P 160. Tail-mounted engine with intake for the inner chamber located at the base of the vertical stabilizer, while the outer chamber drew its air from the intake ringing the aft fuselage. Wingspan was 18.3 m, length 20.5 m, area 47.6 m^2.

	Brief Description	Year of Development
	shown at the 1963 Paris Aérosalon revealed two General Electric CF 700 mounted on either side of aft fuselage, a CF 700 in aft end with intake for inner chamber ahead of vertical stabilizer (ring intake for outer chamber); large supplemental cargo capacity; designed for ranges to 1000km with a capacity of 5500kg (58 passengers, 18600kg takeoff weight); freigher version with 6.5t capacity and 1000km range	
SP 161	VTOL/STOL transport with two thrust vectoring Bristol Siddeley Pegasus 5 engines, each with output of 76.5 kN(7800kp) thrust; 5000kg payload, 21400kg takeoff weight, 850km/h cruising speed; 45.5 m wing area; SP 161-4 with 53.3 m wing area; one engine shut down during cruise flight	1961/1962
SP 162	small commercial airliner/transport with two General Electric CF 700-2B for 20 passengers; blown flaps; 7800kg takeoff weight; SP 162-0 and SP 162-3 variants	1961/1962
Me P 170	(Me P 160 VTOL) - vertical takeoff/landing version of the Me P 160 with two variable diameter retracting rotors of 12.5 m diameter	1963

394

	Brief Description	Year of Development		Brief Description	Year of Development
	(fully extended) and static thrust of 104kN(10600kp) each at maximum rotor blade speed of Mach 0.9; each rotor required 2940kW(4000 hp) or 12.65 kN(1290kp) exhaust thrust directed to the rotor blades, equating to 42.2 kg/s exhaust air mass, by utilizing for example Rolls-Royce RB 185 turbojets with 28.4kN(2900kp) static thrust; takeoff weight would therefore have been 17700kg; rotors connected via air channels (in event of engine failure)		HA-200/ Me 200	two-seat basic and advanced jet trainer with two Turboméca Marboré II, each rated at 3.9 kN(400kp thrust); various versions with different levels of involvement by Messerschmitt	1953/1955
			HA-210	two-seat basic and advanced trainer with two Turboméca Marboré IV, each with 4.7 kN(480kp) thrust	1963
			P 201	two-seat jet trainer/courier plane; developed from Me 200	1955
Me P 180	variant of Me P 123 with VTOL capability and same role, i.e. trainer and commuter for up to five persons (including crew); featured extendable rotor (Rotor-jet concept) and two Daimler-Benz DB TL 6 turbojets, each with 5.1 kN(520kp) static thrust; separately controlled free-running turbine for driving air compressor for second (cold) air flow; use of cyclic-controlled reduced diameter and retractable rotor driven by exhaust air from second airflow from turbines blown over airfoil trailing edges; once attaining pure aerodynamic state, rotor (blade diameter having been reduced to minimun length) would be braked and retracted into upper fuselage, causing no additional drag; blade exhaust flow automatically diverted to wing exhaust flow (boundary layer control); pressurized cabin with oval cross section; wings with small surface area, optimized for cruise flight; 2300kg takeoff weight; 3000km range at 785 km/h cruising speed; two variants	1963	P 204	eight-seat commuter plane/high-speed small transport with two engines outputting 3.9 kN(400kp) to 7.8 kN(800kp); follow-on development from Me 200; 4000kg takeoff weight; depending on powerplant, cruising speed varied between 510 and 680km/h; 1000km range	1956
			P 208	commuter plane or high-speed small transport for six persons including crew; wing identical to that used on HA-200; a "modular design" airplane in the following variants:	
			P 208/OM	six-person (including crew) commuter plane with two piston engines outputting 184kW-294kW(250 hp-400 hp); 3000kg takeoff weight, 380km/h cruise speed, 1600km range	1957
			P 208/TL	six-person (including crew) commuter plane with two turbojet engines outputting 3.9 kN-6.9 kN(400kp-700kp); 2900kg takeoff weight, 550km/h cruise speed, 1000km range	1957
Me P 190	(Me 300 VTOL) VTOL trainer with variable diameter and retractable rotor; two Daimler-Benz DB TL 6; Me 300 VTOL with two General Electric J-85 engines, each rated at 17.2 kN(1750kp) static thrust with afterburner; rotor-I 10 m, 5300kg takeoff weight	1963	P208/PTL	six-person (including crew) commuter plane with two turboprop engines outputting 294kW-515 kW(400 hp-700 hp)	1957
Me P 191	variant of Me P 190 with wing area reduced from 17.12 m to12.75 m	1963	Me P 216	five-seat commuter and business plane; powered by two Turboméca Astazou XII, each rated at 515 kW(700 hp); pressurized, climate-controlled cabin; 2975 kg takeoff weight, 675 km/h cruise speed at 8000 m, 2500km range	1965

	Brief Description	Year of Development
Me P 216 PT/T1	five-seat commuter and business plane; powered by two Turboméca Astazou, each rated at 470kW(640 hp); 3025 kg takeoff weight	1965/1966
Me P 216 J/T1	five-seat (including crew) commuter and business plane; powered by two Turboméca Marboré VI, each rated at 4.7 kN(480kp) static thrust; 2970kg takeoff weight, 750km/h cruise speed at 11000 m, 2150km range	1965/1966
Me 230	six-seat commuter plane; two Turboméca Marboré II of 3.9 kN(400kp) thrust each; extensive use of parts and components from HA 200, but with engines buried in wing roots vice forward fuselage section; 3340kg takeoff weight, 400km/h cruising speed, 1930km range (see also P 1216); HA 230: two Turboméca Aubisque, each rated at 6.86 kN(700kp) static thrust; designed by Max Blumm parallel to Hornung's Me P 124; 3800kg takeoff weight, 680km/h cruise speed, 2400km range (without fuel reserves)	1957/1959
Me P 260	short-range commercial airliner with three aft-mounted E 300 turbojets, each outputting 29.4kN(3000kp) static thrust; low-wing layout with swept, tapered wings and conventional control surface layout; 26000kg takeoff weight, 8250 max. payload or 60 passengers and cargo; 950km/h cruising speed, 1120 km range with 60 passengers and 2500kg cargo	1968
Me P 260/ 2 m	short-range commercial airliner for 45 passengers, with 19000kg takeoff weight and two E 300 turbojets, each rated at 29.4kN(3000kp) static thrust; 760km/h cruising speed, 1020km range with max. payload (5830kg)	1968
Me P 260/ 3 m	short-range commercial airliner for 60 passengers, with 26000kg takeoff weight and three E 300 turbojets, each rated at 29.4kN(3000kp) static thrust; 800km/h cruising speed at 11000	1968

Full-scale mockup of the HA 230/Me 230 in June of 1959 at one of HASA's assembly hangars in Seville.

	Brief Description	Year of Development
	m, 1100km range with max. payload (8350kg)	
Me P 263/ M1	light STOL army reconnaissance aircraft with one double-flow Lycoming KHD ZT 53 turbojet rated at 9.5 kN(970kp) static thrust; blown flaps/ailerons; rough-terrain landing gear with main skid under fuselage and extendable auxiliary skids under wingtip tanks; 2100kg takeoff weight, 780km/h cruising speed at low altitude, 600km range; 350 m takeoff run, 290 m landing run (using blown flaps)	1965
Me P 263	STOL army recce plane with one double-flow turbojet engine; up to 2100kg takeoff weight; dorsal intake behind cockpit; support wheels on wingtip tanks	1965
P 300	VTOL/normal takeoff interceptor powered by Bristol Orpheus 12R afterburning turbojet and two rocket motors, each outputting 9.8 kN(1000kp) thrust	1958/1960
HA 300	light supersonic fighter; preliminary work completed in Germany; initial development in Spain with continuation in Egypt; see chapter	1961/1966
Me P 308	"Jet Taifun" as follow-on development of Me 108 Taifun; 6-10 seat commuter and business plane with pressurized cabin and	1965

A Messerschmitt AG advertisement for the Me 308 Jet Taifun as it appeared in the mid 'sixties.

ME 308 »JET TAIFUN«
DAS 6 SITZIGE GESCHÄFTS-REISEFLUGZEUG VON MORGEN

Bei Start auf Betonpiste mit 4 Personen und Gepäck ca. 4000 km Reichweite. Bei Start auf Sportflugplätzen mit 6 Personen und Gepäck bis ca. 2000 km Reichweite · Reisegeschwindigkeit ca. 710 km/h.

MESSERSCHMITT AG · AUGSBURG
AUGSBURG · POSTFACH 322 · TELEFON 34001 · TELEX 053745

	Brief Description	Year of Development		Brief Description	Year of Development
	two turbojet/double-flow turbojet engines, e.g. General Electric CJ610-4, each rated at 10.7 kN(1090kp) static thrust; variants included freigher/military transport and paradrop variants capable of 14 persons and crew, plus ground attack, NASARR trainer, and medevac plane for two pilots, a medtech, four prone and two sitting injured or sick persons; up to 5380kg takeoff weight, 800km/h cruising speed at 12000 m, and 3100km range			as fighter-bomber missions; two turbojets; use of HA-200 components (fuselage aft section, wings, wingtip tanks); up to 6000kg takeoff weight	
			Me P 400	twin-jet attack trainer	1968
			Me P 408 A/01	Me 308 with 50 percent blown flaps; mixed airflow; two double-flow turbojet engines	1965
			Me P 408 B/01/02/012	Me 308 with 25 percent blown flaps; two double-flow turbojet engines	1965
Me P 362/M1	light, two-seat STOL ground attack and training aircraft with two double-flow turbojets; 25 percent blown flaps; 4800kg takeoff weight, S1, M2 and S2 variants	1965	Me P 408 (RJ 408)	"Rotor-jet" convertiplane; seven-seat VTOL commuter plane with folding rotor; see "Rotor-jet" chapter	1967
P 364/4 m	transport with four turboprop engines; 280 m wing area	1955	Me P 508 (RJV 408 2A)	experimental testbed for "Rotor-jet" convertiplane concept, using fuselage of Mitsubishi MU-2J including landing gear, which would have been moved out to rectangular wings	1977
P 364/2 m	transport with two turboprop engines; 150 m wing area	1955			
Me-400A	two-seat multipurpose aircraft for initial and advanced training as well	1966	(RJ 408 2B)	same, but with T-tail	1977

In Praise of the Messerschmitt Foundation

Characterized by Dr. Gerd Materne in the 12 November edition of the Frankfurter Allgemeine as "a patron of the arts and refined private banker, chairman of the board of the Messerschmitt Foundation and MBB's board of directors, former confidant and friend of Willy Messerschmitt," Dr. Hans Heinrich Ritter von Srbik was at Messerschmitt's side when he established the Willy and Lilly Messerschmitt Foundation in 1969. Upon the death of its benefactor, the Foundation was to have inherited Messerschmitt's primary industrial assets. The childless Messerschmitt originally envisioned the purpose of the Foundation to be the advancement of the rising generation of scientists in the field of aerospace technology. The rationale behind Messerschmitt's subsequent change in the Foundation's role was clearly explained by Hans Heinrich von Srbik in his preface "The Benefactor and the Foundation" to the book "The Epitaphs in the Church of Our Lady in Munich," the first work to be published by the Messerschmitt Foundation. The following paragraphs are excerpts from his preface.

"During our three decades of close friendship, and with me in close consultation with him on matters financial, Willy Messerschmitt and I had many lengthy, detailed conversations," writes Dr. von Srbik, "and as one of whom he requested that he retain the post of chairman of the Foundation for life, I can surmise his motives behind this change: never for a moment did Messerschmitt give up hope of restoring Germany to its place at the peak of the aviation industry, not even during the period of the Allied ban prohibiting Germany from exploring aircraft development even from a theoretical standpoint. His own efforts—despite living in a period of abstinence—were devoted to remaining current with the technological state of affairs in the world, particularly that of the U.S. industry.

In an era of increasingly complicated technology and with the creative power of an individual taking a back seat to the juggernaut that is the electronic age, Messerschmitt was firmly convinced that the German aviation industry should avoid tackling major projects solely by fielding cumulatively larger teams of specialists. According to Messerschmitt, conventional development methods were insufficient to make up for lost advantages in the international market. Messerschmitt felt strongly that the next—and possibly the last—truly decisive step in aircraft development would be made in the direction of vertical takeoff.

With the reestablishment of the Messerschmitt AG, all of Willy Messerschmitt's efforts—and a not inconsequential amount of funds—were focused on this theme; he dedicated all his physical and emotional strength to this end up until the final years of his life.

The fact was, however, that the senior executives within the merged corporation gave little support to these ideas and the 'Rotor-jet' project (an aircraft which would lift off and land using rotors and could transition from a standing hover to forward—even supersonic—flight), a project which Messerschmitt considered to be the way of the future, disappeared into the company's file cabinets. For Willy Messerschmitt, this was the greatest disappointment of the latter part of his life.

As Messerschmitt related to me in many discussions where he revealed how hurt and depressed he felt, this act was the reason which caused him to realize there was no longer any sense in dedicating the Messerschmitt Foundation to promoting aviation and space technology. In the future, the efforts of the Foundation would chiefly serve for the 'care and maintenance of German art and cultural memorials at home and abroad,' as the newly revised charter now read.

Such a theme was also well within the scope of Willy Messerschmitt's interests. A brilliant engineer and designer, throughout his life Messerschmitt was fundamentally driven by artistic impulses. Each of his creations bears this mark. His aircraft exemplified refined aerodynamic beauty of form; they were noted not only for being designed as light as humanly possible, but also by the attention to detail embodied in each of his products, products geared both toward simplicity and performance and optimally suited to the role for which they were created. The artistic element constantly manifested itself in Messerschmitt's works. In addition to the fruits of his own labors, he was keenly interested in the fine arts and, to a great extent, music.

The focus and direction of the Foundation—to preserve German art and cultural treasures—was a natural result of this creative talent and his interest in music, particularly as these were combined with an unbroken national consciousness in the face of countless setbacks and catastrophes.

With the death of Prof. Messerschmitt on 15 September 1978 the Foundation inherited both his ideological legacy as well as his industrial assets. The latter consisted almost exclusively of his significant interests in the Messerschmitt-Bölkow-Blohm GmbH, and the Foundation's board of trustees was able to subsequently diversify the value of these interests by making investments in industry and other fields, thereby expanding the financial base of the Foundation and becoming self-perpetuating.

However, it wasn't until 1980 that the complicated administrative matters had been resolved and the Foundation was able to begin its actual mission. Since then, it has completed several different projects and become the initiator of many new programs, concentrating its efforts in southern German-speaking areas, particularly in Bavaria and North and South Tyrol.

In carrying out these projects it leaned on the expertise of the Offices of Care and Preservation of Memorials in the various regions, ensuring that the donated funds would be utilized in the most effective manner possible.

A group of highly qualified individuals, united into a five-member board of trustees with a function similar to that of a board of directors, supports the chairman, who is the determining body in the selection of projects. (members of the board of trustees initially included Kurt Tank, Bavarian Minister of the Economy Dr. Otto Schell, and Willy Messerschmitt as chairman. At that time it was determined that Gero Madelung would assume the responsibility as chairman of the board of trustees upon the death of Messerschmitt. With the change in the role of the Foundation came two additional members, Bavarian Chief Curator Dr. Michael Petztet, and the chairman of the department for care and preservation of memorials at the Technische Universität Munich, Prof. Dr. Otto Meitinger)

This cooperative effort is guided by the common belief that the danger is great and speed is of the essence if the Foundation's measures are to be effectively implemented.

Two concerns guide the chairman in selecting those projects worthy of the Foundation's support: In addition to the major programs, the care and preservation of which are first and foremost a matter of public responsibility, there is also an urgent need to protect and maintain the treasures found among the medium-sized and smaller artistic and cultural monuments. Despite the fact that interest in this area is on the rise, there is still a lack of adequate funding for such projects. It is these which together form that rich cultural landscape so threatened in our leb*ensraum*. The role of our Foundation is to provide financial assistance in the area of 'cultural environmental protection,' in a very real sense a responsibility no less important than the ecological goals receiving so much attention in today's world.

The Messerschmitt Foundation is the largest of its kind in the Federal Republic of Germany. Its greatest undertaking to date has been the previously mentioned restoration and preservation of the approximately 120 epitaphs (gravestones) on the outer wall of Munich's Church of Our Lady, completed in 1983 at a cost of over a million marks. A significant donation towards the restoration of the control tower at Oberschleißheim, the first Bavarian airfield built, most likely provided the impetus for rebuilding this historically significant site as the new aviation center of the Deutsches Museum. More recently, funds from the interest accrued on the initial moneys deposited when the foundation was established have once again been made available for the promotion and advancement of aerospace technology."

Messerschmitt-Bölkow Merger Paves the Way for Restructuring the Aviation and Space Industry

Since its rebirth after the war, the German aviation industry had a conception of itself developing its own ideas regarding its performance and partnership abilities as well as fostering a sense of independence. By the same token, the industry was forced to accept the fact that, with its wartime prominence and its new-found dependence on the State serving as a background, it would be subject to relentless scrutiny, not only from the various political agencies—chief among these being the German Parliament and the federal government—but also from the public. Nevertheless, all governments, both past and present, have eventually overcome their initial doubts and recognized the need for a national air and space industry, with the latter being included from about 1962 onward. The reasons emphasized for this need were primarily economic, technological, foreign and defense policies. At the same time, however, there lurked in the wings a desire to reduce dependency of the German air and space industry on government contracts and aid through a corresponding increase in commitment from the industry, now growing in its capabilities. The governments in power at the time expressed their views by actively sponsoring the industry on the one hand, while on the other hand forcefully demanding the concentration of the country's growing national air and space potential. This governmental pressure was first felt as early as the late 'fifties, and its intent was to maintain industrial competitiveness within the Federal Republic of Germany.

Christopher Andres, a student at the university in Munich, completed his master's thesis on the subject and viewed the matter thusly:

"The initial groundwork for possible future mergers can be found within individual workers' unions, established for the first license-building programs. One of the first of the large-scale mergers in the northern Federal Republic occured in late 1963, when the two Bremen companies of Weser Flugzeugbau GmbH and Focke-Wulf GmbH consolidated under the new name of 'Verinigte Flugtechnische Werke GmbH' (VFW). In one fell swoop Focke-Wulf and its entire industrual base was absorbed into Weser Flugzeugbau, in which the Krupp Konzern and the American company United Aircraft Corporation had already acquired interest holdings. VFW now employed over 7,000 workers and enjoyed approximately one-third of the entire West German aircraft manufacturing capacity.

In the south, too, attempts were made to unite southern German aviation companies with the establishment in Munich of the 'Entwicklungsring für Luft- und Raumfahrt GmbH' (ELR). The goal here was the establishment of a standardized, single corporation which would be able to compete with VFW. In addition to Siebel ATG, the companies involved in this conglomeration included Bölkow, Messerschmitt, and Heinkel, the three firms which had formerly banded together into the E*ntwicklungsring Süd* for work on the VTOL project VJ 101. During the course of negotiations, however, there was such a great difference of opinion over matters of stock value for the individual companies, who would occupy the chairman's position and the board of directors, and project policies that Heinkel first pulled his development team out of the *Entwicklungsring Süd*, then after much legal wrangling withdrew his shares altogether. In December 1964 the Heinkel family added its shares in the Ernst Heinkel Flugzeugbau GmbH to the Vereinigte Flugtechnische Werke....

The Messerschmitt AG, which pressed on with the Bölkow company in managing the E*ntwicklungsring Süd*, at the same time inherited the Ernst Heinkel Flugzeugbau GmbH's 50 percent shares in the Flugzeug Union Süd

As early as 1953—two years before the ban on construction was lifted—the companies of Daimler-Benz, Dornier, Focke-Wulf, Heinkel, Junkers, and Messerschmitt formed the "Aero-Union" in Stuttgart, the purpose of which was to represent their collective aeronautical interests. Left to right: Fritz Nallinger (Daimler-Benz), Claude Dornier, Ernst Heinkel, and Willy Messerschmitt. The "Aero-Union" was naturally later disbanded due to individual interests.

(FUS). With this move, the FUS became a one-hundred-percent subsidiary of the Messerschmitt AG."

Ludwig Bölkow, whom the Bavarian Minister president Dr. Alfons Goppel later called "the driver behind the merger movement within the German aviation industry," had initially attempted a cooperative venture between his company, the Messerschmitt AG, and the Ernst Heinkel Flugzeugbau GmbH as early as 1963. On 22 November 1963 the fundamental concepts for unification were in place. But Willy Messerschmitt, with a backdrop of lifelong business independence, backed out, and the "merger theme" was put off for the time being.

Despite this, on 3 August 1964 the Bölkow Entwicklungen KG, the Messerschmitt AG, and the Siebelwerke ATG GmbH formed an "Interessengemeinschaft Luft- und Raumfahrt" (IGLR)—an organization with the immediate goal of winning the lucrative contract for license building the Bell UH-1D, which would in theory lead to the subsequent merger of the three companies. But when the competing Dornier company was awarded the license contract these plans fell through again, at least for the time being. Andres addresses the subsequent course of merger talks with the following comments:

"In August 1966 the chairman of the Messerschmitt AG came down decisively in favor of talks with the Bölkow group. But the Messerschmitt AG not only negotiated with Bölkow, it also attempted to deal with Dornier. Ludwig Bölkow, for his part, was also embroiled in talks with VFW....

The starting point for new merger discussions between the Messerschmitt group and the Bölkow group was an agreement signed on 16 November 1966. Both the Office of Finance for the state of Bavaria and the Bavarian Ministry of Economy took part in these negotiations from the outset, mainly acting as intermediaries. The two firms had

Press conference of the IGLR on 25 November 1964 at the Bayerischer Hof in Munich. Left to right: Ludwig Bölkow, Willy Messerschmitt, and Dr. Bernhard Weinhardt.

agreed to the rules for share assessment and formed their own committees to oversee this. By the end of 1967 several sticking points had been cleared up, and the future corporation was to have the legal standing of a joint-stock company, with a goal toward a possible merger with Dornier or the Nord group. One troublesome point was the distribution of shares within the future company. Initially, the breakdown was agreed upon as follows: each of the foreign business partners (Bölkow GmbH, the Boeing Company, and Nord-Aviation) were to receive 22.7 percent each, Ludwig Bölkow himself was to receive 22.7 percent, as well, while the Messerschmitt AG would get 32 percent. Chairman of the executive board for the new stockholders would be Ludwig Bölkow.

This interim step drew much discussion in a meeting of the Messerschmitt AG's board of directors on 14 November 1967. Although Willy Messerschmitt was not happy with the 32 percent share, other members of the board who had taken part in the negotiations felt that this was the optimal solution....

Even the board of executive directors at Messerschmitt AG pushed for the merger (against the backdrop of the Bonn government's position). Unlike Messerschmitt, it considered the planned involvement of the foreign companies as being advantageous 'from a technical standpoint.' Moreover, the board felt that the (planned) distribution of shares with Bölkow, with Boeing or with Nord-Aviation. It would be better, so the board reasoned, if the Messerschmitt AG were to avoid contributing its entire capital to the new corporation; it should instead hold back some of these funds as reserves. Even by contributing its entire capital, it would not be possible for the Messerschmitt AG to corner 51 percent of the shares. Ultimately, Messerschmitt's board of directors accepted the outcome of the previous round of negotiations. Merger negotiations should proceed rapidly, concluding before the end of 1967 if possible....

In order for the merger to take place at all (according to this plan), the Messerschmitt AG would have to first divest itself of a 'large percentage of its operating capital,' spinning off its shares in the Junkers-Flugzeug- und Motorenwerke GmbH, the Hispano Aviacion S.A., the Entwicklungsring Süd GmbH, the Augsburger Flughafen GmbH and the Gesellschaft für Flugtechnik GmbH. It transferred these stocks to its subsidiary Flugzeug-Union Süd GmbH. The Messerschmitt AG itself was not to be drawn into the merger."

In the end, the Messerschmitt/Raulino family consortium took a different position than that of Willy Messerschmitt. He was of the opinion that Messerschmitt should have the financial responsibility and leadership in any merger with Bölkow and it was therefore necessary to bring the weight of the company's entire capital to bear on the venture. Walther Stromeyer, spokesman for the Raulino-Messerschmitt branch of the family, considered this prospect to be too great of a business risk, even given a reduced valuation of the Messerschmitt group's shares. The subsequent falling out over capital saw Lilly Raulino-Messerschmitt and her sons retain the reformed Messerschmitt AG including its real estate holdings, i.e. property capital. The company was later reorganized as the Raulino Treuhand- und Verwaltungs AG. Willy Messerschmitt took over the remaining operations, to be merged into the new corporation, and thus became the sole owner of his company's shares within the Messerschmitt-Bölkow GmbH.

Merger talks initially began full of hope, but broke down shortly before Christmas 1967. As a direct consequence, in February 1968 Franz Sackmann, the state secretary within the Bavarian Ministry for Economy and Trade and chairman of the board of directors for the Bavarian Office of Finance (LfA), and Dr. Hans Peter, president of the LfA, proposed to the Bavarian Minister president Dr. Alfons Goppel that the state of Bavaria should acquire shares in Bölkow GmbH with the intent of smoothing the path to the formation of an air and space center in Ottobrunn. At the same time in Bonn, the scene of the decision-making process for the next generation combat aircraft, which would become the successor to the F-104, was in full swing. The Office of Defense and Acquisition hinted to Ludwig Bölkow that if Messerschmitt and Bölkow could complete the merger by 1 November 1968, the new company would have a good chance of securing the contract. In releasing this information, the government contractor increased pressure on the two firms involved—and on Ludwig Bölkow himself to provide the highly qualified workers within the EWR with adequate work.

> "There are continued efforts to promote reorganization and streamlining of the German air and space industry, if necessary by legal means in the event that the industry does not merge into larger units of its own accord. As the largest customer, the government has the right to exercise such measures," explained the Minister of Economy at the time, Prof. Dr. Karl Schiller, in April 1968 at the opening of the German Aviation Exhibition in Hannover.

Word of the Bavarian government's stock acquisition plans soon got out. Undoubtedly, this made it easier for the representatives of the Messerschmitt AG to resume merger talks again. Chaired by Dr. Hans Pter, who soon proved to be an excellent moderator, the parties met for the first time on 29 May 1968. Walther Stromeyer continued as representative of the Messerschmitt family consortium, while Sepp Hort led the discussions on behalf of the Bölkow GmbH.

Yet to be overcome were numerous differences regarding valuation, leadership hierarchy and company structure, not to mention the emotional barriers. 7 June 1968 saw the signing of an initial agreement, and negotiations continued apace. On 28 October the LfA acquired a portion of the shares of Bölkow GmbH. On 31 October, with the notaries already in the building, the final open points were resolved and negotiations were brought to a close. On 1 November 1968 the business leaders within the new corporation signed the merger agreement in the presence of Minister president Dr. Alfons Goppel.

Sepp Hort later commented:

Upon arriving at the Bölkow factory in Ottobrunn, the Bavarian minister-president Alfons Goppel is greeted by Walther H. Stromeyer. Dr. Bernhard Weinhardt is on the far left, while Sepp Hort is between Stromeyer and Goppel. In the background is K. H. Gierenstein, MdB, and state secretary Erwin Lauerbach, with Ludwig Bölkow on the right.

Ludwig Bölkow (left) and Walther H. Stromeyer sign the merger agreement.

402

The "historic" handshake merging Messerschmitt and Bölkow, symbolically sealed by the Bavarian minister-president Alfons Goppel.

On the periphery of Messerschmitt-Bölkow GmbH's moment of birth: Overseer of the Economy Mrs. Hildegard Fischer in conversation with Dr. Friedrich Dreschler.

"The Federal government never had a consistent plan for merging Germany's aviation and space industry. It was individual personalities who were convinced of the need for larger, more capable companies in this area, if only to improve Germany's standing in the international marketplace in comparison with the U.S. industry and the large European enterprises. Chief among these personalities was the chairman of the Christian-Socialist Union, the Bundesminister and later Bavarian Minister president Franz Josef Strauß, who had clear-cut ideas, as well as the ministerial director in the Ministry of Defense, Albert Wahl, and later, with the merger of MBB and VFW, the mayor of Bremen, Hans Koschnick. The Ministry of Economy was ultimately unable to make up its mind with regard to a decisive policy. It forced mergers under the threat of revoking purchase contracts, yet on the other hand it was prepared to strengthen remaining, smaller competing firms with preferential treatment in handing out contracts. In so doing, the idea was to maintain 'national competitiveness.' Such policies never gave a nod in the direction of international challenges. And in subsequent years this led to problematic situations when determining the German role within the framework of the British and French joint-venture companies and workers' unions which had been set up in the interim."

Figures and Facts Subsequent to the Messerschmitt-Bölkow Merger

Location	Emloyees
Messerschmitt-Werke Flugzeug-Union Süd GmbH, Augsburg	1,750
Bölkow GmbH, Ottobrunn	2,900
Messerschmitt-Werke Flugzeug-Union Süd GmbH, Munich	70
Messerschmitt-Werke Flugzeug-Union Süd GmbH, Manching	1,450
Bölkow Apparetebau GmbH, Schrobenhausen	240
Bölkow Apparetebau GmbH, Nabern	720
Bölkow GmbH, Lampoldshausen Branch	120
Waggon- und Maschinenbau AG, Donauwörth Siebel-Werke ATG GmbH, Donauwörth	2,350
Waggon- und Maschinenbau AG, Laupheim Plant	140
Junkers Flugzeug- und Motorenwerke GmbH, Munich	530
Entwicklungsring Süd GmbH, Munich	1570
Entwicklungsring Süd GmbH, Manching Flight Test Center	180
Total # of employees	12,020
Total sales for 1967:	DM 525,000,000
Business capital:	DM 19,200,000
Business partners:	
Messerschmitt AG, Augsburg	2/6
Ludwig Bölkow, Munich	1/6
Bavarian Office of Finance, Munich	1/6
The Boeing Company, Seattle	1/6
Nord Aviation, Paris	1/6

Having only been established on May 14th, 1969, the newly-founded Messerschmitt-Bölkow-Blohm GmbH first revealed its products to the international aviation and space-minded public at that year's Paris Air Show at Le Bourget. Willy Messerschmitt, Ludwig Bölkow, and Werner Blohm (left to right) at the company's exhibit booth.

70th Birthday Celebrated as Sign of Messerschmitt-Bölkow Merger

Congratulations from around the world, a sea of flowers, hundreds of personal well-wishers, meaningful gifts highlighting old friendships, the warm atmosphere generated by long-time co-workers, famous pilots and engineers, one time rivals and colleagues; these all formed the backdrop to Willy Messerschmitt's 70th birthday celebrations on 26 June 1968 at the Hotel Regina in Munich.

Fritz Rudorf, the chairman of the board at Messerschmitt AG, was the keynote speaker at the festivities, and his remarks included the following: "...I'm glad that, with your vision and the willingness of all those involved, including the Free State of Bavaria, we've succeeded in reaching a conclusive merger agreement between the companies of Messerschmitt and Bölkow which, with the involvement of the Free State of Bavaria, will result in a major company for the aviation and space industry..."

"Thus, at the same time this birthday celebration sealed the agreement for one of the most significant companies in aviation history and the overtures of a hopeful beginning" was how the Bölkow Group's company journal noted the event.

Ludwig Bölkow congratulates his former boss and current partner. Hubert Bauer is in the center.

Fritz Rudorf, chairman of the board of directors at Messerschmitt AG and long-time associate of Willy Messerschmitt, photographed during his speech.

Willy Messerschmitt and his wife listen to the guest speakers. Behind the couple, left to right, are state secretary Erwin Lauerbach, Ludwigh Bölkow, Dr. Hans-Heinrich Ritter von Srbik, and Gero Madelung (with Julius Henrici behind him).

Hubert Bauer relays the best wishes of the Augsburg plant. Dr. Friedrich Drechsler is in the foreground on the left, with Hanna Reitsch standing center left.

75th Birthday in Auditorium of Deutsches Museum

Wilhelm Schwaibold, in writing for the July 1973 edition of the company's "MBB aktuell" newspaper, portrayed the celebrations thusly:

"On 26 June 1973, surrounded by his friends, contemporaries, and old colleagues, Prof. Dr.-Ing. E.h. Willy Messerschmitt accepted the well wishes conveyed to him on his 75th birthday by representatives from the State, the aviation industry, the old and new *Luftwaffe*, and the business and science communities. In ceremonies held in the auditorium of Munich's Deutsches Museum, Ludwig Bölkow congratulated MBB's honorary chairman of the board on behalf of its management and workers. His brief remarks honored the engineer-businessman Messerschmitt, who taught his colleagues to "never give up" when it came to turning technical challenges into reality. Bavaria's Minister-president praised Messerschmitt's devotion to aviation, technology, the economy, and the country, while Prof. August Wilhelm Quick, speaking on behalf of the scientific community, covered the high points of Willy Messerschmitt's life and creations."

In the auditorium of the Deutsches Museum. First row, left to right: Rakan Kokothaki (edge of picture), Graf von Castell, Gräfin von Castell, Luise Ulrich, Professor August Wilhelm Quick, MBB chairman Dr. Hermann Theodor Brandi, Lilly Messerschmitt, Willy Messerschmitt, Bavarian minister-president Alfons Goppel, Ello Madelung, Mrs. Müller, Dr. Müller ("Ochsensepp"), and their daughter Christa Müller, who is today head of Apparatebau Gauting GmbH.

Konrad Merkel (left) conversing with Dr. Müller. In the center is Willy Messerschmitt's nephew Dr. Elmar Messerschmitt.

Ludwig Bölkow congratulates Willy Messerschmitt.

The life and times of the guest of honor are reflected in a Messerschmitt exhibit in the Glass Hall of the Deutsches Museum in honor of Messerschmitt's 75th birthday. Standing at the prop of the M 17 (from MBB-Augsburg) is Willy Messerschmitt, with August Wilhelm Quick (half-hidden), Adolf Galland (former General of the Fighters) behind and MBB chairman Dr. Hermann Theodor Brandi next to him.

Willy Messerschmitt in the cockpit of a Bf 109.

Willy and Lilly Messerschmitt are obviously pleased by the best wishes related by minister-president Alfons Goppel.

With the bust of Hugo Junkers looking over his shoulder, Willy Messerschmitt expresses his gratitude for the honors he's received at his 75th birthday celebration in the auditorium of the Deutsches Museum.

80th Birthday:
"Messerschmitt - More Than Just A Name for Technological Products"

This was one of the core themes of keynote speaker General(ret.) Johannes Steinhoff at the reception given in honor of Willy Messerschmitt's 80th birthday in the Montgelas-Palais at Munich's Hotel "Bayerischer Hof" on the 26th of June, 1978. Figures from the political world, the scientific, research and managerial communities, high-ranking military leaders, former colleagues, friends, contemporaries, and family members all were on hand for the celebrations.

"In honoring the life's work of this 'pioneer of modern aircraft development,'" continued Steinhoff, "we must first understand that Messerschmitt had been building civilian airplanes until well into the 'thirties and that, even later, he was 'more interested in solving technical problems than in how the warplanes were being used.' But in a tragic way, the war itself became a master of events. Messerschmitt the artist, who not only mastered the laws of physics, but also devoted himself to the Muses, was able to create 'flying works of art and brilliant masterpieces.'" The general then addressed the designer directly: "Your aircraft became living beings, with which we pilots enjoyed a special relationship."

In his acceptance speech, Willy Messerschmitt emphasized "that that which today is called Messerschmitt's life work" should be attributed to good fortune and the work of all those who stuck by him over the decades. After 55 years of working in the aviation field, for Messerschmitt it was still the greatest praise to hear someone comment that he'd "made designs which are pleasing to the eye." There are still many frontiers to explore in aviation and space design. "And although my 80 years of age keeps me from becoming actively involved—in my heart I'm always there by your side."

Willy Messerschmitt in the company of his closest colleagues and pilots. Left to right: Kurt Schnittke, Sepp Gerstmayer, Hermann Wurster, Werner Göttel, Willy Messerschmitt, pilot Helmut Kaden, senior foreman Kaspar Meinhardt, Wendelin Trenkle, Lukas Schmid.

Willy Messerschmitt runs into Claudius Dornier, Jr., chairman of Dornier GmbH. In the middle is Wulf Diether, Graf zu Castell, director of the Munich airport (Flughafen München GmbH).

Elly Beinhorn relays her best wishes. On the left is Ingeborg Siebel-Trulson, with MBB businessman Werner Blohm in the background.

Well-wishers also included (left to right): Wolfgang Ruppelt, vice president of Germany's Office of Defense Technology and Acquisition, Professor Dr.-Ing. Hans Scherenberg, member of the board of directors at Daimler-Benz AG, Karl J. Dersch, director of Daimler-Benz AG's Munich branch and now member of the board of directors of Deutsche Aerospace AG.

Ludwig Bölkow and his wife congratulate Willy Messerschmitt. On the left is Dr. h.c. Franz Josef Strauß, at the time the chairman of the CSU and chairman of the board of directors at Deutsche Airbus GmbH.

Standing, right to left: Gerhard Messerschmitt and his wife, and Dr. Eleonore Matouscheck. Seated, left to right: Hilmar Stromeyer, Messerschmitt's stepson, Willy Messerschmitt, Micahel von Dercks, husband of Lilly Messerschmitt's niece Stella.

Some of Messerschmitt's longtime associates: (left to right) Kurt Schnittke, Dr.-Ing. E.h. Kurt Tank, Ferdinand Brandner.

Willy Messerschmitt autographs a publication entitled "Messerschmitt - Six Decades of Aircraft Design," printed by MBB on the occasion of his birthday. The man on the left is Albert Speer.

Celebrations also took place among MBB colleagues at the company's guest casino in Ottobrunn. Enjoying a moment together are: (left to right) Ludwig Bölkow, Hubert Bauer, Willy Messerschmitt, and Sepp Hort.

Talking shop away from the main celebrations in Ottobrunn: Kurt Tank, Willy Messerschmitt, and Ludwig Bölkow.

Gero Madelung (left) and Hubert Bauer in discussion near a model of the Me 264 presented to the guest of honor by the management of MBB.

Epilogue

"Willy Messerschmitt passed away in Munich on 15 September 1978, shortly after his 80th birthday. On 20 September 1978 a memorial service was held in Augsburg's Plant 3, with the final assembly line for the Tornado's center section providing an appropriate backdrop. Willy Messerschmitt today rests in the family cemetery in Bamberg.

As chairman of the board for Messerschmitt-Bölkow-Blohm GmbH, I had the opportunity to greet the many guests who attended the funeral. The Bavarian Minister-president, Dr. Alfons Goppel, noted the following in his remarks:

'Thus we stand in reverence of a life, of the life of the deceased as a highly talented personality, one who was as intelligent as he was loving of the arts, self disciplined and ever industrious, a stranger to tomfoolery, always focused and attentive. He lived and worked and calculated—his designs weren't based on sudden flashes of his genius— and even the results of those labors stemming from urgent, pressing personal and impersonal war and defense crises came from his refined and well-disciplined work and leadership style. He was a fatherly teacher, a masterful instructor, and a leader who was always willing to learn.'

Willy Messerschmitt was actively involved in the 1969 merger of Messerschmitt-Bölkow-Blohm GmbH with the Hamburger Flugzeugbau. Up until the time of his death, he remained honorary chairman of the board at Messerschmitt-Bölkow-Blohm GmbH. Yet he did not live to see the integration of the Vereinigte Flugtechnische Werke (VFW) into the MBB GmbH on 1 January 1980, resulting in the creation of what was then the largest air and space company in Germany.

In the ten years since Messerschmitt's passing, this company—which bears his name at the front—has been able to reestablish ties with the leading air and space enterprises throughout Europe due to its technical advances which have become renowned throughout the world.

At the same time MBB, with its mixed economic corporate structure, found itself pushing its entrepreneurial bounds as an equal leading partner in the expansive European Airbus consortium. It coupled the German partner's relatively late arrival on the large aircraft design scene with considerable amounts of investment in what was an extremely competitive field—particularly from the American aviation industry. Whereas the French Airbus partner, as a State-sponsored company in a nation enthused about aviation, was able to survive the pace of expansion even over long periods of operating in the red, MBB wore itself out in regular penitential processions to plead for assurances of financial support from a government somewhat more critical of aviation. In the land of free market economy the company's entrepreneurial competence and that of its powerful partners increasingly came under scrutiny.

After MBB's business partners declined the proffered but tenuous leadership role, in 1988 the considerable political powers within the Federal Republic were able to persuade the largest privately owned company in Germany— the expansive Daimler-Benz AG—to assume the responsibility of leadership. The Messerschmitt Foundation, in the spirit of its founder, felt obliged to support and encourage the establishment of private-sector management. As the required managerial majority could not be gained without the inclusion of the Foundation's shares, the Foundation decided, reluctantly, to reduce its own shares to a symbolic amount.

Beginning in 1986, it became apparent that the entrepreneurial risks for a non-profit foundation were in fact too great. Reunification in 1990 and the opening of eastern areas of Central Europe, however, have given fresh impetus to the maintenance of German cultural memorials. A plaque was laid in memory of our founder the same year that work began on restoring Frederick's "Belvedere" in the gardens of Potsdam's Sanssoucci palace. And the Foundation will continue in its resolve to keep the name of Willy Messerschmitt alive in the continuing tumultuous world of aviation technology."

Gero Madelung
Chairman of the Advisory Board for the Messerschmitt Foundation

Appendix

Coworkers and Contemporaries

Asam, Moritz (1903-1986), worked at Messerschmitt since 1933. Full of ideas, he worked as a supervisor and shop manager in the research department. Emigrated to the US in 1955. Creator of the Super Guppy, a transport based on the Boeing 377 used by the USAF for carrying rockets (today employed by Airbus for transporting airliner subassemblies).

Baeumker, Adolf (1891-1976), leader in the RLM on matters of aviation research from 1933 on, chancellor of the Deutsche Akademie für Luftfahrtforschung, where Willy Messerschmitt served as vice president. A quote from his "Ein Beitrag zur Geschichte der Führung der deutschen Luftfahrttechnik im ersten halben Jahrhundert 1900-1945"(1971) reads:
"Based on pure intellectual creativity coupled with tangible, measurable results, it seemed to me then, and still does today, that Messerschmitt's performance can be considered 'outstanding to the point of brilliance' ."

Barmeyer, Paul (*1903), from 1929 onward designer of all types from the M 23, eventually working in the department responsible for Me 262 cockpit pressurization.

Bartels, Werner (*1902), pilot, Bf 109 operational pilot, from 1943 commissioner for the Me 262 with the Fighter Armament Staff and for a time the technical director of foreign affairs for the Messerschmitt works.

Bauer, Hubert (1902-1986), came to BFW from Junkers in 1929. Initially operational assistant, then from 1938 director of operations and, to 1945, manager of the company's test design department. From 1955 on was manager of Magister, G.91, F-104G and Tornado assembly programs. Member of the board of directors at Messerschmitt AG from 1957 to 1974. 1970-1971 was director of manufacturing of MBB's aviation division.

Bauer, Richard (1898-1962), employed at Messerschmitt from 1929 on, in 1933 became director of the fabrication section for all significant types up until the beginning of the war. Subsequently supervised the Bf 109 and all further developments. Following the war, worked in cooperation with the Stuttgart engineering office of Bölkow, among others.

Baur, Karl (1914-1963), famous glider and airplane pilot. From 1940 to 1945 test pilot for all Messerschmitt types.

Beauvais, Heinrich (*1908), as an engineer-pilot for the E-Stelle Rechlin, flew and evaluated all Messerschmitt types. After the war was an advisor for the HA 200 high speed test program.

Beinhorn, Elly (*1907), famous stunt pilot and adventuress, often used the Bf 108 and christened it the "Taifun".

Binz, Wilhelm (1908-1986), Messerschmitt design engineer from 1933-1945 and 1947-1972. Recognized undercarriage specialist.

Bitz, Josef (1900-1975), aircraft builder for Messerschmitt (particularly for the Me 323). Established an "oldtimer" reconstruction company after the war.

Blaich, Theo (1900-1975), stunt pilot and farmer in Cameroon. Beginning in 1940 flew a Bf 108 with 'hog's bladders' (making it unsinkable) in the African theater.

Blümm, Max (*1915), aerodynamics engineer at Messerschmitt from 1941 to 1945. From 1955 with HASA in Spain, later as a liaison man with MBB.

Bölkow, Dr.-Ing. E. h. mult. Ludwig (*1912), engineer and businessman. At Messerschmitt from 1939 to 1945 (developmental work on late model Bf 109, Me 262). In 1948 established Bölkow Engineering, in 1965 Bölkow GmbH, which merged with Messerschmitt AG in 1968 and in 1969 linked with Hamburger Flugzeugbau to become MBB (Bölkow was chief executive officer to 1977). Since 1978 has been active in futuristic developments in the energy (solar) and transportation fields.

Bosch, Lorenz (*1906), Messerschmitt design engineer from 1932 onward. From 1941 director of the Messerschmitt branch at Fokker in Amsterdam, eventually manager of production department for fighters and assembly of Enzian anti-aircraft missile.

Braeutigam, Otto (1912-1941), renowned glider pilot. Crashed in bad weather and was killed on 28 May 1941 in Obertraubling while testing the Me 321. Also killed were Bernhard Flinsch, Josef Sinz, and observers Engel and Schwarz.

Brandner, Ferdinand (1903-1986), engine developer at Junkers. Worked in the Soviet Union after the war. From 1961 onward in Egypt developing the E 300 engine for the HA 300.

Brandt, Kurt (*1908), 1935 to 1945 at Messerschmitt in the fabrication sections at Augsburg and Oberammergau. 1948-1949 at "Neue Technik GmbH", from 1950 onward played a major role in the development of sewing machines (co-holder of 60 patents). Subsequently (to 1973) manager of design-liaison.

Braun, Heinz (*1910), at Messerschmitt from 1935. Aerodynamics engineer under Schomerus and Eisenmann (Me 262, among others). Worked on Tornado development at MBB after the war.

Brausewaldt, Horst (1905-1979), worked for Arado and Junkers prior to the war. Wrote Junkers related books. From 1958 to 1971 press advisor and advertising director for Messerschmitt in Augsburg.

Brindlinger, Otto (1895-1975), from 1933 sales manager and public relations worker at BFW. Carried out daily flights between Berlin and Stockholm during the 1936 Olympic Games. In 1938 conducted a seven-month America tour in the Bf 108.

Broschwitz, Dr. Johannes (*1926), 1956-1962 in the Ministry of Economy. From 1963 director of Messerschmitt AG's finance and accounting department. Appointed as member of the board of directors in 1965. From 1970 to 1991 executive director for economy and finance at MBB.

Caroli, Gerhard (1907-1981), from 1938 tasked with expanding and managing flight test program with an average of 60-70 test planes at any given time. Early proponent of nose gear design. Instructed Americans on Me 262s at Lechfeld from May to July 1945.

Chlingesperg, Rolf von (†1945), deputy director in the project section. In 1945, together with Riclef Schomerus, despatched to Japan for license development of Me 163/262 in that country. Perished when submarine sank en route.

Conta, Eberhard von (1895- ?), stunt flyer. Together with Werner von Langsdorff crossed the Central Alps in the M 17 in 1926.

Croneiß, Theo (1894-1942), pilot in WWI and supporter of Messerschmitt in his early years. In 1926 established the Nordbayerische Verkehrsflug regional airlines using M 18s. Special commissioner for aviation in Bavaria. Successful competition flyer and test pilot for Messerschmitt aircraft. In 1936 founded Messerschmitt GmbH Regensburg, later became chief executive officer at Messerschmitt AG, and subsequently chairman of the advisory council and operations manager for the GmbH.

Degel, Wolfgang (*1910), From 1933 to 1945 in managerial positions of increasing importance at Messerschmitt AG, eventually becoming developmental manager for the Me 262 program. From 1950 to 1975 worked in the technical department of Messerschmitt, EWR, and MBB. Finally was manager of F-104/F-4 oversight programs. Was close associate of Willy Messerschmitt throughout his adult life.

Dittmar, Heinrich "Heini" (1911-1960), renowned Rhôn sailplane pilot. Test pilot for the DFS. Along with Alexander Lippisch, went over to Messerschmitt AG in 1939, where he became the test pilot for all Lippisch products. Was the first to break the 1000 km/h barrier in the Me 163 in 1941. Was killed in the crash of the HD-156, a plane of his own design.

Drechsler, Dr.-Ing. E. h. Friedrich (*1906), from 1948 involved in rebuilding the Donauwörth factory of the Waggon- und Maschinenbau AG. From 1957 managing director of WMD/SIAT, from 1965 managing director of Bölkow GmbH. Managing director of Messerschmitt-Bölkow GmbH since its founding in 1968, then at MBB.

Dungern, Wolf Freiherr von (1900-1934), stunt pilot, killed in the crash of Bf 108V-1 while on a training flight for the 1934 Europa Circuit.

Ebner, Georg (1899-1982), from 1931 to 1945 involved as stress analyst with all Messerschmitt types. In 1948 went to work in the studies department, from 1951 with the Messerschmitt team in Spain, from 1959 with EWR.

Faber-du-Faur, Dr.-Ing. Karl-Otto von, from 1943 to 1945 Willy Messerschmitt's assistant in Oberammergau. After the war gave his name to an engineering bureau, which later became Messerschmitt's "technical department".

Friedrich, Oskar (*1935), with Messerschmitt from 1958 as a technical advisor on aircraft powerplant matters, then with EWR from 1959. Worked on the VJ 101, AVS, MRCA. From 1974 program director for the MRCA at MBB, in 1975 became manager of development branch in the aviation department, and from 1979 on was manager of the department itself.

Fröhlich, Josef (1904-1978), came to Messerschmitt from Arado in 1937. Drew up a shop manual for streamlining inventory management and large scale production within the Messerschmitt corporation and companies building its products under license. Management development of the "*Giganten*".

Fuchshuber, Josef (*1922), at Messerschmitt from 1950 onward. Expanded sales. Management of the worker's unions for the F-104 and G.91. Member of board of executive directors for JFM and FUS. Manager of logistics branch in MBB's aircraft department up to 1983.

Geduldig, Wilhelm (*1905), came to Messerschmitt from Arado in 1940. Worked in the project section up to 1945. 1955-1957 in Spain with HASA, from 1958 worked in the studies bureau.

Göttel, Werner (*1911), at Messerschmitt from 1935 to 1945. Management of Bf 109 program in 1938, then in 1940 became technical assistant to Willy Messerschmitt, and in 1943 was appointed as F.W. Seiler's assistant on the board of directors. Eventually worked on the production management team in Augsburg with a focus on dispersed operations. With Bölkow in Ottobrunn from 1958 onward, subsequently as a liaison man to the Messerschmitt studies bureau.

Hackmack, Hans (1898-1928), pioneering sailplane pilot during the early Rhôn years. Crashed and was killed in 1928 while testing the M 20.

Harth, Friedrich (1880-1936), *Regierungsbaumeister*, from 1910 was involved in the Rhôn sailplane flights in the spirit of Lilienthal. Young Willy Messerschmitt assisted him from 1912 on. In 1921 established the company of Segelflugzeugbau Harth-Messerschmitt in Bischofsheim. On 13 September 1921 flew for 21 minutes, setting a world's record. Messerschmitt and Harth parted ways in 1923.

Heinz, Johann (1897-1975), worked at Messerschmitt Bamberg in aircraft construction from 1926 on. Later operations engineer in Augsburg to 1965.

Hentzen, Fritz H. (1897-1978), in 1921 built the groundbreaking "Vampyr" sailplane based on a design by Georg Madelung, in which he set world's endurance records in 1922 with flights over three hours. Following a stint with Fokker, came to BFW in 1934 as the assembly manager and became a member of the board of directors. Was manager of all Messerschmitt production aircraft in the "assembly ring" during the war. After the war, worked on special assignment for Willy Messerschmitt in India and in Egypt.

Hirth, Wolf (1900-1959), received his training on Harth-Messerschmitt aircraft in 1921. Instructor pilot for Harth-Messerschmitt Flugzeugbau. Messerschmitt's fellow student at the TU Munich. Sailplane pioneer. In 1938 founded the Wolf Hirth GmbH. During the war produced wooden components for the Bf 109, Me 163, and Me 323. In 1976 MBB became the majority shareholder.

Hoffmann, August (1909-1987), design engineer at Messerschmitt from 1934 on, later worked in the project section (to 1945). 1952-1958 with Messerschmitt in Spain (general systems). Was operations engineer to 1974, subsequently working on systems for the Tornado.

Hornung, Hans (1910-1978), from 1938 to 1945 director of the preliminary projects bureau at Messerschmitt. With the technical bureau from 1947, and from 1951 to 1956 with the Messerschmitt team at HASA, eventually as a design supervisor in the studies bureau.

Hübsch, Fritz (*1911) In 1942/43 made the longest towed flight, in an Me 321, from Lechfeld to what is now Byelorussia (1000 km). From 1959 to 1975 training supervisor at the Augsburg plant. Set up the flying club MBB-Flugsportgruppe Augsburg and has over 9000 glider training flights to his credit.

Jodlbauer, Dr. Kurt (†1937), company pilot. On 17 July 1937 crashed into the Müritzsee and was killed, diving from 5000 meters in an attempt to determine the point of pullout for a Bf 109 trimmed nose-heavy.

Kaden, Helmut (1910-1992), a design engineer at Messerschmitt from 1935 on. In 1940 became manager for test flying production aircraft at Augsburg when Willi Stör left. From 1956 to 1976 was manager of the Augsburg airports. *Flugkapitän* and *Alter Adler*.

Keilholz, Hans (1902-1967), from 1934 to 1945 marketing director at the Junkers-Werke and general director of production. From 1959 onward worked at setting up the Messerschmitt facilities at Manching. From 1960 to 1966 was marketing director for the Manching division.

Keller, Siegfried (1902-1968), during the war was senior executive officer of the Uher & Co. Gesellschaft für Apparatebau, Munich. From 1949 until his death was on the managing board of directors at Messerschmitt AG as the marketing director, also managing director of several subsidiary companies.

Kempter, Josef (1913-1984), came to BFW in 1927 as an apprentice. Initially in the fabrication section, then from 1935 to 1945 in the project section. Went back to Messerschmitt in 1948. From 1954 to 1957 part of the Messerschmitt team in Spain, then worked in prototype and production at Augsburg until 1976.

Klages, Paul (1899-1959), worked in aeronautics since 1924. Came to Messerschmitt's Spanish team in Seville as chief of design. From 1957 until his death was chief design engineer in the studies bureau.

Knoetzsch, Hans-Dietrich (1907-1957), test pilot for Klemm in 1929, then with the DVL in 1930. At BFW from 1935 to 1936. First pilot to fly the Bf 109 (28 May 1935).

Kokothaki, Peter Rakan (1902-1975), part of BFW from 1926 on. Became marketing director and member of the board of executives in 1932. From 1942 until war's end was successor to Theo Croneiß as operations manager of the Messerschmitt AG. From 1968 to 1969 was member of the board of directors at MB and MBB GmbH.

Krauß, Julius (1894-1972), was senior stress analyst in 1924 at Udet-Flugzeugbau. In 1926 was picked up by BFW. Served as senior stress analyst at Messerschmitt until 1945. In 1936 became a regular lecturer on matters of aeronautical engineering at the TH Munich, where in 1942 he was made professor and dean of the aeronautical design department. After the war acted as a consultant for Messerschmitt (for the HA 200 and HA 300, among others).

Langhammer, Willi (*1914), worked in the project section at Messerschmitt from 1937 to 1945. Test engineer for specialized weapons, including the R4 m rockets for the Me 262. Moved to Messerschmitt's developmental bureau in Munich in 1953, and from 1954 to 1956 was on the staff of the RSM's design and test department. In 1957 oversaw the expansion of the Munich-Riem airport. From 1961 to 1978 was technical director of the Manching Works.

Langfelder, Helmut (1929-1978), worked at Messerschmitt from 1952 on. In Spain from 1954 to 1957, and with EWR from 1959 to 1969 (aerodynamics, new developments). In 1969 became director of systems engineering at Panavia, and in 1970 was appointed Tornado program director at MBB. In 1975 became director of the aviation branch, and from 1 January 1978 until his death (helicopter crash) on 6 April 1978 was chairman of MBB's executive board.

Lauser, Kurt (*1908), was managing director at Heinkel from 1959 to 1965, then became a member of the board of directors at Messerschmitt AG and managing director of EWR. From 1967 on was managing director of Messerschmittwerke FUS GmbH.

Leistner, Rudolf (*1925), aerodynamics engineer. With Messerschmitt from 1951 on; from 1955 to 1957 in Spain. Subsequently with EWR and MBB until 1984.

Linder, Karl, from 1941 was deputy director of operations and managing director of Messerschmitt GmbH Regensburg. Streamlined Bf 109 large scale production. Later director of the Armament Staff's branch overseeing manufacture of the Me 262. From 1948 to 1950 was managing director of the Neue Technik company.

Lindner, Gerd (†1945), test pilot for Me 262 aircraft at Lechfeld up to 1945, subsequently trained Americans on the type.

Lippisch, Dr.-Ing. Alexander (1894-1976), aerodynamics engineer and builder of gliders. Advocate of the tailless (delta, flying wing) design. Worked for DFS. Came to Messerschmitt in 1939. Developed the Me 163, the only rocket-propelled fighter ever produced and flown in combat. Went to the Aeronautical Research Institute in Vienna in 1943. Emigrated to the US after the war.

Lucht, Roluf (1901-1945), chief engineer of the RLM's *Technisches Amt*. In 1942 succeeded Theo Croneiß as operations manager and managing director at Messerschmitt GmbH Regensburg and the Leichtbau Regensburg GmbH.

Lusser, Robert (1899-1969), successful stunt pilot. Joined Klemm, then went to Heinkel. From 1933 on was director of the project section at Messerschmitt AG, influencing all Messerschmitt aircraft up to the Me 262. Returned to Heinkel in 1939, and in 1941 went to work for Fieseler. Worked on the V1 production until the war's end. From 1948 to 1958 helped develop cruise missiles for the US Navy. From 1959 to 1963 was employed by Messerschmitt again, this time as the technical managing director at EWR. Was an expert on matters of structural integrity.

Madelung, Prof. Dr.-Ing. Georg (1889-1972), in 1921 oversaw design work of the "Vampyr". In 1925 became manager of the aircraft division of the DVL. In 1928 became a professor of aviation. Married Messerschmitt's sister Ello. Acted as an advisor to his brother-in-law until long after the war.

Madelung, Prof. Gero (*1928), son of Georg Madelung. Went to work for Messerschmitt in 1952 after working in the US. From 1953 to 1957 was a project advisor and representative for Messerschmitt in Spain. Joined Messerschmitt AG in 1958, and in 1959 became a member of the developmental management team (technical department). Played a major role in the development of the VJ 101. In 1963 became a member of the board of directors. In 1969 became managing director at Panavia. In 1978 was appointed director of MBB's aircraft branch. From September 1978 to 1983 was chairman of MBB's executive board, then became a member of the board of directors. Head of the department of aviation technology at the TU Munich. President of the advisory committee for the Messerschmitt Foundation.

Mallet, Guillermo F., represented Messerschmitt's interests in Madrid vis-a-vis businessmen from HASA and the Air Ministry from about 1951 on.

Mayer, Friedrich, technical managing director at Messerschmitt GmbH Regensburg until 1941, then same position at LBR.

Meinhardt, Kaspar (1900-1987), worked sporadically for Messerschmitt Flugzeugbau, Bamberg from 1924 to 1926, when he became employed full time there. Later operations engineer at Augsburg (to 1968).

Merkel, Konrad (1893-1974), official receiver for BFW AG Augsburg from 1931 to 1933, then member of the board of directors and temporarily operations manager at the Regensburg facility.

Messerschmitt, Ello (1907-1990), youngest sister of Willy Messerschmitt. Assisted in his flying efforts on the Wasserkuppe. Married Georg Madelung. Mother of Gero Madelung.

Messerschmitt, Dr.-Ing. Elmar (*1921), nephew of Willy Messerschmitt (on the side of the painter Pius Ferdinand Messerschmitt). Study and degree at the TU Munich. Inventor of various devices. Pioneer of the screen printing method. Assisted his uncle in reestablishing his business in Munich after the war.

Messerschmitt, Lilly (also Lilli), nee *Freiherrin von Michel-Raulino* (1892-1973), major role in BFW since the early 'thirties. Financial supporter of Willy Messerschmitt. Member of the board of directors at BFW AG/Messerschmitt AG. Married Willy Messerschmitt in 1952.

Miller, Ferdinand (*1906), design engineer in the undercarriage division at Messerschmitt from 1938 to 1945. Came back in 1948, and from 1953 to 1957 workted with Messerschmitt in Spain.

Mohnike, Eberhard (†1930), WWI pilot with nine kills to his credit. Test pilot for the RVM. Killed on 14 October 1930 when propeller of M 22 he was flying disintegrated.

Morzik, Fritz (1891-1985), WWI pilot, instructor for the DVS. Victor of the 1929 and 1930 Europa Circuit with the M 23. During the Second World War was *General der Transportflieger* and was therefore responsible for the "Giants", among other types.

Münemann, Rudolf (†1982), industrial financier, assisted Willy Messerschmitt during his struggle to reestablish himself after the war. Inventor of the "revolving system" (long-term guaranteed credit from short-term loans)

Nerud, Alfred (1913-1986), with the BFW fabrication section from 1934 on. From 1947 with Neue Technik GmbH, then from 1952 with Messerschmitt in Spain. In 1958 joined the design management team at EWR.

Ostertag, Wilhelm (†1943), *Luftwaffe* pilot assigned to Messerschmitt for testing the Me 262. Killed in crash of the V-2 prototype at Lechfeld on 18 April 1943.

Puffert, Hans-Joachim (*1912), aerodynamics engineer from 1938 to 1945 in the project section (director of wind tunnel testing). Went to Argentinia with Tank after the war. Rejoined Messerschmitt in 1955, subsequently moving on to Weserflug and VFW.

Reicherter, Julius (*1913), came to Messerschmitt from Klemm in 1935. Planning engineer to 1945. With Neue Technik GmbH from 1948 to 1950, later with EWR and, until 1978, with MBB.

Reitsch, Hanna (1912-1979), world-famous aviatrix and *Luftwaffe* test pilot. Flew the "giants" and the Me 163, among other types.

Rethel, Walter (1892-1977), came to Messerschmitt from Arado in 1938. Design director (for the Me 321/323, Me 210, among others).

Rothe, Dr. Leo S. (1900-1975), with Henschel from 1933 to 1940, then chairman of the management board at Junkers to 1945. In 1953 became the business director for the Verband zur Förderung der Luftfahrt (Association for the Advancement of Aviation, later BDLI). President of the BDLI to 1963. Chairman of the management board at Messerschmitt AG from 1957 to 1962.

Rudorf, Fritz (1901-1988), chairman of the management board of the Bank der Deutschen Luftfahrt during the 'thirties. After the war was a member of the board of Hamburger Flugzeugbau, Junkers, and Messerschmitt.

Schäffer, Max (1907-1986), with Messerschmitt as a stress analyst starting in 1933, later representative of chief analyst Julius Krauß. In Spain from 1954 to 1957, with EWR and MBB to 1972.

Schmid, Lukas (*1907), with the fabrication section from 1934 to 1937. Following pilot training, Messerschmitt test pilot from 1940 to war's end. Flew the Bf 109 to its ultimate maximum speed of 900 km/h.

Schmidt, Kurt (1905-1944), world-renowned Rhön record-setting glider pilot. As a *Luftwaffe* pilot, assigned to Messerschmitt for flight testing. Killed in crash of Me 262V-6 at Lechfeld on 9 March 1944.

Schmitz, Dr. Gerd (*1923), with Messerschmitt since 1958. Folliing the renaming of the Messerschmitt AG to the Raulino Treuhand- und Verwaltungs AG became the authorized signatory and associate member of the management board for the company.

Schomerus, Riclef (1909-1945), came to Messerschmitt in 1933. Manager of aerodynamics in the project section. In 1945, together with Rolf von Chlingensperg, despatched to Japan for working out the details of that country's Me 163/262 license construction program. Was lost at sea when the submarine transporting them sank enroute.

Schwarz, Alois (*1931), with Messerschmitt from 1957. From 1958 until now has been chairman of the works council, first at the Munich-Riem plant, later at Manching. From 1972 the chairman of the central works council and vice chairman of the board for MBB GmbH. Since 1990 has been chairman of the company works council and vice chairman of the board for Deutsche Aerospace AG.

Swharz, Friedrich (1906-1982), came to Messerschmitt Flugzeug-Bau, Bamberg in 1924. Later, in Augsburg, was a specialist for matters of armament. Following the war was an operations engineer in Augsburg.

Schnittke, Kurt (*1920), relative of Ernst Udet. Participated in the production aircraft test program at Messerschmitt at Augsburg from 1939 to 1940. Accompanied and occasionally flew as Willy Messerschmitt's personal pilot (Bf 108), then for Ernst Udet (Bf 110). In 1941/42 flew combat missions in North Africa. From 1943 to 1945 assigned as personal pilot to the Regensburg factory operations manager *Generalingenieur* Roluf Lucht.

Seifert, Karl (*1907), along with his step-brother Paul Konrad, with the project section from 1927. Responsible for large aircraft such as the Me 261 and Me 264. Worked for EWR and MBB after the war.

Seiler, Friedrich Wilhelm (1895-1979), banker. Along with Konrad Merkel, bailed out BFW following the company's financial problems in 1931/33. Vice chairman of the board of directors at BFW AG, then the same position at Messerschmitt AG. Chairman of the board from 1942 to the war's end. Negotiator in the conflict between Messerschmitt and Milch.

Seywald, H., pilot of the first engine-powered Messerschmitt airplanes, the S 15, S 16, and M 17 (1924 to 1927)

Sido, Franz, a pilot since 1907. BFW chief pilot from 1928 to 1933, flew a majority of Messerschmitt designs up to the M 31.

Srbik, Dr. jur. Hans Heinrich Ritter von (1916-1988), banker. Advisor to Messerschmitt AG from the late 'forties, then from 1958 on the board of directors. Close personal friend of Willy Messerschmitt. From 1969 until his death was a member of the board of directors at MBB. Co-initiator and chairman of the Messerschmitt Foundation for the Advancement of Aeronautical Science, also heavily involved in the care and maintenance of German art and cultural memorials both at home and abroad.

Steigenberger, Dr.-Ing. Otto (*1893), began his career at Junkers in 1920, later becoming the director of the Argus-Entwicklungswerke in Berlin. Played a major role in the development of the As 10C engine (used in the Bf 108, among other types). With Messerschmitt from 1939 to the end of the war, working on coordination of all matters relating to powerplants. Rejoined Messerschmitt in 1952 (automobile engines, and from 1957 assigned to the Mk IV, T-33, Fouga Magister).

Stör, Willi (1893-1977), record-setting pilot. 1926 to 1935 instructor at the DVS in Schleißheim, carried out special tasks for BFW part-time, such as test flying the M 27. In 1935 and 1936 won the German aerobatics championship in his own M 35. From 1935 on was chief of production flight testing at Messerschmitt. From 1941 to 1945, together with his mechanic Herbert Kaden, was stationed in Japan for supervising projects relating to the Bf 109 and Me 210. Interned there until 1947.

Storp, Christine Dora, nee Messerschmitt, adopted Marr (1901-1966), provided Willy Messerschmitt with his first abode in Munich, 1945.

Stromeyer, Walther H. (*1922), son of Otto Stromeyer (chairman of the board at BFW from 1931 to 1932), and Lilly, nee *Freiherrin von Michel-Raulino*. From 1957 managing director of the stockholder's consortium of the Messerschmitt group. Became a member of the management team on 31 October 1968 with the founding of Messerschmitt-Bölkow GmbH, in which capacity he served until 30 June 1969.

Stumm, Hilmar (1908-1981), with Messerschmitt from 1934, initially in the fabrication section (mockup construction for all major types up to the Me 262). From 1951 worked on development of cars and engines. World record-setter with the KR 200. Later worked on the VJ 101 while employed with EWR.

Suciu, Dr.-Ing. Titus (1918-1976), project manager with the Messerschmitt studies department in Munich. Particular focus was on the Rotor-jet.

Tank, Prof. Dr.-Ing. E. h. Kurt (1898-1983), aircraft designer and pilot-engineer. With Rohrbach from 1924 to 1929, then from 1930 to 1931 with BFW (e.g. designed the M 28). Subsequently chief of design at Focke-Wulf until the end of the war. After developing fighters for Argentina and India, renewed cooperative efforts with Messerschmitt in Egypt in 1967 (HA 300).

Thiel, Margarete, company photographer at Augsburg from 1934 to 1945, documented all Messerschmitt types during this period.

Trenkle, Wendelin (*1912), test pilot at Augsburg since 1937, then chief pilot for the new Regensburg facility. Test flew thousands of Bf 109s on their initial checkout flight. With EWR after the war.

Udet, Dr.-Ing. E. h. Ernst (1896-1941), Pour-le-Mérite pilot from the First World War. In 1922 helped to establish the Udet-Flugzeugbau in Munich-Ramersdorf (became part of BFW in 1926). Sport pilot. With the RLM from 1936 on, *Generalluftzeugmeister* from 1939 (responsible for the procurement of all aircraft)

Voigt, Woldemar (1907-1980), with BFW from 1932, expanded the project department, and in 1933 became Robert Lusser's deputy, succeeding him in 1938. Had a major influence on all developments from the Bf 108 to the P 1101. After the war, went to Glenn Martin in the US.

Wendel, Fritz (1915-1975), following pilot training and a stint as an instructor pilot for the *Luftwaffe*, in 1936 became a test pilot and eventually chief pilot at Messerschmitt. Flew all the company's products up to 1945. On 26 April 1939 set the world's speed record for a piston-powered aircraft with 755 km/h in the Me 209, a record which stood for thirty years.

Wurster, Dr.-Ing. Hermann (1907-1985), studied engineering, assistant to Georg Madelung. With the DVL beginning in 1933. Held all pilot's licenses. Test pilot at Rechlin and Travemünde. An engineer-pilot for BFW from 1936 on. On 11 November 1937 captured the first speed record for Germany when he clocked 611 km/h in a Bf 109. Initial test flight of all types up to the Me 210. Project engineer from 1942 on, from 1943 was the developmental director for the Enzian anti-aircraft missile.

Ziegler, Mano (1908-1991), journalist and glider pilot. Received pilot training from the *Luftwaffe*. Flew the Me 163 with *Erprobungskommando 16* from 1943 on (author of the book "Raketenjäger Me 163"). After the war became senior editor of Flug Revue, press secretary for Heinkel, Messerschmitt, and Deutsche Airbus. From the mid-'seventies devoted himself to reopening the assembly lines for the Taifun.

Ziegler, Rudolf (†1952), Akaflieg TH Munich, pilot-engineer for the *E-Stelle* Rechlin. In 1942/1943 was the first to fly the Me 328 with Schmidt-Argus pulse-jet engines.

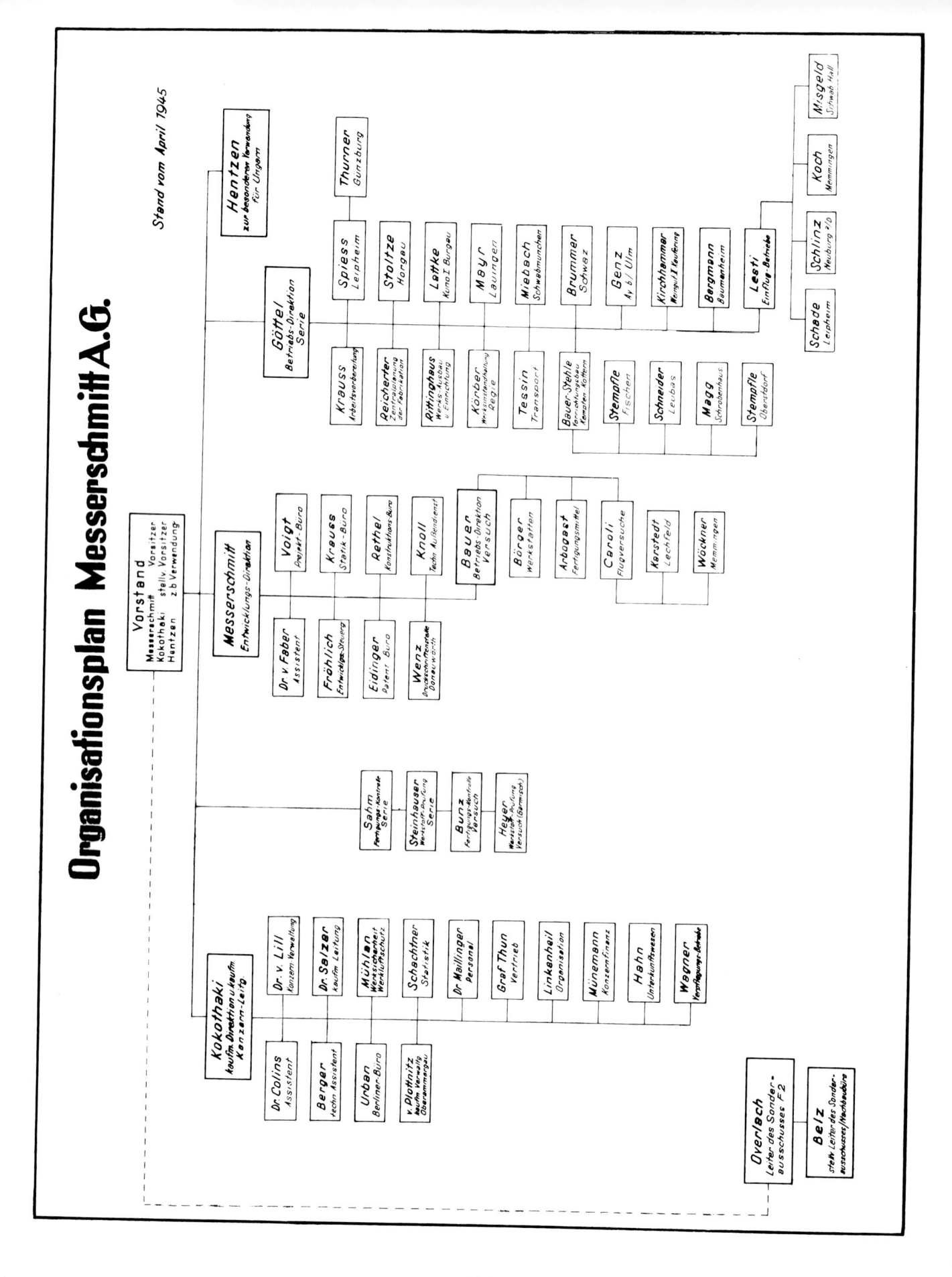

Organisationsplan Messerschmitt A.G.

Stand vom April 1945

Vorstand
- Messerschmitt — Vorsitzer
- Kokothaki — stellv. Vorsitzer
- Hentzen — z.b.Verwendung

Hentzen — zur besonderen Verwendung für Ungarn

Kokothaki — kaufm. Direktion u. kaufm. Konzern-Leitg.

- **Dr. v. Lill** — Konzern Verwaltung
  - Dr. Colins — Assistent
- **Dr. Seizer** — kaufm. Leitung
  - Berger — techn. Assistent
  - Urban — Berliner-Büro
  - v. Plottnitz — kaufm. Verwaltg. Obergaumergau
- **Mühlan** — Werkssicherheit Werkluftschutz
- **Schachtner** — Statistik
- **Dr. Maillinger** — Personal
- **Graf Thun** — Vertrieb
- **Linkenheil** — Organisation
- **Münemann** — Konzernfinanz
- **Hahn** — Unterkunftswesen
- **Wegner** — Verpflegungs-Betriebe

Messerschmitt — Entwicklungs-Direktion

- **Dr. v. Faber** — Assistent
- **Voigt** — Projekt-Büro
- **Krauss** — Statik-Büro
- **Rethel** — Konstruktions-Büro
- **Eidinger** — Patent Büro
- **Knoll** — Techn. Außendienst
- **Wenz** — Organisationszentrale Donauwörth

Bauer — Betriebs-Direktion Versuch
- Börger — Werkstätten
- Arbogast — Fertigungsmittel
- Ceroli — Flugversuche
- Karstedt — Lechfeld
- Wöckner — Memmingen

- Sahm — Fertigungs-Kontrolle Serie
- Steinhauser — Werkstoffprüfung Serie
- Bunz — Fertigungskontrolle Versuch
- Heyer — Werkstoffprüfung Versuch (Germisch)

Göttel — Betriebs-Direktion Serie

- **Krauss** — Arbeitsvorbereitung
- **Reicherter** — Zentralisierung der Fabrikation
- **Rithinghaus** — Werks-Ausbau u. Einrichtung
- **Körber** — Werksinstandhaltung Regie
- **Tessin** — Transport
- **Bauer-Stehle** — Fertigungsbau Kempten Kottern
- **Stempfle** — Fischer
- **Schneider** — Leubas
- **Magg** — Schrobenhausen
- **Stempfle** — Oberstdorf

- Spiess — Leipheim
- Stoltze — Horgau
- Lettke — Kuno I Burgau
- Mayr — Lauingen
- Miebach — Schwabmünchen
- Brummer — Schwaz
- Genz — Ay b/Ulm
- Kirchhammer — Mangold Kaufering
- Bergmann — Baumenheim
- Thurner — Günzburg

Lesti — Einflug-Betriebe
- Schade — Leipheim
- Schlinz — Neuburg a/D
- Koch — Memmingen
- Misgeld — Schwab. Hall

Overlach — stellv. Leiter des Sonderausschusses F2
- Belz — stellv. Leiter des Sonderausschusses / Nachtmuster

416

MESSERSCHMITT KONZERN

Diagram showing the Messerschmitt AG conglomerate structure (early 1944), with Augsburg at center and branches including:

- Reparaturbetreuung La Rochelle / Büro Paris
- Zweigwerk Ratingen
- Konstruktionsbüro Gotha
- Berliner Büro und Vertriebsleitung
- Zweigwerk Eisfeld
- Zweigwerk Marienbad
- Regensburg: Leichtbau Regensburg GmbH, Messerschmitt GmbH
- Zweigwerk Obertraubling
- Zweigwerk Leipheim
- Schwäbische Form-Holz GmbH Ulm
- Werk Gablingen
- R. Messerschmitt K.G. Straubing
- Konstruktionsbüro Wiener Neustadt
- Lizenzbetreuung Györ
- Werk Kaufbeuren
- Büro Budapest / Ungar. Flugzeugarmaturenfabrik A.G.
- Zweigwerk Dachau
- "Eiso"-Schrauben GmbH München
- UHER u. Co. Ges. f. Apparatebau
- Büro Kronstadt
- Büro Bukarest
- Zweigwerk Kottern
- Zweigwerk Oberammergau
- Messerschmitt GmbH Tirol / Kematen
- Zweigwerk Innsbruck
- Büro Sevilla
- Büro Zürich
- Büro Como

Legend:
MESSERSCHMITT A.G.
Zweigwerke u. Werke davon
Aussenstellen
Konstruktionsbüros
VOLLBETEILIGUNGEN
Zweigwerke davon
MEHRHEITSBETEILIGUNGEN
Zweigwerke davon
MINDERHEITSBETEILIGUNGN
Zweigwerke davon

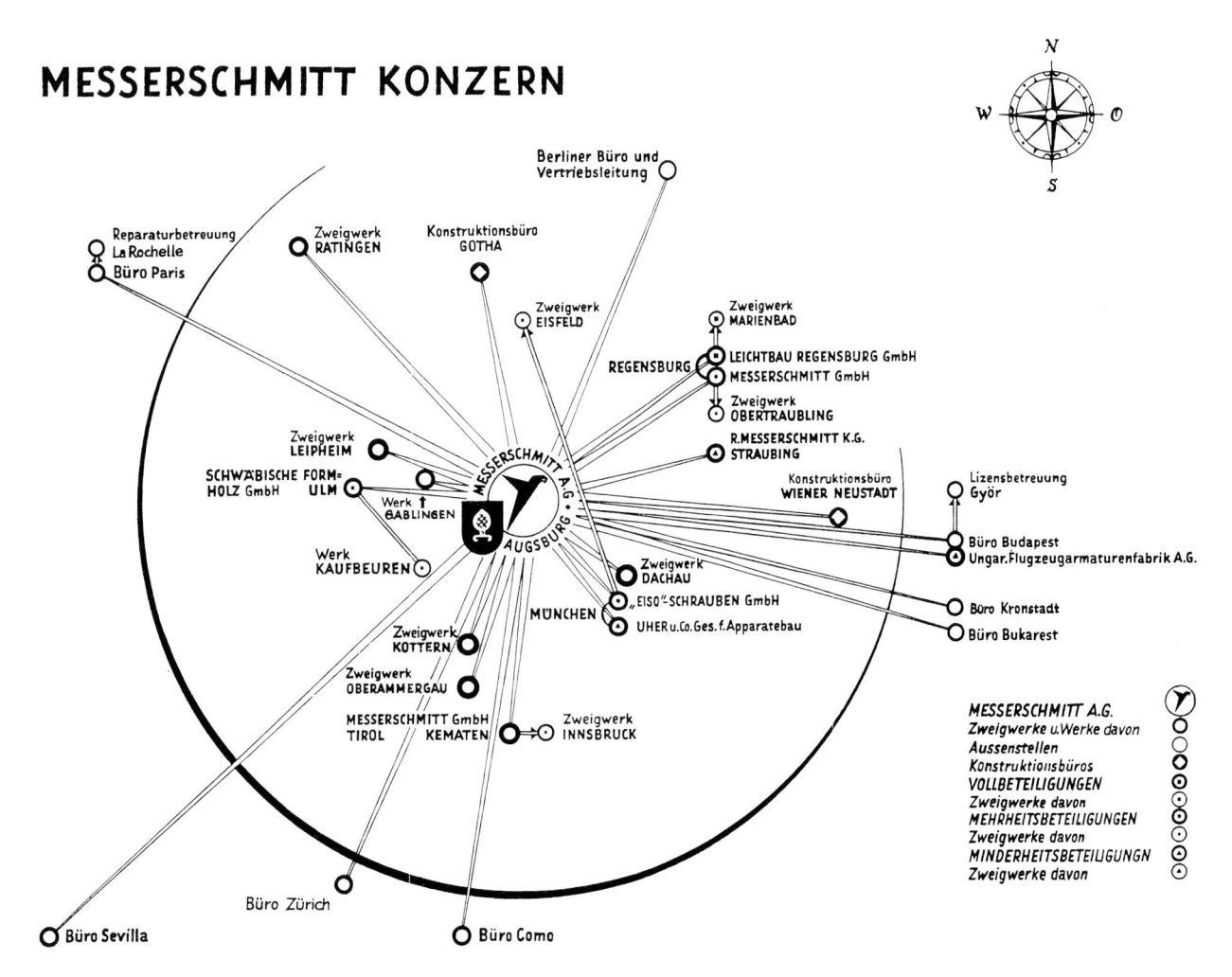

Within a decade of 1934, Messerschmitt's Augsburg facility had blossomed into an entire conglomerate of companies, as shown in this diagram from early 1944. In addition to the establishment of the Regensburg works in 1937, Messerschmitt also had minority and majority interests in metals industry companies. In establishing these interests, the company's senior management desired to make the company somewhat less reliant on supplier firms (Kematen, Eiso, and Uher works). In the case of the Leichtbau Regensburg GmbH (LBR), established in 1939, Willy Messerschmitt created an innovations company which would apply experience gained from aviation technology to other technological fields, e.g. applying alloy construction methods to traincar manufacturing, shipbuilding, or housebuilding. The program met with success in the area of radar systems manufacturing: the weight of the "Würzburg" radar was reduced from 17 metric tons to 4.5 tons.

The branch facilities generally were set up during the war due to production capacity (Leipheim, Obertraubling) and for reasons of safety (Oberammergau, Kottern). Additional disbursements occurred particularly during the last nine months of the war and included bunker complexes such as at Kaufering.

Those foreign companies shown appeared early in the war and were established for planning and executing license-manufacturing programs (Bf 108 in France, Bf 109 in Spain, Hungary, and Romania).

The Messerschmitt GmbH Regensburg facilities under Karl Lindner went through a similar reorganization, as this production center also operated several dispersed sites which were winding down production of the Bf 109 and replacing it with the Me 262. The transportation systems linking these production facilities suffered increasingly from Allied bombing raids and strafing attacks, so that in the end production virtually came to a standstill.

Left: When compared to earlier plans, the April 1945 organizational plan for the Messerschmitt AG at Augsburg shows the greatest changes taking place in the area of production. The *Sonderausschuß F 2* special department - previously headed by Fritz H. Hentzen - now fell under the administration of the Armament Staff, i.e. outside the control of the company. Internal management of the numerous dispersed production sites - now almost exclusively devoted to Me 262 production - was retained by Werner Göttel.

Technical Data for Harth- BFW- and Messerschmitt Aircraft Flown Between 1911 and 1945

Type	**Harth S 1**	**Harth S 2**	**Harth S 3**	**H&M S 4**	**H&M S 5**	**H&M S 6**	**H&M S 7**	**H&M S 8**
Role	glider	glider	glider	glider	glider	glider	glider	sailplane
Design	high-wing canard, open wood fuselage frame, cloth-covered wood wings	high-wing, standard control sfcs, open wood fuselage frame, cloth wrapped wood wings	high-wing, twist control, fixed rudder, single-skid, wood truss fuselage	high-wing, similar to S 3	high-wing, similar to S 3, built solely by Mtt from Harth plans	high-wing, wing controlled, fixed rudder, steel tube fuselage frame	high-wing, wing controlled, fixed rudder, steel tube fuselage frame	high-wing, wing controlled, no rudder, steel tube fuselage frame
Crew+passengers	1	1	1	1	1	1	1	1
Wingspan (m)	12	7	9	8	8	12	11	11
Length (m)		4.5	4.5	4.5	4.5	5.00 apprx.	4.35	4.5
Wing chord (m)	1.5	1.35	1.65	1.65	1.65	1.9	1.6	1.5
Wing area (m2)	18	9	14.5	14.5	14.5	22	16.5	15.35
Weight equipped (kg)	56	75	35	50.4	32	51.3	50	48
Load (kg)	75	75	75	74.6	75	74.7	75	65
Takeoff weight (kg)	131	150	110	125	107	126	125	113
Wing loading (kg/m2)	7.3	16.6	7.5	15-16	7.3	5.73	7.57	7.7
First flight	1911	1912	1913		1915	1916	1918	1920
No. produced	1	1	1	1	1	1	1	1
Notes	short hops only	unmanned flights only	first free flights to 120 m	destroyed before first flight	tail-heavy	3.5 min flight at Heidelstein summer 1916	1918 trials with flaps in place of rudder	Harth sets endurance record on 9/13/1921

Type	**Mtt S 9**	**Mtt S 10**	**Mtt S 11**	**Mtt S 12**	**Mtt S 13**	**Mtt S 14**	**Mtt S 15**	**Mtt S 16b**
Role	sailplane	training glider	sailplane	sailplane	sailplane	sailplane	powered sailplane	powered sailplane
Design	high-wing canard, open wood fuselage frame, cloth-covered wood wings	high-wing, standard control sfcs, open wood fuselage frame, cloth wrapped wood wings	high-wing, twist control, fixed rudder, single-skid, wood truss fuselage	high-wing, similar to S 3	high-wing, similar to S 3, built solely by Mtt from Harth plans	high-wing, wing con-trolled, fixed rudder, steel tube fuselage frame	aileron control using wing warping, but with normal elevator and rudder control; single stick control, both aircraft powered by 2-cyl. 700 cm2 22.6 hp Douglas engine	
Crew+passengers	1	1	1	1	1	1	1+1	1+1
Wingspan (m)	12	14	14	14	14	13.80	14.60	14.40
Length (m)	3.20	4.5	5.0	4.40	4.90	5.50	5.00	5.00
Wing chord (m)	1.17	1.00	1.50	1.35	1.30	1.36	1.05	0.95
Wing area (m2)	14.00	19.00	19.00	19.00	18.00	18.80	15.40	14.00
Weight equipped (kg)		80	80	100	100	105	170	218
Load (kg)		75	75	75	75	75	130	122
Takeoff weight (kg)		155	155	175	175	180	300	340
Wing loading (kg/m2)		7.9	7.9	8.9	9.5	9.3	21	24
First flight	1921	1922	1922	1922	1923	1923	1923	1924
No. produced	1	4(?)	1	1	2	2	1	1
Notes	unstable in all three axes	stable trainer for wing-controlled gliders		535 profile. Wolf Hirth crashed and injured on first flight	Control shaft broke on first flight, crashed and destroyed	Hans Hackmack won alt. prize at 1923 Rhön competition	First flown in June 1924 at Bamberg by Seywald	First flown in 1924 by Seywald from the Rhön
Max. speed(km/h)							105	115
Cruise speed (km/h)							90	95
Landing speed (km/h)							50	60
Range (km)							400	

Type	BFW 1	BFW 3a	U 12b	M 17	M 18b	M 19	M 20b	M 21a
Role	trainer	trainer	trainer and aerobatic	sportplane and trainer	passenger	advertising and sportplane	passenger	trainer, aerobatic and sport
Engine	Siemens Sh 12, 79/92 kW (108/125 hp)	Siemens Sh 11, 63 kW (86 hp)	Siemens Sh 12, 79 kW (108 hp)	Bristol Cherub, 19 kW (26 hp)	Siemens Sh 12, 79 kW (108 hp)	Bristol Cherub, 21 kW (29 hp)	BMW VI 5.5Z, 368 kW	Siemens Sh 11, 62 kW (84 hp)
Design	biplane, steel tubing, composite	biplane, wooden	biplane, wooden	cantilever, mid-wing, wooden	cantilever, mid-wing, alloy	cantilever, low-wing, wooden	mid-wing, alloy	biplane, steel tubing
Crew+passengers	1+1	1+1	1+1	1+1	1+4	1	2+8 to 10	1+1
Wingspan (m)	10.00	10.00	10.00	11.60	15.60	9.60	25.50	10.00
Length (m)	7.11	7.50	7.46	5.85	8.05	5.40	15.90	7.32
Height (m)	2.82	2.60	2.80	1.50	2.25	1.50	4.80	2.80
Wing area (m2)	24.00	24.00	24.00	10.40	24.80	7.90	65.00	20.80
Weight equipped (kg)	666	529	530	186	600	140	2800	460
Load (kg)	234	271	300	184	600	200	1800	280
Takeoff weight (kg)	900	800	800	370	1200	340	4600	740
Wing loading (kg/m2)	37.5	33.3	33.3	35.6	48.4	43	70.8	35.6
Max. speed (km/h)	137	136	145	140	145	145	205	145
Cruise speed (km/h)	110		115	125	130	120	165	130
Climb to 1000 m (min)	7.2	9	9.6	12.5			5.5	9.5
Service Ceiling (m)	3600	3300	3300	4000	2700	850	4000	3300
Range (km)	400	500	450	600	700		1000	500
First flight	11927	1927	1925	1925	1926	1927	1928	1928
No. produced	1	1	150 + license (a and b var.)	at least 5	approx 30 (a-d var.)	2	15 (a-b var.)	2 (a-b var.)

Type	M 22	M 23b	M 24	M 26	M 27	M 28	M 29	M 31
Role	nightfighter, night recce	trainer and sportplane	passenger	commuter	sportplane	high-speed mailplane	advertising and sportplane	sportplane
Engine	2xSiemens Jupiter, 2x368 kW (2x500 hp)	Siemens Sh 13 (or equiv) 50 kW (68 hp)	BMW Hornet (or equiv.), 368 kW (600 hp)	Siemens Sh 14, 68 kW (92 hp)	Argus As 8, 74 kW (100 hp)	BMW Hornet, 368 kW (500 hp)	Argus As 8R (or Siemens Sh 14a), 96 kW (130 hp)	Hirth HM 60, 44 kW (60 hp) or BMW Xa, 29 kW (40 hp)
Design	biplane, fabric covered steel tubing	cantilever low-wing, wooden	cantilever, mid-wing, alloy	cantilever, mid-wing, alloy	cantilever, low-wing, composite	all-metal low-wing	low-wing, composite	low-wing, composite
Crew+passengers	3	1+1	2+8	1+2	1+1	2	1+1	1+1
Wingspan (m)	17	11.8	20.6	12.4	12	15.5	11	12
Length (m)	13.6	6.5	12.8	7.15	7.9	10	7.75	7.85
Height (m)	4.8	2.3	4.2	2.4	2.4	3	2	2.2
Wing area (m2)	63.2	14.4	43	14.3	14.5	25.6	14.5	17.24
Weight equipped (kg)	2900	330	1480	480	430	1160	390	350
Load (kg)	900	265	1520	420	300	1590	310	300
Takeoff weight (kg)	3800	595	3000	900	730	2750	700	650
Wing loading (kg/m2)	60	41	70	63	50.4	107.4	48.3	37.7
Max. speed (km/h)	220	160	220	168	200	260	262	175
Cruise speed (km/h)	185	130	195	150	170	220	225	150
Climb to 1000 m (min)	2.4	5.5	4	9.5	4.5	3	3	5.8
Service Ceiling (m)	6200	4700	5500	2700	5200	5200	6000	3800
Range (km)		800	800	850	700	2450	700	700
First flight	1930	1928	1929	1930	1931	1931	1932	1932
No. produced	1	over 80 (a-c variants)	4 (a-b var.)	1	12	1	6	2

Type	M 35	M 36	Bf 108B-1	Bf 109B-1	Bf 109C-1	Bf 109D-1	Bf 109E-3	Bf 109T
Role	sport and aerobatic	passenger and cargo	touring and liaison	light fighter---				carrier-based fighter
Engine	Siemens Sh 14a, 96 kW(130 hp)	Armstrong Siddely Serval 4, 257 kW (350 hp)	Argus As 10C, 177 kW (240 hp)	Junkers Jumo 210D, 500 kW (680 hp)	Jumo 210G	Jumo 210D	Daimler-Benz DB 601A, 809/758 kW (1160 hp)	Daimler-Benz DB 601N, 864 kW (1175 hp)
Design	low-wing, composite	mid-wing, composite	all-metal, low-wing monocoque ---------------------------------------					
Crew+passengers	1+1	2+6	1+3	1	1	1	1	1
Wingspan (m)	11.57	15.40	10.62	9.90	9.90	9.90	9.90	11.08
Length (m)	7.48	9.80	8.30	8.70	8.70	8.70	8.76	8.76
Height (m)	2.75	2.80	2.02	2.45	2.45	2.45	2.45	2.60
Wing area (m2)	17.00	30.50	16.40	16.35	16.35	16.35	16.35	17.50
Weight equipped (kg)	500	1320	880	1580			2060	2250
Load (kg)	300	930	500	380			550	830
Takeoff weight (kg)	800	2250	1380	1960	2170	2170	2610	3080
Wing loading (kg/m2)	47.0	73.8	84.2	119.9			159.7	176.0
Max. speed (km/h)	230	235	305	460	470	460	570	570
Cruise speed (km/h)	205	215	265					
Climb to 1000 m (min)	3.3	4.	3.15	6.85 to 5000 m	6.2	7.0	5 to 5000 m	
Service Ceiling (m)	5300	4500	5000		9500	9500	11000	1050
Range (km)	1000	700	1000	450	450	450	560	700
First flight	1933	1934	1934	1935	1937	1937	1939	1940
No. produced	15	1	over 1000 (a-d variants)	total production of all Me 109 variants from 1935 to 1955 is approx 35000 (variants A-Z and foreign models)				
Armament				2-3xMG17	4xMG17	4xMG17	2xMG17 2xMGFF	2xMG17 2xMGFF

Type	Bf 109F-4	Bf 109G-6	Bf 109H-2	Bf 109 k-4	Bf 110G	Bf 161	Bf 162	Bf 163
Role	light fighter		high-alt ftr.		twin-engined heavy ftr., nightfighter, recce	strategic recce	bomber	liaison, tactical recce
Engine	Daimler-Benz DB 601E, 993kW(1350 hp)	Daimler-Benz DB 605A, 1085 kW (1475 hp)	DB 605E, 993kW (1350 hp)	DB 605D, 1324kW (1800 hp)	2xDB 605, 1085 kW (2x1475 hp)	2xJumo 210E or 2x DB 600	2xDB 600, 736 kW (2x1000hp)	Argus As 10C, 177 kW (240hp)
Design	all-metal low-wing, monocoque				all-metal low-wing, monocoque			braced high-
Crew+passengers	1	1	1	1	2	2-3	3	3
Wingspan (m)	9.92	9.92	13.26	9.92	16.27	16.69	16.20	13.58
Length (m)	8.94	8.94	9.02	9.02	12.07	12.85	12.30	9.75
Height (m)	2.60	2.50	3.33	2.68	4.13		4.12	
Wing area (m2)		199.8			193	127	163	57.5
Weight equipped (kg)	2255	2680		2755	5850		4360995	995
Load (kg)		520			1560		1920	315
Takeoff weight (kg)	2980	3200	3900	3400	7410	4890	6280	1310
Wing loading (kg/m2)		199.8			193	127	163	57.5
Max. speed (km/h)	635	630@70000 m		680	561	440@40000 m	480	170-200
Cruise speed (km/h)		590@60000 m			525			
Climb to 1000 m (min)	6	10.5 to 8400 m		10	4.5 to 4000 m			
Service Ceiling (m)	11600	1200	1400	11800	11100	8100		
Range (km)	650	650		700	900			450
First flight	1941	1942	1944	1944	1936	1938	1937	1938
No. produced					over 6000 (B-G var.)	2 (V-1-V-2)	3 (V-1-V3)	1 (V-1)
Armament	2xMG17 1xMG151	2xMG131 1xMG151	2xMG131 1xMK108	2xMG131 1xMK108	4xMG17 2xMG151 1xMG81Z	2xMG15	2xMG15 500 kg bombs	

Type	Me 163A	Me 163B	Me 208	Me 209V-1	Me 209V-5	Me 210A-1	Me 261V-1	Me 262A-1
Role	rocket powered testplane	rocket powered interceptor	liaison and touring	record-setting plane	fighter	heavy ftr, bomber, recce	courier, strategic recce	jet powered fighter
Engine	Walter R II 203 rocket, 7.35 kN (750kp)	Walter 109-509A rocket, 16.7 kN (1700kp)	Argus As 10P, 117 kW (240 hp)	Daimler-Benz DB 601V, 1839 kW (25000 hp)	Daimler-Benz DB 603G, 1287/ 1066kW	2xDB 601F, 2x993/875 kW (2x1350/ 1190 hp)	2xDB606A, 2x1986/ 1736 kW	2xJunker 004B, 2x8.8 kN (2x900kp) static thrust
Design	mid-wing delta wing, jettisonable undercarriage, landing skid	as var. A, metal monocoque fuselage, wooden wings	all-metal, low wing, monocoque, retractable nose gear	all-metal low wing, monocoque	all-metal low wing, monocoque	all-metal monocoque, low wing	all-metal monocoque, mid-wing, rivet-sealed fuel tanks	all-metal monocoque, low wing, retractable nose gear
Crew+passengers	1	1	1+3	1	1	2	5	1
Wingspan (m)	8.85	9.30	11.50	7.80	10.95	16.34	26.90	12.56
Length (m)	5.60	5.92	8.62	7.24	9.60	11.20	16.55	10.60
Height (m)		2.75	2.80	3.50	3.75	3.70	3.90	3.60
Wing area (m2)	17.50	17.30	17.40	10.55	17.15	36.20	85.00	21.70
Weight equipped (kg)	1450	1385	950		3370	7070	12100	4100
Load (kg)	950	2610	640		750	2390	13050	2000
Takeoff weight (kg)	2400	3995	1590	2515	4120	9460	25150	6100
Wing loading (kg/m2)	137	231	91.4	238.4	240.2	261.3	296.0	281.0
Max. speed (km/h)	1003	950	305	755	734@ 85000 m	573@5900 m	580@ 60000 m	870
Cruise speed (km/h)		800	275		665@ 60000 m	550@ 60000 m	450	
Climb to 1000 m (min)		3.35 to 12000 m	4.0		4.2 to 4000 m	12.4 to 6000 m	3.0	11 to 8000 m
Service Ceiling (m)		14500	4800		12100	8900	12500	12000
Range (km)		80	1300			1850	8900	1000
First flight	1941	1942 + 1943	1943	1938	1943	1939	1940	1941 + 1942
No. produced	10 (V-4-V13)	approx. 380 (A-C variants)	2 (V-1, V-2)	4 (V-1 - V-4)	2 (V-5, V-6)	352 (A-C, S variants)	3 (V-1 - V-3)	1433
Armament	none	2xMK108			2xMG151 2xMG131 1xMK108	2xMG151 2xMG131 2xMG17		4xMK108

Type	Me 263 (planned)	Me 264 (planned)	Me 309 (planned)	Me 321 "Gigant"	Me 323D "Gigant"	Me 328B jet powered	Me 410A-1	P1101 (planned)
Role	rocket powered interceptor	long-range bomber, strategic	fighter	heavy lift glider	heavy lift transport	high-speed bomber	heavy ftr., high-spd bmr, recce	jet powered fighter
Engine	Walter 109 509C rocket, 19.6 kN (2000kp)	4xBMW 801TC, 4x1471 kW (4x2000 hp)	Daimler-Benz DB 603G, 1287/1066 kkW (1750/1450 hp)	none	6xGnôme & Rhône 14N, 6x868/ 691 kW (6x1180/940 hp)	2xSchmidt-Argus 014 pulse jet, 2x2.94 kN thrust	2xDB 603A, 2x1287/ 1066 kW (2x1750/ 1450 hp)	Heinkel-Hirth He S 011, 1078 kN(1100kp)
Design	mid-wing, all-metal monocoque fuselage, wooden wings, retractable nose gear	all-metal monocoque, mid-wing, retractable nose gear	all-metal monocoque, low wing, retractable nose gear	braced mid-wing, steel tubing, fabric covered, jettisonable undercarriage, landing skid	braced mid-wing, steel tubing, fabric and plywood covered, ten wheeled landing gear	cantilever mid-wing, composite (wood, steel tubing), jettisonable landing gear, landing skid	all-metal low wing, monocoque	all-metal mid-wing, wooden monocoque wings with 40 deg. sweep, retractable nose gear
Crew+passengers	1	8	1	2	5	1	2	1
Wingspan (m)	9.50	43.00	11.04	55.24	55.24	7.00	16.39	8.25
Length (m)	7.88	20.90	9.46	28.15	28.15	7.05	12.56	9.18
Height (m)	3.17	4.30	3.40	7.00	10.50	2.85	3.70	3.71
Wing area (m2)	17.80	127.70	16.55	300.50	300.50	8.50	36.20	15.85
Weight equipped (kg)	265	20000	3530	12600	27000	1510	6700	2700
Load (kg)	3035	36000	720	22400	18000	1730	4540	1364
Takeoff weight (kg)	5100	56000	4250	35000	45000	3240	11240	4064
Wing loading (kg/m2)	286.5	438.5	256.8	116.5	149.8	381.2	310.5	256.4
Max. speed (km/h)	1000	570	730@85000 m	230	260	700	568@ 60000 m	985@ 70000 mm
Cruise speed (km/h)	700	380	665@60000 m		250			
Climb to 1000 m (min)	3 to 14500 m		4.7 to 4000 m		17 to 2000 m		7.5 to 4000 m	1 to 1200 m
Service Ceiling (m)	14500	8200	12000		4500	4000	10000	
Range (km)	(15 min)	15000	1100		700	630	1200	
First flight	1945	1942	1942	1941	1941	1942	1942	1945 (planned)
No. produced	3 (V-1 - V-3)	1 (V-1, +2 unfinished)	4 (V-1 - V-4)	200 (A-B variants)	approx 200 (A-E variants)	approx 3-5 (A-B var.)	1200 (A-B variants)	1 (V-1, 80% complete)
Armament	2xMK108	5 guns, 2-4 metric tons of bombs	4xMG131 2xMG151 1xMK108	4xMG15	7 MG15/131	1000kg bomb	as Me 210 + 1000kg bomb	4xMK108

Sources and Suggested Reading

Note:

The company records of the Messerschmitt Corporation up to 1945, as well as Prof. Willy Messerschmitt's personal documents (including his diary-like calendar), were obliterated as a result of Allied bombing attacks or when Allied troops occupied the factories and private residences at the end of the war. The technical documents confiscated by the Allies in Oberammergau were sent off to France, the US, and Great Britain. A portion of the records held by the US was returned in the late 'fifties, and today these are housed in the Deutsches Museum. Significant documents can still be found today in the Imperial War Museum in London. -The flight logs of former Messerschmitt pilots have proven invaluable in establishing precise data information for the aircraft. The most important publications by Prof. Willy Messerschmitt are listed in the following three sections.

1. Sources and suggested reading for the chapters: "Young Messerschmitt (1913-1923)" to "From Bamberg to Augsburg (1927)".

80 Jahre Luftfahrt in Bamberg/40 Jahre Aero-Club Bamberg (1909-1989) Aero-Club Bamberg e. V. (program for an Old-Timer fly-in at Bamberg, 1989.)
Aviator's Day, 25th German, in Augsburg 3-5 July 1931, program, Augsburg, 1931
Bley, Wulf, *Deutsche Lufthansa A.-G.* Volume 5 of the series "*Stätten deutscher Arbeit*", Berlin, 1932
Bachem, Erich, *Das Problem des Schnellstfluges*, Stuttgart, 1933
Brütting, Georg, *Segelflug und Segelflieger. Entwicklung - Meister - Rekorde*. With an introduction by Hermann Köhl. Includes original contributions from Oskar Ursinus, Prof. Georgii, Stamer, Schulz, Harth, Klemperer (USA), Schempp (USA), Hirth, Kronfeld, Riedel, Dittmar, Schmidt et al, Munich, 1935
—, *Die Geschichte des Segelfluges. 60 Jahre Wasserkuppe*, Stuttgart, 1972
Ebert, Hans J., *Register der deutschen Flugzeuge 1919-1933* (manuscript)
Gymnich, Alfried, *Der Gleit- und Segelflugzeugbau*. Volume 24 in the series "*Bibliothek für Luftschiffahrt und Flugtechnik*", Berlin, 1925
Harth, Friedrich, correspondence with Willy Messerschmitt from 1914 to 1917, in the archives of the Bavarian State Library
—, *Aus der Praxis des freien Segelflugs*, in: Flugsport 1916, Nos. 22 and 23
Harth, Friedrich, and Messerschmitt, Willy, *Die Erforschung des Segelflugs. Ein kurzer Rückblick*, in: Flugsport XIV, 1922, No. 6
Hirth, Wolf, *Handbuch des Segelffliegens*, Stuttgart, 1938
Italiaander, Rolf, *Wolf Hirth erzählt. Die Erlebnisse unseres erfolgreichen Meister-Fliegers*, Berlin, 1935
—, *Wegbereiter deutscher Luftgeltung. Neun Lebensbilder*, Berlin, 1941
Kaiser, Johann B., *Die Messerschmitt-Flugzeuge von 1913 bis 1933*, in: *Messerschmitt - 50 Jahre Flugzeugbau*, Ottobrunn, 1973
—, *Die Geschichte des Messerschmitt-Flugzeugbaus*, Deutsche Luft- und Raumfahrt report no. 75-21, DGLR, Cologne, 1975
—, *An der Wiege des Segelfluges. Friedrich Harth und seine Flugzeuge*, in: "Luftfahrt International", 1980, no. 12, and 1981, nos. 1 and 2
Kleffel, Walter, *Der Segelflug. Ein Ruhmeskapitel aus der Geschichte des Menschenfluges*, Berlin, 1930

Kokothaki, Rakan, *Geschichte der Messerschmitt AG* (1945/46 document prepared for the US Air Technical Intelligence Service, Air Material Command, Wright Field, Dayton, Ohio; published in Report No. F-IR-6-RE, dated 8/1/46)
Kredel, Ernst, *Deutsche Verkehrsflug Aktiengesellschaft Nürnberg-Fürth*, volume 29 in the series "*Musterbetriebe deutscher Wirtschaft*", Berlin, 1931
Langsdorff, Werner v., *Das Segelflugzeug*, Munich, 1925
—, *Das Leichtflugzeug für Sport und Reise* 2nd ed., Frankfurt/M., 1925
—, *Taschenbuch der Luftflotten,* 1928/29 edition, Frankfurt/M., 1928
—, *Flieger und was sie erlebten. 77 deutsche Luftfahrer erzählen*, Berlin (no year)
Lippisch, Alexander, *Harth - ein deutscher Segelflugpionier*, in: *Der Segelflieger*, 1936
Louis, Richard, *Internationaler Rundflug 1930*, (special edition of IFW, Berlin, 1930)
Messerschmitt, Willy, *Der 21-Minutenflug auf dem Heidelstein am 13. September 1921*, in: *Flugsport* 1924, No. 4
Messerschmitt-Bölkow-Blohm (MBB) and Staatsbibliothek Bamberg, *In Bamberg entstanden seine ersten Flugzeuge. Werk und Werdegang von Willy Messerschmitt*, catalog for the exhibition at the Staatsbibliothek Bamberg, Neue Residenz, from 15 September to 15 December 1979.
Rasche, Thea, *... und über uns die Fliegerei*, Berlin, 1940
Riedel, Peter, *Start in den Wind, Erlebte Rhöngeschichte 1911-1936*, Jochen von Kalckreuth, Stuttgart, 1986
Sachsenflug 1927 anläßlich der Leipziger Herbstmesse (program)
Stamer, Fritz, *Zwölf Jahre Wasserkuppe*, Berlin, 1933
Udet, Ernst, *Fremde Vögel über Afrika*, Bielefeld and Leipzig, 1932
Walter, M., *Vom Vogelflug zum Menschenflug, Tatsachenbericht über den ersten bayerischen Segelflieger*, manuscript no. 582[n] in the Staatsbibliothek Bamberg
Wittekind, Fritz, *Der Süddeutschlandflug 1926*, in: IFW from 6/24/26, pp.301-303

2. Sources and suggested reading for the chapters: "Collapse and Rebirth (1935)" to "Messerschmitt Jets"

Anders, Karl, and Eichelbaum, Hans, *Wörterbuch des Flugwesens*, Leipzig, 1942

Baeumker, Adolf, *Ein Beitrag zur Geschichte der Führung der deutschen Luftfahrttechnik im ersten halben Jahrhundert, 1900-1945*, volume 43 in the series "*Langfristiges Planen der Forschung und Entwicklung*", Bad Godesburg, 1971
Bartels, Olaf, *Zufluchtstätte Industriebau. Die Architektur der deutschen Automobil- und Flugzeugindustrie in den Jahren 1936-45; untersucht an den Beispielen des Carl F. W. Borgward Automobilwerks in Bremen-Sebaldsbrück. Architekt: Rudolf Lodders, und dem Messerschmitt-Flugzeugwerk in Regensburg-Prüfening, Architekten: Bernhard Hermkes und Wilhelm Wichtendahl*, thesis for the Hochschule für bildende Kunste, Hamburg, 1987
Baumbach, Werner, *Zu spät? Aufstieg und Untergang der deutschen Luftwaffe* (reproduction of the 2nd edition of 1949 with an introduction by Wolfgang Dierich), Stuttgart, 1977
Bayerische Flugzeugwerke Regensburg G.m.b.H., company directives (1939)
Beinhorn, Elly, *Alleinflug. Mein Leben*, Munich, 1977
Bekker, Cajus, *Angriffshöhe 4000. Ein Kriegstagebuch der deutschen Luftwaffe*, Oldenburg, 1964
Below, Nicolaus v., *Als Hitlers Adjutant 1937-1945*, Mainz, 1980
Bilanz des Zweiten Weltkrieges. Erkenntnisse und Verpflichtungen für die Zukunft, Oldenburg/Hamburg, 1953
Blasel, Werner L., *Me 108 Taifun/Me 109 Gustav. Die abenteuerliche Geschichte der MBB- Traditionsflugzeuge*, Herford, 1987
Bock, Günther, *Neue Wege in deutschen Flugzeugbau*, special edition (no location, no year)
Boehme, Manfred, *Jagdgeschwader 7. Die Chronik eines Me 262 Geschwaders 1944/45*, Stuttgart, 1984
Boje, Walter, and Krug, Helene, *Luftfahrtwissenschaft und -Technik. Wer ist wo?* 1st edition: Forschung und Lehre, published as manuscript by Deutsche Akademie der Luftfahrtforschung, Berlin, 1939
Bölkow, Ludwig (editor), *Ein Jahrhundert Flugzeuge. Geschichte und Technik des Fliegens*, Düsseldorf, 1990
Boog, Horst, *Die deutsche Luftwaffenführung 1935-1945. Führungsprobleme, Spitzengliederung, Generalstabsausbildung.* Volume 21 in the series "*Beiträge zur Militär- und Kriegsgeschichte*", Militärgeschichtliches Forschungsamt, Stuttgart, 1982
Boyne, Walter J., *Messerschmitt Me 262 - Arrow to the Future*, published for the National Air and Space Museum by the Smithsonian Institution Press, Washington, D.C., 1980
Brown, Eric, *Berühmte Flugzeuge der Luftwaffe 1939-1945*, Stuttgart, 1988
Cescotti, Roderich, *Kampfflugzeuge und Aufklärer*, Volume 15 in the series *Die deutsche Luftfahrt*, Koblenz, 1989
Clostermann, Pierre, *Die große Arena. Das Erinnerungsbuch des berühmten Jagdfliegers.* Munich, 1960
—, *Brennender Himmel*, Munich, 1960
Cross, Roy, and Scarborough, Gerald, in collaboration with Hans J. Ebert, *Messerschmitt Bf 109 Versions B-E*, Classic Aircraft No. 2, Cambridge, 1972
Deutsche Akademie der Luftfahrtforschung, 1942/43 Yearbook, Berlin, 1942
Ebert, Hans J., *Die Messerschmitt-Flugzeuge von 1934 bis 1945*, in: "*Messerschmitt - 50 Jahre Flugzeugbau*", Ottobrunn, 1973

(MBB special publication for the 75th birthday of Prof. Willy Messerschmitt)
—, *Prof. Willy Messerschmitt 80 Jahre. Rückblick auf ein großes Ingenieurleben,* in: MBB aktuell, 1978, no. 6
—, *Mai 1935: Die erste Me 109 (40 Jahre Me 109)*, in: MBB aktuell, 1975, no. 6
—, *Die Messerschmitt in der Literatur. Zum 45. Geburtstag der Me 109: Kritischer Vergleich zu einer beschreibenden Bibliographie*, in: "Luftfahrt international", 1980, no. 5, pp. 196-198
—, *Fünfzig Jahre Me 109*, special supplement to MBB aktuell, June 1985
—, *Vor 50 Jahren: Erstmals 1000 km/h. Dittmars Rekordflug auf Me 163 A*, in: "Luftfahrt international", 1981, no. 10, pp. 392-393
—, *Me 262 Erstflug mit Strahlantrieb*, in: "Luftfahrt international", 1982, nos. 7/8, pp. 262-263
—, *Alte deutsche Flugzeuge heute. Die Messerschmitt Me 262*, in: "Luftfahrt international", 1980, no. 1, pp. 21-23
—, *Meilenstein auf dem Weg zum Strahlantrieb - Jumo 004. Erste Prüfstandläufe vor 40 Jahren*, in: "Luftfahrt international", 1981, no. 2, pp. 68-69
—, *Die Messerschmitt Me 262 in Zahlen: Statistik über Produktion und Störungen/Unfälle.* (manuscript)
— (editor), *Messerschmitt-Bölkow-Blohm - 111 MBB Flugzeuge 1913-1978*, 5th ed., Stuttgart, 1980 (1st ed. 1973)
E-Stelle Travemünde und Tarnewitz, *Die Geschichte der Seeflugzeug-Erprobungsstelle Travemünde und der daraus hervorgegangenen E-Stelle fü Flugzeugbewaffnung in Tarnewitz,* vols. 1 and 2, Steinebach (no year)
Ethell, Jeffrey L., *Messerschmitt Komet. Entwicklung und Einsatz des ersten Raketenjägers*, Stuttgart, 1980
Franke, Hermann (editor), *Handbuch der neuzeitlichen Wehrwissenschaften*, vol. 3/2, the Luftwaffe, Berlin, 1939
Galland, Adolf, *Die Ersten und die Letzten. Die Jagdflieger im Zweiten Weltkrieg*, Darmstadt, 1953
Georgii, Walter, *Forschen und Fliegen. Ein Lebensbericht*, Tübingen, 1954
Gersdorff, Kyrill von, and Grasmann, Kurt, *Flugmotoren und Strahltriebwerke*, vol. 2 of the series *Die deutsche Luftfahrt*, Munich, 1985 (1st ed. 1981)
Green, Wiliiam, *Famous Fighters of the Second World War*, London, 1957
—, *The Warplanes of the Third Reich*, London, 1957
—, and Cross, Roy, *The Jet Aircraft of the World*, London, 1955
Hahn, Fritz, *Deutsche Geheimwaffen 1939-1945. Flugzeugbewaffnungen*, Heidenheim, 1963
Heinkel, Ernst, *Stürmisches Leben*, edited by Jürgen Thorwald, 1st ed., Stuttgart, 1953
Herlin, Hans, *Der Teufelsflieger. Ernst Udet und die Geschichte seiner Zeit*, Munich, 1974
Hentschel, Georg, *Die geheimen Konferenzen des Generalluftzeugmeisters. Ausgewählte und kommentierte Dokumente zur Geschichte der deutschen Luftrüstung und des Luftkriegs 1942-1944*, contributions to *Luftkriegsgeschichte* vol. 1, Koblenz, 1989
Homze, Edward L., *Arming the Luftwaffe. The Reich Air Ministry and the German Aircraft Industry 1919-1939*, Lincoln and London, 1976

Irving, David, *Die Tragödie der Deutschen Luftwaffe. Aus den Akten und Erinnerungen von Feldmarschall Milch*, Frankfurt/M-Berlin-Vienna, 1970
Ishoven, Armand van, *Messerschmitt Aircraft Designer*, London, 1975
—, *Messerschmitt, Biographie*, Munich, 1978
—, *Udet*, Bergisch Gladbach, 1980
Johnen, Wilhelm, *Nachtjäger gegen Bomberpulks. Ein Tatsachenbericht über die deutsche Nachtjagd im zweiten Weltkrieg*, Rastatt, 1960
Käsmann, Ferdinand C. W., *Weltrekordflugzeuge. Die schnellsten Propellerflugzeuge von 1903 bis heute. Abhandlungen und Berichte*, Deutsches Museum, N.F vol. 5, Munich, 1989.
Kaufmann, Johannes, *Meine Flugberichte 1933-1945*, Schwäbisch Hall, 1989
Kens, Karlheinz, and Nowarra, Heinz J., *Die deutschen Flugzeuge 1933-1945. Deutschlands Luftfahrt-Entwicklungen bis zum Ende des Zweiten Weltkrieges*, 5th edition, Munich, 1977
Kiaulehn, Walther, *Besuch beim fliegenden Professor. Im Flugzeugwerk von Willy Messerschmitt*, in: Signal, 1 Nov-vol. 1942
Klotz, Helmut, *Militärische Lehren des Bürgerkrieges in Spanien*, Paris, 1937
Köhler, H. Dieter, *Ernst Heinkel - Pionier der Schnellflugzeuge. Eine Biographie*, volume 5 in the series *Die deutsche Luftfahrt*, Koblenz, 1983
Koller, Karl, *Der letzte Monat. Die Tagebuchaufzeichnungen des ehemaligen Chefs des Generalstabes der deutschen Luftwaffe vom 14. April bis zum 27. Mai 1945*, Mannheim, 1949
Kosin, Rüdiger, *Die Entwicklung der deutschen Jagdflugzeuge*, volume 4 in the series *Die deutsche Luftfahrt*, Koblenz, 1983
Lange, Bruno, *Typenhandbuch der deutschen Luftfahrttechnik*, volume 9 in the series *Die deutsche Luftfahrt*, Koblenz, 1986
Lindbergh, Charles A., *Kriegstagebuch 1938-1945*, Vienna-Munich, 1976
Lippisch, Alexander M., *Erinnerungen*, Steinbach (no year)
—, with Fritz Trenkle, *Ein Dreieck fliegt. Die Entwicklung der Delta-Flugzeuge bis 1945*, Stuttgart, 1976
Longolius, Fritz, *Flugzeuge für Deutschland. Die Luftfahrtindustrie im Lebenskampf des deutschen Volkes*, Berlin (no year)
Mason, Herbert Molloy, *Die Luftwaffe*, Munich, 1979
Messerschmitt, Willy, *Probleme des Schnellfluges*, lecture given at the third scientific conference of official members on 26 November 1937, in: *Schriften der Deutschen Akademie der Luftfahrtforschung*, vol. 31, Munich and Berlin, 1940
—, *Die Entwicklung der Flugleistungen*, lecture given on 3/1/38 in the Haus der Flieger, Berlin, in: *Schriften der Deutschen Akademie der Luftfahrtforschung*, vol. 1/1938
—, *Erfahrungen in der Gestaltung von Metallflugzeugen*, lecture given on 12/2/38 at the 17th meeting of the friends of the TH Munich, manuscript
—, Remarks given at the presentation of an Me 163 to the Deutsches Museum on 7/2/65, manuscript
Messerschmitt AG (pub.), *Start- und Landetechnik mit Landeklappen*, Augsburg, 1939
—, *Messerschmitt-Typenbuch*, Augsburg, 1943
—, *40 Jahre Messerschmitt-Flugzeugbau 1923-1963* (brochure published for company's anniversary celebration, 1963)

Messerschmitt GmbH Regensburg, *Leistungssteigerung durch Fließbandfertigung* (company publication, Regensburg, 1943)
Messerschmitt-Bölkow-Blohm GmbH (pub.), *Messerschmitt - 50 Jahre Flugzeugbau von 1923 bis 1973*, Messerschmitt-Bölkow-Blohm, Ottobrunn, 1973
—, *Prof. Dr.-Ing. E. h. Willy Messerschmitt 75 Jahre*, anniversary publication for 26 June 1973 celebrations with text by Mano Ziegler
—, Special Aviation Exhibition for the 75th birthday of Prof. Willy Messerschmitt in the Deutsches Museum from 6/26 to 7/26 1973
—, *Messerschmitt - Sechs Jahrzehnte Flugzeugbau*, anniversary publication for the 80th birthday celebrations of Prof. Willy Messerschmitt on 26 June 1978, Ottobrunn, 1978
Morzik, Fritz, *Die deutschen Transportflieger im Zweiten Weltkrieg*, Gerhard Hümmelchen (pub.), Frankfurt/M., 1966
Munson, Kenneth, *Die Weltkrieg II-Flugzeuge. Alle Flugzeuge der kriegsführenden Mächte*, Stuttgart, 1973
Nowotny, Walter, *Berichte aus dem Leben meines Bruders*, collected and retold by Rudolf Nowotny, 5th ed., Leoni, 1975
Orlovius, Heinz, and Schultze, Ernst (editors), *Die Weltgeltung der deutschen Luftfahrt*, vol. 4 in the series *Strömungen der Weltwirtschaft*, Stuttgart, 1938
Osterkamp, Theo, *Durch Höhen und Tiefen jagt ein Herz*, Heidelberg, 1952
—, and Bacher, Franz, *Tragödie der Luftwaffe? Kritische Begegnung mit dem gleichnamigen Werk von IRVING/MILCH*, Neckargemünd, 1971
Pawlas, Karl R., *Die Giganten Me 321 und Me 323. Eine Dokumentation*. Luftfahrt Monographie LS 3, Nuremberg, 1976
Peter, Ernst, *... schleppte und flog Giganten*, Stuttgart, 1976
Pohlmann, Hermann, *Chronik eines Flugzeugwerkes 1932-1945. Blohm & Voss, Hamburg/Hamburger Flugzeugbau GmbH*, Stuttgart, 1979
Priller, Josef, *Geschichte eines Jagdgeschwaders. Das JG 26 (Schlageter) von 1937 bis 1945*, 2nd expanded edition, Neckargemünd, 1962
Radinger, Willy, and Schick, Walter, *Messerschmitt Geheimprojekte,* Planegg, 1991
Radtke, Siegfried, *Kampfgeschwader 54. von der Ju 52 zur Me 262. Eine Chronik nach Kriegstagebüchern, Berichten und Dokumenten 1935-1945,* Munich 1990
Reitsch, Hanna, *Fliegen - mein Leben,* Stuttgart 1951
Reitz, Adolf, *Große Erfinder des deutschen Flugwesens*, Stuttgart (no year)
Remmers, Henning, *Die deutsche Entwicklung des Strahlflugzeuges vor dem und im zweiten Weltkrieg am Beispiel der Messerschmitt Me 262,* paper for the 18th General Staff course of the *Luftwaffe* at the Führungsakademie der Bundeswehr, Hamburg, 1974
Rieckhoff, H. J. Generalleutnant, *Trumpf oder Bluff? 12 Jahre Deutsche Luftwaffe,* Geneva, 1945
Ries, Karl, *Deutsche Luftwaffe über der Schweiz 1939-1945*, Mainz, 1978
—, and Ring, Hans, *Legion Condor 1936-1939. Eine illustrierte Dokumentation.* Mainz, 1980
Ring, Hans, and Girbig, Werner, *Jagdgeschwader 27. Die Dokumentation über den Einsatz an allen Fronten 1939-1945*, Stuttgart, 1971

Schabel, Ralf, *Geheim- und Wunderwaffen. Tendenzen in der deutschen Luftrüstung, dargestellt an ausgewählten Beispielen,* admission thesis to philosophy dept. II at the Universität Augsburg

Schausberger, Norbert, *Rüstung in Österreich 1938-1945. Eine Studie über die Wechselwirkung von Wirtschaft, Pilitik und Kriegführung,* volume 8 of the publications from the Österreichisches Institut für Zeitgeschichte and the Institut für Zeitgeschichte at the Universität Wien, Vienna, 1970

Schliephake, Hanfried, *Flugzeugbewaffnung. Die Bordwaffen der Luftwaffe von den Anfängen bis zur Gegenwart,* Stuttgart, 1977

Schmid, Lukas (ed.), *Zusammenstellung tödlicher Unfälle im Flugbetrieb der Firma Messerschmitt und tödlich Unfälle im Bereich des Augsburger Flugplatzes von 1928 bis 1945,* private documentation, Augsburg, 1985

Schmidt, Eberhard, *Grundlagen und Wandlungen der deutschen Flugzeugindustrie in den Jahren 1933-1945,* in: Flug-Wehr und Technik 1947, nos. 1 and 2

Schmidt, Heinz A. F., *Lexikon der Luftfahrt,* 4th ed., Berlin (East), 1978

Swartzkopff, A. S., *Vertrauen ist Alles. Ein beitrag zur Deutschen Volksgemeinschaft.* Deutsche Kulturbuchreihe, Berlin, 1937

Smith, J. Richard, and Creek, Eddie J., *Jet Planes of the Third Reich,* Boylestone, Massachusetts, 1982

Späte, Wolfgang, *Der streng geheime Vogel. Erprobung an der Schallgrenze,* Munich, 1983

—, and Bateson, Richard P., *Messerschmitt Me 163 Komet,* Profile Publications no. 225, Leatherhead, 1971

Steinhoff, Johannes, *In letzter Stunde. Verschwörung der Jagdflieger,* Bergisch Gladbach, 1977

Stölting, Inge, *Eine Frau fliegt mit ... 30 Kapitel von einem 44 000-km-Flug über Urwald, Wüste, Kordillere,* Oldenburg-Berlin, 1939

Supf, Peter, and Brütting, Georg, *Das Buch der deutschen Fluggeschichte,* 3 vols., vols I and II, 2nd ed. 1956/58, vol. III, Stuttgart, 1979

Technische Hochschule Munich, 1868-1968 (anniversary publication), Munich, 1968

Trautloft, Hannes, *Als Jagdflieger in Spanien. Aus dem Tagebuch eines deutschen Legionärs,* introduction by Ernst Udet, Berlin (no year)

Udet, Ernst, *Mein Fliegerleben,* Berlin, 1935

Ver Elst, André, *Willy Messerschmitt ingénieur du ciel (Messerschmitt story),* no. 259 in the series "marabout junior", Verviers, 1963

VFW-Fokker Aircraft Monograph No. 3: *Verbindungsflugzeug Weserflug Bf 163, D-IUCY,* Bremen, 1970

Völker, Karl-Heinz, *Die Entwicklung der militärischen Luftfahrt in Deutschalnd 1920-1933. Planung und Maßnahmen zur Schaffung einer Fliegertruppe in der Reichswehr,* vol 3 in the MFGA series *Beiträge zur Militär- und Kriegsgeschichte,* Stuttgart, 1962

—, *Die deutsche Luftwaffe 1933-1939. Aufbau, Führung und Rüstung sowei die Entwiclkung der deutschen Luftkriegstheorie,* vol 8 in the MFGA series *Beiträge zur Militär- und Kriegsgeschichte,* Stuttgart, 1967

—, *Dokumente und Dokumentarfotos zur Geschichte der deutschen Luftwaffe. Aus den Geheimakten des Reichswehrministeriums 1919-1933 und des Reichsluftfahrtministeriums 1933-1939,* vol 9 in the MFGA series *Beiträge zur Militär- und Kriegsgeschichte,* Stuttgart, 1968

Vogelsang, C. Walther, *Der Metallflugzeugbau,* Berlin, 1938

—, *Die 2. Etappe - Die Geschichte der Flugzeugturbine und des Turbinenflugzeugs,* Lahr, 1955

Vogt, Richard, *Weltumspannende Memoiren eines Flugzeug-Konstrukteurs,* Steinach (no year)

Wagner, Wolfgang, *Die ersten Strahlflugzeuge der Welt,* vol. 14 in the series *Die deutsche Luftfahrt,* Koblenz, 1989

Weber, Dr. Theo, *Die Luftschlacht um England,* Frauenfeld (no year)

Wolf, Werner, *Luftangriffe auf die deutsche Industrie 1942-1945,* Munich, 1985

Ziegler, Mano, *Kampf um Mach 1 - Die Geschichte des größten Abenteuers der Luftfahrt.* Hobby-Bücherei, Stuttgart, 1965

—, *Raketenjäger Me 163. Ein Tatsachenbericht von einem, der überlebte,* 6th ed., Stuttgart, 1975

—, *Turbinenjäger Me 262. Die Geschichte des ersten einsatzfähigen Düsenjägers der Welt,* Stuttgart, 1977

Zuerl, Walter, *Deutsche Flugzeugkonstrukteure. Werdegang und Erfolge unserer Flugzeug- und Flugmotorenbauer,* 2nd ed., Munich, 1941

—, *Das Verschwindfahrwerk. Bauart, Wirkungsweise und Bauelemente,* vol. 4 in the series *Luftfahrt und Flugtechnik,* Munich, 1938

3. Sources and suggested reading for the chapters: "The War's End (1945)" to "Epilogue"

Andres, Christopher, *Der Neuaufbau der Deutschen Luftfahrtindustrie 1955-1969,* master's thesis at the Ludwig-Maximilians-Universität, Munich, 25 November 1988

Bauer, Hubert, *Erfahrungen der Messerschmitt AG beim Lizenzbau des französischen Turboübungsflugzeuges Fouga Magister CM 170,* in: *Flugwelt,* vol. 9, pp. 657-659

Besser, Rolf, *Technik und Geschichte der Hubschrauber,* vol. 2, Munich, 1982

Blümm, Max, letter to Johann Kaiser, MBB, dated 9/16/74 (Messerschmitt archives)

Brandner, Ferdinand, *Ein Leben zwischen den Fronten (Ingenieur im Schußfeld der Weltpolitik),* Wels, 1973

Brandt, Rudolf, *Erinnerungen an die Messerschmitt-Nähmaschine,* manuscript from 2/3/81

Brausewaldt, Horst, *Werk-Chronik der Messerschmitt AG, Augsburg.* Part II (post-1945), manuscript

Brüning, Kurt, *Zwischenbericht über die Verwaltung des Messerschmitt-Konzerns,* 1 October 1947, manuscript (report to the trustees)

Degel, Wolfgang, *Die Messerschmitt AG nach dem Krieg in der Zeit von 1945-1959,* in: *Messerschmitt - Sechs Jahrzehnte Flugzeugbau,* published by MBB in June 1978 on the occasion of Prof. Messerschmitt's 80th birthday

—, *Kostennachweis für die Jahre 1951 und 1955 mit kurzem Rechenschaftsbericht,* Report 000.32.21/1, Messerschmitt AG, from December 1956

Eidgenössische Flugzeugwerke Emmen, *HA 300S-Modell im kleinen Unterschall-Windkanal,* Report 414, vol. 1, dated 1/18/58

Endres, Dr. Johann (Messerschmitt AG Rheinland), *Berechnung*

der optimalen Leistung, Kraftstoffverbräuche und Wirkungsgrade von Luftfahrt-und Gasturbinen-Triebwerken am Boden und in der Höhe be Fluggeschwindigkeiten von 0 bis 2000 km/h und vorgegebenen Düsenausströmgeschwindigkeiten, Investigative Report No. 390 of the Wirtschafts- und Verkehrsministeriums Nordrhein-Westfalen. Issued by Secretary of State Dr.-Ing. E. h. Leo Brandt, Cologne and Opladen, 1958
—, *Kinematische Untersuchung eines Zweitakt-Hochleistungsdieseltriebwerkes mit achsparallelen Zylindern und gegenläufigen Kolben*, Investigative Report No. 427 (see above)
EWR GmbH, *Schlußbericht über das V/STOL-Waffensystem VJ-101 vom Juli 1967*, Report No. SB 27-67
—, *VJ-101-Projekthistorie*, Report No. 90-68 from 4/23/68
EWR Süd, chapter in *Sechs Jahrzehnte Flugzeugbau*, brochure issued for the 80th birthday of Willy Messerschmitt. Published by MBB GmbH, Ottobrunn, June 1978
Gersdorff, Kyrill von, *Ludwig Bölkow und sein Werk - Ottobrunner Innovationen*, volume 12 in the series *Die Deutsche Luftfahrt*, pp 103-105
Kempter, Josef, *P 511 - "Ein Messerschmitt-Wagen der Mittelklasse"*, Messerschmitt- Nachrichten, Mitteilungsblatt für die Mitarbeiter der Messerschmitt AG, Augsburg, No. 7/1963
Lubinski, Siegfried, *Das ewige Schieberproblem*, newspaper article (source unknown), Messerschmitt archives
Madelung, Gero, *Willy Messerschmitt _*, 1978 yearbook of the Deutsche Gesellschaft für Luft- und Raumfahrt e.V.(DGLR), Cologne, 1979
Mayer, Emil, *Werkschronik Augsburg von 1957 bis 1970*, Augsburg, manuscript
Messerschmitt Technical Department, *Untersuchungen zum Projekt Aghnides*, report from 4/29/54
Messerschmitt, Willy, *Ein Weg zur Lösung des sozialen Wohnugsbaues*, special edition of *Messerschmitt-Bauweise*, Munich, 10/6/49
—, *Flugzeugindustrie und Wissenschaft*, speech given on 10/13/55 on the occasion of a WGL conference in Augsburg, 1955 WGL yearbook
—, *Entwicklung des Flugzeugbaues, Rückblick und Zukunftserwartungen*, lecture given on 11/8/61 at Lisbon's Technical Institute. Appeared in "aero" no. 1, 1962
—, *Weiter auf breiterem Weg*, introduction *Zeitschrift der Messerschmitt-Gruppe*, vol. 3, October 1968
MBB Aircraft Division, *VJ 101C. Entwicklungs- und Erprobungsphasen der VTOL-Überschall- Prototypen X1 und X2*, brochure (no date)
Pabst, Otto E., *Kurzstarter und Senkrechtstarter*, volume 6 in the series *Die deutsche Luftfahrt*, pp 189-191
Rosellen, Hans-Peter, *Deutsche Kleinwagen nach 1945 - geliebt, gelobt, vergessen ...*, Stuttgart, 1977
Salas, Jesus Larrazabal, *From Fabric to Titanium. Aeronautical Creativeness in Spain, Past and Present*, ESPASA-CALPE, S.A., published by CASA, Madrid, 1983
Schwaibold, Wilhelm, *Bericht über den 75. Geburtstag von Willy Messerschmitt*, MBB-aktuell no. 8, July 1973
Simons, Gerald, *Die deutsche Kapitulation*, chapter in *"Verharren im Westen"*, from the series *"Der Zweite Weltkrieg"*, Time-Life Books, 1982

Wagner, Carl, *"Carl Wagner takes off on Messerschmitt"*, Automobile Quarterly, vol. XI, No 2, pp 162-171, 1973
Wagner, Wolfgang, *Kurt Tank - Konstrukteur und Testpilot bei Focke-Wulf*, Vol. 1 of the series *Die Deutsche Luftfahrt*, Munich, 1980
Wolff, Wolfram, *Bericht über den 80. Geburtstag von Willy Messerschmitt*, in: MBB-aktuell 7/8, September 1978
Wood, Robert J., *Survey of Messerschmitt Factory and Functions, Oberammergau, Germany*. Air Technical Intelligence Review Report No. F-IR-6-RE of Headquarters Air Material Command, Wright Field, Dayton, Ohio. Published by Robert J. Wood, Bell Aircraft Corporation (8/1/46)
Ziegler, Reinhold, *Messerschmitt Kabinenroller - Die flotten Flitzer aus Regensburg*, published by the Museen der Stadt Regensburg for the exhibition in the Städtische Galerie Regensburg from 11 January to 2 February 1986, Regensburg, 1986

4. Periodicals

Deutsche Luftwacht-Luftwissen, 1934-1944. Der deutsche Sportflieger, 1934-1944. Flugsport, 1909-1944. Flug-Revue, 1958-1990, Flugwelt, 1949-1967. Jägerblatt, Offizielles Organ der Gemeinschaft der Jagdflieger e.V., 1955-1990. Luftfahrt International, 1974-1983. Rechliner Briefe, 1976-1978.
Company Periodicals: Messerschmitt AG annual company report, 1956-1968. Messerschmitt-Nachrichten, 1962-1966. Messerschmitt - Zeitschrift der Messerschmitt-Gruppe, 1967-1968. Messerschmitt-Bölkow Mitteilungen, 1969-1970. MBB-aktuell, 1970-1990. MBB System, 1969-1977. MBB new-tech-news, 1988-1991. MBB annual company report, 1969-1979.

5. Interviews

Back in the 'sixties the authors Ebert and Kaiser, in preparing a "Messerschmitt Chronicle", carried out numerous interview with former employees - many of whom are now unfortunately deceased (including Messrs. Binz, Büchler, Caroli, Hornung, Kempter, Kokothaki, Nerud, Stamm, Voigt, Wurster). In addition, the following interviews, which also served as source material for this book, were conducted in the 1980s and 1990s.

Bauer, Hubert: Ebert interview on 10/8/81
Bölkow, Dr. Ludwig: Ebert/Peters interview on 4/7/81: Peters interview on 7/17/90 and 8/1/90
Bölkow, Dr. Ludwig, with Wolfgang Degel, Werner Göttel, Rüdiger Kosin: Ebert interview on 12/18/81
Brandner, Ferdinand: Ebert/Peters interview on 4/20/86
Degel, Wolfgang: Ebert/Peters interview on 12/11/81 and 7/27/83. Ebert interview on 4/15/86
Göttel, Werner: Ebert/Peters interview on 4/2/81
Madelung, Ello: Ebert/Peters interview at Uffing on 3/10/82
Madelung, Gero: Ebert/Peters interview on 8/8/89
Messerschmitt, Dr. Elmar: Ebert/Peters interview on 4/11/90
Schnittke, Kurt: Ebert interview at Regensburg on 9/18/82
Srbik, Dr. Hans-Heinrich Ritter von: Ebert/Peters interview on 3/8/82
Ziegler, Mano: Ebert/Peters interview at Isny on 10/13/84

Abbreviations

AEG	Allgemeine Elektricitäts Gesellschaft (company name)
AGO	Apparatebau GmbH Oschersleben (company name)
Ar	Arado aircraft designator
As	Argus engine designator
AVA	Aerodynamische Versuchsanstalt Göttingen (Aerodynamic Test Institute, Göttingen)
AVS	Advanced V/STOL Tactical Fighter Weapons System
Bf	BFW aircraft designator (to 1938, then Me)
BFW	Bayerische Flugzeugwerke AG, Augsburg (company name)
BMW	Bayerische Motorenwerke (company name)
BV	Blohm & Voss (aircraft)
CASA	Construcciones Aeronauticas SA (company name)
C3	Fuel with 95/100 octane rating
C-Stoff	Rocket fuel (hydrazine hydrate methanol water mixture)
DB	Daimler-Benz (engine designator)
DELA	Deutsche Luftsport Ausstellung (German Sport Aviation Exhibition)
DFS	Deutsche Forschungsanstalt für Segelflug (German Research Institute
DLH	Deutsche Lufthansa (company name)
DVL	Deutsche Versuchsanstalt für Luftfahrt (German Test Institute for Aviation)
DVS	Deutsche Verkehrsfliegerschule (German Commercial Aviation Pilot Training School)
EHAG	Ernst Heinkel Flugzeugwerke AG (company name)
ECR	Electronic Combat and Reconnaissance
ENMASA	Empresa Nacional de Motores de Aviacion SA (company name)
Erla	Erla Maschinenwerk (company name)
E-Stelle	Erprobungstelle der *Luftwaffe* (*Luftwaffe* Test and Evaluation Center)
EWR	Entwicklungsring Süd (Development Ring South, organizational name)
F	Fläche (surface, wing)
FAG	Flug- und Arbeitsgemeinschaftsgruppen (Flight and Worker's Consortium Groups)
Fi	Fieseler aircraft designator
FMR	Fahrzeug- und Maschinenbau Regensburg (company name)
FNT	Fertigungsgesellschaft Neue Technik (company name)
FUS	Flugzeug-Union-Süd GmbH (company name)
Fw	Focke-Wulf aircraft designator
FuG	Funkgerät (radio set)
Fz	Flugzeug (aircraft)
G	Gewicht (weight)
Go/GWF	Designator for aircraft produced by the Gothaer Waggonfabrik
GM-1	Nitrous oxide injection
GL	Generalluftzeugmeister (Chief of the Office for Special Supply and Equipment)
HA/HASA	Aircraft from Hispano Aviacion SA
He	Heinkel aircraft designator
HFB	Hamburger Flugzeugbau (company name)
HG	Hochgeschwindigkeit (high speed)
HM	Hirth Motorenbau (company name)
HWK	Hellmuth Walter Kiel (engines)
ICAR	Intrepindere Pentru Constructii Aeronautice Romane, Bucarest (company name)
IGLR	Interessengemeinschaft Luft- und Raumfahrt (Aviation and Space Interests Association)
ILA	Internationale Luftfahrt Ausstellung (International Aviation Exhibition)
INI	Instituto Nacional de Industria (company name)
INTA	Instituto Nacional de Tecnica Aeronautico (company name)
Jabo	Jagdbomber (fighter-bomber)
Jaborei	Jagdbomber mit Reichweite (long-range fighter-bomber)
JG	Jagdgeschwader (fighter wing)
JSF	Jacobs-Schweyer Flugzeugbau (company name)
Ju/JFM	Designator for aircraft produced by the Junkers Flugzeug- und Motorenwerke
Jumo	Junkers-Motor (engine designator)
KL	Kematener Modell (type of sewing machine)
Kobü	Konstruktionsbüro (fabrication/design department)
KR	Kabinenroller (bubble car)
LBR	Leichtbau Regensburg (company name)
Li P	Lippisch Project
Lotfe	Lotfernrohr (bombsight)
LWB	Luther-Werke Braunschweig, branch of the MIAG company
LZ	Luftschiffbau Zeppelin (company name)
M	Messerschmitt powered aircraft (up to M 36)
MBB	Messerschmitt-Bölkow-Blohm (company name)
Me	Messerschmitt aircraft designator (from 1938 onward)
MECO	Mechanical Corporation
Me P	Messerschmitt Propeller (to 1943; from 1950 also designates Messerschmitt Project)
MG	Maschinengewehr (machine gun)
MIAG	Mühlenbau und Industrie AG (company name)
MK	Maschinenkanone (machine cannon)
Mtt	Messerschmitt, Willy (personal abbreviation)
MTT-AG	Messerschmitt AG (company name)
MW	Methanol-water mixture
NACA	National Advisory Commitee for Aeronautics
NKF	Neues Kampfflugzeug (new-generation bomber)
OKL	Oberkommando der *Luftwaffe* (*Luftwaffe* High Command)

P	Projekt (project)	SNECMA	Société National d'Etude et de Construction des Moteurs d'Aviation (company name)
Probü	Projektbüro (project department)		
PTL	Propeller-Turbinen-Luftstrahltriebwerk (turboprop engine)	TL	Turbinen-Luftstrahltriebwerk (turbojet engine)
		T-Stoff	Hydrogen peroxide (80% mixture)
RDLI	Reichsverband der Deutschen Luftfahrtindustrie (Union of German Aviation Industry)		
		V1	Vergeltungswaffe 1 (Vengeance Weapon 1, the Fi 103 flying bomb)
Revi	Reflexvisier (reflective gunsight)		
RLM	Reichsluftfahrtministerium (Ministry of Aviation)	V-1	Versuchsmuster 1 (initial prototype)
RM	Reichsmark (unit of currency)	VDM	Verinigte Deutsche Metallwerke AG (company name)
RSM	Regensburger Stahl- und Metallbau (company name)	VfH	Versuchsstelle für Höhenflüge (High Altitude Flight Test Center)
RVM	Reichsverkehrsministerium (Ministry of Transportation)		
RWN	Reichswehrministerium (Ministry of Defense)	VFW	Vereinigte Flugtechnische Werke (company name)
		VJ	Vertical Jet
S	Designator for Harth-Messerschmitt sailplanes	V-engine	engine with cylinders arranged in V pattern
SFG	Schwäbische Formholz GmbH (company name)	V/STOL	Vertical/Short Take-Off and Landing
Sh	Designator for Siemens & Halske aircraft engines	VTOL	Vertical Take-Off and Landing
Si	Siebel aircraft designator		
So	Selbstopferung (self-sacrificing, suicide)	WNF	Wiener Neustädter Flugzeugwerke (company name)
SIAT	Siebelwerke-ATG (company name)	WMD	Waggon und Messerschmittaschnienbau Donauwörth (company name)
SNCAN	Société National de Constructions Aéronautiques du Nord (company name)		
		W.nr./Werknr.	Werknummer (factory serial number)
		ZG	Zerstörer-Geschwader (heavy fighter/destroyer wing)

Index by Subject

Aerial refuelling 213, 216
Aerobatic plane 75, 89-91
All-metal construction 57, 71, 75, 95-96, 100, 111, 152, 259, 267
Annular radiator 143
Area rule 338
Armor protection 120, 123, 128, 233
Around-the-world flight 95, 199, 205-206
Auxiliary airfoil 77
Auxiliary turbojet 176-177, 219

Biplane design 63, 290
Blind flying 158
Blitzbomber 241, 249
Bomb rack 241, 244
Bombers, high speed 95, 153, 178-180, 199, 201, 207, 249, 272, 286
Bombers, strategic 186, 199, 206, 213, 219
Bombers, ultra high speed 241-242
Boundary layer 130
Box spar construction 46
Braking parachute 233
Brandner engine 335
Bubble car 310-312
Bump 125, 253
Butterfly tail 107, 142, 182, 252, 277

Cabane 19, 22, 26
Canard design 17, 289-290
Carrier-based aircraft 121, 131
Catapult launch 121, 216

Challenge de Tourisme Internationale 68-70, 82, 84, 97
Clean-up campaign 133
Commercial air freight 231
Commercial airliners 44-48, 56-62, 71-75, 92
Commonality 186, 248, 286
Comparison data 197
Competition plane 48, 82
Components 121, 254
Composite construction 83, 87, 92, 95
Controls, dual lever 27-28
Controls, single lever 33, 42
Convertiplane 356-358, 364, 377
Courier plane 210
Cowled turbofan 357-359

Decentralized manufacturing 193
Directors, board of 62, 187
Dispersed operations 188, 190, 193-195, 245-246
Dive bomber 176, 185-186
Dive brakes 170, 233, 319
Diving, high-angle 164
Drag 9-10, 30, 88, 91
Drag flaps 27
Ducted radiator 124

Ein-TL-Jäger fighter 275-277, 286-288, 290
Ejection seat 147-148, 279
Electron 85, 89, 91
Elevators, balanced 169, 171, 173

European Combat Aircraft 346
Evaporation cooling 112, 137
Exhaust system 321, 324
Exhaust pipe 131, 234-235, 269-271, 286-287, 324
Experimental plane 279
Export 101, 199, 254
Extension shaft 179-180

Fighter-bomber 129, 239, 242
Fighter Staff 197, 244-245
Flutter 244, 338
Flying wing 179
Folding rotor system 359-361, 362, 375, 377
Folding wings 62, 83, 90, 98, 111, 121
Fowler flaps 98, 204
Freighter 79
Fuel injected engine 116-118
Fuel tanks, jettisonable 120, 125, 130, 155-156, 279
Fuselage, lengthening of 165, 174-176

Glide bomb 273, 283
GM-1 fuel injection system 125, 129, 131, 133-134
Ground attack aircraft 154, 342-343
Ground looping 117, 127, 170

Heavy lift glider 220-222, 231
Heavy lift transport 226-228
Heavy transporter 366-368

Heavy fighter 150, 152, 154, 156, 176, 178-179, 185, 203
Helicopter flight 256-257
High-speed flight 252, 258, 276
High-altitude fighter 131-132, 185-186, 286
High-lift systems 99, 203

Industry advisory board 169, 196
Interceptor 343
Interzeptor 248

Jet deflection 347, 353, 370-371, 374
Jet deflector 350
Jet engine 131, 233, 235, 245
Jet trainer 321-323
Jet fighter 232
Jig construction 152, 188, 195, 239

Kematener Modell 303-304
Kippnase 234
Kitplane 87
Kurvenleger 315

Laminar profile 147
Landing flaps 83, 98, 203-204
Landing gear, cantilever type strut 82, 90, 98
Landing gear, jettisonable 257, 261
Landing gear, retractable 10, 89, 96, 99, 113
Landing gear, shock absorber strut 83
Landing gear, single strut 10, 84
Landing gear, skid type 260-261, 288
Landing gear, tricycle 13, 104, 107, 148-149, 233, 237-238, 318, 339, 350
Large scale production 44, 195, 197
Liaison plane 95, 203
License manufacturers 114-115, 118, 191-192
License manufacturing 75, 77, 92, 134, 254
License manufacturing program 343-345
Lift engine 347, 349, 353, 355
Lightweight airplane 40, 42, 44
Lightweight construction method 112
Lightweight fighter 111
Load factor 48-49
Long-range aircraft 88, 213-214, 219, 286-288

M-planes 40-92
Mach number 251, 258, 272, 276
Mailplane 78-79
Mailplane, high-speed 79
Manufacturing improvements 44, 67, 73
Material, maximum exploitation of 11
Metal construction method 44
Minicar 313-315
Mistel combination 242, 272
Modular airplane 151, 184-185, 286, 317, 324, 366
Monocoque fuselage design 96, 98, 109, 111
Multiple exploitation 11
MW 50 fuel injection system 128-129, 131, 134

NACA shroud 85, 91
Negative wing sweep 176
Night fighters 155, 157-158, 162, 185-186, 248-249, 255

Open wing spar construction 44
Operational testing 246-248, 267

Parasite drag 114
Parasite fighter 216, 247, 272, 286-288
Patents 12
Performance review 49, 51, 65
Plywood hull 33, 36
Powered gliders 36-40
Pusher propellers 183, 219
Pressurized cockpit 13, 148, 176, 266-267, 279, 324
Propeller developments 292
Pulse-jet engine 226

Railtrack gauge 228
Recirculation 353
Reconnaissance aircraft 136, 185, 249
Reconnaissance aircraft, strategic 95, 176, 186, 199, 209-210, 212
Reconnaissance aircraft, tactical 201, 247
Record breaking flights 103, 118, 137-139, 258
Rocket armament 131, 135, 247, 265
Rocket-assisted takeoff 216, 224-226, 243, 279, 283
Rocket-powered fighter 255-257
Rocket propulsion 235, 255-257, 284, 288

Safety factor 12
Safety loading 50
Sailplanes 18-36
Scissor-wing 290-291
Sewing machines 303-304
Single-axle car 309
Single-skid design 18, 26
Slats 96-98, 101, 105, 203, 243, 245
Sonderausschuss F2 196
Spin handling 78, 91, 111, 113-114, 171-172
Split flaps 96, 319
Spoiler 97
Sportplanes 64-70, 77-78, 82-86
Stalling 105, 123, 157, 167, 172, 176
Standard fighter 113, 141, 188, 191
Stationary vibration testing 327
Steam turbine 218, 226
Steel tubing construction 53, 56, 62-63, 77, 83, 86, 90, 223
Stress test airframe 112, 152
Strike fighter 186
Structural limitations 40, 49
Structural soundness testing 79
Structural weight 9, 10
Suicide missions 273
Supersonic flight 285, 288
Supersonic fighter 326-328

Tailseater 347, 369
Tailskid 42
Tailless design 27, 179-181, 255-257
Tailplane 19-20, 24
Tailplane, all-moving 12, 82
Tailwheel, retractable 134
Thrust, loss of 280
Thrust vector engine 347, 349, 359, 369-370
Torpedo bomber 185
Torsion-resistant leading edge 11, 86
Torsion tube 27, 33, 36
Touring plane 75-77, 96-98
Tow system 224-225, 247, 288
Towed takeoff 257, 270-271
Towing, air-to-air 220-221
Trailing edge flaps 152
Trainers 54, 62, 64, 77, 86-87, 94, 122, 248-250, 340-341
Transport helicopter
Transport plane 186
Transport plane, high speed 186
Troika-tow 224-225
Tropicalized 107, 120
Turbojet engine 183, 186, 232-234
Twin tail design 151, 211-212
Twin engine 179, 211-212

Upward-firing armament 158

Variable pitch propellers 101, 104, 147, 292
Variable sweep (in air) 280
Variable sweep (on ground) 281, 353
VTOL plane 347-349

Walter rocket engine 256, 261-262, 284
Weak point research 12
Weight ratio 48
Wheel braking 73, 83, 99, 114
Wheel steering 74
Wide track undercarriage 147-148
Wind tunnel testing 12, 119, 165, 222, 234, 256, 273, 318, 321-322, 328, 338, 354, 363
Wing, all-through 65
Wing, sealed 11, 139, 209
Wing control 9, 15-16, 19-20, 26, 28, 31-33, 37-38, 40
Wing, folding 281
Wing, monospar 9, 11-12, 30, 42, 59, 69, 73, 99, 152
Wing, swept 147, 218, 233-234, 251-252, 275-277, 280-282
Wing, tapered 9, 114, 152
Wing warping 19
Wood construction method 100, 143, 257, 259, 269, 283-285

Zwillingsflugzeug 149, 183-184, 186

Index by Persons

(Roman numerals indicate color pages)

Adenaur, Paul 311
Agello, Francesco 137
Aghnides, E. P. 309
Aichele, Erwin 82-83
Althoff, Karl 233, 239
Andrès, Christopher 380-381
Antz, Hans 256
Arnold, Karl 304
Arritio, Don Pedro 317
Asam, Moritz 243, 281, 391

Bader, Paul 257
Baeumker, Adolf 226, 255, 356, 391
Barmeyer, Paul 391
Bartels, Werner 391
Bauer, Hubert 63, 95, 98, 137, 165, 190, 321, 340, 342-343, 351, 355, 384, 389, 391
Bauer, Richard 75, 82, 95, 98, 112, 121, 142-143, 147, 152, 164, 204-205, 391
Baumgärtner 18
Baur, Karl 148, 174, 210-211, 216-217, 223, 226, 235, 269, 391
Beauvais, Heinrich 141, 149, 174, 236, 238, 257, 326, 391
Behrbohm 282
Beinhorn, Elly 67, 101-102, 106, 108, 292, 388, 391
Below, Nicolaus von 206-207, 240-241
Belz, H. 301
Benecke, Theodor 348-349
Berger, Traudel 330
Besser, Rolf 344
Betz, Albert 275
Binz, Wilhelm 294, 310, 324, 327, 350-351, 391
Bischoff 59
Bismarck, Alexander von 55, 68
Bitz, Josef 391
Blaich, Theo 106, 391
Blank, Theodor 318
Blasel, Werner 13, 337
Blenk, Hermann 48
Blocher, Martha 327
Blohm, Werner 383, 388
Blümm, Max 318, 338, 391
Bock, Günther 167
Bölkow, Ludwig 82, 108, 132-133, 143, 167, 196, 222, 234, 275, 299, 329, 348-349, 355, 364, 381-382, 384, 386, 388-389, 391
Bönsch, Helmut Werner 311
Bosch, Lorenz 294, 317-318, 391
Braeutigam, Otto 391
Brandi, Hermann Th. 385-386
Brandner, Ferdinand 331, 335-337, 389, 391

Brandt, Kurt 303, 391
Brandt, Leo 304
Braun, Heinz 391
Brausewaldt, Horst 301-302
Bravo, Claudio 329-330
Brenner, Paul 111
Bright, George 351
Brindlinger, Otto 80, 100, 106, 204, 208, 391
Bringewald, August 255
Brising, Lars 282
Broschwitz, Johannes 391
Brown, Eric 125, 128, 248
Brüning, H. 301
Büchler, Peter 252
Busemann, Adolf 275

Caroli, Gerhard 104, 211, 213, 243, 391
Caspar, Victor 255
Castell, Wulf-Diether Graf zu 385, 388
Chanute, Octave 17
Chlingensperg, Rolf von 186, 254, 391, 394
Clostermann, Pierre 248
Conta, Eberhard von 41, 43, 49-50, 391
Croneiß, Carl 43-44
Croneiß, Theo 40-41, 44, 46, 49-51, 56, 58, 62, 65-66, 71, 77-78, 95, 110, 160, 171, 174, 180, 187, 191-193, 206, 391, 393

Degel, Wolfgang 107, 213, 233, 293-294, 307, 309, 313, 339, 352, 392
Dercks, Michael von 388
Dercks, Stella von 388
Dersch, Karl J. 388
Dewoitine 10
Diels, Rudolf 294
Dieterle, Hans 137
Dietrich, Christian 106
Ditmar 38
Dittmar, Heinrich (Heini) 255-256, 258, 261-262, 264-265, 283, 392
Doetsch, Karl H. 167
Dogchel 315
Dornier 183
Dornier, Claude 216, 380
Dornier, Claudius 388
Douhet, Guilio 150
Drechsler, Friedrich 383, 392
Dronsek, Max 346
Dürr, Ludwig 228
Dungern, Wolf von 100, 392

Ebner, Georg 317, 330, 392
Ecenarro, Luis M. Balparda 329
Eckener, Hugo 216, 228-229
Eidenschink, Georg 304, 317
Eisenmann, Walter 391
Empacher, Hans 355
Engel 391

Endres, Johann 305
Erhard, Ludwig 294
Espenlaub, Gottlieb 255
Esteva, Francisco Salom 322

Faber-du-Faur, Karl-Otto von 294, 296, 392
Falck, Wolfgang 158
Feilcke, Fritz 204
Fend, Fritz 310-311, 315, X
Ficht, Reinhold 108, 372
Fiebrich, H. 346
Fieseler, Gerhard 67
Finsterwalder, Sebastian 35
Fischer, Hildegard 383
Fitzek 66
Finsch, Bernhard 223, 391
Föppl, Ludwig 36
Förnzler 211
Föttinger 305
Francke, Carl 97, 100, 114, 116-117, 169, 172, 174, 223
Franco, General 115
Franz, Anselm 235
Frêne, Martial 313
Freyberg, Baron von 32-33
Friedrich, Oskar 346, 392
Fröhlich, Josef 222-223, 228-229, 392
Frydag, Karl 343
Fuchshuber, Josef 329, 392
Gablenz, Freiherr von 213
Galland, Adolf 116, 119, 128, 134, 143-144, 236-238, 241-242, 247, 386
Gasperi, Capt. 204
Gaus, Günter 346
Geduldig, Wilhelm 326, 333, 392
Gentzen, Hannes 106
Georgii, Walter 221
Gerstmayer, Sepp 387
Gierenstein, K. H. 382
Göring, Hermann 93, 95, 110, 119, 144, 159, 162, 172, 174, 213, 236, 238-240, 246
Göttel, Werner 133, 387, 392
Gonzales, Emilio 329
Goppel, Alfons 381-383, 385, 390
Graichen, Max 37-38
Grande, Munoz 318
Greenamayer, Darryl E. 141
Greim, Ritter von 114
Gronau, Wolfgang von 106
Grozea 68
Günter, Siegfried 347, 354
Gurevich 268
Haase, Gottfried 327
Haberkorn, Erich 347, 354
Hackmack, Hans 34, 57, 392
Hall, Paul J. 86-87
Hamilton, Graf 34
Handley-Page 99

Harth, Friedrich 9, 15-29, 31-32, 34, 40, 392
Hartmann, Erich 134
Heinkel, Ernst 87, 92, 118, 137-138, 213, 256, 322-323, 326, 328, 331, 380
Heinrich, Prinz von Preußen 39
Heinz, Johann 52, 392
Hellmann 62
Hellmann, Herbert 329
Hempel 167
Hentzen. Fritz H. 144, 154, 171, 187, 192, 196, 206, 216, 259, 329, 351, 392
Henrici, Julius 284
Herbst, Wolfgang 346
Herlitzius 251
Hermkes, Bernhard 192
Herring 17
Hermann, Hans 52-53
Hertel, Heinrich 267
Heß, Ilse 162
Heß, Rudolf 161-162
Heusinger, Adolf 317
Hierl, Alfred 157
Hille, Fritz 81
Hills 228
Hirth, Wolf 28-32, 34-35, 81, 392
Hitler, Adolf 144, 161-162, 164, 206-208, 213, 219-220, 232, 238-240, 242, 244, 247, 249, 254, 269
Höllein, Peter 296
Hofmann, Ludwig 251
Hoffmann, August 317, 392
Hornung, Hans 107, 181, 276, 286, 291, 294, 317, 320, 338-339, 347, 354, 365, 392
Hort, Sepp 382, 389
Horten, Walter 155
Hubert, J. 266
Hubrich, Gerhard 204-205
Hübsch, Fritz 392
Hügelschäffer, Fritz 132, 143
Hughes, Howard 137
Huth, Edgar 302

Ihlefeld, Herbert 116
Illg, Hermann 103, 106

Jacobs, Hans 220
Jodhbauer, Kurt 114, 392
Johnen, Wilhelm 158, 162-163
Junck, Werner 100
Junkers 124
Junkers, Hugo 9, 386

Kaden, Helmut 161-162, 387, 392
Kaden, Herbert 254, 394
Kaiser, Johann B. (Hans) 161-162, 208, 266, 316
Kamil, Hassan Sayed 331, 335
Kammhuber, Josef 158, 349
Kármán, Théodore von 329
Kaufmann, Johannes 155, 171

Kaul 12
Keilholz, Hans 392
Keller, Siegfried 302, 304, 392
Kempter, Josef 326, 392
Kensche, Heinz 273
Kielmannsegg, Graf 317
Kistner 38
Klages, Paul 327, 392
Klein, Gerald 63
Klotz, Fritz 326
Knemeyer, Siegfried 216-218
Knoetzsch, Hans Dietrich 113, 152, 392
Knott, Valentin 311
Köppen, Joachim von 64, 113
Kogon, Eugen 294
Kokothaki, Rakan 144, 146, 187, 192, 196-197, 299, 393
Konrad 38
Konrad, Dr. Walter 284
Konrad, Paul 208, 213, 394
Koschnick, Hans 383
Kosin, Rüdiger 112
Krauß, Julius 9, 35, 52, 54, 62-63, 316-318, 340, 393
Küßner 12
Kurth 284

Lässig, Theo 204
Langfelder, Helmut 323-324, 326-327, 377, 393
Langhammer, Wilhelm 310, 393
Langsdorff, Werner von 37, 43, 51, 391
Lattre de Tassigny, Jean de 293
Lauerbach, Erwin 382, 384
Lauser, Kurt 355, 393
Leistner, Rudolf 326, 393
Lennartz, Helmut 8, 246
Lent, Helmut 158
Liebe, Wolfgang 130
Liebmann, Fritz 327
Lilienthal, Otto 16-17
Lindbergh, Charles 101-102
Linder, Karl 125, 143, 183, 185-186, 192, 245, 296, 393
Lindner, Gerd 240, 243, 247, 249, 251, 393
Lippisch, Alexander 34, 178-180, 182, 205, 255-259, 261, 266, 272, 283, 299, 392-393
Lösch 226
Lorenz, Hermann 256
Lorenzo, Raffael Vellido 321
Lubinski, Siegfried 307
Lucht, Roluf 114, 174, 193, 393-394
Ludowici 32
Ludwieg, Hubert 275
Lusser, Robert 95, 98, 111, 137, 151-152, 164, 199, 201, 203, 208, 232, 286, 350, 354, 393
Luukkanen, Eino 128
Madelung, Ello 385
Madelung, Georg 28, 48, 88, 318, 320, 392-393, 395

Madelung, Gero 108, 318, 320, 327, 339, 348-349, 354-355, 372, 379, 384, 389-390, 393
Maier 43
Maiershofer, Karl 160
Mallet, Guillermo F. 316-317, 328-329, 331, 393
Malz 169, 171
Marienfeld, Walter 242
Marr, Christine Dora 394
Masching 38
Materne, Gerd 378
Mathis 268
Mathy, H. 301
Matouscheck, Eleonore 388
Maximovic, Peter 329
Mayer, Friedrich 192, 393
Mayr, Josef 187, 191
Meier, Hans Justus 205
Meinhardt, Kaspar 393
Meitinger, Otto 379
Menzel, H. 301
Merkel, Konrad 82, 93, 187, 296, 302, 304, 386, 393
Messerschmitt, Andreas 14
Messerschmitt, Anna Maria 14
Messerschmitt, Elisabeth (Ello, Betti) 14, 36, 393
Messerschmitt, Elmar 294, 299, 386, 393
Messerschmitt, Ferdinand 14, 36
Messerschmitt, Ferdinand Baptist 14
Messerschmitt, Gerhard 388
Messerschmitt, Johann-Baptist 294
Messerschmitt, Lilly 393
Messerschmitt, Maria (Maja) 14, 36
Messerschmitt, Pius Ferdinand 14
Messerschmitt, Rudolf (Bubi) 14, 36
Messerschmitt, Wilhelm (Willy) 14, 24
Michel-Raulino, Freiherrin Lilly von 302, 316, 329-330, 352, 382, 386, 394
Mikoyan 286
Milch, Erhard 61, 92-93, 95, 114, 144, 160, 174, 178, 206, 213-214, 236-238, 240, 244, 261, 272, 394
Miller, Ferdinand 318, 324, 350, 393
Mittelholzer, Walter 69, 73
Mölders, Werner 116, 119-120
Mösinger, H. 301
Mohnike, Eberhard 63-64, 393
Monge, Louis de 315
Morzik, Fritz 68-70, 82, 84, 230, 393
Müller, Christa 385
Müller, Josef 385-386
Münemann, Rudolf 294-296, 302, 393

Naguib, Ali Mohammed 331
Nasser, Gamal Abdel 331-332
Nallinger, Fritz 117, 380
Nauschütz, Peter 284
Neidhart, Gustl 165, 169, 171
Nerud, Alfred 318, 393
Niel, Captain 301

Nitschke, Gerhard 114
Nowotny, Walter 246-247

Oesau, Walter 116
Oestrich, Hermann 233
Oerzen, Otto 263
Oettil, Franz 161
Ohain, Hans von 232
Opitz, Rudolf 258, 261
Orduna, Fernando Gomez 329
Osterkamp, Theo 100, 154
Ostertag, Wilhelm 236-237, 243, 393
Oswatitsch, Klaus 327
Otani, General 146, 150, 254

Pabst, Otto 348-349, 355
Padua, Paul Mathias 15
Pantazi, Hauptmann 68
Patch, Alexander 293
Pauker 226
Peter, Ernst 224
Peter, Hans 382
Peters, Heinz 265, 267
Petersen 213-214, 238
Petzet, Michael 379
Piaggio, Rinaldo 313
Pintsch, Karlheinz 162
Pleines, E. W. 321
Pöhlmann 211
Pohl, Hermann 52
Pohlmann, Hermann 133
Porro, General 204
Prasthofer, Frederico 316
Prause, Robert 292
Price, Nathan 322
Priller, Josef 134
Puffert, Hans Joachim (Jochen) 167, 326, 393

Quick, August Wilhelm 167, 385-386

Regelin, Hans 207
Reicherter, Julius 239, 394
Reitsch, Hanna 223, 261-262, 273, 384, 394
Reitz 307
Rentel, Rudolf 266
Rethel, Walter 164-165, 207, 222, 279, 394
Riedel, Hans Otto 355
Riedel, Peters 38-39
Rieppel, Paul 62
Rittner, Günter XV
Rodig, Helmut 133
Rohrbach, Adolf 204
Rosemeyer-Beinhorn, Elly 101, 106
Rothe, L. S. 301, 328, 348, 394
Rougeron, Camille 150
Rowehl 212
Ruden, Paul 326, 337
Rudorf, Fritz 384, 394
Rühler 38

Ruppelt, Wolfgang 388

Sackmann, Franz 382
Salas, Larrazabal Jesus 316, 333
Salomon, Horst von 106
Salvador, Julio 320
Santa Cruz, Pedro Barcelo 328
Saur, Karl Otto 197, 242-243
Scancony, V. 62
Schäffer, Max 326, 337-338, 394
Schairer, George 282
Schaller, Anna Maria 14
Schedl, Otto 379
Scheibe 217
Schelp, Helmut 256
Schenck 242
Scherenberg, Hans 388
Scheuermann, Erich 53, 58, 70
Schiel 68
Schiller, Karl 382
Schinzinger, Reginald 268
Schmid, Lukas 387, 394
Schmidt, Erich 318
Schmidt, Kurt 240, 394
Schmitt, Dieter 102
Schmitz, Gerd 394
Schnabel, Herbert 327
Schnaller 284
Schnause, H. 301, 302
Schnaufer, Heinz-Wolfgang 158
Schnittke, Kurt 106, 162, 181, 193, 387, 389, 394
Schönbaumfeld 321
Schönebeck, Carl-August von 37, 39
Schomerus, Riclef 167, 234, 254, 276, 391, 394
Schrüffer, Alexander 53, 62
Schwärzler, Karl 348, 350
Schwaibold, Wilhelm 385
Schwarz, Friedrich 38-39, 41, 149, 394
Schwarzkopff, A. S. 187
Sebald, H. 346
Seewald, Friedrich 300
Seidemann, Hans 106, 116-117
Seifert, Karl 208, 394
Seiler, Friedrich-Wilhelm 144, 174, 187, 192, 196, 244, 392, 394
Seitz, Rudolf 233, 269, 272-273, 286
Seywald, Heinz 38-41, 43, 394
Sido, Franz 63, 71, 77, 394
Siebel, Friedrich-Wilhelm 164
Siebel-Trulson, Ingeborg 388
Sinz, Josef 391
Späte, Wolfgang 236-237, 265, 270
Speck von Sternberg, Freiherr 106
Speer, Albert 197, 244, 389
Speidel, Hans 317
Srbik, Hans-Heinrich Ritter von 316-317, 331, 335, 355, 372, 378, 384, 394
Standfuß 34

Stehle, Max 195
Steigenberger, Otto 307-308, 314, 394
Steinhoff, Johannes 247, 387
Steininger 195-196
Stender, Walter 180-181
Stölting, Inge 106
Stör, Willi 67, 78, 89, 91, 113, 153, 161, 204, 254, 346, 392, 394
Storp, Dora 294, 394
Straßl, H. 275
Strauch, Hans 304, 311
Strauß, Franz Josef 343, 347-349, 354, 383, 388
Streib, Werner 158
Stromeyer, Eberhard 293, 340
Stromeyer, Hilmar 388
Stromeyer, Karin 330
Stromeyer, Lilli 187, 192
Stromeyer, Otto 62, 81
Stromeyer, Walther 355, 382, 394
Stromeyer-Raulino 61
Stumm, Hilmar 313, 330, 394
Stürmer 20
Suciu, Titus 363-364, 394

Tank, Kurt 79-80, 108, 123, 144, 282, 286, 331, 336, 356, 379, 389, 394
Tein, Volker von 346
Teufel, Adolf 360
Thalau, Karl 340, 348, 356, 394
Thiel 284
Theil, Margarete 120, 138, 204, 394
Thierfelder, Werner 243, 246
Thomsen, Otto Robert 34, 106
Thun, Graf 284
Tracinski 34
Trautloft, Hannes 115
Trenkle, Wendelin 192, 387, 394
Triebnigg 305

Udet, Ernst 39, 53, 68, 75, 86, 93, 95, 106, 111, 113-114, 117, 138, 140, 162, 164, 169, 180, 183, 206-207, 213, 221-222, 225, 257-259, 261, 394

Valiente, Julian 323
Vogel, Ludwig 315
Vogt, Richard 133
Voigt, Woldemar 147, 178, 180-181, 213-214, 222, 226, 232, 272, 276, 394
Vorwaldt 171
Voß 211
Voy, Karl 263
Vuillemin, General 160

Wahl, Albert 383
Walter, Fritz 343
Walter, Hellmuth 256
Wassermann, Karl 326
Watson, Harold E. 252

433

Wegner 211
Weinhardt, Bernhard 355, 381-382
Weissenberger, Theodor 247
Welter, Kurt 248
Wendel, Fritz 8, 106, 132, 134, 137-139, 144-145, 148, 182, 204, 210, 234-236, 248, 346, 395
Wenz 63
Wever, Walter 150, 207
Weyermann, R. 41, 50
White, Robert M. 324
Wichtendahl, Wilhelm 192
Wiesner, Reinhold 327
Wimmer, Major 111
Wimmer, Thomas 297

Woods, Robert J. 281
Wischhöfer, Gert 318
Wright Brothers 17
Wurster, Hermann 101, 113-114, 117, 120, 123, 130, 137-138, 141, 152, 165-166, 167-168, 174, 178-180, 200-202, 204, 206, 283-284, 292, 387, 395

Yeager, Charles 256

Zeiler, Curt 223
Ziegler, Mano 108, 263-264, 372, 395
Ziegler, Reinhold 311
Ziegler, Rudolf 269, 395

The Authors

Left to right: Klaus Peters, Johann B. Kaiser, Hans J. Ebert (1988)

Hans J. Ebert was born in Stettin in 1937. A book dealer by trade, from 1955 to 1964 he worked in sorting, importing and wholesale in Luneburg, Cologne and Hamburg. His interests lie with aviation history and sport flying. In 1965 he changed jobs to work in the archival center at Messerschmitt AG, where he began building up the historical Messerschmitt archives. He subsequently migrated to the literature/archives department at EWR, then to MBB's aircraft division. From 1975 to 1990 he worked here as the group director for the largest (as of this writing) private archives, at MBB. Since 1991 Ebert has been assigned to the scientific-technical information department at MBB with the responsibility of building up and maintaining the historical archives. He has been a member of the advisory board for air and space at the Deutsches Museum since 1979.

Since 1962 Ebert has written on the subject of historical aviation for various magazines and books, with regular articles appearing since 1969 in the company's "MBB aktuell" and "newtech news" periodicals. In addition, Ebert has contributed to company information and anniversary publications and exhibits.

In 1973 he authored the book "Messerschmitt-Bölkow-Blohm, 111 MBB-Flugzeuge 1913-1973/78" (five editions).

Johann B. (Hans) Kaiser was born in 1911 in Oberstdorf. In 1934 he completed his machine engineering studies at the Technical University in Augsburg and subsequently went to work for MAN. In 1936 Kaiser was hired by BFW and initially worked in the technical publications department, where he was responsible for the Bf 109 and Bf 110 aircraft handbooks. From 1939 onward he worked in the project section under Robert Lusser and Friedrich Schwarz, where he dealt with armament matters to include live testing of the Bf 110's weapons at Tarnewitz. In 1939 he became the technical assistant to Woldemar Voigt in the project section. During the war, among other things he was the coordinator between Mtt-AG and Jacobs-Schweyer on the development of the Me 328 and supervised the Me 163B prototype program in Laupheim as well as the program's move to Kronach.

From 1945 to 1961 Kaiser was a machine design engineer in the Ruhr district, among other areas. In 1962 he again became involved in the aviation industry with the Messerschmitt team at Development Ring South, where he established the technical writings and historical archives department. With the merger, this department was transferred to MBB's aircraft division. he retired in 1975. Kaiser worked closely with Friedrich Harth on Messerschmitt history with particular focus on the early period. His first major publication "*Friedrich Harth und seine Flugzeuge*" (appearing in Luftfahrt International, Nos. 12/80 through 2/81) stemmed from this work.

His writings have also appeared in Messerschmitt's 1973 and 1978 anniversary editions as well as in the DLR-Mitteilung 75-21 (1975) under the caption "*Die Geschichte des Messerschmitt-Flugzeugbaus.*"

Hans Kaiser died on 12 November 1991 as this book was going to press, having written the first portion up to the 1934 period.

Klaus Peters was born in Bernburg/Saale in 1938. He and his wife have three children. Peters studied machine engineering at the Rheinisch-Westfälische Technische Hochschule in Aachen. He was given practical training at Dornier in Munich, at Fokker in Dordrecht an Schiphol, and Sud Aviation and Breguet in Toulouse. In 1966 he completed his thesis work in aeronautical engineering and subsequently went to work at the Avicentra-Archiv company in Cologne. Over time, he became the editor for several aeronautical publications, including "deutscher aerokurier", Avicentra-Nachrichten Ausrüstung," "Weltraumfahrt/Raketentechnik" and the aviation and space section of the journal "Soldat und Technik." In 1971 he joined the Messerschmitt-Bölkow-Blohm firm, where he worked in the public affairs division as an advisor for the technical press. In 1983 he became the chief of records. In 1988 Peters was appointed the deputy director, and in 1989 the provisional director of the public affairs branch. In 1990 Peters joined the Deutsche Aerospace company, where he works as chief of records. Peters is a member of the Deutsche Gesellschaft für Luft- und Raumfahrt (since 1966), the Luftfahrt-Presse-Club (since 1966), the Deutsche Gesellschaft für Wehrtechnik, the advisory council for the Deutsches Museum's air and space department, and the Royal Aeronautical Society, Munich branch.

Notes

Notes

Notes

Notes

Notes